Process Geomorphology

Frontispiece, G. K. Gilbert
standing by perched granite
boulder in Yosemite National Park,
1908.

Process Geomorphology

Second Edition

Dale F. Ritter
Southern Illinois University at Carbondale

Wm. C. Brown Publishers
Dubuque, Iowa

wcb
group

Wm. C. Brown
Chairman of the Board
Mark C. Falb
President and Chief Executive Officer

wcb
Wm. C. Brown Publishers, College Division

Lawrence E. Cremer   *President*
James L. Romig   *Vice-President, Product Development*
David A. Corona   *Vice-President, Production and Design*
E. F. Jogerst   *Vice-President, Cost Analyst*
Bob McLaughlin   *National Sales Manager*
Catherine M. Faduska   *Director of Marketing Services*
Craig S. Marty   *Director of Marketing Research*
Marilyn A. Phelps   *Manager of Design*
Eugenia M. Collins   *Production Editorial Manager*

Book Team

Edward G. Jaffe   *Executive Editor*
Lynne M. Meyers   *Associate Editor*
Nova A. Maack   *Associate Editor*
Mark Elliot Christianson   *Design Supervisor*
Vickie Blosch   *Production Editor*
Mary M. Heller   *Photo Research Editor*
Vicki Krug   *Permissions Editor*

Cover photograph: Frost melting on sand dunes at sunrise at Monument Valley, Arizona. Setting and colors are natural. © Kathleen Norris Cooke

The credits section for this book begins on page 569, and is considered an extension of the copyright page.

To my family.

Contents

4

Physical Weathering, Mass Movement, and Slopes 109

5

The Drainage Basin—Development, Morphometry, and Hydrology 153

6

Fluvial Processes 205

7

Fluvial Landforms 255

8

Wind Processes and Landforms 303

9

Glaciers and Glacial Mechanics 335

10

Glacial Erosion, Deposition, and Landforms 363

11

Periglacial Processes and Landforms 405

12

Karst—Processes and Landforms 445

13

Coastal Zones—Processes and Landforms 481

Preface

Geomorphology has undergone a drastic change in scope and philosophy during the last several decades. In the past, the discipline was primarily concerned with the evolutionary development of landscapes under a wide variety of climatic and geologic controls. More recently, geomorphologists have recognized the need for an applied rather than a historical emphasis. This change in philosophy has placed geomorphology at an interface with many other disciplines. Today's geomorphologist must relate to problems that face hydrologists, engineers, pedologists, foresters, and many other types of earth scientists. The bond that unites geology and geomorphology with so many apparently diverse disciplines is the common need to understand the processes operating within the Earth's surficial systems. Thus, although the historical aspect of landscapes remains important, it is absolutely essential for earth scientists to have a basic understanding of surface mechanics and, in addition, of how those process mechanics are reflected in the landforms they create. This edition of *Process Geomorphology,* like its predecessor, is an attempt to satisfy those needs. The prime purpose of the book remains as it was, to provide undergraduate students with an introductory understanding of process mechanics and how process leads to the genesis of landforms.

A wealth of new information concerning surficial processes has emerged since the first edition was completed, and many new techniques to analyze process have been developed. In most chapters new data and interpretations have been assimilated within the format of the first edition. Some chapters, however, have been changed significantly. This is especially true in the introductory chapters (chapters 1 and 2) and in the treatment of the drainage basin (chapter 5) and coastal processes (chapter 13). A lengthy bibliography is again presented so that students wishing to pursue a particular topic in greater depth will find a ready nucleus of source material. Most references cited were purposely selected from journals and books that will most likely be found in libraries of North American colleges and universities. There is less mathematical treatment in the revised edition, and such an approach is used primarily to clarify concepts that are particularly complex.

I wish to acknowledge the help and guidance I received from numerous colleagues in the geomorphological discipline. I am especially indebted to Steven P. Esling, Ronald C. Flemal, Thomas W. Gardner, Andre K. Lehre, R. Craig Kochel, Frances J. Hein and Arthur N. Palmer who reviewed parts or all of the revised text. Their constructive advice and criticism were instrumental in the completion of the text, and their conscientious efforts are deeply appreciated. Shortcomings and errors in the book are, of course, mine.

D. F. R.

Process Geomorphology
An Introduction

1

Introduction

One of the remarkable aspects of planet Earth is the infinite variety of its surface forms. It is probably safe to assume that as humans became aware of their physical environment, landscape was the first geologic characteristic they noted. Familiar surface features guided their travels and established their territorial boundaries. As time passed, people learned how best to utilize regional characteristics for different purposes, such as agriculture, trade, and military adventure. They also learned that some landforms possess certain peculiarities that somehow, almost imperceptibly, set them apart from others. Gradually these isolated observations grew into an organized collection of knowledge, and a separate branch of science was born.

Geomorphology is best and most simply defined as the study of landforms. Like most simplistic definitions, this does not do justice to a discipline that can be exciting to even the uninspired and challenging to anyone who enjoys science. Historically, landforms have been analyzed in a variety of ways because different students seek from the landscape different information and different truths. For example, since people live on landforms, geographers may justifiably be concerned with how landscapes affect human events. Engineers, on the other hand, examine surface forms to select the best construction sites or to control the physical environment in the most advantageous manner. While engineers and geographers may look at the same landscape, they probably never ask the same questions about it.

Landform data come from widely divergent disciplines. Synthesizing the facts into a cohesive picture of the Earth's surface, therefore, becomes a monumental task. The diverse nature of the data may explain the appearance of subdisciplines such as dynamic geomorphology or climatic geomorphology (Büdel 1968) as well as the difficulties geomorphology has always had in finding a definite academic home. Today in the United States, geomorphology is still taught in both geology and geography departments, and the subject matter becomes the responsibility of anyone who will properly adopt it. The stepchild existence between geology and geography has created in the minds of some the undeserved image that geomorphology is not clearly defined as a science or based on scientific facts.

It is true that traditional geomorphology has been excessively descriptive. Much emphasis in the past was given to placing landforms, both regional and local, into some evolutionary model, so that the field was concerned primarily with historical interpretations. In recent years, however, the discipline has become more quantitative, and research has shifted to studies with a more practical value. Modern geomorphologists often deal with problems that link them directly to other professionals working at the Earth's surface. Obviously, geomorphology is more than any definition can adequately express. Although it has identity, its boundaries are ill-defined and more certainly ephemeral.

More important than a precise definition is the fact that geomorphology is and probably always will be a field-oriented science. Map and photo analyses are necessary first steps to good geomorphic work, and laboratory data support interpretations. But the real test of geomorphic validity is outdoors, where all the evidence must be pieced together into a lucid picture showing why landforms are the way we find them and why they are located where they are. A prime requisite for a geomorphologist is to be a careful observer of relevant field relationships. This trait cannot be easily taught, and truly outstanding geomorphologists usually develop it by learning from their own mistakes. Geomorphic processes are remarkably subtle, and minor changes of basic controls can result in an infinite array of landforms. Invariably, the person with the greatest experience under varied conditions will make the most viable geomorphic interpretations. Thus a geomorphologist, like any other scientist, must learn the trade. There are no shortcuts that produce geomorphic insight. It must be acquired gradually through long field experience.

This book will concentrate on processes that create the features we see at the surface of the Earth. **Process** can be defined as the action produced when a force induces a change, either chemical or physical, in the materials or forms at the Earth's surface. In simpler terms, process may be thought of as the method by which one thing is produced from something else. It may not be clear why this approach is more beneficial than using some other criterion, such as climate or time, as a central theme. As we have said, geomorphology stands at the interface between geology and many other disciplines that deal with surficial phenomena. Today geomorphologists must be aware of the problems facing hydrologists, civil engineers, pedologists, foresters, urban planners, and other specialists. And since those scientists are working in an environment underlain and partly controlled by the geologic fabric, they must be concerned with geologic concepts and problems. It follows that there must be a common interest uniting these apparently diverse fields, since they all function in the same place at the same time. It is the universal need to understand processes that is basic to all surficial disciplines.

Understanding what process is also serves as a basic component of other scientific disciplines. For example, we now know that application of our knowledge about geomorphic processes is basic in the field of environmental science. Every surface environment is controlled by process. We have known for years that human intervention into surface environments causes rapid changes in processes (Gilbert 1917) and invariably requires adjustments in the environment itself.

A good example of human influence on geomorphic processes is occurring today in San Diego County, California (Kuhn and Shepard 1983). Here the effects of cyclic climate change are beginning to produce accelerated erosion of the bluffs overlooking the Pacific Ocean. In the past several decades, wave action has not been severe because the prevailing dry climate during that interval created very few major storms. Beaches, shorelines, and sea cliffs were relatively stable. This led to large-scale urbanization along the coast, and with it excessive watering of lawns, irrigation, septic tanks, leach lines, and cesspools. The extensive use of water has caused a steady rise in the water table,

Street

Scarp

Scarp

Bulldozers

Ocean

Figure 1.1.
Large landslide and earthflow
along the California coast.

which is a prime culprit in slope failure (see chapter 4). Thus, as the climate
change has produced more precipitation and more erosive storm waves, the
sea cliffs have been primed for failure by human activities. Landslides and
other mass movements are now more numerous, and blocks of the coastal bluffs
(often supporting homes) are slipping downward into the ocean (fig. 1.1).
Clearly, we are geomorphic catalysts. Therefore, prior to its inception, any
major surficial project requires a detailed understanding of geomorphic processes in order to predict how those processes will respond to our activities
(Coates 1976).

Another discipline directly dependent on a knowledge of process is plan-
etary science (Baker 1981). There is little doubt that recognition of landforms
is a key factor in interpreting the surface domain of our sister planets (fig.
1.2). However, simple landform identification is not enough. Understanding
the genesis of those features requires knowledge of how processes function in
analogous Earth environments and, equally important, how processes might
function in conditions that are alien to anything known on Earth (for example,
Komar 1979).

Figure 1.2.
Large landslides on Mars.
Compare with figure 1.1.

Finally, an understanding of process is critical in geoscience itself. Cause and effect are essential components in the events that document geologic history. Our reconstructions of history suffer, however, because we lack an explicit understanding of what effects will arise from particular causative processes. We know, for example, that Holocene climate changes were severe enough to upset the delicate geomorphic balance. What is confusing is the diverse geomorphic responses resulting from the same climatic trends. We must conclude that, for the very recent past, we commonly understand the cause better than the effect. How then can we confidently infer how processes functioned 300 million years ago when, in fact, we cannot always predict their responses to modern stimuli? Certainly our insight does not increase as we contemplate rocks because what we see are not processes but the results of processes. We oversimplify the system in order to make any interpretation at all. There is nothing wrong with this practice as long as we admit that our models are based on what we *think* about processes, not what we know. What geology needs is a precisely defined understanding of modern processes; until geomorphologists provide it, our reconstruction of past events will be at best educated guesses.

The Basics of Process Geomorphology

Assuming that our focus on process is a viable way to examine geomorphology, we must identify those concepts that, when integrated, constitute the basic principles of process geomorphology. They are listed here and discussed in detail on the following pages.

1. A delicate balance or equilibrium exists between landforms and processes. The character of this balance is revealed by considering both factors as systems or parts of systems.
2. The perceived balance between process and form is created by the interaction of energy, force, and resistance.
3. Changes in driving force and/or resistance may stress the system beyond the defined limits of stability. When these limits of equilibrium (**thresholds**) are exceeded, the system is temporarily in disequilibrium and a major response may occur. The system will develop a different equilibrium condition adjusted to the new force or resistance controls, but it may establish the new balance in a complex manner.
4. Various processes are linked in such a way that the effect of one process may initiate the action of another.
5. Geomorphic analyses can be made over a variety of time intervals. In process studies the time framework utilized has a direct bearing on what conclusions can be made regarding the relationship between process and form. Therefore, the time framework should be determined by what type of geomorphic analysis is desired.

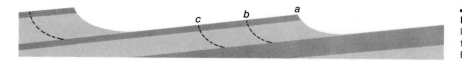

Figure 1.3.
Interpretation of slope adjustment
to geology by G. K. Gilbert.
Equilibrium slope developed at *a*
is maintained at times *b* and *c*.

The Delicate Balance

The idea that some form of balance or equilibrium exists between landforms and the processes that create them is not new. It was clearly expressed by G. K. Gilbert during the latter part of the nineteenth century in his classic reports on the geology in the western United States (see Gilbert 1877). Essentially, Gilbert believed that under any given climate and tectonic setting, landforms reflect some unique accommodation between the dominant processes and the local geology. He often used the terms "dynamic adjustment" and "balanced condition" to describe this relationship. An example of Gilbert's perception of equilibrium is shown in figure 1.3. Here we see a series of slopes that are adjusted to alternating weak and resistant rock layers. The slopes developed on the different units are produced and maintained by the interaction of geology and processes such as mass movement, sheet wash, and river flow. Importantly, Gilbert believed that continuous erosion would not change the slope angles as long as the processes and their climatic and tectonic controls remained constant. Thus, the slopes at times *b* and *c* will be a mirror image of the slopes at time *a* because the process types and rates have not changed through time. If tectonic or climatic controls change, processes will also change, and new slope characteristics will develop in an adjustment to the altered processes.

In the first half of the twentieth century, Gilbert's ideas were pushed aside when geomorphologists espoused the concept developed by W. M. Davis that landscapes change continuously with time and progress through distinct stages that can be identified by regional geomorphic characteristics. It was not until after World War II that the equilibrium approach was revitalized in a number of papers reemphasizing the importance of the adjustment between process and form (Horton 1945; Strahler 1950, 1952a; Leopold and Maddock 1953). This shift in emphasis resulted in the **dynamic equilibrium concept** in which J. T. Hack (1960b) essentially brought back Gilbert's approach as a philosophical framework for geomorphic analyses. Dynamic equilibrium suggests that elements of landscape rapidly adjust to the processes operating on the geology, and thus process and form reveal a cause-and-effect relationship. The forms within a landscape maintain their character as long as the fundamental controls do not change.

Many workers believe that the balance between form and process is best demonstrated by considering both factors as systems or parts of systems. A *system* is simply a collection of related components. For example, suppose we define a drainage basin as a system and consider its measurable parts to be basin area, valley-side slopes, floodplains, and stream channels. The balance or equilibrium condition within our system is revealed by statistical relationships between the various parameters; i.e., basin area may be directly related to total channel length, etc.

The systems approach has become highly sophisticated (Chorley 1962; Chorley and Kennedy 1971), and different types of systems have been identified and used in geomorphology (Schumm 1977). For our purposes, it is best to consider landforms and processes as part of the same open system in which energy and/or mass are continually added or removed. Any flux in energy or mass requires that the processes and their statistically related landforms adjust to maintain balance in the system.

The systems approach has these advantages:

1. It emphasizes the intimate relationship between process and form.
2. It stresses the multivariate nature of geomorphology.
3. It reveals that some forms may not be in balance because they owe their character to relict conditions. Some glaciated regions, for example, may have landforms that were adjusted to geomorphic controls different from those of the present.

As stated above, equilibrium implies that landforms (and presumably processes) exist in some type of unchanging condition. In theory this requires that factors which ultimately control landforms and process (such as climate and tectonics) must also remain unchanged. In reality changes do occur in the controlling factors with time. Thus, the true meaning of equilibrium depends on the time interval over which our balance is being considered. Schumm and Lichty (1965) argued that different time intervals, which they called *cyclic, graded* and *steady,* are critical to our understanding of process and landform development, and the distinction of these is extremely important in our perception of equilibrium. Indeed this insight was followed by the further suggestion (Chorley and Kennedy 1971) that different kinds of equilibrium are related to each particular interval of time (fig. 1.4). **Static equilibrium** is that which exists over the short steady-time interval (days or months). In this framework of time, landforms do not change, and therefore they are truly time-independent. In **steady-state equilibrium,** landforms and/or processes are considered over graded time, perhaps 100 to 1000 years (Schumm 1977). The equilibrium demonstrated in this interval is one in which changes do occur, but their offsetting effects tend to maintain the system in a constant average condition (fig. 1.4). In contrast, **dynamic equilibrium** must be considered over cyclic time, perhaps millions of years (Schumm 1977). In this case, even though fluctuations of variables occur, they are not offsetting and the average condition of the system is progressively changing (fig. 1.4).

With the foregoing perspective of time, it is apparent why the concept of equilibrium has been difficult to define or understand. Time is a major factor in the sense of equilibrium, and effective use of the concept in geomorphology demands that the time framework be specified. We will examine the time factor in process geomorphology later in the chapter.

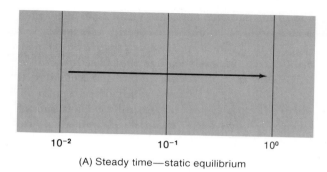

(A) Steady time—static equilibrium

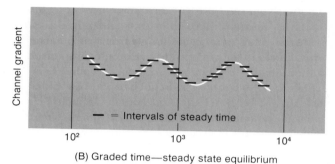

— = Intervals of steady time

(B) Graded time—steady state equilibrium

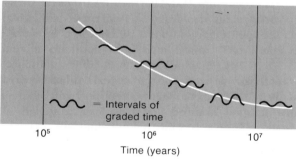

⌇⌇ = Intervals of
  graded time

Time (years)

(C) Cyclic time—dynamic equilibrium

Figure 1.4.
Different time intervals and associated equilibrium in geomorphic analyses. (A) Steady time (static equilibrium). No change in channel gradient over short periods. (B) Graded time (steady state equilibrium). Constant average channel gradient with periodic fluctuations above and below the average condition. Measurements made during intervals of steady time within the graded time period may show no change in channel gradient. (C) Cyclic time (dynamic equilibrium). Gradual lowering of the average channel gradient over long time intervals. Intervals of graded time and steady state equilibrium exist within the cyclic time scale. (Adapted from Schumm 1977)

Force/Process/Resistance

We know from Newtonian physics that only a small number of fundamental quantities are needed to explain mechanics. One of the mechanical quantities identified by Newton is *force,* which—on the basis of his laws of motion—can be loosely defined as anything that changes or tends to change the state of motion in a body. In more specific terms, Newton defined force as a function of mass times acceleration,

$$F = ma;$$

since acceleration is a vector parameter having both magnitude and direction, force also is a vector quantity.

Table 1.1 Common systems of units used in mechanical analyses.

	Length	Mass	Force	Time
Systems		**Units**[a]		
cgs	centimeter	gram	dyne	second
fps	foot	slug	pound	second
mks	meter	kilogram	newton	second

[a] 1 slug = 1 lb $\sec^2$ft^{-1}; 1 dyne = 1 g cm $\sec^{-2}$; 1 newton (N) = 10^5 dynes.

The measure of force is *weight*. Therefore, the standard units of force are pounds, dynes, or newtons depending on what system of units is being used (table 1.1). Another fundamental quantity, *mass,* is directly related to force as can be seen in the equation on the preceding page. In fact, by substituting weight and acceleration of gravity into that equation, it can be expressed as

$$W = mg \text{ or } m = \frac{W}{g},$$

where W is weight, m is mass, and g is the acceleration due to gravity. This demonstrates the interchangeable relationship between force and mass in mechanical analyses.

Force is also related to energy, and in geomorphology we can think of landforms and processes as resulting from the application of energy. *Energy* is defined as the capacity for doing *work*. It can neither be destroyed nor created, but it can exist in many forms and can be changed from one form into another. *Kinetic energy* of an object is energy derived by virtue of its motion. *Potential energy* stems from the position of an object. Any change in the kinetic or potential energy of a body is equal to the work done on that body to produce the change. Therefore, units of energy are the same as the units in measuring work.

For example, consider a 100-pound steel ball carried vertically to the top of a building 30 feet high. The work required to lift the ball to a higher elevation represents a change in potential energy due to the increase in elevation. The amount of work is defined as the product of force and the displacement of the body in the direction of the force such that

Work $= Fs,$

where F is force and s is distance of displacement. In the fps system of units (table 1.1), the unit of work is the *foot-pound*. In other systems, work is expressed in units such as the *erg* (dyne-centimeter), the *joule* (10^7 ergs), or the *newton-meter*.

Thus, in our example above

Work $= PE$

$PE = Fs$

$PE = 100 \text{ lb} \times 30 \text{ ft} = 3000 \text{ ft lb},$

where PE is the gain in potential energy.

This discussion is meant to show that energy and force are not the same but that they are related through the concept of work. Since energy is the capacity for work and work is a function of force, mental substitution suggests to us that (1) force is a major component of energy measurement and (2) force in mechanics can be thought of as the application of energy, a phenomenon that we will refer to as **driving force.**

In process geomorphology landforms represent interaction between driving forces and resistance. Driving forces in geomorphology are climate, gravity, and other forces generated inside the Earth. Resistance is provided by the geologic framework. The link between these two components is process. Thus, as stated earlier, process may be considered as the method by which one thing is produced from something else, and as the vehicle by which a quantity of one system is transferred into, and participates in, the mechanics of another system.

In general, processes are either exogenic or endogenic. *Exogenic* processes operate at or near the Earth's surface and are normally driven by gravity and atmospheric forces. *Endogenic* processes are different because the energy that initiates the action is located inside the Earth. The processes themselves may operate at the surface, but their energy source is usually well below the surface. Both types of processes may sometimes be involved in the development of the same landform. For example, the shape of a volcanic cone is the product of both endogenic volcanism and normal exogenic slope processes.

In sum, we suggest that geomorphology can be examined by using physical concepts that revolve around the application of force on surface materials. In our model the effect of processes depends on how vigorously the forces drive them and how strongly their action is resisted by the geological framework. Process, in this sense, allows us to explain the incredible variety of landforms at the Earth's surface.

Driving Forces   Having suggested that energy is exerted on Earth materials as a driving force, we should briefly examine the major forces in our systems. Although each of these has been detailed after long and careful study, we will treat them only briefly to fit our specific needs.

Climate   Radiation emitted from the sun is the major source of energy needed to drive exogenic processes. Radiation is expressed in terms of heat, a form of energy possessed by molecules of matter because of their motion. Heat could be expressed in normal units of energy, but it has historically been measured in the special, more convenient units of calories or British thermal units (Btu). These are simply measurements of the amount of heat required to raise the temperature of a specified mass or weight of water one degree.

If an imaginary plane were placed at the outer limit of the atmosphere, perpendicular to the incoming rays of sunlight, it would receive 2.0 cal/cm²/ min of radiant energy over its entire surface. This value, called the *solar constant,* represents the small fraction of the estimated 100,000 cal/cm²/min of energy produced by the sun that survives the long journey to the Earth. The solar constant averaged over the entire surface is only 0.5 cal/cm²/min. As

Table 1.2 Annual heat balance and the transfer of heat in different latitude zones.

Zones of Latitude (degrees)	Fraction of Total Area	Short-Wave Radiation Absorbed (cal/cm²/min)	Long-Wave Radiation Emitted (cal/cm²/min)	Poleward Transport of Heat Across Latitude Parallels (cal/min)
0–20	0.34	0.39	0.30	
				57×10^{15} (20°)
20–40	0.30	0.34	0.30	
				77×10^{15} (40°)
40–60	0.22	0.23	0.30	
				50×10^{15} (60°)
60–90	0.14	0.13	0.30	
Weighted mean		0.30	0.30	

sunlight passes through the atmosphere, another segment of the radiation is reflected back into space by clouds, particulate matter in the atmosphere, and the Earth's surface. Therefore, the amount of energy absorbed in the system (*insolation*) and actually available for work averages about 0.30 to 0.35 cal/cm²/min over the entire globe.

The average insolation value varies greatly, however, with latitude (table 1.2) and with the seasons. Since the total heat budget does not change, the earth-atmosphere system must return to space as much heat as it receives, which it does in the form of long-wave, blackbody radiation. It is significant that although absorbed radiation decreases with increasing latitude, heat loss is fairly constant (table 1.2). This produces an obvious temperature differential between the equator and the poles, which drives a poleward transfer of heat in the oceans and, even more, in the atmosphere. The transfer of solar energy demands a series of complex processes that generate the various components of our weather. These processes tend to establish reasonably well-defined temperature and precipitation patterns for all portions of the Earth's surface. The average weather conditions at any place, considered over a long period of time, are called *climate*. Climate represents the net result of how solar energy is distributed in the earth-atmosphere system. In combination with the surface material of a region, climate determines vegetation, weathering and soil-forming processes, and the hydrology. Combined with gravity, it controls glaciation, mass wasting, and fluvial processes.

Some of the heat absorbed at the surface is transferred by *conduction* into exposed Earth materials (rock, soil, etc.) and into the air above the surface. How much heat is conducted into each medium and how deeply it penetrates depend on the physical properties of the air and surface matter. As Petterssen (1964) points out, these conductive properties can be linked to several important facts about thermal distribution. First, the high heat capacity of water makes large lakes and the oceans natural storage bins for heat. Second, the temperature ranges over continents are significantly greater than over the oceans. Third, the depth of heat penetration will be much larger in air or water than in solid, denser materials.

Considered separately in terms of radiation and of conduction, the Earth's surface and the atmosphere do not have balanced heat budgets. The surface gains more heat than it loses by these methods; the atmosphere loses more than it gains. What balances the thermal ledger on a local scale is another process of heat transfer, *convection,* which causes hotter and lighter air (or water) to move toward zones of lower temperature. Air near the Earth's surface is warmer and so will rise into the cooler zones higher in the atmosphere. Some heat absorbed at the surface is used in the process of evaporation and transferred into the atmosphere in a latent form, along with the ascending water vapor. About 600 cal of heat is absorbed by air when one gram of water is evaporated. It is subsequently released in the atmosphere as sensible heat during condensation and precipitation. This heat usually is liberated at some distance from the point of evaporation, and therefore the redistribution of heat depends partly on air motions and the conditions needed to produce condensation.

The combined effects of these factors help explain the great differences between oceanic and continental climates. Continents heat and cool faster than oceans, with more extreme variations in temperature. Since the thermal character of the surface controls the heating of the adjacent air, we can expect temperatures over land and water to function the same way as the surface itself. The seasonality of mid-latitude continental climates can also be understood in terms of the relative rates of heating and cooling and the associated pressure changes of the land-water settings.

On a larger scale, the inequality of heat with latitude (table 1.2) requires a transfer of heat from the equatorial region to the poles. The precise mechanics of this transfer involves a series of complexly interrelated processes that are a basic concern of meteorology. Variables include world circulation patterns of air masses, vertical and horizontal pressure distribution, fronts, rotation of the Earth, and the distribution of landmass and oceans. In general, heat is transferred poleward by air motions, controlled by the average air pressure and wind patterns of the Earth.

As figure 1.5 shows, at latitudes near 30° (north and south), high-pressure centers dominate the circulation pattern and drive the persistent trade winds at low latitudes. Where the trade winds converge, an equatorial trough is created, and humid low-pressure air forms the rising limb of a giant convection cell. The air moves poleward and sinks, as the descending limb of the convection cell, along the zone of high pressure at 30° latitude. Sinking high-pressure air inhibits cloud formation and causes the predominance of desert conditions in this region. Thus, at low latitudes most heat it transferred poleward by convection, circulation, and the evaporation-precipitation process.

In the middle latitudes, heat migrates in association with cyclone and anticyclone wind circulation in a zone of unsteady westerly air flow ("prevailing westerlies"). Here cold polar air meets the air moving poleward from lower

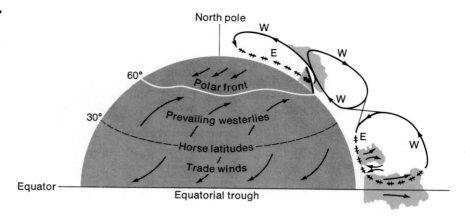

Figure 1.5.
Prevailing wind patterns and cellular air motions in the Northern Hemisphere. (Adapted from Rossby 1941)

latitudes. Instead of mixing, the different air masses remain clearly defined and are turned into warm low-pressure masses rotating counterclockwise and cool high-pressure masses rotating the opposite way. In the middle latitudes, much of the poleward heat transfer is associated with atmospheric disturbances.

At higher latitudes, warmth is delivered poleward in the upper air by horizontal mixing, called *advection,* when warm air rides over southward-flowing polar air at the polar front.

Ocean currents also are important dispensers of heat. Warm ocean currents moving toward the poles flow under cold air and give off energy as latent heat that ultimately warms the overlying atmosphere. The distribution of heat is not complete in the sense that all regions attain equivalent temperatures. It is effective, however, in stabilizing temperature conditions on a regional basis, providing the Earth with a rather well-defined temperature pattern.

Precipitation is also controlled by the same processes that distribute temperature, and the pattern of precipitation generally follows the major zones of atmospheric circulation. Precipitation is greatest near the equatorial trough and least near the 30° sinking limb of the low-latitude convection cell (fig. 1.6). Secondary peaks of precipitation occur in the middle latitudes with the unsteady pressure cells of that region. Precipitation differs from temperature, however, in that it can only occur when large air masses are cooled. There must be some triggering mechanism that causes moisture-laden air to ascend and be cooled by the decreasing temperatures at higher elevations. The principal mechanisms that lift air masses and thereby initiate precipitation are convection, orographic effects, and frontal mechanics. In convection, warm air, being lighter than surrounding air, rises until condensation forms the familiar bulging shape of cumulus clouds. The orographic effect occurs when air masses are forced to rise over high mountain ranges. Frequent rain occurs on the windward side of the mountains, with a characteristic rain shadow on the leeward side. Frontal activity involves the interaction of low-pressure and high-pressure zones. A *front* is simply the line of contact between the moist, warm air of the low-pressure cells and the cool, dry air associated with high-pressure cells. At the frontal contact, warm air is forced to rise up and over the cooler air and so is cooled; the result is precipitation.

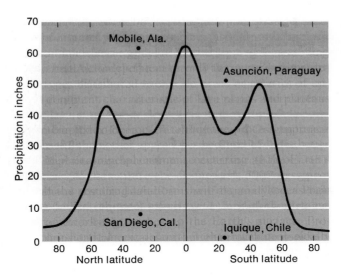

Figure 1.6.
Generalized world distribution of
average precipitation according to
latitude.

Gravity   The second major driving force, gravity, manifests itself in a myriad of both endogenic and exogenic geomorphic processes. Combined with the climatic engine, gravity determines the rigor of fluvial power, mass wasting, glaciation, tidal effects on coastal processes, and the movement of ground water. Internally, gravity bears directly on the process of isostasy, which tends to control the distribution of Earth materials of different densities, ultimately powering regional uplift. Gravity is ubiquitous, affecting all substances. The force of gravity is applied continuously in every system at, above, and beneath the surface and so can never be completely ignored in any consideration of process.

Sir Isaac Newton's classic work on the force of gravity was published in 1687, introducing his law of universal gravitation. Simply stated, the law says that there exists between any two objects a mutual attractive force that is a function of the two masses (m_1 and m_2), the distance separating them (r), and the universal gravity constant (G):

$$F = G\,\frac{m_1\,m_2}{r^2}.$$

Thus, gravity attraction between two objects is an action-reaction phenomenon. Each body exerts a force on the other that is equal in magnitude but oppositely directed along a straight line joining the two bodies. Our main interest in gravity is how it affects geomorphology, especially surficial matter. The gravitational force exerted on surface materials is measured in terms of the amount of acceleration that the force imparts to any freely falling particle having mass. It is normally expressed by the equation

$$g = \frac{GM}{r^2},$$

where M is the mass of the Earth. In most scientific work g is assumed to be constant, having a value of 980 gals (a *gal* being a unit of gravity, which is 1 cm/sec/sec).

From this equation, it is obvious that *g* in fact cannot be a constant, as we normally assume, because it depends on several variable factors. The distance (*r*) changes because of topographic irregularities and because the Earth is not a perfect sphere. The density of the Earth materials is not evenly distributed and so may vary along the line connecting the masses. In addition, the rotation of the Earth introduces a counteracting force and causes a distinct latitudinal variation in gravity. Therefore, *g* is not distributed regularly at the Earth's surface. This fact is a justifiable concern of geophysicists because slight variations in gravity have real significance, especially as an exploration tool. However, the variation in gravity at the Earth's surface is so small compared with the total magnitude that for most exogenic analyses *g* can reasonably be considered to be constant, and this is normal practice in process analyses. On the other hand, the minor variations that reflect internal density or mass differences are extremely important in endogenic processes. We will discuss gravity again in chapter 2 when considering isostatic adjustments as a geomorphic process.

Internal Heat   Thermal energy is generated inside the Earth, primarily by radioactive decay and secondarily by friction caused by earth tides and rock deformation. The exact amount of heat available for geologic work is unknown because thermal characteristics deep within the Earth must be estimated from other physical properties (density, pressure, gravity, etc.) that have been determined from analyses of seismic waves. Internal heat can be measured directly only in deep wells or mine shafts; any postulated thermal distribution below the outer fringe of the crust is based on assumption, not observation. Not only are we uncertain about the physical and chemical properties of subcrustal rocks, but hypotheses about temperature distribution tend to involve us in consideration of how much heat the Earth obtained during its formation and early history. Because of the ambiguities involved, estimates of thermal gradients within the Earth vary considerably (see Wyllie 1971). Temperatures proposed for a depth of 1000 km, for example, differ by as much as 1500°C; even at a relatively shallow depth of 100 km, estimated temperatures vary by approximately 600°C.

Regardless of the many problems inherent in determining the vertical distribution of temperature, it is a fact that the Earth transmits to the surface about 2.4×10^{20} cal each year of its internal heat. The total amount of heat is minor compared with the heat received at the surface from solar radiation, but it does indicate that heat, no matter what its origin or gradient, is being transferred from place to place within the Earth. The mechanics of heat transfer is significant since the energy distributed drives internal geological processes. Like its atmospheric counterpart, internal heat is transferred by several methods. Conduction is very slow because of the low conductivity of silicate minerals, but nonetheless it is the dominant transfer mechanism in the crustal layers. Convection as a method is still hypothetical since it cannot actually be seen, but its presence is supported by observed tectonic features, such as the evidence for seafloor spreading, that are virtually inexplicable without some type of convective overturn. Theoretically, convection is caused by temperature differences at depth (presumably in the mantle) that heat rocks locally and thereby create a less dense mass. The hot, light rocks rise toward the surface as cool, denser rocks are simultaneously sinking to replace the ascending mass. In this way, rock materials of different heat and density are continuously exchanged, following the path of a large convective "cell." The excessive heat at depth is carried along with the rising rock masses and released closer to the surface, efficiently transferring heat.

Measurements of heat reaching the Earth's surface are difficult and costly, and often they are affected by secondary factors such as ground water, variations in conductivity, and recent volcanism. In addition, measurements are not randomly spaced but tend to be concentrated in areas of some specific interest so that large regions exist for which little or no data are available. Nonetheless, the development of sophisticated instrumentation and the current interest in ocean tectonics have produced a storehouse of information that is beginning to yield a reasonable picture of surface heat flow. Except for local abnormalities, heat emerges from all parts of the Earth in amazingly equal amounts, with average continental and oceanic values differing by only 0.2 μ cal/cm^2/sec (Wyllie 1971). If radioactivity is the major source of internal heat, the equality of heat flow from continents and ocean floors would require an unusual distribution of radioactive minerals beneath the two environments unless the thermal condition were balanced by a convective process. Such a process may be demonstrated by examining heat flow for major physiographic regions of the Earth, as presented by Lee and Uyeda (1965) and shown in table 1.3. Note that heat flow from ocean ridges and trenches differs considerably from average values for entire ocean basins; ridge crests are abnormally high and trenches notably low. Heat may be actively rising under ocean ridges as part of a convective overturn, while the low heat values beneath the trenches presumably represent the descending limbs of the overturning cells. On continents, as one would expect, the lowest heat flow values occur in the very stable shield areas and the highest in the most recent orogenic belts and their associated regions of Cenozoic volcanism.

The transfer of internal heat plays a significant role in determining the major topographic framework of Earth. Heat transfer drives processes beneath the surface causing uplift and deformation, distributes rock masses of varying resistance, and controls the volume of ocean basins, thereby influencing the position of sea level. Precisely how or if heat flow relates to gravity distribution is debatable, but certainly the two forces combined represent a major geomorphic element.

The Resisting Framework   As pointed out earlier, landforms reflect a balance between the application of driving forces and the resistance of the material being worked on. Having reviewed the salient features of the driving forces in our systems, we should now examine the resisting elements, but exactly how to do so is rather perplexing. It is tempting simply to state that the

Table 1.3 Heat flow values from major geologic features.

Geologic Feature	Average Heat Flow (μcal/cm^2/sec)
Land Features	
1. Precambrian Shields	0.92
Australian Shield	1.02
Ukrainian Shield	0.69
Canadian Shield	0.88
S. African Shield	1.03
Indian Shield	0.66
2. Post-Precambrian	
Nonorogenic areas	1.54
Europe	1.67
Interior Lowlands, Australia	2.04
Interior Lowlands, N. America	1.25
S. Africa	1.36
3. Post-Precambrian	
Orogenic areas[a]	1.48
Appalachian area	1.04
E. Australian highlands	2.03
Great Britain	1.31
Alpine system	2.09
Cordilleran system	1.73
Island arcs	1.36
4. Cenozoic volcanic areas[b]	2.46
Ocean Features	
1. Ocean Basins	1.28
Atlantic	1.13
Indian	1.34
Pacific	1.18
Mediterranean seas	1.20
Marginal seas	1.83
2. Ocean Ridges	1.82
Atlantic	1.48
Indian	1.57
Pacific	2.13
3. Ocean Trenches	0.99
4. Other ocean areas	1.71

From Lee and Uyeda, *Geophysical Monograph 8*, 1965. Copyrighted by American Geophysical Union. Reprinted by permission.
[a]Excluding Cenozoic volcanic areas.
[b]Excluding geothermal areas.

resistance in geomorphic systems is geology—the geologic affect on geomorphology is so pervasive and so varied that any brief review of its role in determining process and form must be inadequate. A complete discussion of the geological control of geomorphology would require an analysis of every possible geologic framework in every possible climatic and tectonic regime. Although such an effort is impossible here, some general examples will show how geological resistance manifests itself in landforms.

Lithology   The resisting force in geomorphology is implemented through the two major geologic variables, lithology and structure. The diverse origins of rocks create lithologies at the surface that differ vastly in their chemical and mineralogic compositions, textures, and internal strengths. In geomorphology

Table 1.4   Weight percent of common elements in Earth's lithosphere.

	Continental Crust			Oceanic Crust		Total Lithosphere	
	Polder-vaart 1955	Pakiser and Robin-son 1966	Ronov and Yaro-shevsky 1969	Polder-vaart 1955	Ronov and Yaro-shevsky 1969	Polder-vaart 1955	Ronov and Yaro-shevsky 1969
SiO_2	59.4	57.8	61.9	46.6	48.7	55.2	59.3
TiO_2	1.2	1.2	0.8	2.9	1.4	1.6	0.9
Al_2O_3	15.5	15.2	15.6	15.0	16.5	15.3	15.9
Fe_2O_3	2.3	2.3	2.6	3.8	2.3	2.8	2.5
FeO	5.0	5.5	3.9	8.0	6.2	5.8	4.5
MgO	4.2	5.6	3.1	7.8	6.8	5.2	4.0
CaO	6.7	7.5	5.7	11.9	12.3	8.8	7.2
Na_2O	3.1	3.0	3.1	2.9	2.6	2.9	3.0
K_2O	2.3	2.0	2.9	1.0	0.4	1.9	2.4

Table 1.5   Abundance of rock and mineral types in the Earth's crust.

Rocks	% of Crustal Volume	Minerals	Modal %
		Quartz	12
Sands	1.7	K-feldspar	12
Clays and shales	4.2	Plagioclase	39
Carbonates	2.0	Micas	5
Granites, gneiss and crystalline schist	36.9	Amphibole	5
		Pyroxene	11
Granodiorite and diorite	11.2	Olivine	3
Syenite	0.4	Clay	4.6
		Calcite and dolomite	2.0
Basalt, gabbro, amphibolite, eclogite	42.5	Magnetite	1.5
Peridotite, dunite	0.2	Others	4.9
Total	99.1	*Total*	100.0

Modified from Ronov and Yaroshevsky, *Geophysical Monograph 13*, pp. 37–57, 1969. Copyrighted by American Geophysical Union. Reprinted by permission.

we are concerned with the modern resisting framework, regardless of its history. It is important to gain an overall picture of the crustal and surface rock distributions as they presently exist.

Table 1.4 synthesizes several estimates of the bulk chemical composition of the Earth's lithosphere. As expected, the chemistry of continental crust is higher in silica and K_2O than that of oceanic crust, and lower in CaO, MgO, and total iron. Such a chemical distribution can be converted into reasonable estimates of the volume-percentage of common rock types and their modal mineral composition (table 1.5). The significance of these analyses is to emphasize that the resisting framework in geomorphology basically entails only

Table 1.6 Volume percentage and chemical composition of silicic and mafic crust in the United States.

Western Provinces	Silicic	Mafic			Western	Eastern
California coastal			SiO_2		60.0	57.1
region	75	25	TiO_2		1.1	1.3
Sierra Nevada	50	50	Al_2O_3		15.1	15.2
Pacific NW (coastal)	28.6	71.4	Fe_2O_3		2.3	2.3
Columbia Plateau	22.3	77.7	FeO		4.9	5.7
Basin and Range	66.7	33.3	MgO		4.5	5.6
Colorado Plateau	62.5	37.5	CaO		6.3	7.5
Rocky Mountains	62.5	37.5	Na_2O		3.0	3.0
			K_2O		2.0	2.1
Average	56.4	43.6				
Eastern Provinces				*Total*	99.2	99.8
Interior Plains						
and Highlands	40.0	60.0				
Coastal Plain	57.2	42.8				
Appalachian Highlands						
and Superior Upland	37.5	62.5				
Average	42.6	57.4				
Total United States	46.3	53.6				

From Pakiser and Robinson, *Geophysical Monograph 10*, pp. 620–26, 1966. Copyrighted by American Geophysical Union. Reprinted by permission.

two igneous and metamorphic rock suites, and approximately ten mineral varieties. The crust consists primarily of a silicic assemblage (granites, gneisses, schists, granodiorites, and diorites) that makes up 48 percent of the crustal volume and a mafic association that constitutes about 43 percent. Obviously the silicic group is plutonic or metamorphic in origin and is dominantly continental; the mafic types are overwhelmingly volcanic and rooted beneath the oceans.

The crust beneath the conterminous United States, however, is more mafic than one might guess (table 1.6). Pakiser and Robinson (1966) point out that, based on seismic velocities, the total U.S. crust is 54 percent mafic by volume (55 percent by weight). In addition, they show that the mafic content is considerably greater in the provinces of the eastern United States. In general, the eastern regions have a crust that is predominantly mafic, and the western provinces a crust that is mostly silicic.

If Pakiser and Robinson are correct, it is even more interesting to examine the igneous rocks exposed at the surface in the Appalachian and Cordilleran regions (table 1.7). In the Appalachians, where the crust is predominantly mafic (as noted in table 1.6), the surface igneous rocks are overwhelmingly calc-alkalic, plutonic rocks. Of the rocks of this type indicated in table 1.7 (84.5 percent of the total), 96 percent of the plutons are granites. In contrast, the igneous rocks exposed in the Cordilleran system are mainly extrusive (63.6 percent), and of these 77 percent are basaltic or andesitic in composition. The thick mafic crust in the eastern United States supports a surface rock assemblage that is dominantly granitic, in contrast to a silicic crust supporting mafic surface rocks in the west.

Table 1.7 Area of different igneous rocks exposed in the Appalachian and Cordilleran regions of the United States (percent).

	Cordilleran	Appalachians
Plutonic Rocks		
Calc-alkalic Rocks (granite, granodiorite, quartz monzonite, quartz diorite, diorite, gabbro, anorthosite)	33.6	84.5
Alkalic Rocks (syenite, monzonite, others)	0.4	neg.
Ultramafic Rocks (periodotite, pyroxenite)	0.5	neg.
Hypabyssal Intrusives		
Calc-alkalic Rocks (porphyries, quartz diabase, diabase)	1.5	7.4
Alkalic Rocks (porphyries)	0.3	0.2
Extrusive Rocks		
Calc-alkalic Rocks (basalt, dacite, andesite, rhyolite)	63.4	7.9
Alkalic Rocks (trachyte, latite, phonolite, others)	0.2	neg.
Total	99.9	100.0

Adapted from Daly, P., (ed.), *Igneous Rocks and the Depths of the Earth.* © 1933 McGraw-Hill Inc. Used with permission of McGraw-Hill Book Company.

Table 1.8 Rocks exposed at the surface of the North American continent (expressed as % of area).

	Gilluly 1969	Blatt and Jones 1975
Sedimentary	61.5	52
Volcanic	8.2	11
Plutonic	3.8	6
Metamorphic and total PC	26.5	31

In North America, sedimentary rocks make up most of the exposed materials (table 1.8) even though they are only a minor constituent of the total crustal volume. Their ultimate source, however, is older igneous, metamorphic, and sedimentary rocks, and so their chemistry and mineralogy reflect changes induced by exogenic geomorphic processes. Geomorphology, therefore, becomes an important link in the rock cycle.

The wide areal distribution of sedimentary rocks undoubtedly causes a surface mineral composition different from that shown in table 1.5. At the surface, quartz and feldspars are dominant and probably exist in equal amounts (feldspar 30 percent, quartz 28 percent); calcite and dolomite increase to about 9 percent; and clay minerals and micas become much more significant, rising to approximately 18 percent of the surface material (Leopold et al. 1964).

In any given climate each rock type will respond to the processes of weathering and erosion in a different manner and at a different rate. With time and tectonic stability, high-standing landmasses commonly will be underlain by resistant rocks, and low-standing regions will be formed from rocks that are more susceptible to weathering and erosional attacks. These effects of differential weathering and erosion in landscape development are stressed in every introductory course in the basics of geology. In fact, we are conditioned early in our geological training to view regional topography as a mirror of gross lithology, tectonics, and geologic history. For example, the concept of physiographic provinces stresses this approach, causing us to think of geological controls in geomorphology as regional phenomena. It is worthwhile to emphasize, however, that geomorphic processes will accentuate lithologic differences on many scales. Mega-scaled differentiation produces regional features such as mountains and plains (fig. 1.7), and can be utilized in an erosional topography as a first approximation of the gross lithologic distribution. Within any large region of similar rock type, small lithologic discrepancies will also surrender to geomorphic processes and appear at the surface as minor landform deviations. These tiny blips in the general landscape provide critical information about geological history and exert important controls on subsequent geomorphic developments (figs. 1.8 and 1.9). Even on a microscopic scale, lithologic variations may have a distinct effect on the style of weathering (fig. 1.10), ultimately causing subtle topographic differences (Eggler et al. 1969). It seems certain that even long periods of erosion cannot completely erase the

Figure 1.7.
Mountains and surrounding plains, looking west-southwest from a point about 1.6 km northeast of Boulder at an elevation of 2,152 meters. Boulder County, Colo., ca. 1934.

Figure 1.8.
A topographic irregularity caused by differences in lithology—West Spanish Peak, from the northwest. Dikes cutting flat-lying Eocene strata. Spanish Peaks quadrangle, Huerfano County, Colo.

Figure 1.9.
Variations in lithology as evidenced in cuestas formed by hard sandstones north of Galisteo Creek, N.M. The rocks in succession from left to right are Mansamo red beds, Morrison, Dakota, and Mancos (Galisteo Creek and the Santa Fe Railroad in foreground).

influence of minor lithologic abnormalities from the landscape (Flint 1963), although their appearance may be greatly subdued.

Lithologic diversity must be considered on a variety of levels. Large areas underlain by crystalline rocks or sedimentary rocks may develop a distinct regional character, but smaller variations within the region are revealed in subtle topographic changes that often provide significant geologic and geomorphic information. The geomorphologist must be able to "read" these subtle topographic modifications in order to present a coherent interpretation of history and process.

Structure   Geologic structures that influence landforms also range in magnitude from large, areawide tectonic styles to minor features that exert only local control (Lattman 1968). Structural influence is readily apparent only when the rocks and climate involved are conducive to differential weathering and erosion. In depositional environments, structures may be buried by thick accumulations of sediment that mask the surface expression of the underlying structure. Comparably, the internal structure may not be immediately evident in erosional topography formed in areas with distinctly similar lithology, such as shields or crystalline mountain cores, but minor structures still may produce a discernible topographic control (Flint 1963). Spacing of joints, for example, is recognized as a prime factor in the development of the longitudinal "staircase" profiles that characterize glaciated valleys in mountains held up

Figure 1.10.
Disintegration of granitic boulders due to expansion of biotite grains. Boulders in terrace gravel near Red Lodge, Mont.

by rocks of uniform lithology. The most likely lithologic environment to display structural control is a sedimentary sequence with alternating resistant and nonresistant units, such as the Valley and Ridge province of the Appalachian Mountains. There resistant sandstone and conglomerate layers form ridges that are separated by intervening valleys underlain by easily eroded shales and limestones. The regional topography reveals the pervading structure of plunging anticlines and synclines because the ridges cross the countryside in a sinuous pattern that shows the character of the underlying folds.

Thresholds and Complex Response

The third basic principle of process geomorphology involves the **threshold concept.** It may or may not be apparent to you that any concept proposing equilibrium inherently implies a contrasting state of disequilibrium. If variations in controlling factors demand a response within the system, there must be a period of readjustment during which process and form are out of equilibrium. Landslides, subsidence, and gulley erosion are examples of disequilibrium generated when the variables of force and/or resistance are altered so they can no longer maintain a balanced relationship (fig. 1.11). They represent events that occur as systems attempt to reestablish a new equilibrium. Such events can happen suddenly or can proceed toward equilibrium over a long period of time, depending on how great the disequilibrium is and how much energy is involved.

Schumm (1973) recognized that if a system in equilibrium can be defined by real parameters, it follows that there must be parameter values that represent the limits of the balanced condition. If these limiting values are exceeded, the system enters a condition of disequilibrium. The limits of equilibrium are critical conditions called thresholds (Schumm 1973).

If parameter values are pushed to the limiting condition by variations of external controlling factors, the threshold is known as an *extrinsic threshold.* Examples are numerous in nature; geologists will be most familiar with threshold velocity in streams, at which sediment movement begins. The change in external variables (in this case the force) causes instability of the channel

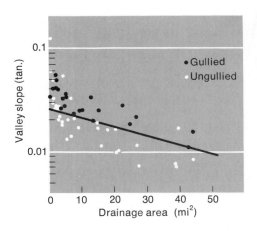

Figure 1.11.
Threshold relationship between gullied and ungullied valley floors in several drainage basins of northwest Colorado.

sediment. Other examples are found in responses to the fluctuating climate that characterized much of the Pleistocene epoch. A more subtle type of threshold, however, is the *intrinsic threshold,* where instability and failure of a system occurs even though external variables remain relatively constant. The threshold conditions develop in response to gradual, often imperceptible, changes within the system. In many cases the threshold represents a deterioration of resistance rather than an increase in driving forces. For example, a region characterized by periodic heavy rains may have stable slopes for a long time, but continuous freeze and thaw or other soil-forming processes gradually reduce the cohesion of slope material. Eventually one storm, no more severe than thousands that have preceded it, triggers slope failure.

A special type of threshold, called a *geomorphic threshold* (Schumm 1973, 1980), essentially refers to the stability of a landform itself. Originally Schumm (1973) considered this to be a type of intrinsic threshold; i.e. progressive alteration of the landform, especially the slope factor, eventually reaches a stability limit and demands a rapid adjustment in the character of the feature. The significance of the concept is that abrupt changes may be a normal part of landscape development and do not always require a change in external controls to precipitate the event. Schumm (1980) has since expanded the concept to include externally generated changes in landforms.

It can be demonstrated that a threshold response often occurs as a series of reactions called **complex response.** The sequence of events happens because all processes and components of a system may not reach the threshold condition at the same time. This phenomenon was demonstrated experimentally by Schumm and Parker (1973) in a study of an artificial drainage basin. An induced base level decline at the mouth of the basin caused downcutting of the trunk river at that point and the formation of terraces. At the same time, however, tributary channels were unaffected and remained in their equilibrium state. With time, the site of channel incision migrated progressively upstream until the base level of each tributary was lowered and channel entrenchment ensued. The tributary incision, however, provided so much sediment to the trunk river that aggradation began at the basin mouth because the stream was incapable of transporting the increased load derived from entrenchment of the tributary channels. Clearly, the processes functioning in different parts of the systemic basin were out of phase.

It is not difficult to imagine the same sequence of events occurring during major glacial stages when sea level declines dramatically. The effect of that base level decline will be initially felt at the mouths of major rivers such as the Mississippi. Tributaries in that huge basin may not experience the expected incision until long after the initiating event.

Actually, natural complex responses similar to the experimental study have been documented. For example, the changes in fluvial systems in response to hydraulic mining for gold on the west flank of the Sierra Nevada produced a sequence that was generally the same but diametrically opposed in detail to the Schumm and Parker study. Gilbert (1917) was able to show that as mine tailings were released from the mountain deposits, the coarse fraction gradually invaded the channels of the major downstream rivers as sand and gravel bedload. The rivers, unable to transport such an overwhelming load, adjusted by drastically raising their channel bottoms as the material was deposited. As each segment filled, the gradient increased so that the river acquired the capacity needed to transport the sediment farther down the valley. The rise in channel level stopped at different times in each segment of each river, depending on the distance from the source and the amount of load. For example, the channel of the Yuba River at Marysville (Cal.) rose about 6 m (19.1 ft) between 1849 and 1905, when it reached its highest level. The Sacramento River at Sacramento (Cal.) elevated about 3 m (10.8 ft), attaining its highest level in 1897. It is interesting to note that both rivers continued to fill even after mining ceased in 1884. This happened because the upstream reaches, no longer receiving great volumes of sediment, had excess energy on their steepened gradients and therefore entrenched their channels. The sediment from the entrenching process was transported to the lower river segments. Thus, part of the fluvial system was filling while, at the same time, other parts were entrenching.

The complex response observed by Gilbert differs from that of Schumm and Parker because (1) the force initiating the threshold occurred near the basin divides rather than at the trunk stream mouth, (2) the driving force was associated with human activity, and (3) the initial response was aggradation, the site of which proceeded downvalley.

Thresholds in geomorphology were first demonstrated in fluvial systems. They have since been recognized in almost every aspect of the discipline (for examples see Coates and Vitek 1980), and innovative methods are being devised to indicate when a system is tending toward a threshold condition (Bull 1979, 1980). In addition, it is increasingly clear that the systemic stress that produces instability is commonly generated by human activity. Thus, the widespread applicability of the concept has prompted the suggestion that it should be the primary working model of geomorphology (Coates and Vitek 1980). It is difficult to argue with that suggestion because it is when stability limits are exceeded that things begin to happen, and many deleterious events at the surface may be nothing more than nature's way of responding when a threshold is passed. With that consideration, it becomes critical for geomorphologists to define threshold values for every environment and for all conceivable combinations of process and geology. Such information would be extremely important for future land management and could be the foundation for identifying natural hazards and predicting imminent disasters.

The Principle of Process Linkage

Complex adjustments to altered conditions often involve a chain reaction of responses that we will call process linkage. **Process linkage** essentially operates on the domino principle; it means that the changes that occur in one process or landform during an adjustment period often initiate subsequent responses in totally different processes and/or landforms. Linkage works because a driving force can transfer from one process type to another as its effect filters through a system, or it can even shift to processes operating in totally different systems. Thus, a myriad of different processes can be involved in the response to a single threshold-inducing force.

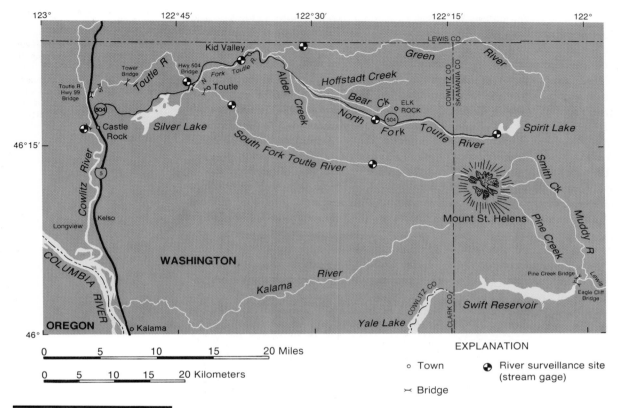

Figure 1.12.
Map showing location of river surveillance sites in the lower Toutle and Cowlitz river systems. (From Lombard et al. U.S. Geological Survey Circular 850, 1981)

A recent case history exemplifies how process linkage works. On May 18, 1980, Mount St. Helens in southwestern Washington was blown apart by a violent volcanic eruption. The widespread effects of the eruption have been documented in a series of short papers published as U.S. Geological Survey Circular 850. The initial process response occurred during the eruption as a massive debris avalanche that deposited enormous volumes of rock, ice, and other debris in the upper 17 miles of the North Fork Toutle River valley (fig. 1.12). The deposits are up to 600 feet thick at places. Physical, chemical, and biological characteristics of lakes close to the eruption were drastically altered, and benthic faunas in the adjacent rivers were destroyed.

Immediately following the avalanche, snow and ice that had melted during the eruption provided enough water to generate a mudflow in the same valley. In addition to environmental damage, the mudflow deposited about 25,000 acre feet of sediment in the Cowlitz River channel (fig. 1.12). This elevated the channel floor and decreased its cross-sectional area, making the valley bottom prone to more frequent flooding (fig. 1.13). Furthermore, a significant volume of sediment reached the Columbia River, where it created a shoal area that blocked the channel used for shipping (see Lombard et al. 1981).

The Mount St. Helens catastrophe involved a number of process links which demonstrate that the location of the dominant response shifted progressively downstream. In addition, the single driving force in this case was

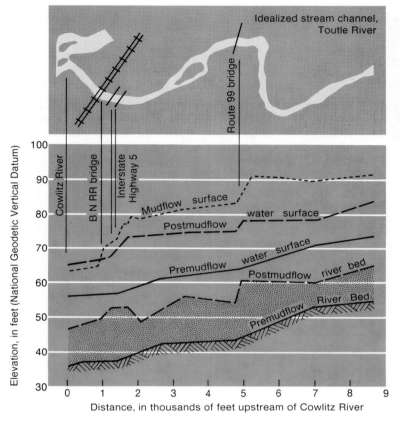

Figure 1.13.
Channel bottom and surface elevation of the lower Toutle River prior to and after the mudflows of May 18, 1980. Pre- and post-eruption water surfaces are based upon a flow of 38,000 cubic feet per second. (From Lombard et al. U.S. Geological Survey Circular 850, 1981)

produced inside the Earth. The results of volcanic action, however, were quickly transferred into slope processes (debris avalanche and mudflows) and from there into fluvial, hydrologic, and lake systems and processes. Physical changes even altered the biological balance.

The remarkable ability of process action to shift from one form to another in response to a single impetus is a critical ingredient in process geomorphology. It often provides the only explanation for isolated or apparently unprovoked geomorphic actions.

The Time Framework

Earlier in the chapter we saw that geomorphology can be considered over different time intervals, which Schumm and Lichty (1965) called cylic, graded, and steady. The question before us now is which of these time spans is most conducive to demonstrating the mechanics of process geomorphology?

To visualize the time factor more clearly, consider a hypothetical drainage basin and the component subsystems (rivers) within it. If we observe a single cross section of any river, we can describe the channel morphology at that point by a group of parameters such as width, depth, slope, and shape. Any single flow event in the river may bring about temporary scouring or filling, which causes immediate changes of the channel parameters but does not affect them permanently; they will return to their previous state when the flow event passes. Measurements taken an hour or day apart will show different values for the variables, but they will always be internally consistent and apparently adjusted to their external controls. Significantly, observed sediment and water discharges through the cross section are dependent variables when viewed in this time sense because they are modified by changes in the channel configuration. For example, if the river scours its bed, the moving load will increase and the channel area will expand, allowing a greater volume of water to be held within the banks. That is, both water and sediment discharge are temporarily affected by the channel changes. However, we are observing the river for only a very short period of time, and any equilibrium that we define is almost instantaneous (or "steady" in the terminology of Schumm and Lichty). Since no permanent changes can be expected on this temporal scale, we can justifiably say that processes and landforms are time-independent when considered in this sense (see fig. 1.4).

Over longer periods, the steady-time channel measurements will vary with mean values of sediment and water discharge; the statistical relationship between the external and internal variables defines the equilibrium state for graded time (fig. 1.4). On this time scale (unlike steady time), sediment and water discharge function as independent variables to determine the morphologic and flow characteristics of the stream reach. In graded time, changes in external variables due to modifications of climate or base level may require a new set of equilibrium conditions within the channel. If and when thresholds are exceeded, some significant change in the river is necessary and (again in contrast to steady time) the adjustment will be permanent unless still another change occurs in the external controls. Commonly the response takes the form of channel incision or aggradation, pattern adjustments, or modification of sinuosity.

The point is that the system is temporarily out of balance, and there exists a certain time interval during which the river is approaching a new equilibrium state. The time involved is intermediate between long-term cyclic time and instantaneous steady time. Absolute time in years is not the important element in recognizing graded time as a valid geomorphic concept. Its significance lies in the fact that, once equilibrium is achieved, disruption of the balance will be counteracted by each subsystem's ability to reestablish a new equilibrium quickly (in the geologic sense); graded time thus does not involve continuous and progressive change in the landscape. In addition, all parts of the regional system may not be affected simultaneously or in precisely the same way.

If we consider our basin over cyclic time, the balance (or temporary imbalance) seen in steady or graded time becomes irrelevant. The inexorable loss of sediment and energy from the basin suggests that the system must be continuously approaching, but perhaps never attaining, an ultimate equilibrium condition. In this time framework, landforms within the basin should be progressively losing relief in phase with the deteriorating systemic energy.

Clearly a hierarchy of time exists; how one is able to interpret geomorphology depends, in a real sense, on which time scale is used. In process geomorphology, analyses are best made on a steady- or graded-time scale, assuming landforms and processes to be time-independent or only temporarily time-dependent phenomena.

Summary

In this chapter we have attempted to demonstrate that the process approach in geomorphology is more useful than other approaches because it relates better to a variety of disciplines that examine phenomena occurring at the Earth's surface. A set of basic principles constitute the framework of process geomorphology. Processes and the resulting landforms will be analyzed as balances between the driving forces (climate, gravity, etc.) and the resistance offered by the geologic framework that makes up the Earth's surface. Processes will be considered on short time scales rather than over geologically significant intervals.

Suggested Readings

The following references will help you understand the approach used in this book.

Chorley, R. J., and Kennedy, B. A. 1971. *Physical geography.* London: Prentice-Hall International.

Coates, D. R., and Vitek, J. D., eds. 1980. *Thresholds in geomorphology.* London: Allen and Unwin Ltd.

Hack, J. T. 1960. Interpretation of erosional topography in humid temperate regions. *Am. Jour. Sci.* Bradley Vol. 258-A: 80–97.

Schumm, S. A. 1973. Geomorphic thresholds and complex response in drainage systems. In *Fluvial geomorphology,* edited by M. Morisawa, pp. 299–310. S.U.N.Y., Binghamton: Pubs. in Geomorphology, 4th Ann. Mtg.

———. 1977. *The fluvial system.* New York: John Wiley & Sons.

Schumm, S. A., and Lichty, R. W. 1965. Time, space, and causality in geomorphology. *Am. Jour. Sci.* 263:110–19.

Climate and Internal Forces

2

Introduction

One of the major principles of process geomorphology is that spasmodic disruptions of equilibrium are significant components of surficial mechanics. In many cases the cause of instability is external to the geomorphic system being affected. In this chapter we will examine the primary external controls on geomorphic systems.

In theory, exogenic processes unimpeded by opposing forces will gradually reduce the landscape to a rather dull, featureless surface with only minor topographic irregularities to interrupt its sameness. However, as the Earth's surface does possess relief, and has done so throughout geologic time, the Earth must be constructed in such a way that exogenic processes are not always preeminent. The surface is not a static environment but a locale where denudational processes have been repeatedly counteracted as new mass is elevated or created by endogenic forces. Each influx of mass not only brings with it escalated relief but also new potential energy that is available to accelerate or change the character of the exogenic processes. The introduction of new mass and energy occurs primarily through volcanic activity and vertical uplift of the land surface.

In addition to pulses of endogenic input, we know from the glacial and interglacial episodes of Quaternary history that the climatic regime is not a static phenomenon. Thus, although the equilibrium condition in geomorphic systems is adjusted to climate, changes in climate are equally significant because they serve as spark plugs of temporary instability.

The importance of major climate change and significant endogenic input is that they tend to occur as irregular events separated by long periods of relative constancy. This explains why these factors are able to create conditions whereby threshold responses in geomorphic systems are required. Furthermore, periods of climate change and endogenic action are the ingredients needed to decipher the sequential nature of geomorphology. Therefore, the major external controls are the bases for *Quaternary geomorphology,* another subarea of the discipline, in which the determination of history rather than process is a common goal of geomorphic investigation. In this book we will occasionally examine various aspects of how geomorphology may be used in Quaternary studies. However, we will take these excursions into Quaternary geomorphology cautiously so that we do not subvert our primary process theme.

The Endogenic Effect

The fact that endogenic forces are an integral part of what we observe at the Earth's surface is beyond question. Volcanic activity and seismicity are unmistakable manifestations of how internal energy and force can affect landforms and processes. Often, however, exertion of internal force results in equally important though less dramatic geomorphic effects. For example, figure 2.1 shows that most land surface exists at two persistent levels, which correspond to the continents and the ocean floors. The Earth's hypsometric curve (fig. 2.1) also indicates the physiographic importance of continents and the ocean floor; surfaces at elevations between sea level and 1 km account for 20.8 percent of

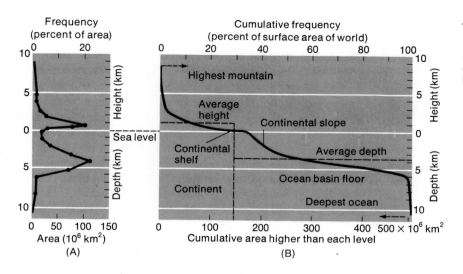

Figure 2.1.
Distribution of areas of the Earth
standing at different elevations.
(A) Distribution frequency. (B) The
Earth's hypsometric curve.

the world's total area, and ocean bottoms between 3 and 6 km in depth comprise about 52.4 percent of the total surface area. The average elevation of continents is about 0.875 km, and the average depth of the ocean floor is approximately 3.729 km.

The rocks that make up continents are drastically different in mineralogic and chemical composition from those that form the ocean basins. They are richer in silica and less dense than the ocean rocks, so that continental crust is thicker and stands higher than the ocean floors. Exactly how this monumental difference between oceanic and continental crust developed is not certain because theories concerning the origin of continents lack hard data. The reconstruction of events that led to the formation of protocontinents depends entirely on indirect evidence of the Earth's early history. Nonetheless, the maintenance of high continents and low ocean basins represents a balance perpetuated by endogenic mechanics. In addition, as figure 2.2 shows, with the exception of Antarctica, larger continents stand at higher elevations, a fact that must reflect some primary systemic operation. Our task, then, is to decipher what internal process is responsible for this topographic phenomenon.

Epeirogeny

The processes by which the Earth's crust is deformed are collectively known as *diastrophism* or *tectonism* and are commonly divided into two distinct tectonic styles. *Orogenic* processes culminate in the formation of structural mountains. Such mountain systems are typified by intense disruption of the included rocks due to folding and overthrusting; the effects are usually localized in narrow, elongated belts. In contrast, *epeirogenic* processes cause uplift or depression on a regional scale and proceed without internal disruption of original rock structures. Response to driving forces is rather passive in this type of deformation. Although gentle tilting of strata may accompany the vertical displacements, folding and thrusting are absent during the movement.

Figure 2.2.
Relationship between area and
elevation of the continents.

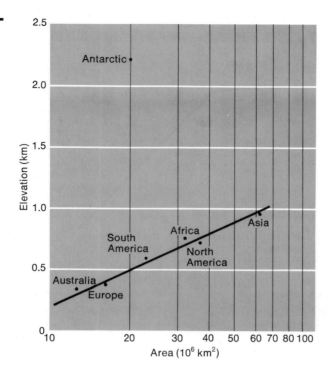

Mountain systems, however, are also affected by epeirogeny *after* the orogeny
that formed them has ended. Such vertical displacement of rocks and surfaces
is driven by the fundamental gravitational force.

In chapter 1 we briefly examined the force of gravity and the various factors that determine its effect on a body at any location. The net force on any
mass is the vector sum of all gravitational attractions acting on it. Each body,
therefore, possesses a discrete amount of potential energy because mutual attractions can be transferred into a kinetic form that is capable of doing work.
Unfortunately, all the attracting elements influencing any particular mass are
not applied in the same direction; thus, precise calculations of gravity should
be resolved into separate components operating in the orthogonal (x, y, z)
directions. This complicated procedure can be simplified by viewing gravity
as an energy field consisting of horizons of equal potential (U) in which the
attractive force is defined by

$$F = - \text{ grad } U.$$

In this field, F is everywhere normal to a series of surfaces, each of which
includes only points with equal potential and therefore a constant value of U.
In such a model the value of F can be expressed in terms of energy, and its
magnitude, perpendicular to the equipotential surfaces, is

$$F = - \frac{dU}{dr} = \frac{Gm}{r^2},$$

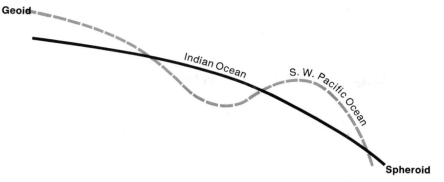

Figure 2.3.
Generalized relationship between the spheroid and the geoid along a line between east Africa and the Pacific Ocean near Australia.

where dU is the change in potential over the distance dr. By integration,

$$U = \frac{Gm}{r},$$

where U goes to zero when r is infinity.

In this model, sea level becomes an extremely important equipotential surface in the Earth's gravitational field. Even though the surface may be slightly distorted because of local factors, the inflections are small in amplitude compared with the radius of the Earth and limited in areal extent.

The sea level equipotential surface is called the **geoid,** which on land is usually defined by the water level in a series of imaginary canals cut through the solid mass. The Earth's surface topography is referred to the geoid because any elevation is determined by extending upward a succession of planes that are parallel to sea level. At any point to be measured, a surveying instrument is set tangent to the geoid, with the tangential line being the perpendicular to a vertical plumb line over the site. Like any body, however, the direction of the plumb bob itself is the vector sum of all the gravitational forces acting on it and so may not be perfectly normal to the Earth's center of gravity. To resolve this complication, geodesists utilize a second surface called the **spheroid,** which is a mathematical representation of sea level with all irregular influences removed. Essentially, the spheroid is the *hypothetical* sea level surface of an Earth with no lateral variations in density or topography and with a vertical change in density that is uniform from the center of gravity to the surface. With such a mass distribution, gravity would vary consistently from pole to equator, and its theoretical sea level values can be easily calculated. The differences between the predicted values of gravity, calculated under the above assumptions, and the actual measured values are called **gravity anomalies;** they indicate the departure of the geoid from the spheroid (fig. 2.3).

Because few gravity measurements are made at sea level, most observations must be reduced into separate components indicating the portion of the measurement produced by mass and the portion due to distance. Corrections for each must then be made before a gravity measurement can be compared with the spheroidal value. There is a general tendency for anomalies produced by mass (*Bouguer anomalies*) to be strongly negative in mountains and increasingly more negative with higher elevations, demonstrating a most important principle in geomorphology—that surface topography is somehow related to the internal distribution of mass.

The idea that topography is influenced by the distribution of mass within the Earth is not new. It was expressed by Leonardo da Vinci, and its concrete formulation as a hypothesis arose from analyses of data obtained in the mid-nineteenth-century land surveys in India. C. E. Dutton, working in the Colorado Plateau, introduced the term **isostasy** to define the internal process involved. In essence, the result of isostatic adjustment is a condition in which large, elevated regions such as continents or mountain ranges are compensated by a mass deficiency in the crustal rocks beneath them. The process of isostasy requires that at some depth beneath sea level the pressure exerted by overlying columns of rock will be the same, regardless of how high the various columns stand above sea level. Mountains, ocean basins, shields, etc., are balanced with regard to the total mass overlying each area at some internal level called the *depth of compensation*. This isostatic equilibrium is probably controlled by lateral variations in rock density (Hsu 1965; McGinnis 1966) and maintained by internal adjustments that are not clearly understood, and in fact, the mechanics may differ depending on the scale of topography being supported at the depth of compensation (Officer and Drake 1982).

A correction of measured gravity due to isostasy can be calculated, and when this is subtracted from the Bouguer anomaly, a residual value called the *isostatic anomaly* remains. If the isostatic anomaly is zero, the system is in perfect balance. We know, however, that the equilibrium condition is easily upset so that negative or positive anomalies are not unusual. A negative isostatic anomaly indicates deficiency of mass at the locality, and the surface should have a tendency to rise because more matter must be added at depth to establish the equilibrium state. Similarly, positive isostatic anomalies should portend sinking since they indicate an excess of mass beneath the surface.

Because most topographic blocks, local or regional, are not perfectly equalized with respect to one another, vertical movements of crustal segments are inherent in the attempt to establish equilibrium. When isostatic balance is disrupted by erosion, thick sediment deposition (example, Lake Mead), or tectonics, a counteraction by isostasy is required to restore equilibrium. It is also known that the accumulation of massive glaciers is accompanied by depression of the surface; conversely, when the weight is removed as the ice disappears, the surface will rise to reestablish the isostatic balance. This response to glacial and interglacial conditions is called *glacio-isostasy* and represents one of the most important geomorphic processes in high latitudes of the world (Andrews 1974).

The significance of this is that isostasy is the endogenic process that causes epeirogenic diastrophism and is responsible for maintaining the topographic relationship between large blocks of the Earth's crust. The relief between ocean basins and continents probably reflects the isostatic balance established because the crustal thickness and density of rocks underlying the two areas are different. The same process, however, functions on scales smaller than continents or ocean basins.

The fact that isostasy works is uniquely important in geomorphology because its balancing act requires vertical motion of the Earth's surface. When the movement is upward, isostasy produces potential energy that is available for use in exogenic processes. Geologists intuitively recognize this effect because enormous rates of deposition have been observed in the geologic record during and following a major mountain-building episode. What is difficult to ascertain from these events, however, is how much uplift of the surface is associated with the diastrophism. The enigma is due to the fact that the geologic time scale lacks the precision needed to calculate a meaningful uplift rate, and the geologic evidence is often ambiguous. For example, assume that a shallow-water marine deposit containing fossils of Late Pliocene age is now exposed at an elevation of 3000 m. We may state confidently that 3000 m of uplift has occurred since the Late Pliocene, even though we cannot be certain how far below sea level the animals lived or whether the Pliocene sea level was exactly the same as sea level today. More troublesome is the question of how to translate the assumed 3000 m uplift in terms of rate with units of velocity such as m/yr. Two problems are inherent in such a calculation. First, how many absolute years are represented by the Late Pliocene age of the marine fossils? Is the deposit 2 m.y. old or perhaps 5 m.y. (for example, see Yeats 1978)? Obviously the calculated rate of uplift depends on one's interpretation of an inadequately known time interval. For the second problem, assume that an absolute age of 2.2 m.y. is given for the deposit by radiometric dating of interbedded volcanics. Here again large plus-or-minus errors must be considered. But assuming the date is accurate, the calculated rate of uplift then becomes 3000 m in 2.2 m.y. or .00136 m/yr. Although the numbers may seem precise, they say very little about the actual rate of uplift because there is no way of knowing whether the vertical motion was continuous over the entire time interval or whether it occurred in one catastrophic spurt sometime between the Late Pliocene and the present. The calculated rate assumes a constant and continuous uplift for 2.2 m.y. and so is a minimum value. If the entire uplift was accomplished during a limited segment of the total time, the rate may have been much higher.

In spite of the problems discussed above, it may be possible to determine precise uplift rates if the environmental setting is proper and the time span being considered is relatively short. This is especially true where glacio-isostatic uplift has affected coastal regions (Andrews 1974; Ten Brink 1974; Hillaire-Marcel and Fairbridge 1978). For example, Ten Brink (1974) was able to derive detailed uplift curves (fig. 2.4) by correlating the ^{14}C age of marine shell samples with the elevation of the strandlines in which they were preserved. *Strandlines* are essentially old shorelines formed by bodies of water, such as a lake or ocean, that are now elevated above the present water level. In Ten Brink's study, the oldest strandlines, dated by the related fossils, are higher than subsequent strandlines that formed as the Greenland coast emerged during deglaciation. The curve shown in figure 2.4 is a clear indication of how uplift proceeded in the postglacial adjustment.

Figure 2.4.
Uplift curve along the Greenland coast based on ^{14}C dates of marine fossils in emerged strandlines.

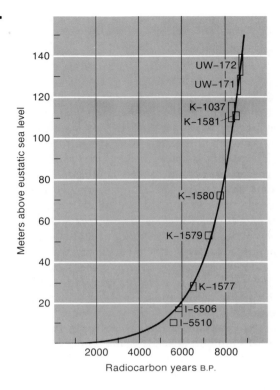

Meaningful uplift rates are important in geomorphology because they determine whether a surficial system can remain in equilibrium during the uplift event. If rates of uplift exceed by far the prevailing rates of denudation, the system will cross a threshold and enter into disequilibrium. How long it will take to establish a new balance depends on how radical the difference is between the rates of uplift and of denudation. Schumm (1963c) discusses the complications involved in making such an analysis because although uplift rates are significantly higher than denudation rates, the uplifts occur in short, spasmodic bursts rather than as long, continuous events.

Because of the problems inherent in determining a precise uplift rate from long-term data, table 2.1 is presented to demonstrate rates of vertical displacement measured in the modern setting or based on data from late glacial and postglacial times. The data presented were chosen at random and should not be considered complete. Some of the represented movement is caused by isostatic adjustment to the unloading of ice and/or water or the loading of sediment and water. Other motion is generated in regions that presumably are experiencing the effects of active tectonism. On the basis of this limited sample, it is interesting to note that the rates of vertical displacements are high regardless of the tectonic environment. For example, the uplifts in Fennoscandia

Table 2.1   Rates of vertical displacement.

Area	Time	Rate (cm/1000 yr)
Fennoscandia[1]	Recent	(+) 1100
Hudson Bay[2]	Recent	(+) 1700
Lake Superior[1]	Recent	(+) 500
Lake Mead[3]	Modern	(−) 1200
Lake Bonneville[4]	Late glacial—Modern	(+) 1200–60[a]
East Coast U.S.[5]	Modern	(+) 200 − (−) 500
California Coast Ranges[6]	Late glacial—Recent	(+) 500–800
Alaska[7]	Modern	(+) 2400
Greenland[8]	Modern	(+) 1000–1700

(1) Gutenberg 1941 (2) Walcott 1972 (3) Longwell 1960 (4) Crittenden 1963 (5) Fairbridge and Newman 1968 (6) Bandy and Marincovich 1973 (7) St. Amand 1957 (8) Ten Brink 1974
[a]Represents declining rates during last 20,000 years.

and central North America, undoubtedly isostatic rebounds initiated by deglaciation, are within an order of magnitude of those in the California and Alaskan regions, which are probably undergoing active deformation. It also appears that all vertical movement is considerably greater than maximum rates of denudation, which will be discussed later. The values presented in table 2.1 are also complicated by the fact that rates for isostatic rebound decline progressively as the region gets closer to equilibrium. They will be very high immediately following the removal of the excess weight of ice or water but may be very low when equilibrium is nearly reestablished (see Gutenberg 1941; Crittenden 1963). Schumm (1963c) suggests that a reasonable average rate of uplift is about 750 cm/1000 yr.

The significant geomorphic aspects of isostatic uplift can be summarized briefly:

1. Almost all regions within the continents are in or near some form of isostatic equilibrium.
2. Structural features and initial relief are formed by vertical epeirogenic movements associated with density variations in the basement rocks. Presumably postorogenic uplift in mountains is also related to isostatic compensation.
3. Redistribution of mass at the surface by erosion and deposition, glacier development, thrusting, etc., requires vertical movement of the underlying rocks to reestablish isostatic equilibrium.
4. The driving force behind isostasy is gravity, which is responsive to a heterogenous distribution of rock density. The cause of density variations and the precise mechanics of isostatic compensation are very poorly understood, but the process and its effect on geomorphic systems are real.
5. Rates of uplift are normally high compared with rates of denudation.

Orogeny and Tectonic Geomorphology

Scientists have pondered the origin of mountains ever since they first recognized that rocks in mountain belts were structurally different from those in other areas. The intense folding and overthrusting displayed within mountainous regions led geologists to realize that significant crustal shortening was involved in their formation, but learning what caused the deformation was hindered by our ignorance of the Earth's interior. Initially, a progressively cooling and shrinking Earth was suggested to explain the needed compressional stress, but this idea was rejected after the perception of continuous addition of internal heat by radioactivity and the acceptance of a cold origin for the Earth. Other proposals met an equally unsatisfactory fate. The advent of the plate tectonic theory, however, forced geologists to reevaluate mountain building in light of the new global model. A detailed discussion of plate tectonics is beyond the scope of this book, and the basic concepts of the plate model can be found in any textbook of physical geology. For our purposes it is sufficient to say that mountains and ocean features such as island arcs and trenches are intimately associated with the seismicity and volcanism found at plate margins where the lithosphere is being actively consumed.

The effect of orogeny on geomorphic process is less tangible than that produced by epeirogeny because the act of mountain building occurs below sea level. Thus, it is not the rocks or region squeezed in the mountain-building event that shows geomorphic response during the application of orgenic forces. As indicated earlier, these regions are usually affected after orogeny is over, when isostasy raises the thick pile of low-density and highly deformed rocks. Nonetheless, geomorphic systems adjacent to the orogenic zone may be thrown into disequilibrium during the orogeny. The driving forces that cause the changes are found in secondary phenomena that are ancillary to the orogenic force. The most prominent of these are seismicity and volcanism.

Seismic activity transfers internal energy into exogenic process in several ways. First, earthquake motion increases driving force while simultaneously reducing the resisting strength of materials. This commonly results in sudden slope failures (landslides) that represent threshold responses to seismic activity in regions close to an active orogenic zone (see Garwood et al. 1979). Second, movement along faults that cause the earthquakes sometimes displaces the surface; upward displacement produces the same geomorphic responses as those occurring in epeirogenic movement, but they are usually less dramatic.

The fact that deformation leaves an imprint on landscapes is one of the oldest tenets in geomorphology. Initially, regional tectonics were used to explain diversity of character in large-scale topography. For example, geomorphologists recognized that the block mountains and intervening basins of the Basin and Range province in the western United States, and the sinuous valleys and ridges in the Folded Appalachians of Pennsylvania were reflections of significantly different tectonic styles. Such observations are interesting in themselves, but the utility derived in demonstrating the relationships is primarily physiographic and does little to further our understanding of process.

More recently, the relationship between form and tectonics has become the basis for much more detailed and sophisticated geomorphic studies and has led to the development of a special branch of the discipline known as tectonic geomorphology. **Tectonic geomorphology** as now practiced deals with how tectonic activity affects processes and morphology in geomorphic systems and, conversely, how tectonically controlled landforms can be used to assess Quaternary tectonic activity (Bull and McFadden 1977). Most studies have been directed at understanding the geomorphic response to vertical and/or horizontal displacement along faults and the tilting or warping associated with broad uplifts. Significantly, in many studies landforms are used to provide insight into the style and rate of tectonic processes, therefore revealing a shift in emphasis from the explanation of form to the analysis of what form tells us about process.

Tectonic geomorphology has been used for a variety of scientific purposes, and studies range in scale from contrasting the tectonic style of Earth, Venus, and Mars (fig. 2.5; Arvidson and Guinness 1982) to the analysis of a single landform such as an alluvial fan (Keller et al. 1982). The approach has tremendous implications in environmental planning (Schowengerdt and Glass 1983), especially with regard to landform stability (Bull and McFadden 1977), seismicity and earthquake risk (Bull 1974), and prediction of fault-movement periodicity (Wallace 1977, 1978; Buckman and Anderson 1979; Nash 1980; Colman and Watson 1983).

Figure 2.5.
Tectonic features on Mars. Linear depressions on left of photo are interpreted as large grabens.

20 KM

Geomorphic response to faulting constitutes one of the basic components in tectonic geomorphology. Faulting produces adjustments in process and form that can be used as clues to present and past tectonic activity and, additionally, to the rates and magnitudes of displacements. For example, Bull and McFadden (1977) were able to characterize the activity of Quaternary tectonism in subareas associated with the Garlock fault in southern California. The degree of tectonic activity was recognized on the basis of two geomorphic parameters: (1) the *mountain-front sinuosity,* which is the ratio of the length measured along the junction of the mountain and piedmont to the total straight-line length of the mountain front, and (2) the ratio between valley floor width and valley height in major river valleys measured at a given distance upstream from the mountain front.

Theoretically, a tectonically inactive region shows mountain fronts with high sinuosity because continuing erosion and deposition would progressively shift the position of the mountain front away from the range-bounding fault. Valleys in the mountains should be wide and shallow under tectonic stability because lateral erosion controlled by a stable base level would dominate the system. Conversely, in tectonically active areas the mountain-front sinuosities are low and the mountain valleys are narrow and deep. This follows because repeated vertical movements preserve the coincidence between the bounding fault and the mountain front, and the rivers upstream from the front are spurred into downcutting on the upthrown side of the fault. The basis for this expected response has been established by countless observations that river gorges are related to enhanced stream power brought on by rapid, fault-related uplift (for example, see Jackson et al. 1982).

Horizontal displacement on faults can also be documented by variations in geomorphic landforms. Keller et al. (1982), for example, recognized numerous tectonically produced landforms along a branch of the San Andreas fault in the Indio Hills of southern California. Offset drainages, stream diversions, and an alluvial fan-pediment complex demonstrated approximately 0.7 km of horizontal displacement. Other features were recognized as products of vertical displacement along the fault. Perhaps the greatest importance of this study is the fact that Keller and his colleagues were able to explain almost all of the fault-related features by simple shearing associated with a bend in the trace of the fault. Therefore, landforms here were used as primary evidence to explain the mechanics of local deformation.

Evidence of regional, large-scale tilting and warping are also present in the elements of landscapes, especially in the profiles of planar surfaces or stream channels. Marine terraces and wave-cut platforms (fig. 2.6) are excellent indicators of these tectonic movements because they originate as nearly horizontal surfaces related to sea level (Bradley and Griggs 1976; Pillans 1983). Thus, any divergence from their original state is a manifestation of tectonic influence. In both studies referred to above, the researchers were able to demonstrate distinct shore-parallel warping of the coastline based on differential uplift of the platforms and terraces. In addition, seaward tilting of the planar features and associated river terraces indicates that the tectonic motions are domal type regional uplifts.

Longitudinal profiles of river terraces, alluvial valley floors, and stream channels are also sensitive indicators of regional tectonism. For example, repeated geodetic surveys have shown clearly that two areas in Louisiana and Mississippi are sites of rapid modern uplift. Burnett and Schumm (1983) examined the characteristics of fluvial features crossing these uplifts (fig. 2.7) and found considerable geomorphic evidence to substantiate the geodetic findings. Terrace and valley floor profiles show pronounced convexities where they cross the axes of the uplifts, being abnormally gentle upstream from the axes and oversteepened downstream from the axes (fig. 2.7). In addition, river reaches upstream and downstream from uplift axes develop markedly different channel patterns, depths, and sinuosities, which are probably controlled in part by the orientation of the underlying bedrock structure (Gardner 1973). Precisely how much response will occur in modern rivers experiencing neotectonism depends on the river size. Larger rivers may possess enough energy to keep degradational pace with the rate of uplift, and therefore, they may experience little change. Their valley floors and terraces, however, may show considerable change in character at the uplift axes. Nonetheless, fluvial features clearly reflect neotectonism and should be considered as prime criteria in tectonic geomorphology.

Figure 2.6.
Uplifted marine terraces (wave-cut platforms) along the central California coast.

Figure 2.7.
(A) Longitudinal profiles of terraces, floodplain, and a projected channel profile of the Pearl River, a major river crossing the Wiggins uplift. (B) Longitudinal valley profiles of streams crossing the Monroe uplift. Macon Ridge, a remnant of a deformed Pleistocene Mississippi River terrace, is also shown; *MSL*, mean sea level.

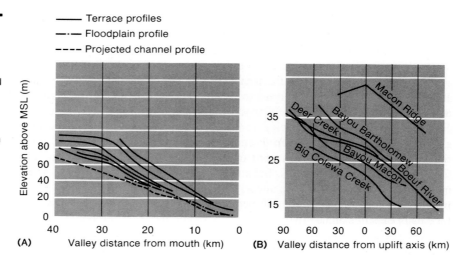

The selected cases just presented are only a few examples of how the relationship among landform, process, and tectonics can be employed to gain insight in numerous geological and environmental investigations. The practice of tectonic geomorphology will certainly increase in the future, and every geomorphologist should become familiar with its basic premises and uses (for an excellent review see Bull 1984).

Volcanism

Although some (but not all) volcanic activity is associated with orogenic events, volcanism is such a major input of endogenic force that it deserves separate treatment. Volcanism is nothing more than a surface manifestation of the internal processes that create and mobilize magma. Although volcanoes are spectacular in eruption and unique in topographic form, describing examples of active volcanoes is not necessary here. Most physical geology texts do this, and excellent treatises on volcanoes and volcanic landforms are available (Bullard 1962; Ollier 1969; Green and Short 1971; Macdonald 1972). It may be pertinent, however, to examine briefly how internal variables control the type of volcanic eruption and the ensuing topographic form.

The violence of a volcanic eruption is determined mainly by the composition of the magma and the amount of gas in it. By affecting the viscosity of the magma, these factors influence the observed differences between continental and oceanic volcanism. Highly fluid, basaltic magmas are produced in both continental and oceanic environments. The more viscous, high-silica lavas that crystallize as andesites, dacites, and rhyolites are generally restricted to continents or marginal island arcs. In orogenic zones, where most viscous lava occurs, the erupting magma also contains more gas then do the true oceanic types. The combined effect of higher gas content and greater viscosity creates a tendency toward more explosive eruptions. This trend is magnified in continental volcanoes where violent explosions are commonplace, and solid volcanic ejecta, called *tephra,* is the dominant extrusive material rather than flowing lava.

The major volcanic landforms consist of three types: plateaus or plains, cones, and calderas. *Lava plains* and *plateaus* are extremely flat surfaces, both continental and oceanic, that have been aggraded by overlapping flows of fluid lava with a mafic composition. Although the nature of the vents is not clear, they probably are fissure types distributed over wide areas or a series of unconnected pipe vents. The dominant characteristic of lava plains and plateaus is the enormous volume of lava extruded and spread over a vast surface area. For example, in the Columbia River Plain of Washington and Oregon, areas greater than 200,000 km^2 are known, and in the Deccan Plain of India, more than 500,000 km^2. Oceanic plains can be even more extensive (Kuno 1969). Commonly the total thickness of the extrusive rocks exceeds 2000 m—great enough to bury a mountainous terrain. In fact, in the Columbia River Plain peaks of buried mountains (called steptoes) may project through the flat surface as isolated "islands" of older rock.

The surfaces of lava plains and plateaus show minor perturbations where broad, low cones rise 30–60 m above the general level. However, individual flows, 2 to 50 m thick, can extend for hundreds of kilometers from the cones, blending imperceptibly into lava issued from fissured vents to create a surface that normally slopes less than 1°. Lava plains are almost always composed of basaltic rocks because silicic magma tends to be too viscous for the long distance of flow necessary to create a planar topography.

Some silica-rich volcanics do underlie flat plains, but these rocks did not crystallize from lava and the surface extends over much smaller areas. Silicic plains develop when incandescent volcanic glass is erupted within dense clouds of gas capable of flowing for considerable distances. The welded tuffs (ignimbrites) resulting from the ash flows are common in the cordilleran of the United States, especially in the southwest. Field evidence there suggests that the ash was vented from both linear and arcuate fissures and central pipes.

The second major volcanic landform occurs as the topographic expression surrounding a single vent. These surfaces, called *cones,* have a variety of shapes depending on the type of volcano and the predominant mode of eruption. *Shield volcanoes* develop cones that are typical of those seen in Hawaii and Iceland. The sides of these cones slope from 2° to 10° and merge gradually, at their base, with the adjacent ocean floor (fig. 2.8). The massive Hawaiian shields rise at least 4800 m above their bases, and some (Mauna Loa, Mauna Kea) have an absolute relief greater than 9 km, making them among the largest topographic mountains of the world.

Shield volcanoes are always built up from fluid, basaltic magma; tephra is only a minor part of the erupted material, and explosive eruptions are rare. The magma in the Hawaiian chain moves surfaceward in rift zones that generally parallel the Hawaiian Ridge (topographic high in the central Pacific). The rift zones usually contain hundreds of fissures that the ascending magma uses as vents for eruption. The Hawaiian rifts extend discontinuously for 3500 km across the center of the Pacific plate to a point where they abruptly bend to join the Emperor Chain. This linear group of seamounts, similar in all aspects to the Hawaiian volcanoes, continues northward for another 2500 km.

The Hawaiian eruption rates, estimated between 0.05 and 0.1 km³ a year, (Moore 1970; Swanson 1972) are the greatest known on Earth. There seems to be little question that the primary Hawaiian magma originates in the upper mantle (Eaton and Murata 1960) and, furthermore, that no genetic relationship exists between the volcanoes and the age or structure of the adjacent sea floor (Dalrymple et al. 1973). The plate-interior location of the island chain precludes an orogenic origin for the volcanic activity. This apparent lack of coincidence between plate tectonics and volcanism has prompted the hypothesis that volcanic action here occurs over a roughly circular hot spot in the upper mantle, about 300 km in diameter. Volcanism occurs as the Pacific plate rides over the heated region, and successive generations of volcanoes are carried away from the heat source.

Figure 2.8.
Profile and crater area of shield volcanoes, Hawaii. Summit caldera of Kilauea volcano in foreground. Mauna Kea in background shows gentle slopes typical of a shield volcano.

The Icelandic shields commonly are smaller in all dimensions than the Hawaiian volcanoes. Although they originate in a plate-margin environment, they are not related to orogenic events. The magma probably rises along the mid-Atlantic ridge in tensional fractures associated with the forces that move the oceanic plates apart.

Most other volcanoes take the form of a **composite cone,** so named because the cone is constructed of interbedded lava flows and layers of tephra. The flows are usually blocky and the tephra mostly cinder or ash, but the characteristics of all components may vary with the viscosity of the ascending magma. Composite cones, like the one in figure 2.9, are more peaked than shield volcanoes, commonly rising several thousand meters above a narrow circular base. Most develop from a single pipe vent. Side slopes are typically

Figure 2.9.
Steep slopes of composite volcanoes. Mount Rainier, Wash., in foreground.

high, ranging from 10° to 35°, because the lack of fluidity in the magma leads to a more localized deposition. Most of the world's continental volcanoes are of this type; in North America famous examples are Mt. Rainier (Washington), Mt. Shasta (California), and Mount St. Helens (Washington).

The third major landform associated with volcanic activity is known as a caldera. **Calderas** are large depressions in volcanic regions that result where eruption spews forth large quantities of material, creating an empty space in the underlying magma chamber. This results in an inward collapse of the upper part of the volcanic cone, often along fractures that develop following the eruption. The caldera depression, therefore, has a larger diameter than the original crater, and in fact, the primary distinction between a normal volcanic crater and a caldera is size. Depressions larger than 1.6 km (1 mile) in diameter are usually accepted as being calderas (Macdonald 1972) because they almost certainly result from subsidence rather than incremental construction of a cone.

The largest calderas are associated with volcanoes that produce tephra sheets. For example, the caldera basin buried under the thick ignimbrite sheets of the Yellowstone Plateau is 70 km by 45 km. In addition, some of the most well-developed calderas, such as the one occupied by Crater Lake, Oregon, occur on the summits of composite cones, and their origin is directly associated with the occurrence of widespread ash falls.

Volcanic activity produces a myriad of smaller landforms that, like the major forms, also relate to the viscosity of the involved magma. The interested reader is referred to Rittmann (1962) for discussions about the classification and origin of these forms.

Climatic Geomorphology

Climate, Process, and Landforms

Because of climate's vast effect on many surface phenomena, it was inevitable that scientists would attempt to group climates into some useful classification. Temperature and precipitation commonly are the prime variables used to define climatic boundaries, although vegetation and soil assemblages may serve as auxiliary factors. The most successful classification is an empirical system devised by Köppen in which five major climate groups are distinguished on the basis of observed temperature and precipitation values. Each major group can be divided into progressively smaller units with local variations such as seasonality. The Köppen approach has the great advantage of relying on measured data; climates can be defined by precise physical characteristics, and maps based on this system can provide rather detailed information about local or regional climates. The Köppen classification, however, tells us nothing about what causes the climate and therefore provides no genetic information. More recent attempts at classification employ variables that indicate the origin of climatic zones rather than simply placing them in numerical pigeonholes (Strahler 1965). Geomorphologists should be more concerned with how well energy in the climatic regime is utilized in geomorphic work than with how the climate is classified. In other words, we need to know which landforms will develop most efficiently under any prevailing climate.

The relationship between landforms and climate is the basis for a major philosophic approach in geomorphology known as **climatic geomorphology,** which has been most forcefully championed by European scientists (Tricart and Cailleux 1972; Büdel 1982). Essentially, the underlying premise of this approach is that geomorphic mechanics vary in type and rate according to the particular climatic zone in which they function. If that assumption is correct, landforms produced from these mechanics will be different from region to region and will reflect the dominant climate.

Büdel (1982) suggests that modern processes in mid-latitudes are too weak to remove features that formed prior to the Holocene. As a result, 95 percent of the topography found in those regions developed under a former, more rigorous climate, and landforms there are not related to modern processes. Thus, the framework of climatic geomorphology becomes evident only if we consider landforms that developed from processes operating in climatic zones established during the Holocene. If that basic rule is followed, landforms will be uniform over large areas called *morphoclimatic zones,* and their character will change drastically as climate varies from one zone to another. Forms that do not fit in a specific zone are considered as relict features that developed under climatic conditions that no longer exist in the region.

Although most geomorphologists recognize some validity in the tenets of climatic geomorphology, the approach suffers because we do not yet understand how landforms are affected by climate changes of different magnitude and/or frequency (Stoddart 1969). In addition, the approach has no process base, and its practitioners have not applied experimental methods to climogenetic problems (Derbyshire 1976). Thus, the descriptive and qualitative associations between climate and topography proposed in climatic geomorphology have simply not been adequately tested.

In spite of such problems, there is little doubt that some relationship exists between the type of dominant process in a region and the prevailing climate. This relationship has been expressed by Peltier (1950), who designated certain areas as *morphogenetic regions.* In these regions, dominant processes can be ascertained and related to a specific climatic regime that is broadly defined by its prevailing temperature and precipitation. This concept has been followed by others, although the climatic variables used to distinguish the regime are sometimes different. L. Wilson (1968) objects to the term "morphogenetic regions" because it usually has been used to refer to a process and not to a tangible region of the earth. He suggests that the relationship between climate and process be called **climate-process systems,** and the relationship between climate, process, and landforms be called **morphogenetic systems.** The term "morphogenetic region" would be restricted to the case where an actual region is being considered.

Regardless of semantics, it is valuable in geomorphology to establish the relative importance of a particular process under a set of temperature-precipitation conditions. The concept is still useful even when it is recognized that landscapes are controlled by factors other than climate and that the details of morphogenetic systems are poorly understood. It serves as a viable first approximation of what landforms we can expect to find in any environment. As

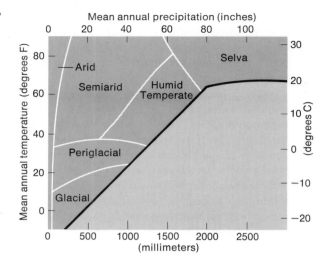

Figure 2.10.
Six possible climate-process systems as suggested by Wilson (1968). Each set of temperature-precipitation values tends to drive processes that function most efficiently under those climatic conditions.

Büdel (1982) would suggest, it is helpful in detecting relict features, forms that developed under an earlier and different climate and have not yet adjusted to the present climatic regime. For example, moraines in central Illinois reflect a morphogenetic system that no longer exists; the landforms have not completely readjusted to coincide with modern conditions.

Figure 2.10 is a diagrammatic representation of six climate-process systems that ideally relate climate to a set of paramount processes. The diagram was derived by combining the relationships of several specific processes and climates; the arbitrary system names are meant to be descriptive of the climatic type. For instance (other factors being equal), we can expect that a region with a mean annual temperature of 0°C and mean annual precipitation of 500 mm will function as a periglacial climate-process system, and those processes that function most efficiently in such a climate will prevail. Each climate-process system in the figure can be converted into a morphogenetic system by defining the landforms that most commonly result from the processes involved, as is done in table 2.2.

In reality, minor heterogeneity of geology may produce subtle variations in the topographic outcome. In addition, relict landforms, and seasonally changing climates may complicate the expected form. Nonetheless, the morphogenetic approach does provide a reasonable framework for the analysis of landforms. Absolute servility to this approach, however, leads us away from the study of processes, since the method passes quickly from climate to form. This nonchalant leap from climate to landform does not mean that we understand the intermediate phase of process.

In addition to isolating the predominant processes, it is important to understand how climatic variables influence process mechanics. In most cases, solar energy must be converted into another form of energy to do geomorphic work. For example, although precipitation releases stored heat, it is not the temperature of raindrops that drives hydrologic processes. Evaporation and

Table 2.2 Simple morphogenetic systems and their landscape characteristics.

System Name	Equivalent Köppen Climates[a]		Dominant Geomorphic Processes[b]	Landscape Characteristics[c]
Glacial	EF	Icecap	Glaciation Nivation Wind action (freeze-thaw)	Glacial scour Alpine topography Moraines, kames, eskers
Periglacial	ET EM D-c	Tundra Humid microthermal	Frost action Solifluction Running water	Patterned ground Solifluction slopes, lobes, terraces Outwash plains
Arid	BW	Desert	Desiccation Wind action Running water	Dunes, salt pans (playas) Deflation basins Cavernous weathering Angular slopes, arroyos
Semiarid (subhumid)	BS Cwa	Steppe Tropical savanna	Running water Weathering (especially mechanical) Rapid mass movements	Pediments, fans Angular slopes with coarse debris Badlands
Humid temperate	Cf D-a	Humid mesothermal	Running water Weathering (especially chemical) Creep (and other mass movements)	Smooth slopes, soil covered Ridges and valleys Stream deposits extensive
Selva	Af Am	Tropical Monsoonal	Chemical weathering Mass movements Running water	Steep slopes, knife-edge ridges Deep soils (laterites included) Reefs

[a]Equivalents are approximate, and not exhaustive.
[b]Processes are listed in order of relative importance to landscape (not in order of absolute magnitude). List is abbreviated.
[c]Both erosional and depositional forms are included. List is neither comprehensive nor definitive, merely suggestive.

rising air provide water vapor with potential energy that is transformed to kinetic energy when rain begins to fall. A maze of reactions changes energy into various forms that ultimately drive the different geomorphic processes.

Climate Change and Geomorphic Response

Regardless of the obscure relationships existing among climate, process, and landforms, the fact that climate changes is an irrefutable cornerstone in geomorphology. In modern times, for example, we know from direct observation that average air temperatures in the Northern Hemisphere increased from the late 1800s to 1940 and since then have been steadily decreasing (Budyko 1977; Lockwood 1979). Additionally, overwhelming evidence of past climate change has been documented in such diverse fields as pedology (soil science), palynology (study of pollen), archeology, paleontology, oceanography, and many others. The most dramatic manifestation of climate change is, of course, that several widely spaced episodes of prolonged glaciation have occurred in the Earth's history. The most recent of these, encompassed by the Pleistocene epoch, was characterized by distinct alternations of glacial and interglacial stages when climate was patently different from the present. Thus, the ephemeral nature of climate is beyond question. The *reasons* for climate change, however, are open to considerable debate because they are tucked away in the complex maze of interactions that control climate itself.

Detailed examination of the dynamics of climate change are well beyond the purposes of this book and are best left to the climatologists. In fact, some climatic variations are presently inexplicable. In general, however, the variables that trigger climate change exist in several major groups; significantly, the variables in each category seem to function over different time intervals. *First,* variations in the composition of the atmosphere may alter the amounts of solar radiation reaching or escaping the Earth's surface. This is especially true when changes in the CO_2 content alter the Earth's greenhouse effect or when recurring volcanism introduces large volumes of fine-grained ejecta into the atmosphere. Variations in atmospheric factors can produce tangible climate change in a matter of years or decades.

Second, astronomical motions may produce changes in the pattern and intensity of solar radiation. For example, parameters such as the tilt of the Earth's axis and the orbital path around the Sun vary from maximum to minimum values in specific intervals of time. Periodically, a number of such factors coincide to produce minimum solar radiation, enough decrease in radiation to induce an episode of glaciation. The astronomical motions have a periodicity such that their effect on climate would occur in a time span ranging from 20,000 years to 100,000 years.

Third, many scientists believe that a cause-and-effect relationship exists between climate change and variations in the elevation and distribution of continents. The concept implies that the major cooling needed to produce glaciation occurred when landmasses were high. It also has been suggested that the distribution of landmass and ocean basins associated with the mechanics of polar wandering and plate tectonics is important in controlling climate. Landmass movements probably influence climate over cyclic time intervals, perhaps millions of years.

In geomorphology the cause of climate change is less important than the adjustments demanded within surficial systems in response to the change. Space prevents us from examining the effect of climate in every geomorphic system. It may be instructive, however, to consider several examples of threshold-producing phenomena that serve as intermediary links between the variables of climate change and the process response in geomorphic systems.

Sea Level Fluctuation   One major intermediary factor related to climate change and responsible for generating geomorphic adjustments is the phenomenon of sea level change. As stressed earlier, sea level is a most important horizon in geomorphology. In addition to serving as the datum for relative gravity analyses, sea level represents the ultimate base level for rivers draining the land and the theoretical end point of continental erosion. We know from

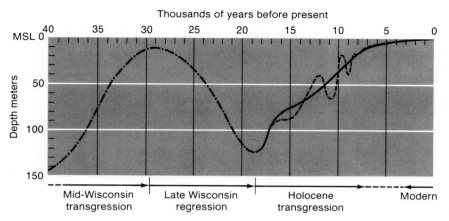

Figure 2.11.
Late-Quaternary fluctuations of sea level, from compilation of published and unpublished radiocarbon dates and other geologic evidence. Dotted curve estimated from minimal data. Solid curve shows approximate mean of dates compiled. Dashed curve slightly modified from Curray (1960, 1961). Probable fluctuations since 5000 B.P. not shown here.

geologic history that sea level is not constant relative to the adjacent land-masses. This relative position has shifted through time because (1) the continents are subject to isostatic uplift or depression and (2) the water level itself may rise or fall, an adjustment known as *eustatic* change. Eustatic sea level change occurs on a worldwide basis in contrast to isostatic movements, which are determined by the local or regional gravity environment.

Although there are a number of ways to produce eustatic sea level change, the accumulation and removal of glaciers is by far the most obvious and effective. Glaciers are interruptions of the hydrologic cycle because water locked up in ice represents precipitation that is not being returned to the ocean. As a result, when the volume of ice held on the land increases during a period of glaciation, the ocean volume shrinks and its surface is lowered. Conversely, if all modern glaciers were melted, sea level would rise by an amount comensurate with the volume of water released from storage. Clearly, the growth and wastage of glaciers attendant to glacial and interglacial stages of the Quaternary are major progenitors of eustatic sea level change.

Exactly how much eustatic fluctuation occurred in Pleistocene cycles is difficult to ascertain because all evidence used to make such estimates is subject to error (see Flint 1971). The most reliable evidence is the presence of near shore (littoral) marine fossils that are datable by radiocarbon and are now covered by about 100 m of ocean. These indicate that the shoreline was approximately 100 m lower at the time of the last glacial maximum. Similarly, emerged strandlines bearing datable shells are now preserved well above present sea level and, if isostasy is neglible, they mark sea level during interglacial times. The use of such evidence has allowed the construction of curves showing the rise in sea level since the last glacial maximum (fig. 2.11). These curves, corroborated in numerous studies, indicate a reasonable steady increase in ocean level from about 18,000 years B.P. (before present) to about 5000 years B.P. when sea level apparently stabilized near its present configuration.

It is ironic that the most accepted evidence of eustasy is also the evidence used to document isostasy. This presents Pleistocene geologists with a nasty interpretive problem because vertical change in relative sea level is produced by the mechanics of two different processes that are both based on the same

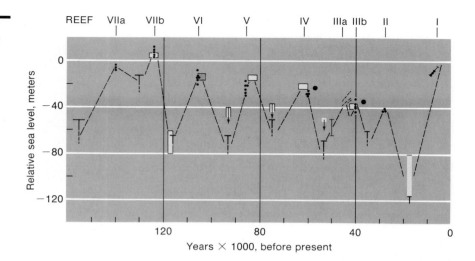

Figure 2.12.
Late Quaternary paleosea levels based on estimates from New Guinea and elsewhere.

Years × 1000, before present

• New Guinea reef crests, dated, each dot is from reef heights on traverse.

⊤ New Guinea reef complexes, low sea level maxima, undated (Chappell, 1974)
⊥

⊥ New Guinea reef complexes, dated (Veeh and Chappell, 1970)

☐ Compilation of Steinen, *et al.* (1973)—primarily Barbados data
Sea level minimum not known

– – – Diagrammatic paleo sealevel trace

• Sea level estimate using premature emergence correction

evidence. Therefore, sea level change attributable to eustasy cannot be accurately determined without knowledge of isostatic movement during the same interval. In addition, the depth measurements of preserved fossils are often made suspect because of subsidence of the sea floor under the weight of water added during the interglacial stage (Clark and Bloom 1979). The problems associated with interpretation of sea level change are amazingly complex (see Morner 1980 for discussions), but detailed studies in widespread locations are beginning to unravel the complicating factors (Bloom 1967, 1980; Bloom et al. 1974; Bender et al. 1979; Cronin 1981; Cronin et al. 1981; and many others).

Notwithstanding the interpretive problems, numerous studies, especially those in Barbados and New Guinea (Matthews 1973; Steinen et al. 1973; Bloom et al. 1974; Chappell 1974), have resulted in a reasonable curve of eustatic sea level for the last 150,000 years (fig. 2.12). The data, based on $^{230}Th/^{234}U$ dates of coral reef fossils, show that the highest sea level in that period was about 6 m higher than present sea level during the Sangamon interglacial stage. Other high sea level stands occurred during interstadials of the Wisconsin stage, but none of these was as high as the present level. The lowest sea level, probably somewhat over 100 m lower than present, occurred during the last glacial maximum.

Sea level fluctuations have a direct influence on a number of shoreline processes and features (beaches, barrier islands, etc.) These will be discussed

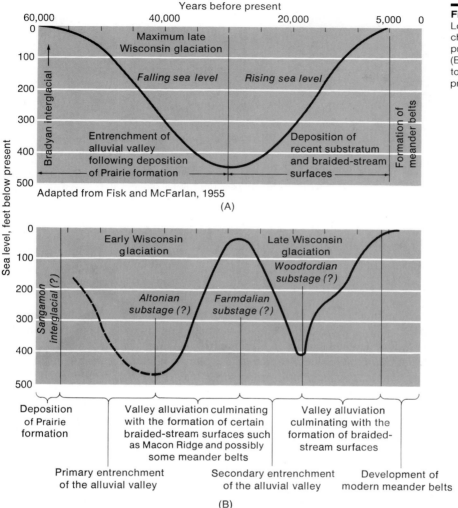

Years before present

Maximum late
Wisconsin glaciation

Falling sea level   *Rising sea level*

Bradyan interglacial

Entrenchment of
alluvial valley
following deposition
of Prairie formation

Deposition of
recent substratum
and braided-stream
surfaces

Formation of meander belts

Adapted from Fisk and McFarlan, 1955

(A)

Sea level, feet below present

Early Wisconsin
glaciation

Late Wisconsin
glaciation

*Woodfordian
substage (?)*

Sangamon
interglacial (?)

*Altonian
substage (?)*

*Farmdalian
substage (?)*

Deposition
of Prairie
formation

Valley alluviation culminating
with the formation of certain
braided-stream surfaces such
as Macon Ridge and possibly
some meander belts

Valley alluviation
culminating with the
formation of braided-
stream surfaces

Primary entrenchment
of the alluvial valley

Secondary entrenchment
of the alluvial valley

Development of
modern meander belts

(B)

Figure 2.13.
Lower Mississippi Valley
chronological concepts (A) as
proposed by Fisk (1944), and
(B) as applied by Saucier (1968)
to sea level fluctuation curve
proposed by Curray (1965).

in a later chapter. Sea level change also permeates into the fluvial system and through it into other systems. For example, in coastal regions river terraces commonly result when alternating cutting and filling are initiated by fluctuating sea level. Theoretically, entrenchment should accompany glacial expansion (when sea level is decreasing), while filling would take place during the waning phase of the glacial cycle (when sea level is rising). Fisk (1944) interpreted the depositional terrace sequence in the lower Mississippi Valley as being related directly to waxing and waning glaciations and their effects on eustatic sea level change. Although the details of aggradation and degradation were modified as more data became available, the relationship between sea level change and depositional terraces in the lower Mississippi basin is generally recognized as being real. Figure 2.13 shows these interpretations.

Figure 2.14.
Location of the Pomme de Terre River in relation to the Missouri–Mississippi river drainage basin, and the limits of continental glaciation.

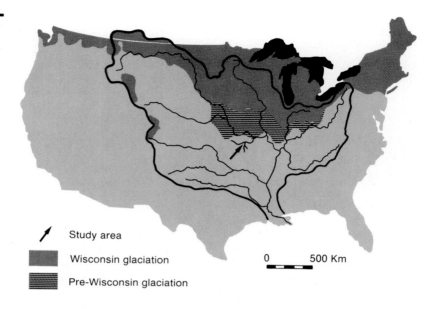

Study area

Wisconsin glaciation

Pre-Wisconsin glaciation

0   500 Km

The downcutting produced during glaciation may be gradually propagated upstream, causing a similar erosional response in each tributary basin. At that time, slopes may be regraded to new levels and groundwater tables may be lowered. Thus, by process linkage, one external change in climate begins a chain reaction of adjustments throughout the geomorphic regime. How far responses will be propagated upstream in a system as large as the Mississippi basin depends on the erosive power of the river and how long conditions remain stable in the interior reaches. We know, for example, that some tributaries in central Missouri (fig. 2.14) were apparently not affected by the lower Mississippi sequence recognized by Fisk (Brakenridge 1981). In fact, Brakenridge (1981) suggests that the terrace and floodplain sequence along the Pomme de Terre river shows no clear correlation with either sea level history or glacial chronology in the upper Missouri–Mississippi basin. Aggradation occurred during glacial, nonglacial, and interstadial periods. Terrace formation by river downcutting proceeded in the late Pleistocene, as well as in the Holocene when major glaciation and eustatic base level changes were absent. Presumably this indicates that alternating filling and cutting can be directly induced by climate change without glaciation or base level change. This supports the contention by Knox (1976) that Holocene downcutting might be caused when variations in atmospheric circulation produce conditions that lead to an increase in river energy. Downcutting, therefore, would represent a response to a threshold of stream power (Bull 1979). In the following section we examine how such responses might occur and whether the relationship between climate and threshold response is as straightforward as it appears.

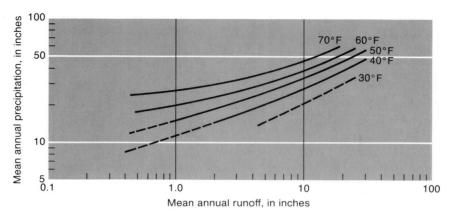

Figure 2.15.
Curves illustrating the effect of temperature on the relation between mean annual runoff and mean annual precipitation.

Geologic and Vegetal Screens   The preceding discussion reveals that it may be more difficult to explain the climatic effect on landforms located far from the site of sea level change. We can use the same landform as above to exemplify this point. River terraces in the interior of continents have historically been attributed to the fluctuating climate accompanying glacial and interglacial conditions. The swing from arid to humid as glaciation begins (and vice versa) affects not only the prevailing discharge but also the amount and type of sediment delivered to the rivers. Such changes in fundamental river controls logically produce the trenching and filling. Pleistocene and Holocene climates in North America are reasonably well known. In general terms, average temperatures in the Pleistocene were probably about 7°C ($\approx$ 13°F) lower than present during glacial stages and 3°C ($\approx$ 5°F) higher than present during interglacials. Average annual precipitation was as much as 25 cm (10 in.) greater during glaciation than now and 12 cm ($\approx$ 5 in.) lower during the warmer periods. These values vary depending on latitude and the technique used to make the estimate (Flint 1971; Budyko 1977).

Knowing this, it is ironic that little agreement exists as to which climate produced the filling and which caused the trenching in the process of terrace formation. To complicate matters, different interpretations of the cause and effect relationship in nonglaciated regions of the United States seem to be supported by field evidence (for a discussion see Flint 1971, pp. 304–307). The ambiguity revealed here relates to the second intermediary link between climate change and response that rests in how climatic ingredients are filtered into geomorphic systems through geologic and vegetal screens.

A major climatic influence in geomorphic systems occurs because temperature and precipitation are fundamental controls on mean annual runoff and the magnitude of erosion. As you might expect, runoff increases with higher annual precipitation, but at constant precipitation the runoff will decrease with higher temperature because evapotranspiration is known to be greater in warmer regions (fig. 2.15). The amount of sediment yielded from basin slopes is an indication of erosion and is also a function of temperature and precipitation (fig. 2.16).

Figure 2.16.
Curves illustrating the effect of
temperature on the relation
between mean annual sediment
yield and mean annual
precipitation.

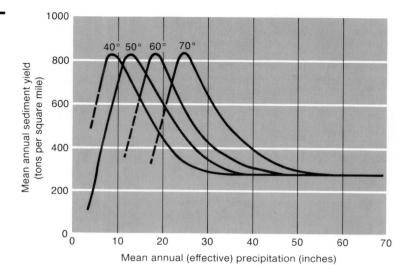

Figure 2.17.
Average annual sediment yield as
it varies with effective
precipitation and vegetation.

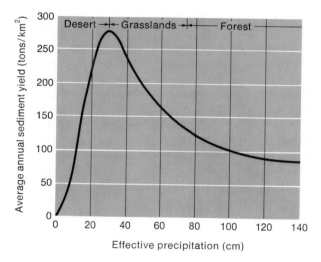

The curves shown in figure 2.16 were derived by using data from basins
in the western United States averaging 3900 km² (1500 mi²) in area and may
not be suitable for other settings. Significantly, however, they show that the
influence of climate on erosion is filtered through a vegetal screen. For example, figure 2.17 represents the 50°F curve, where effective precipitation is
the mean annual precipitation adjusted for that prevailing temperature
(Langbein and Schumm 1958). Under the stated conditions, maximum sediment yield occurs at approximately 30 cm (12 in.) of effective precipitation
because the density and type of vegetation developed at that precipitation and
temperature is most conducive to water-related erosion. Where precipitation
is lower than 30 cm, not enough water is available to erode the slopes. Above
30 cm of precipitation, the vegetal cover becomes more dense and changes

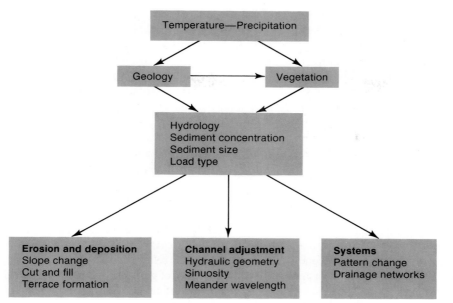

Figure 2.18.
Hypothetical flow chart showing
how climatic variables exert an
influence on rivers. A change in
climate alters sediment
concentration, sediment size, or
load type, requiring a response by
the river system. Responses vary
depending on local conditions.

from desert brush to grasses. The ubiquitous root systems tend to fix sediment
on the slopes, and therefore, with increased precipitation sediment yield becomes progressively lower until it nearly stabilizes under a forest cover.

The Langbein-Schumm analysis will be discussed further in chapter 5.
For now, it is important for you to recognize that climatic change does not
affect geomorphic systems directly but is passed through geological (rock type,
soils) and vegetal screens that determine how much water and sediment get
to a river channel. Variations in those two factors require threshold adjustments that can occur in a number of different ways (fig. 2.18).

Equally important is the fact that the same climatic change may prompt
entirely different sediment yields and, therefore, different geomorphic responses. For example, considering figure 2.17 again, assume that a 15-cm decrease in precipitation occurs in a particular drainage basin that had an effective
precipitation of 45 cm prior to the change. The reduced precipitation will result in a greatly increased sediment yield. However, the same 15-cm decrease
in a basin having a 35-cm annual precipitation prior to the change will produce
a major reduction in sediment yield. Theoretically, then, the same climate
change may result in cutting by one river and filling by another because the
type and amount of sediment yielded during adjustment to the new climate is
oppositely directed. What this tells us is that the effect of climate change may
be highly dependent on the antecedent values of temperature and precipitation. If that is a true statement, knowledge of preexisting climate may be as
important in understanding how systems respond to climate change as knowing
the magnitude of the change itself.

In sum, the relationship among climate, process, and landform is not easily determined because the effect of change is sidetracked into ancillary factors. The adjustments of these factors in the new climate provide variable conditions of load and water which spur responses that are not predictable under the present state of our knowledge. This suggests that we do not have clear-cut relationships between climate and landforms because we are far from understanding the climatic geomorphology scheme.

Summary

In this chapter we briefly examined climate and endogenic factors as major external controls on geomorphic systems. The endogenic influence occurs primarily through the addition of mass and energy by volcanism and tectonic activity. The most important tectonic processes are those producing vertical movements of the Earth's surface. One of these is the isostatic adjustment required when the internal mass balance is upset. Other vertical movements, associated with faulting and warping, are integral parts of a subdiscipline known as tectonic geomorphology in which the relationship among tectonics, processes, and landforms is utilized in a variety of geologic and environmental studies. In most cases, vertical displacements induce threshold conditions and responses in the effected surficial systems.

Climatic geomorphology explores the relationship between prevailing climates and the landforms expected to form under those conditions. Of greater significance is the fact that climate change is a threshold-producing phenomenon, although the cause-and-effect relationship may pass through intermediary factors such as eustatic sea level fluctuations or vegetal and geological screens.

Suggested Readings

The following references provide greater detail concerning the concepts discussed in this chapter.

Andrews, J. T., ed. 1974. *Glacial isostasy.* Stroudsburg, Penn.: Dowden, Hutchinson and Ross.

Bloom, A. L.; Broecker, W. S.; Chappell, J. M. A.; Matthews, R. K.; and Mesolella, K. J. 1974. Quaternary sea level fluctuations on a tectonic coast: New $^{230}Th^{234}U$ dates from the Huon Peninsula, New Guinea. *Quat. Res.* 4:185–205.

Bull, W. B. 1984. Tectonic geomorphology. *Jour. Geol. Educ.* 32:310–24.

Derbyshire, E., ed. 1976. *Geomorphology and climate.* London: John Wiley & Sons, Ltd.

Keller, E. A.; Bonkowski, M. S.; Korsch, R. J.; and Shleman, R. J. 1982. Tectonic geomorphology of the San Andreas fault zone in the southern Indio Hills, Coachella Valley, California. *Geol. Soc. America Bull.* 93:46–56.

Macdonald, G. 1972. *Volcanoes.* Englewood Cliffs, N.J.: Prentice-Hall.

Ollier, C. D. 1981. *Tectonics and landforms.* New York: Longman.

Chemical Weathering and Soils

3

Introduction

We are now ready to focus on the exogenic processes that mold the geomorphic framework into recognizable topographic forms. The first step is to recall that most of the Earth's surface is not composed of solid rock but is underlain by the unconsolidated remains of thoroughly altered rock. The fresh rocks and minerals that once occupied the outermost position reached their present condition of decay through a complex of interacting physical, chemical, and biological processes, collectively called *weathering*. Weathering progressively alters the original lithologic character until what finally remains in the space of the former rock is an unconsolidated mass consisting of (1) new minerals created by the weathering processes, (2) minerals that resisted destruction, and (3) organic debris added to the weathered zone.

Since every mineral species has, by definition, a unique chemical composition or atomic arrangement, it is not surprising that each type resists or responds to weathering in a special way. Considering the wide variety of climates driving the processes and the almost endless array of rock structures and mineral types, the precise mechanics of weathering could easily be beyond our comprehension. Nature, however, has simplified our task because, as indicated earlier, the bulk of the Earth's crust is composed of a surprisingly small number of mineral varieties with an equally limited chemistry. With this advantage, we can explore weathering efficiently even though the systems involved are exceedingly complex.

Weathering is usually divided into separate domains of chemical processes (*decomposition*) and physical processes (*disintegration*). The distinction between the two is real because the processes of disintegration involve no chemical reactions but simply produce smaller particles from larger ones. Nonetheless, the two realms of weathering operate simultaneously and, in fact, each may directly affect the character and rate of the other. For example, breaking a large rock into smaller particles increases the total surface area and thereby accelerates the chemical attack on the material. Conversely, expansion of minerals by chemical processes may exert enough internal stress to hasten the disintegration of the rock. Realistically, then, the physical and chemical functions of weathering may be so intimately intertwined that to consider them as unique processes is mostly a matter of convenience. In this chapter we will deal with only the chemical and biological aspects of weathering, which are dominant in the development of soils. Disintegration will be examined in the next chapter when we consider the stability of slopes, because many of the processes that break rocks apart are also important in the erosion of unconsolidated slope material by mass wasting.

A *soil* is the residuum that results from the application of weathering over an extended period of time. Given the proper conditions, a distinct layering, called the *soil profile,* will develop in the residual material, and its characteristics will directly reflect the weathering and soil-forming processes. Although both geologists and pedologists use soil profiles as a basis for classification, other scientists use the term "soil" in different ways. To an engineer, for example, a soil consists of any accumulation of unconsolidated debris; it may include material such as alluvial deposits that have not experienced the effects of weathering. Agricultural experts may be interested in only the upper part of a soil profile, which is the segment that supports vegetation. Clearly, the definition and classification of soil depend on what information an investigator wishes to gain from it and on what ways such data will be used. The farmer, of course, wants increased crop production, and the engineer requires knowledge of physical properties, such as bearing strength, to design buildings and other constructions judiciously. The geologist, however, utilizes soils and other weathering phenomena as clues to the intricacies of geologic history and the relative age of unconsolidated deposits. Although most soils are young in the geological sense, some profiles have been preserved in the older record, and these provide critical evidence about environmental conditions at the time of formation. Mineral compositions of many sedimentary rocks reflect the combined tectonic and climatic conditions during their origin, and a correct interpretation of such rocks requires a knowledge of weathering and soil-forming processes. Landforms are often recognized as relict features because their soil character is inconsistent with the prevailing modern climate. Many other examples could show that we utilize an understanding of soils in many aspects of geology. The study of chemical weathering and soils is not simply an adventure into esoteric geomorphology, for a working knowledge of weathering processes is essential to any scientist interested in the surface environment.

Decomposition

Rocks and minerals are usually not in equilibrium with conditions that exist at or near the Earth's surface. As a result, **decomposition** can be viewed as a group of processes that attempt to create substances that are more nearly stable in that environment. This march to equilibrium is accomplished by alteration of original materials and/or production of new mineral types.

The most important agent of the weathering regime is water introduced as rain. Because rainwater is usually mildly acidic, a significant part of chemical weathering can be visualized as a process whereby minerals assimilate hydrogen ions and/or water and release cations to the soil liquid. Complicating this simple model are two facts: (1) water entering the ground is not chemically pure but contains a variety of ions captured from the atmosphere and introduced from surface materials; and (2) organic processes, involving metabolism of microorganisms and decay of vegetal matter, add gases and organic acids to the system. These organic functions are of such importance that some authors (Carroll 1970) suggest that chemical weathering proceeds in two stages. The first stage, driven primarily by inorganic processes, is called *geochemical weathering* and produces rotten rocks or *saprolites*. The second stage, called *pedochemical weathering,* leads to the formation of soils from the saprolitic material; it is chiefly a biologically controlled phenomenon.

Figure 3.1.
Ion exchange and chemical bonding take place as surface of orthoclase feldspar comes in contact with a solution containing dissociated H^+ and OH^- ions.

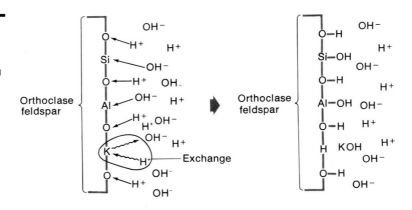

As rainwater percolates into exposed rock material, the first important response commonly is the breaking apart of the structures of the parent minerals. As water surrounds a mineral or penetrates its structure through micro-openings and cleavages, chemical reactions occur that tend to disrupt the mineral's orderly atomic arrangement. Jenny (1950) pointed out that atoms along exposed mineral surfaces are not satisfied electrically and so may attract the dipolar water molecules. If the attraction causes water to dissociate into H^+ and OH^-, these will bond to the exposed ions of the mineral, as in figure 3.1. Hydrogen is then in the proper position to replace mineral cations, thereby releasing them to the surrounding fluid. This process may have a profound effect on the pH of the liquid, as hydrogen is progressively depleted and hydroxyls are concentrated. Experiments observing the change in pH when different minerals are pulverized in distilled water and carbonic acid (Stevens and Carron 1948) show that an equilibrium pH is eventually attained. Presumably this condition is reached when the number of cations removed is balanced by an equal number reentering the mineral structure. These *abrasion pH* values are significant in that they help determine the fluid's effectiveness in attacking other minerals. The values of abrasion pH for common minerals (table 3.1) demonstrate that minerals with alkali or alkaline earth elements in their composition tend to produce high pH liquids, while other minerals yield acidic fluids.

The original mineral is not necessarily completely destroyed. The process may proceed layer by layer from the external surface, or it may break the mineral into many small pieces, each retaining the structure of the original material. Weathering of minerals often occurs at sites of excess energy on the mineral surface (Berner and Holdren 1977, 1979), and the position of these sites is probably determined by the internal properties of the mineral itself. As a result, the mineral surface is not attacked uniformly but, instead, develops numerous widely spaced etch pits (fig. 3.2) where weathering is active (Parham 1969; Keller 1976, 1978; Berner and Holdren 1977, 1979). Thus, the end product of weathering may be directly influenced by the manner in which the minerals are destroyed. Some silicate minerals, for example, break into molecular chains that are easily recombined with available cations to form layered clay minerals.

Table 3.1   Abrasion pH values for common minerals.

Mineral	Composition	Abrasion pH
Kaolinite	$Al_2Si_2O_5(OH)_4$	5, 6, 7
Boehmite	$AlO(OH)$	6, 7
Gibbsite	$Al(OH)_3$	6, 7
Gypsum	$CaSO_4 \cdot 2H_2O$	6
Hematite	Fe_2O_3	6
Montmorillonite	$(Al_2Mg_3)Si_4O_{10}(OH)_2 \cdot nH_2O$	6, 7
Quartz	SiO_2	6, 7
Zircon	$ZrSiO_4$	6, 7
Chlorite	$(Mg,Fe)_2Al_4Si_2O_{10}(OH)_4$	7, 8
Muscovite	$KAl_3Si_3O_{10}(OH)_2$	7, 8
Biotite	$K(Mg,Fe)_3AlSi_3O_{10}(OH)_2$	8,9
Calcite	$CaCO_3$	8
Anorthite	$CaAl_2Si_2O_8$	8
Orthoclase	$KAlSi_3O_8$	8
Albite	$NaAlSi_3O_8$	9, 10
Dolomite	$CaMg(CO_3)_2$	9, 10
Augite	$Ca(Mg,Fe,Al)(Al,Si)_2O_6$	10
Hornblende	$Ca_2Na(Mg,Fe^{+2})_4(Al,Fe^{+3},Ti)_3$ $Si_6O_{22}(O,OH)_2$	10
Olivine	$(Mg,Fe)_2SiO_4$	10, 11

Data from Stevens and Carron 1948.

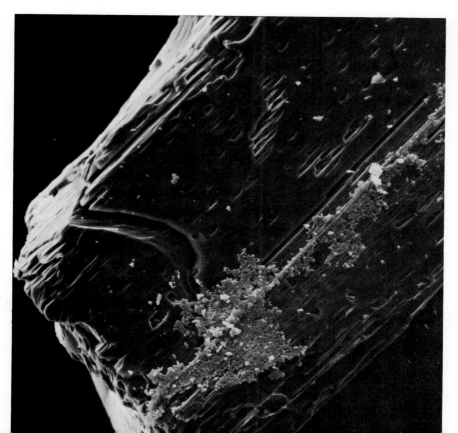

Figure 3.2.
Scanning electron microscope photograph of hornblende grain showing weathering of etch pits that are elongated along cleavage planes.

The second important response in chemical weathering relates to how easily ions are released from the parent structure and how mobile or immobile they are after their liberation. Considerations here are whether released ions or groups of ions are readily fabricated into new, stable minerals, or whether they can be removed in solution by percolating soil fluids. The ultimate fate of released particles depends on what they are and on the characteristics of the fluid into which they are released. The equilibrium state, for example, can only be attained in a closed system, for continuous addition and removal of water surrounding the decomposing mineral will alter the pH, carry some of the released ions away, or provide new elements that can combine chemically with those escaping the mineral. The final result is determined by a myriad of interreactions, depending on how the original structure breaks apart and how mobile the ionic or molecular particles are under the physical and chemical constraints within the weathering zone.

Processes of Decomposition

As just suggested, the magnitude and direction of chemical weathering depend on the composition and structure of the minerals being attacked, how they break apart, and how mobile the constituents are in the weathering environment. Various combinations of mineral types and chemical processes can result in a variety of final products. The common chemical reactions involved in decomposition are *oxidation and reduction, solution, hydrolysis,* and *ion exchange.* Each process plays a particular role in the overall scheme of chemical alteration, although all function simultaneously in the weathering zone.

Oxidation and Reduction   Oxidation occurs when an element loses electrons to an oxygen ion. The process tends to occur spontaneously above the water table where atmospheric oxygen is readily available; therefore, most elements at the Earth's surface exist in an oxidized state. Below the water table the environment is generally reducing (Loughnan 1969); however, high concentrations of organic matter may cause local reducing conditions to occur above the water table.

The ease of oxidation depends on the redox potential (Eh), the magnitude of which is controlled by the abundance of organic matter and the accessibility of free oxygen. In most soils, Eh values range from -350 to $+700$ (fig. 3.3), sufficient to keep the majority of common elements in their oxidized state. However, some elements, such as aluminum, change from a reduced to an oxidized form with considerable difficulty. The most common elements affected by fluctuations of Eh are iron, manganese, titanium, and sulfur. Iron is easily oxidized to the ferric (Fe^{+3}) state:

$$2Fe^{+2} + 4HCO_3^- + \tfrac{1}{2}O_2 + 2H_2O \rightarrow Fe_2O_3 + 4H_2CO_3.$$

Its appearance as crustations on grains or as reddish-brown stains along fractures usually signifies the first tangible evidence of decomposition. The redox phenomenon exerts a significant control on the mobility of certain elements as they are affected by decomposition. This role of the EH factor will be discussed later in the chapter.

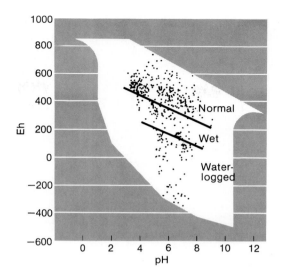

Figure 3.3.
Eh-pH characteristics of soils. Eh values in millivolts.

Solution   The process of solution is critical in chemical weathering because when atoms are dissolved from a mineral, the structure becomes unstable. However, the precise way in which a mineral collapses varies with its crystalline structure and the mobility of its constituent ions. When an atom is removed from its parent mineral, it may remain in solution and be taken completely out of the system by the downward moving fluids. This depends on the concentrations of reactable ions in the fluid and on other chemical characteristics of the medium.

Solution is usually exemplified by the reaction of calcite and acid:

(calcite) (carbonic acid)

$$CaCO_3 + H_2CO_3 \rightarrow Ca^{+2} + 2(HCO_3)^-.$$

The role of CO_2 in the solution of calcite is critical and is especially significant in the development of karst topography. The CO_2 factor will be discussed in chapter 12 as we examine the processes that form karst regions. Here we simply emphasize that most common elements and minerals are soluble to some degree in normal groundwaters, where pH values usually range from 4 to 9 (fig. 3.4). In the weathering of silicate minerals, note that silica is soluble under all normal groundwater conditions. The degree of solubility is rather low ($\approx$ 6 ppm) when the silica is contained in quartz (Morey et al. 1962) but increases considerably ($\approx$ 115 ppm) in amorphous silica (Morey et al. 1964). Aluminum oxides are virtually insoluble under normal groundwater conditions, and iron in the ferric state can be dissolved only by rather acidic fluids. It is not surprising, therefore, that mature weathering profiles in humid climates should be characterized by the presence of abundant ferric iron and aluminum. Other cations are readily soluble, and their removal to the water table concentrates the least soluble constituents in the weathered zone.

Figure 3.4.
The relationship of solubility of common substances to various pH conditions.

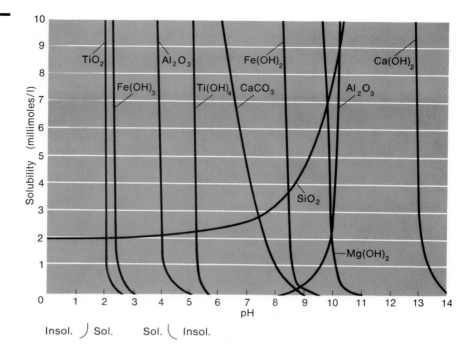

Insol. ⌡ Sol.   Sol. ⌡ Insol.

Hydrolysis   The reaction described on the preceding page between mineral elements and the hydrogen ion of dissociated water is called hydrolysis. Chemically, *hydrolysis* involves a reaction between a salt and water to produce an acid and a base; it is probably the most important mechanism in breaking apart structures of the silicate minerals. During the process, metallic cations are separated from the mineral structure and replaced by H^+, which is held in the original aluminosilicate complex (fig. 3.1). Most of the replaced cations are soluble in natural waters.

In addition to freeing cations, hydrolysis usually produces H_4SiO_4, HCO_3^- and OH^-, all of which can be considered to be in solution (Birkeland 1974). In the common case where water and acid attack an aluminosilicate mineral, the reaction leaves the H^+ embraced in segments of the original structure, and these are subsequently recombined into clay minerals. The hydrolysis of orthoclase feldspar shown here and in figure 3.1 demonstrates this response:

(orthoclase)   (kaolinite)

$$2KAlSi_3O_8 + 2H^+ + 9H_2O \rightarrow H_4Al_2Si_2O_9 + 4H_4SiO_4 + 2K^+.$$

The removal of the metallic cation will proceed as long as free hydrogen ions are available, easily replaced cations are present, and the solvent has not reached saturation with respect to the ion being liberated. The continuous introduction of fresh water and the development of organic acids will assure a ready source of H^+. Since OH^- is carried downward to the groundwater table by percolating water, it seems unlikely that highly alkaline water can ever be produced in open systems with constantly moving water. Therefore, the normal product of a continuously leached system is a residue in which all mobile cations have been freed from the original mineral. After the cations are released, they either remain in solution or attach to the surfaces of minerals held in colloidal suspension. Eventually they are carried to the water table and delivered to the regional streams.

Clearly, the effect of hydrolysis decreases as clays depleted of cations become the dominant aluminosilicate in the weathering zone. In fact, as more clays form, they commonly become colloidally suspended in the fluid and may adsorb H^+ to their surfaces. This may cause the liquid to become more acidic rather than more alkaline as hydrolysis proceeds.

Ion Exchange   Ion exchange is the substitution of ions in solution for those held by mineral grains. Although all minerals possess some capability for ion exchange, the process is most effective in clay minerals. The ions to be exchanged are held on the surfaces of clays because unsatisfied charges, exposed hydroxyl groups, and isomorphic substitutions such as Al^{+3} for Si^{+4} have given the clays an overall negative charge. Cations are held at the mineral surfaces by adsorption. In soils, "adsorption" usually refers to the attraction of ions and water molecules to the surfaces of colloidal clays in an attempt to neutralize their negative charges. The adsorbed ions are not held too tightly and, therefore, are susceptible to replacement by or exchange with other cations.

Each clay species has a different propensity for adsorbing cations, called its *cation exchange capacity* (c.e.c.), which is expressed as the number of milliequivalents per 100 grams of clay. When the adsorbed ion is hydrogen, the c.e.c. directly influences the pH value that the clay assumes. Kaolinite, for example, takes on a pH of 4–5 under complete adsorption, while H^+-montmorillonite attains a pH value as low as 3. Colloidal suspensions of these clays create acids that are capable of attacking other minerals. In soils, colloids of organic compounds produce similar acids because the c.e.c. of organic matter is usually rather high, ranging from 150 to 500 (Birkeland 1974).

It should also be noted that cations other than H^+ may be adsorbed by clays and that the prevailing environment controls which cation types are more likely to be adsorbed. For example, in humid regions colloidal clays will adsorb H^+ and Ca^{++} more readily than Mg^+, Na^+, or K^+. In well-drained arid soils, Ca^{++} and Mg^{++} are usually the most prominent exchangeable ions, and H^+ is the least common. In poorly drained arid soils, the number of adsorbed sodium ions often equals or exceeds the calcium (Buckman and Brady 1960).

Ion exchange is governed by the composition and pH of the interstitial water as well as the type of ion in the exchangeable position. In general, strongly acidic water allows H^+ to replace metal cations of the parent minerals, but this tendency changes as the water becomes neutral. At higher pH values, the mineral cations may remain in the exchangeable position or, in fact, H^+ may be replaced by metallic cations. In soils, the pH of the soil-water mixture is an indicator of the number of cations held in the exchange position by clays.

This characteristic is commonly referred to as a *percentage of base saturation,* meaning the percentage of exchange sites occupied by cations other than hydrogen. The higher the percentage of base saturation, the higher the pH of the soil-water complex because less H^+ is held by the clays.

Mobility

The extent to which chemical weathering will alter the parent mineralogy depends largely on the relative mobilities of the constituent ions. Some ions are easily removed (high mobility) from the weathering system under normal groundwater conditions, while others are relatively difficult to remove (high immobility). The presence of highly mobile ions in a mature weathering zone indicates that some factor is impeding the transfer of the ion from the system. Orthoclase feldspar, for instance, will hydrolyze to kaolinite (as shown earlier) if all the potassium is lost in the process. If some potassium is retained, the clay product will be illite rather than kaolinite, as shown in the following reaction:

(orthoclase)   (illite)

$$3KAlSi_3O_8 + 2H^+ + 12H_2O \rightarrow KAl_3Si_3O_{10}(OH)_2 + 6H_4SiO_4 + 2K^+.$$

Accordingly, then, since K^+ is a mobile ion, its immobile behavior in the formation and preservation of illite must be linked to the prevailing character of the fluid or to an incomplete breakdown of the orthoclase that traps the K^+ in molecules of the original structure. In either case, the easily removed potassium is rendered immobile and remains in the weathered zone.

The relative mobilities of common cations are as follows, in order of decreasing mobility:

$$(Ca^{+2}, Mg^{+2}, Na^+) > K^+ > Fe^{+2} > Si^{+4} > Ti^{+4} > Fe^{+3} > Al^{+3}.$$

Such a mobility distribution is probably related to a parameter known as the *ionic potential,* which is expressed as the ratio of the valence (Z) to the ionic radius (r). In general, Z/r is a useful first approximation of mobility because very mobile ions have Z/r values less than 3; those forming immobile precipitates have values between 3 and 9.5; and those forming soluble, complex anions are usually greater than 9.5 (fig. 3.5). However, as each cation can be immobilized by external factors, those factors in a sense greatly influence how effectively the major processes of chemical weathering will work.

For example, Shoji and his colleagues (1981) found a different mobility sequence than the normal one just shown. They concluded that the volcanic glass parent rock and the secondary minerals derived from its weathering exerted a strong control on the ion mobilities. Nonetheless, if all processes do function without hindrance, the mobile ions will be depleted from the system and the immobile ions will be progressively concentrated. The major external factors that control mobility include leaching, pH, Eh, fixation and retardation, and chelation.

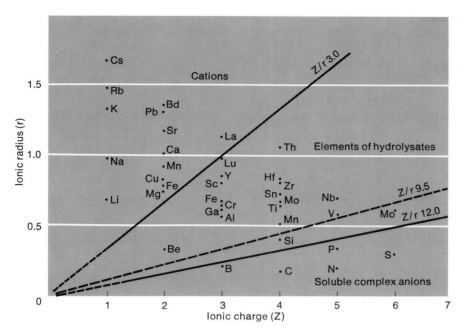

Figure 3.5.
Grouping of some common elements according to their ionic radius (*r*), ionic charge (*Z*), and ionic potential (*Zr*). (From Gordon et al. 1958)

Leaching The most important factor influencing ion mobility is the amount of water leaching the weathering zone. How frequently rainwater flushes through the system, and in what quantity, depends mostly on the climate but also on the permeability of the material. The significance of leaching is severalfold: (1) it removes in solution the constituents that have been separated from minerals by hydrolysis and ion exchange and, by providing new hydrogen, allows these processes continuously to alter the original material toward an ultimate degraded condition; (2) it directly affects the pH of fluids surrounding minerals and thereby helps determine which elements will remain in solution; and (3) it provides the mechanism by which dissolved ions and clays are transferred from higher levels to lower levels in the weathering zone. This introduces new components to each level and so may engender the chemical environment needed to precipitate new minerals.

In general, the absence of continual leaching makes the weathering zone function like a closed system. Since ions removed from the parent minerals are held in the surrounding fluid, the exchange processes will continue only until an equilibrium condition is established between the ions in the fluid and those in the mineral. For all practical purposes, further chemical weathering is impossible beyond this stage, and mobile ions and easily decomposed minerals will remain in the system. In this way, the retardation of leaching, usually due to insufficient rainfall, is a prime factor causing immobility of ions in a weathering system. It is not surprising, then, that mobile constituents are abundant in arid-climate soils. On the other hand, in humid regions where leaching is continuous, the same mobile substances are removed entirely. Even

Figure 3.6.
Characteristic mineralogy of weathering zone formed on kaolinitic sandstone under intense chemical weathering. Note decrease in quartz and increase in stable clays at top of profile. (From Loughnan and Bayliss 1961, *American Mineralogist* 46:211)

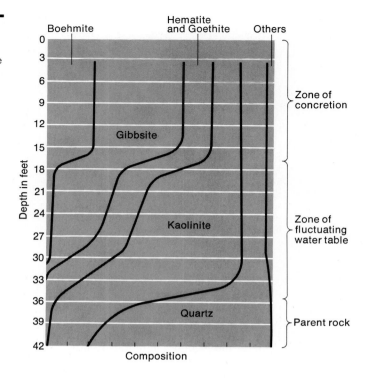

very resistant parent materials such as quartz can be chemically altered to a drastic extent if high-volume leaching continues over a long period of time. Loughnan and Bayliss (1961) reported a truly amazing example of how a kaolinitic sandstone with 90 percent quartz was altered under a tropical climate to a residuum that contained only 5 percent quartz (fig. 3.6).

pH   In addition to its previously discussed influences, H^+ concentration (pH) also plays an important role in the solubility of most common elements (fig. 3.4). Stated in another way, the mobility of ions is partly controlled by pH. Although rainwater reaching the surface is slightly acidic, processes operating beneath the surface can considerably alter the original pH value. Thus, pH is not an independent variable in weathering but may, in fact, be dependent on the type and degree of inorganic and organic processes within the weathering zone.

In geochemical weathering the pH is influenced by leaching, by the c.e.c. of residual minerals, and by the composition and structure of the parent minerals (Loughnan 1969). The effect of leaching and cation exchange capacity can be shown diagrammatically by considering the progressive weathering of a basalt, shown in figure 3.7. With continued leaching, the pH is related almost inversely to the type of clay produced during various stages of the weathering. Assuming complete mobility of Ca, Mg, and Na, the first clay developed

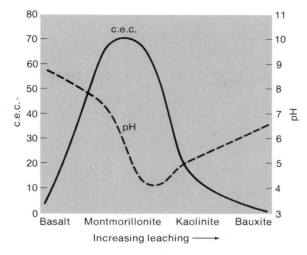

Figure 3.7.
Relationship of cation exchange capacity (c.e.c.) and pH with continuous weathering of a basalt. Transition from basalt to bauxite represents gradual change in mineralogy with increased leaching.

is an H^+-montmorillonite with a high c.e.c. value (70–100 meq/100 g). The high c.e.c. allows the clay to adsorb many of the hydrogen ions made available by the continuous leaching, and the colloidal acid formed causes a pronounced drop in pH. As leaching continues, however, montmorillonite is degraded to kaolinite, which becomes the dominant clay mineral. The much lower c.e.c. of kaolinite permits less hydrogen adsorption, and the pH rises. The final product, bauxite, can adsorb virtually no H^+, and the pH climbs to an almost neutral condition. Obviously, pH is dependent on the other factors.

It is important to remember that organic substances also form colloidal suspensions with very high cation exchange capacities. Although the values vary with climate and the type of organic matter, it is not uncommon for these humic acids to lower the pH below 4.

The Eh Factor   As suggested earlier, the redox potential exerts a control on ion mobility. The effect of Eh on ion mobility is due to the fact that oxidation-reduction reactions are reversible. When soils become waterlogged, free oxygen is excluded and strongly reducing anaerobic bacterial action begins. This generates lowered Eh values that, in turn, cause some elements to revert to their reduced forms. In some cases, ease of reduction is also dependent on pH (Connell and Patrick 1968). Such transitions have a direct influence on the mobility of certain elements that may be relatively insoluble in one form and easily dissolved in the other. Removal of iron oxides attached to clays, for example, can result from only a minor lowering of Eh, which transforms ferric iron to the more soluble ferrous type (Carroll 1958). Apparently, a fluctuating water table can have important ramifications on the redox potential (Eh) and with it the relative mobility of important ions.

The significance of Eh in decomposition and ion mobility can be briefly stated as follows:

1. Ferrous iron (Fe^{+2}) commonly binds silica tetrahedra in the structures of silicate minerals. The oxidation of the iron to the ferric state requires impossible internal adjustments, and the lattice structure is destroyed (Carroll 1970).
2. The by-products of the oxidation process may facilitate the decomposition of other, more stable, minerals. For example, the oxidation of pyrite produces sulfuric acid, as shown:

$$4FeS_2 + 14H_2O + 15O_2 \rightleftharpoons 4Fe(OH)_3 + 8H_2SO_4.$$

This acid lowers pH and will react with any nearby mineral susceptible to attack by acid. Groundwater in regions near zones of sulfide mineralization is usually very acidic.
3. As indicated before, the solubility or insolubility of some elements in groundwater having a normal pH is directly controlled by whether the elements exist in oxidized or reduced form. This is especially true for iron and titanium.

Fixation and Retardation   Some cations, especially potassium, have a tendency to be retained or fixed in the weathered zone, so that their mobility is considerably lower than might be expected. The fixation of potassium may account for its low concentration in seawater and the widespread distribution of illite in sedimentary rocks. The precise mechanism involved is poorly understood, but it is probably related to the fact that potassium silicates (muscovite, K-feldspars) are more resistant to chemical weathering than are other minerals with similar lattice structures. It is known that potassium is most readily fixed in clay minerals with expandable lattices such as chlorite, illite, micas, vermiculite, and montmorillonite. Why this should be or why potassium is so much more susceptible to this phenomenon than other ions is not clear, but Wear and White (1951) suggest that the unique size of the K^+ ion provides the most stable structure when combined with oxygen in layered silicates.

Ion mobility may also be slowed or retarded without complete fixation. Wollast (1967) demonstrated that the initial stage of orthoclase hydrolysis is accompanied by the creation of a thin surface layer of amorphous $Al(OH)_3$ and silica. Although these substances are transferred from the coating layer to the surrounding liquid, the fluid is quickly saturated with respect to aluminum. Silica, and presumably other ions, continue to diffuse through the layer, but the rate diminishes because the distance from fresh feldspar to the fluid increases continuously as the aluminum-enriched sheath thickens. The retarding coat is not formed at pH values less than 5 because $Al(OH)_3$ is soluble under those conditions.

Chelation   The process of chelation represents one of the most dramatic effects on the mobility of ions. Lehman (1963) defines chelation as "the equilibrium reaction between a metal ion and a complexing agent, characterized by the formation of more than one bond between the metal and a molecule of the complexing agent and resulting in the formation of a ring structure incorporating the metal ion." To those of us unversed in organic chemistry, this means that metallic ions that are extremely immobile under normal conditions can be mobilized by reacting with complexing agents and be vertically transported as part of the compound.

Most complexing agents involved in chelation are organic compounds nurtured in soils by alteration of humus into a plant acid called fulvic acid (Wright and Schnitzer 1963), although other organic processes also produce chelating complexes. Lichens, for example, secrete such materials (Schatz 1963). The complexing agent ethylenediaminetetraacetate (EDTA) is by far the most completely understood chelator, and its structure (fig. 3.8) shows how the metallic ion is bonded and held.

Although the chemistry of the chelating process is far from clear, its importance in weathering has been recognized for several decades (Schatz et al. 1954). During this time several excellent studies have demonstrated that EDTA and other chelators may play a dominant role in mobilizing iron and aluminum under pH conditions that are not conducive to Fe and Al solution (Atkinson and Wright 1957; Wright and Schnitzer 1963; Schalscha et al. 1967). Using several different chelating agents Schalscha and his co-workers (1967) were able to extract iron from a variety of minerals. Significantly, their study revealed no correlation between pH and the amount of liberated iron. EDTA released more iron from magnetite and hematite, for example, than did hydrochloric acid, even though the HCl solution was more acidic. However, in other studies (Tan 1980) fulvic acid was capable of extracting ten times more silica and six times more aluminum when pH was decreased from 7 to 2.5.

In soils, chelating agents become soluble in water as they are oxidized (Wright and Schnitzer 1963). Iron and aluminum are locked in the ring structure and carried downward with the percolating water. The downward movement continues until the entire mass is flocculated because of small changes in ionic content of the soil water, or until the complex is broken by microbial action. In either case further downward movement is curtailed, and the iron and aluminum are redeposited.

Even before soils begin to develop, chelation can be important in the alteration of the bare rock. Jackson and Keller (1970) showed that the presence of the lichen *Stereocaulon vulcani* accelerated the chemical weathering of volcanic rocks in Hawaii. Rocks covered by lichen growth had a thicker weathering crust than lichen-free rocks, and the crust was enriched in iron and depleted in titanium, silicon, and calcium. Rocks devoid of lichens were measurably less affected by chemical weathering.

The Degree and Rate of Decomposition

To utilize soils in geological studies, it is important to know what the end products will be in a completely altered system. Geologists are, therefore, concerned about what material will remain when no more chemical reactions are possible, given the constraints of climate, vegetation, and rock type. If such a condition can ever be reached, a steady state will have been established between the driving forces (decomposition) and the resisting materials, and although the weathering zone may get progressively thicker, the upper parts of

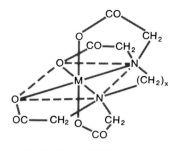

Figure 3.8.
Structure of complexing agent EDTA. Metallic ion (M) is bonded within structure. (Reproduced from *Soil Science Society of America Proceedings*, Vol. 27, 1963, p. 169, by permission of the Soil Science Society of America)

Table 3.2 Weathering stability of the common silicate minerals.

Olivine, Anorthite	Least Stable
Pyroxenes, Ca-Na Plagioclase	
Amphiboles, Na-Ca Plagioclase	↑
Biotite, Albite	│
K-Feldspars	↓
Muscovite	
Quartz	Most Stable

After Goldich 1938. © by The University of Chicago Press. Used with permission.

the profile are degraded to their ultimate form. In thoroughly leached soils, all minerals remaining presumably will be stable and all mobile ions will be gone. Thus, if we have some idea as to what final products are expectable in any climate, we can estimate the degree of weathering (how far the material has progressed toward the steady state) by observing the mineral and chemical composition of the soil.

Mineral Stability   In considering the degree of weathering, a logical first question to ask is what minerals will be most rapidly destroyed in a thoroughly leached open system and, conversely, which types will remain if a steady state is attained. Perhaps the first analytical attempt to answer that question was provided by Goldich in 1938. In a detailed study of the weathering of several varieties of igneous and metamorphic rocks, he suggested that mineral stability of the rock-forming silicates is directly related to their order of crystallization as determined earlier by Bowen (1928). Quartz, being the last to crystallize, forms under the lowest temperature conditions and therefore should be most stable in the surface environment. High-temperature minerals such as olivine and pyroxene are least stable and weather most rapidly. With the exception of muscovite and the plagioclase feldspars, the Goldich stability series (table 3.2) reflects the number of oxygen atoms shared in the silicate structure. Quartz and orthoclase, for example, have four shared oxygens in their structure, making their internal bonding strength much higher than that of a mineral like olivine, which shares no oxygens. Keller (1954) in fact suggested that bonding energy may provide a useful basis on which to estimate mineral stability.

Since Goldich's work, other guides to mineral stability have been developed (for example, Pettijohn 1941; Reiche 1943), and studies using the various techniques have led to the significant conclusion that the extent of chemical weathering can generally be estimated from the mineral assemblage contained in a soil as compared to the minerals of the parent material. Furthermore, this conclusion provides a basis for the determination of relative ages of different soils, which is a major concern of most Quaternary geomorphologists.

In detail, the correlation between the different methods used to determine mineral stability is far from perfect, and the factors that control mineral stability are not clearly defined. For example, the relationship between stability and silicate structure recognized by Goldich is not as strong as we often think. Plagioclase feldspars show considerable variation in stability depending on the mineral variety, even though they all possess the same framework structure. Micas show the same type of discrepancy. In addition, zircon appears to be an extremely persistent mineral under weathering, yet it is structurally the same as olivine, a notably unstable mineral species. Despite these apparent difficulties, the use of mineral stability estimates is still an acceptable technique to gauge the rate and degree of decomposition.

Secondary Minerals   The minerals just discussed are components of the original rock, commonly referred to as *primary minerals.* It is well to remember, however, that the processes of chemical weathering play a dual role. In addition to destroying primary minerals, they also create new minerals by recombining or reprecipitating materials liberated from the parent rocks. These *secondary minerals,* born within the weathered zone, are distinctly more stable than their primary ancestors because they reach equilibrium in the temperature-pressure environment of the soil rather than in the magmatic or metamorphic conditions that created the original crystals. The most common secondary products of weathering are clay minerals and amorphous hydrous oxides of iron, aluminum, silica, and titanium.

Clay Minerals   The advent of X-ray diffraction and electron microscopy provided scientists the capability to examine the actual lattice structures of clay minerals. As a result, clay mineralogy has developed into a separate scientific discipline, the details of which are well beyond the purposes of this book. Nonetheless, clay minerals are important indicators of the degree and character of weathering, and they are significant components in the physical and chemical attributes of soil. As such they deserve mention, albeit brief and oversimplified.

Clay minerals are aluminum silicates in which silica tetrahedra and aluminum octahedra are bonded together into a layered atomic structure. A silica tetrahedron has one silicon atom surrounded by four oxygen atoms; an aluminum octahedron consists of a single aluminum atom bonded to six oxygen atoms. In clays the individual tetrahedrons and octahedrons are linked together in planes, forming distinct sheets or layers typified by either a tetrahedral structure (silica sheet) or an octahedral structure (alumina sheet).

Most clays are either a 1:1 layer silicate or a 2:1 layer silicate (table 3.3). A 1:1 structure has as its fundamental building block one layer of silica tetrahedra and one layer of aluminum octahedra. In 2:1 mineral structures, one aluminum octrahedral layer is positioned between two layers of silica tetrahedra. Other clay varieties do form, but they either represent combinations of the basic types (mixed-layer) or originate under very special weathering conditions (chain silicates).

Table 3.3 Classification of the clay minerals.

	Structure	Common Minerals
External adsorptive surfaces Silica sheet — Crystal unit Alumina sheet Distance fixed — Little or no internal adsorption Silica sheet — Crystal unit Alumina sheet Lattice structure of kaolinite	1:1	Kaolinite Dickite Nacrite Halloysite (4H$_2$O) Allophane
External adsorptive surfaces Silica sheet Alumina sheet — Crystal unit Silica sheet Distance variable — Internal adsorptive surfaces Silica sheet Alumina sheet — Crystal unit Silica sheet Lattice structure of montmorillonite	2:1	*Mica group* Muscovite, 2M Illite Glauconite Biotite Vermiculite *Smectite group* Montmorillonite Many other types *Chlorite group* Chlorite Chamosite

The mineral *kaolinite* represents the 1:1 clay mineral group and has the greatest importance in soils. As shown in table 3.3, kaolinite is constructed of crystal units consisting of one tetrahedral sheet (silica sheet) and one octahedral sheet (alumina sheet). These two layers are held together by ions that are mutually shared by aluminum and silicon atoms in the separate sheets. Importantly, the bonding between adjacent, two-sheet crystal units is also very strong. The bonding strength between the units is significant because it keeps the lattice spacing fixed, thereby preventing expansion of the structure when the clay is wetted. As a result, cations and water do not penetrate between the kaolinite crystal units. This factor explains the relatively low cation exchange capacity of kaolinite because the ion exchange process is restricted to the external surfaces of the mineral. The fixed structure also accounts for the low-plasticity (the capacity to be molded) and swelling characteristics of the mineral.

Of the minerals possessing a 2:1 structure, the smectite group and the mica group have the greatest significance in weathering and soil development.

The *smectite* group is characterized by ion substitutions that occur primarily in the octahedral layer. This leads to wide variations in chemical compositions of the group minerals. *Montmorillonite,* the most common type of the smectite group, has the lattice structure shown in table 3.3. The obvious difference from kaolinite is that montmorillonite has a three-sheet crystal unit. Like kaolinite, the three layers comprising the unit (two tetrahedral sheets and one octahedral sheet) are bound tightly together by shared atoms. In contrast to kaolinite, however, the bonding between adjacent crystal units is notably weak, and the mineral lattice is capable of expansion upon wetting. Therefore, cations and water molecules easily penetrate the mineral interior where cation exchange takes place along the surfaces of the crystal units. The result is a clay with greater plasticity and swelling and a cation exchange capacity that is much higher than can be generated when only external exchange sites are available.

Illite is the most common mica-group mineral found in soils. Because its lattice is 2:1, illite is structurally similar to montmorillonite. It differs from montmorillonite, however, in that some of the silicon in the tetrahedral sheets has been replaced by aluminum. This creates an unbalanced charge in the silica sheets, a charge that is primarily satisfied by potassium ions placed between the crystal units. As a result, the bonding between the units is stronger in illite than in montmorillonite. Therefore, the values of c.e.c., swelling, and plasticity in illite are lower than those found in smectite clays, but they still exceed those found in kaolinite.

The mineral *vermiculite,* also part of the mica group, is found in many soils. It differs from illite because it normally has little K^+ in the interlayer zone. Instead, that position is more likely to be occupied by Ca^{++} or Mg^{++}. These cations do not provide the strong bonding noted in illite and mica; therefore, vermiculite is prone to have some expansion.

Other clays found in soils have variations on the structures we have been discussing. *Chlorite,* for example, has a crystal unit consisting of alternating 2:1 layers and octahedral layers. Therefore, it is often designated as a 2:1:1 clay. Bonding in chlorite is strong because ion substitutions create opposite charges in the layers, making the lattice nonexpanding. In addition, *mixed-layer* clays can form when different crystal units interstratify with one another or with hydroxides of magnesium or aluminum. As you might expect, the physical and chemical properties of mixed-layer clays are quite variable.

Differences in stability are present even within the realm of clay minerals. Jackson and his colleagues (1952) proposed a distinct sequence of clay mineral development in which muscovite or illite is the initial product of weathering and, assuming effective leaching, progressively degrades through montmorillonite to a final kaolinitic clay (fig. 3.9). The suggestion that kaolinite should be the most common end product under normal conditions has received support from evidence other than mineral indices. Feth and his co-authors (1964) found that the composition of groundwater draining the granitic rocks of the Sierra Nevada was chemically in equilibrium with kaolinite. This suggests that other clay minerals in the weathered residuum are unstable and will ultimately change to kaolinite.

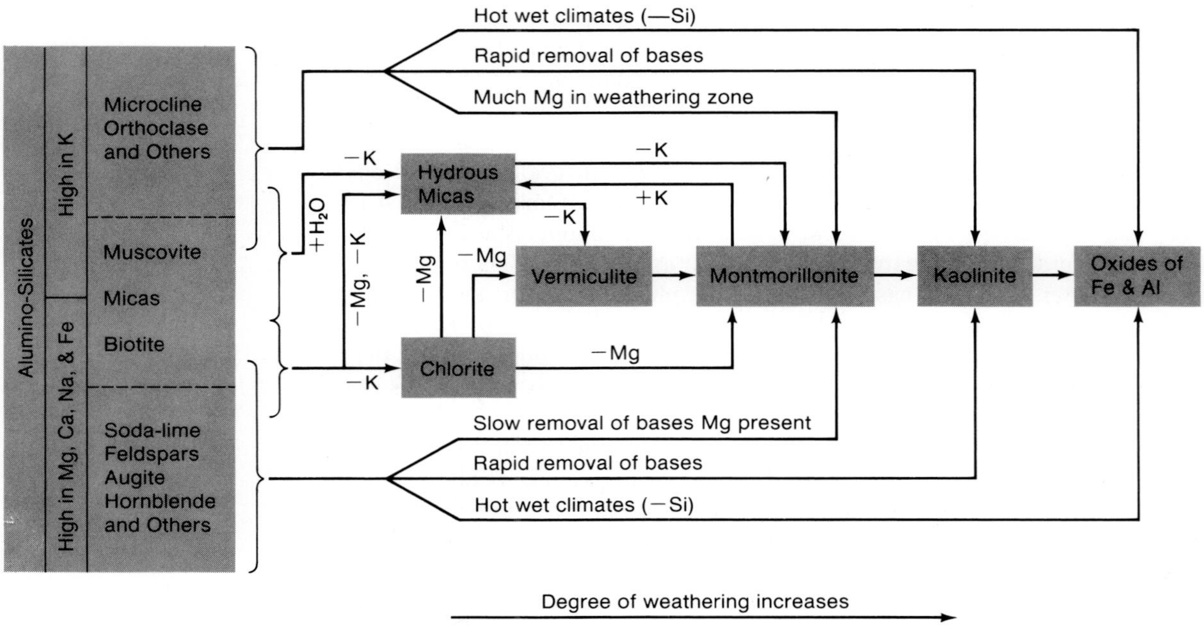

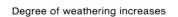

Figure 3.9.
Diagram showing the general
conditions for the formation of the
various silicate clays and the
oxides of iron and aluminum. In
each case, genesis is
accompanied by the removal of
soluble elements such as K, Na,
Ca, and Mg.

It should be recognized, however, that kaolinite itself is a complex mineral having a variety of origins that differ from the evolutionary process earlier suggested. Kaolinite may derive directly from the weathering of feldspar without being the end stage in the degradation of other secondary minerals (Keller 1978, 1982). Additionally, when kaolinite forms directly from feldspar, it seems certain that the mineral develops in an intervening solution phase rather than by solid-state transformation or replacement. Nonetheless, assuming that a sequential development of clay types is possible, the type and amount of clay should provide a reasonable basis for estimating the degree of weathering in a soil; such estimates have in fact been made (Barshad 1964). It should also be apparent that specific clay minerals will tend to be dominant in mature soils developed under a particular set of climatic conditions.

The apparent sequential development of clays emphasizes the important point that clay minerals themselves are highly reactive and readily alter from one form to another. In most weathering situations, the composition of the initial clay developed is usually controlled by the ions available in the parent rock, and many of these ions may be highly mobile. With time, however, climate and other factors operating within the weathering zone exert a greater influence, and clays will alter to forms that are more nearly in equilibrium with the chemical environment.

A note of caution is perhaps needed at this point. In some situations all clay minerals in a soil are not necessarily derived directly from weathering of the parent materials. Colman (1982) examined the clay-sized material in the weathered outer fringe (called a *rind*) of basalt and andesite boulders contained in clay-rich soil horizons at least 100,000 years old. The material in

Table 3.4 Gains or losses of chemical constituents in a hypothetical example of weathering.

	Original Rock	Saprolite	Adjusted %[a]	Loss or Gain[a]
SiO_2	50.3	41.3	24.78	− 25.52
Al_2O_3[b]	18.3	30.6	18.3	0
Fe_2O_3	2.5	11.3	6.78	+ 4.28
FeO	9.4	1.2	0.72	− 8.68
MgO	5.5	0	0	− 5.5
CaO	12.6	0	0	−12.6
TiO_2	1.1	0.1	0.06	− 1.04
H_2O	0.2	14.6	8.76	+ 8.56
Total	99.9	99.1	59.40	−40.5

[a]To obtain loss or gain multiply each value in the saprolite column by 0.6. Subtract the adjusted percentage from the original % in the fresh rock.
[b]Al_2O_3 (fresh)/Al_2O_3 (saprolite) = 18.3/30.6 = 0.6

the rinds is predominantly disorganized or amorphous allophane and iron oxides or hydroxides. However, in the surrounding soil itself the clay fraction consists of highly crystalline, well-developed clay minerals. Because of the rind composition, it appears likely that the clay minerals in the soil horizon are not the result of simple transitions from weathering of the boulders. Instead, the clay minerals were probably introduced to the soil as fully developed crystalline forms by some external process.

Hydrous Oxides   In addition to clay minerals, a number of oxide and hydroxide compounds are formed in the weathering zone as important secondary minerals. The oxides are normally sesquioxides of iron and aluminum occurring mainly as *hematite,* Fe_2O_3, or its hydrated form *limonite,* $Fe_2O_3 \cdot H_2O$, and *gibbsite,* $Al_2O_3 \cdot 3H_2O$ [or preferably $Al(OH)_3$]. In general, these minerals are crystalline, but when a poorly ordered ion arrangement exists in the structure, they are considered to be amorphous. The sesquioxides are usually stable end products of weathering unless the chemical environment becomes strongly acidic.

Estimates Based on Chemical Analyses   The degree of chemical weathering has also been gauged on the basis of total chemical analyses. The most common approach utilizing such data is a direct comparison of the fresh parent rock with the saprolite or soil derived in situ from it. To make an analytical comparison of the two, Al_2O_3 is usually considered to be constant since its mobility is extremely low under normal conditions. The weight percent values of oxides in the soil are recalculated by using the conversion factor %Al_2O_3 (fresh rock) / %Al_2O_3 (weathered material) as a multiplier of each constituent (Birkeland 1974). Table 3.4 demonstrates this method by showing a hypothetical comparison of a saprolite developed from a parent metagabbro. The recalculated values indicate the degree of weathering, and relative gains and losses from the original composition can be used to compare nearby exposures or different horizons within the same soil profile. Maximum errors in this method occur when the assumption of a completely constant aluminum value is incorrect. As mentioned earlier, aluminum may be mobilized by chelators or humic acids, and so it is likely that some aluminum will be lost from the upper weathering zones and some gained in the lower horizons. Nonetheless, the method provides a good first approximation of the relative degree of weathering, even though absolute gains and losses are suspect because of the aluminum problem.

The rate of chemical weathering is also commonly estimated by chemical analyses of water that has filtered through the rock and soil system. In general, these studies rely on the assumption that the parent material is the only source of dissolved constituents in the water. However, because rainwater is not chemically pure, some knowledge of its composition is required before valid conclusions about solution within the weathering profile can be made. Cleaves and his colleagues (1974), for example, estimate that about 37 percent of dissolved solids in a small Maryland stream were introduced into the system as part of the precipitation. Even when the chemistry of rainwater is known, other factors may lead to faulty estimates of solutes produced by weathering. Cyclic wetting and drying of soils in arid climates may cause partial re-solution of salts that earlier were precipitated in the soil (Drever and Smith 1978). Additionally, ions are often brought in or removed from a soil by atmospheric dry fall (deposition not associated with precipitation), vegetation, and land uses such as irrigation, fertilization, and timbering. Therefore, to be totally valid the water chemistry approach should be comprehensive in scope, with all factors causing input or removal of elements being considered (Likens et al. 1977; Reid et al. 1981).

If corrections can be made for precipitation input and effects of the other factors just mentioned, the dissolved load in rivers should represent the weight loss of rock material from a drainage basin. Weights can be converted into volumetric terms, and when considered over a reasonable period of time, rates of chemical weathering can be expressed in terms of surface lowering or in soil development (Hembree and Rainwater 1961; Cleaves et al. 1974; Owens and Watson 1979). In larger areas the rate of chemical denudation also appears to be a function of rock type and climate (Livingstone 1963; Judson and Ritter 1964; Strakhov 1967), increasing in magnitude in warm-humid regions. Rapp (1960) demonstrated, however, that solution may be the dominant erosional process even in arctic regions, and therefore, chemical weathering is a factor that simply cannot be ignored in the overall scheme of landscape development.

Soils

Weathering processes that continue over an extended period of time result in an unconsolidated mass of soil that is measurably different from the original rock in its physical and chemical properties. Soils also contain a significant amount of organic material that is added progressively to the system as decaying vegetation and microorganisms. In addition, pronounced layering develops in the weathered mass during the transition from being simply decomposing rock or alluvium to being a true soil. The vertical arrangement of the layers constitutes a diagnostic property of soils known as the soil profile. The profile extends from the surface downward to the fresh parent material. The time in absolute years needed to form a soil profile, as well as the perfection of the profile's development, varies widely with the intensity of the weathering processes and the character of the original material. Nevertheless, where the soil profile is well developed, its character reflects the environment under which it formed and serves as the basis for classification of soils and interpretation of paleo-conditions.

The Soil Profile

In its simplest form, the soil profile can be visualized as consisting of three main layers, usually designated as the A, B, and C horizons. The A horizon is normally considered to be the thin, dark-colored surface layer where organic matter is concentrated and where clays and mobile components are continuously leached downward or *eluviated*. The C horizon is usually thought of as the underlying parent material which is essentially unmodified by the soil-forming processes. The B horizon, therefore, becomes the transitional zone between the A and C horizons and historically has been considered as an *illuviated* zone, i.e., a zone of accumulation and concentration of the material brought down from the A horizon. Each horizon, defined on the basis of its physical and chemical properties, shows enough internal variation to require subdivisions indicating some special trait is present.

A variety of properties can be used to distinguish the horizons and zones of a soil profile. In addition to characteristics already discussed, such as pH, Eh, and c.e.c., the most important criteria are color, texture and structure, organic content, and moisture characteristics (Birkeland 1974). *Color* in soils is an indicator of high organic content (black, dark brown), ferric iron (yellow-brown to red), or a concentration of SiO_2 or $CaCO_3$ (light grey to white). Small amounts of a pigmentor can cause rather intense discoloration, however, and so color alone may be a poor index of the total quantity of the pigmenting substance. *Texture* is simply the relative proportions of different particle sizes in a soil horizon, analogous to the property of sorting as used by geologists (fig. 3.10). *Structure* in soils, however, is a unique characteristic in that it designates the shape developed when individual particles cluster together into aggregates called *peds* (fig. 3.11). In clay-rich soils, the openings between peds may play an extremely important geomorphic role by providing the only avenues for downward percolation through an otherwise impermeable soil. *Organic matter* in soils consists mainly of dead leaves, branches, and the like, called *litter*, and the amorphous residue, called *humus*, that develops when litter is decomposed. Litter may form at mean annual temperatures as low as freezing, but its optimum production occurs at about 25°–30°C and decreases rapidly above those levels. Microorganisms that convert litter to humus begin to function at temperatures slightly above freezing (5°C), but the optimum

Figure 3.10.
Percentages of clay (< 0.002 mm), silt (0.002–0.05 mm), and sand (0.05–2.0 mm) in basic soil textural classes as defined by the U.S. Department of Agriculture. (From Soil Survey Staff 1951)

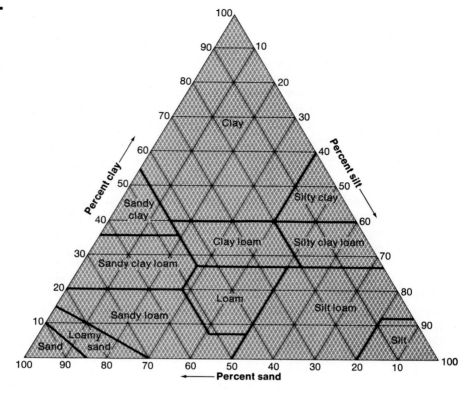

Figure 3.11.
Major types of soil structure: (A) prismatic; (B) columnar; (C) angular blocky; (D) subangular blocky; (E) platy; (F) granular. (From Soil Survey Staff 1951)

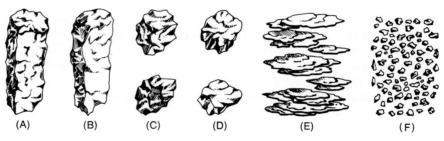

temperature for their life activities may be as high as 40°C (fig. 3.12). It is significant that at temperatures between 0° and 25°C, humus is produced in abundance, but above 25°C little if any humus is accumulated. Humus has an important effect on soil formation because it includes chelators and increases water absorption. In addition, the development of humus releases CO_2 in high concentrations, leading to unusual amounts of carbonic acid within the humic zone and an associated lowering of the pH.

The total quantity of water that can be held in a soil is the *available water capacity* (AWC). This characteristic is very difficult to estimate in the field, and the lack of a consistent laboratory procedure for its computation has led to some confusion as to its relevance (Salter and Williams 1965). Nonetheless, by combining this parameter with the bulk density (dry weight of soil/unit

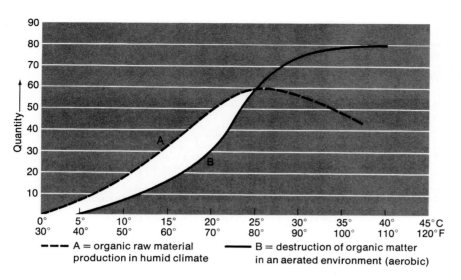

Figure 3.12.
Production and destruction of organic matter in humid climates. The difference between the amount produced and the amount destroyed by microorganisms controls the accumulation of humus.

- - - A = organic raw material
production in humid climate

B = destruction of organic matter
in an aerated environment (aerobic)

volume), an estimate of the depth of wetting can be made (see Birkeland 1974). Such information is significant in that it relates to many soil properties, especially those affected by redistribution of material during downward percolation of water.

Available water capacity can be calculated if the moisture content at an upper limit (*field capacity*) and a lower limit (*permanent wilting point*) are known. The field capacity is determined by allowing a saturated sample to drain by gravity for at least 48 hours, by which time the remaining water content is held by adhesion to mineral and organic particles. After field capacity is reached, water can still be taken from a soil by the normal functions of plants. If plants continue to extract water to the extent that no more can be removed, the vegetation wilts. The water remaining in the soil is held with tensional stresses too high for the plants to break, and this amount of water represents the permanent wilting point. Both field capacity and permanent wilting point are expressed as a weight percentage according to the following equation:

$$P_w = \frac{W_s - W_d}{W_d} \times 100$$

where P_w is moisture percentage, W_s is total soil weight, and W_d is weight of soil after drying at 105°C. The available water capacity is simply the difference between the moisture content at field capacity and that at the permanent wilting point.

Assuming that some criteria are distinct enough to identify a soil horizon or zone, the pertinent consideration then becomes what nomenclature should be used to convey that information. In recent years the simple designation of A and B horizons as zones of eluviation and illuviation has been seriously questioned. Soil scientists have found that some soils contain more than one eluvial or illuvial horizon, obviously necessitating that either an eluvial zone be placed in the B horizon or an illuvial zone be placed in the A horizon. In addition, it

Table 3.5 Nomenclature of soil horizons.

Horizon[a]	Characteristics
O	Upper layers dominated by organic material above mineral soil horizons. Must have > 30% organic content if mineral fraction contains > 50% clay minerals, or > 20% organics if no clay minerals.
A	Mineral horizons formed at the surface or below an O horizon. Contains humic organic material mixed with mineral fraction. Properties may result from cultivation or other similar disturbances.
E	Mineral horizons in which main characteristic is loss of silicate clay, iron, or aluminum, leaving a concentration of sand and silt particles of resistant minerals.
B	Dominated by obliteration of original rock structure and by illuvial concentration of various materials including clay minerals, carbonates, sesquioxides of iron and aluminum. Often has distinct color and soil structure.
C	Horizons, excluding hard bedrock, that are less affected by pedogenesis and lack properties of O, A, E, B horizons. Material may be either like or unlike that from which the solum presumably formed.
R	Hard bedrock underlying a soil.

Adapted from the Soil Survey Staff (1960, 1975, 1981).
[a]Horizons can be divided into subhorizons indicated by Arabic numbers such as B1, A2, B12, etc.

is now clear that concentrations of clays or sesquioxides are not always caused by illuviation but may result as a lag when other materials are preferentially removed. Because of these and other complications, soil processes and characteristics are often incompatible with the general sense of the A, B, C nomenclature; in response, the Soil Conservation Service (S.C.S.) of the U.S. Department of Agriculture proposed new terms to indicate the diagnostic soil horizons (Soil Survey Staff 1960, 1975). The horizons established by the S.C.S. (table 3.5) may not correlate with the older system; in fact, their definitions are so precise that laboratory analyses may be required before certain zones or horizons can be identified.

Surface master horizons are called O or A depending on the amount of organic matter they contain. The A horizon can be further defined as mollic, umbric, or ochric on the basis of criteria established in the S.C.S. system. *Mollic* A horizons are dark colored, contain more than 1 percent organics, have a base saturation of more than 50 percent, and usually develop under a grassland vegetation. *Umbric* A horizons are similar to mollic types but form beneath forest zones and have less than 50 percent base saturation. *Ochric* A horizons are light colored, contain less than 1 percent organics, and normally develop under semiarid vegetation.

The E horizon underlies the O or A horizon. It is characterized by intense leaching that removes Fe^{+3} or organic coatings from the mineral particles and, therefore, is usually bleached gray in color. The B horizon has characteristics that may reflect any or all of the weathering processes discussed earlier. It commonly has a high clay content due to illuviation or to mineral growth in situ, reddish hues, iron and aluminum concentrations as sesquioxides, and stable

Table 3.6　Some common descriptive symbols to be used in conjunction with major soil horizons.

Symbol[a]	Meaning
b	buried soil horizon
g	strong gleying
h	illuvial humus
ir	illuvial iron
k[b]	accumulation of alkaline earth carbonates, commonly $CaCO_3$
m	strong cementation
p	plowing
t	illuvial clay
x	fragipan character

From Soil Survey Staff, 1975, 1981.
[a]Symbols used with other profile designations. For example, B2t, B1h, Cca.
[b]Formerly designated as ca, a term many publications still use.

primary minerals. The properties of the B horizon, however, can change markedly with variations in the fundamental soil-forming controls such as climate and parent material. The fact that the B horizon has such variable traits prompted the descriptive subterms *argillic* (high clay content), *natric* (high exchangeable sodium concentration), *spodic* (translocated organics and sesquioxides), *oxic* (hydrated oxides of Fe and Al and secondary silicates including 1:1 clays), and *cambic* (well-developed structures with intense oxidation and red color common).

The lowermost master horizons, C and R, exist below the B horizon. The C horizon has no properties typical of the overlying horizons, but it has been affected by weathering, as features such as oxidation show. It is composed of unconsolidated material that may or may not be like the material from which the soil presumably formed (Birkeland 1974). The R horizon is simply consolidated bedrock beneath the soil.

In his most recent edition, Birkeland (1984) retains the K horizon to indicate a subsurface horizon characterized by extreme carbonate accumulation. In addition, he suggests a number of formal subhorizons of the A, B, and C horizons that have designations such as Bt, Bh, Cox, etc. These are not reproduced here, but those of you interested in soils should refer to that work for greater detail.

Three kinds of symbols are used to denote horizons and layers in a soil profile. Capital letters as shown in table 3.5 designate master horizons. Lowercase letters are used as suffixes to indicate specific characteristics of layers in the master horizon (see table 3.6), and numbers are used as suffixes to connote vertical subdivision within a horizon or layer. In addition, numbers are prefixed to the master horizon designations to indicate a significant change in particle size or mineralogy within the soil. These signify a difference in the material from which the horizons have formed. In 1975 the S.C.S. used roman numerals as the prefix but have since changed to arabic numerals (Soil Survey Staff 1981). The number 1 is never used because it is implied to represent the

Figure 3.13.
Hypothetical soil profile.

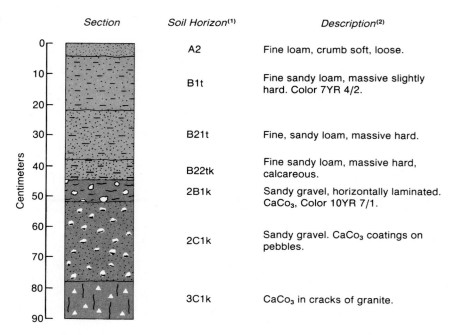

Section	Soil Horizon[1]	Description[2]
	A2	Fine loam, crumb soft, loose.
	B1t	Fine sandy loam, massive slightly hard. Color 7YR 4/2.
	B21t	Fine, sandy loam, massive hard.
	B22tk	Fine sandy loam, massive hard, calcareous.
	2B1k	Sandy gravel, horizontally laminated. $CaCO_3$, Color 10YR 7/1.
	2C1k	Sandy gravel. $CaCO_3$ coatings on pebbles.
	3C1k	$CaCO_3$ in cracks of granite.

(1) Lowercase k may now be used in place of ca.
(2) Color from Munsell color charts (see Soil Survey Staff, 1975, pp. 463–69 for discussion).

material in the surface mineral zones. Therefore, if no changes occur downward into the profile, prefix numbers are not needed. Transitional zones between the master horizons are indicated by the use of both capital letters. For example, a zone transitional between the A and B horizons may have characteristics typical of both. It is designated as AB or BA, the first letter depending on which master horizon it most closely resembles. The B2 zone is reserved for that portion of the horizon where the diagnostic properties identifying the B horizon are most prominently displayed. A second number is sometimes utilized to stress minor variations in the soil. Figure 3.13 represents a hypothetical soil profile using the current nomenclature. However, because the S.C.S. designations seem to be in a perpetual state of flux, descriptive notations shown in figure 3.13 may be changed in the future.

Soil Classification
The classification of soils is, like all classifications, simply an attempt by someone systematically to group together and name soils that exhibit pronounced similarities. The trait or traits being classified serve as the philosophic basis for the groupings. The choice of diagnostic traits depends on what the classifier deems important, and since individuals disagree on that point, no classification will satisfy all members of the pedologic community.

The first real attempt at soil classification came from Russian soil scientists in the late nineteenth century, especially from the work of Dokuchaiev and his students. Although fraught with inconsistencies, Dokuchaiev's scheme had great influence on American pedologists. His groups were defined in part by climate and vegetation, and until recently these genetic criteria were accepted as the basis of all soils classifications used in the United States. In fact, many Russian terms are still integrated in American soils nomenclature.

In the United States, C. F. Marbut was the leading force in efforts to systematize soils. Marbut's classification, developed over a span of years in the 1920s and 1930s, was based on characteristics that are present only in "mature" soils. The system, therefore, had no place for soils that were not fully developed. A more refined classification, designed to rectify many of the problems inherent in Marbut's system (Baldwin et al. 1983; Thorp and Smith 1949), became the most extensively used soil classification in the United States until a dramatic shift in fundamental emphasis was introduced. In 1960 the Soil Conservation Service completely revised the descriptive nomenclature and the classification of soils (Soil Survey Staff 1960). It is important to note that this revision is not simply a formulation of new class names (although that occurred) but represents a fundamental disagreement with the philosophical basis of earlier classifications. Until the new system was devised, all classifications were essentially genetic in scope; that is, the major soil classes were based on climatic and vegetal factors. Although the subdivisions were linked to observable aspects of the profile, these were not explicitly defined, and soil scientists inescapably allowed their knowledge of climatic and vegetation distributions to influence decisions about placing a soil in a particular group. In many cases, therefore, pedologists were classifying the genetic factors and not the tangible resulting soil. The S.C.S. system is nongenetic. It requires no climatic interpretation but is based on very specific, often quantitative, criteria. We will describe it briefly, without becoming mired in its details.

Soils in the S.C.S. system are grouped into 10 *orders* distinguished by the major horizons in their profiles (tables 3.7, 3.8; fig. 3.14). The orders are subdivided into *suborders* (table 3.7), defined by a diagnostic physical or chemical soil property that in some cases may require quantitative laboratory data to be recognized. The terms used to designate a particular suborder represent the combination of two syllables; the prefix indicates the diagnostic property of the suborder and the suffix reveals the order, since it is composed of several key letters of the order name (tables 3.7, 3.9). An Argid, for instance, is an Aridisol with an argillic horizon, and a Udent is a moist Entisol with a low to moderate organic content. Further subdivision into *great groups* and *subgroups* is made on the basis of even more detailed properties than those differentiating the suborders. For example, within the Udent suborder, a great group characterized by a low temperature is a Cryudent, the prefix "cry-" inserted to indicate coldness. Subgroup nomenclature requires an additional descriptive word placed before the great group name. A Cryudent with a weakly developed spodic horizon becomes a Humodic Cryudent.

Table 3.7 Orders and suborders of the Soil Conservation Service classification.

Order	Suborder	Order	Suborder
ENTISOL	Aquent Psamment Ustent Udent	SPODOSOL	Aquod Humod Orthod Ferrod
VERTISOL	Aquert Ustert	ALFISOL	Aqualf Altalf Udalf Ustalf
INCEPTISOL	Aquept Andept Umbrept Ochrept	ULTISOL	Aquult Ochrult Umbrult
ARIDISOL	Orthid Argid	OXISOL	
MOLLISOL	Rendoll Alboll Aquoll Altoll Udoll Ustoll	HISTOSOL	

From Soil Survey Staff 1960.

Table 3.8 Formative elements in names of soil orders in the Soil Conservation Service classification.

No. of Order[a]	Name of Order	Formative Element in Name of Order	Derivation of Formative Element	Mnemonicon and Pronunciation of Formative Elements
1	Entisol	ent	Nonsense syllable.	recent.
2	Vertisol	ert	L. *verto*, turn.	invert.
3	Inceptisol.......	ept	L. *inceptum*, beginning.	inception.
4	Aridisol	id	L. *aridus*, dry.	arid.
5	Mollisol	oll	L. *mollis*, soft.	mollify.
6	Spodosol.......	od	Gk. *spodos*, wood ash	Podzol; odd.
7	Alfisol............	alf	Nonsense syllable.	Pedalfer.
8	Ultisol	ult	L. *ultimus*, last.	Ultimate.
9	Oxisol	ox	F. *oxide*, oxide.	oxide.
10	Histosol	ist	G. *histos*, tissue.	histology.

From Soil Survey Staff 1960.
[a]Numbers of the orders are listed here for the convenience of those who became familiar with them during development of the system of classification.

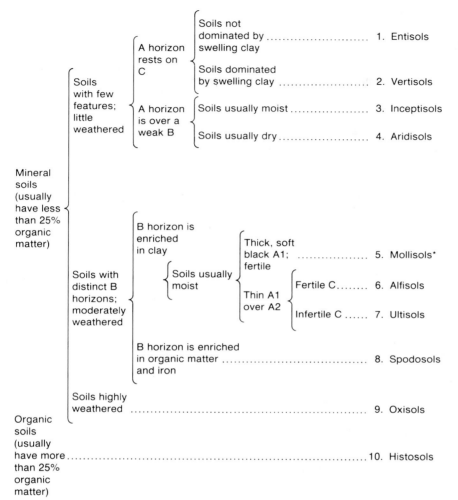

Figure 3.14.
Simple classification of mineral and organic soils of the world based on *Soil Taxonomy* by the USDA. (Reprinted by permission from *Soil Science Simplified* by Milo I. Harpstead and Francis D. Hole. © 1980 by The Iowa State University Press, 2121 South State Avenue, Ames, Iowa 50010)

*Some Mollisols have a weak B horizon; not enriched by clay.

It is obvious even from a brief description that the S.C.S. system allows for extremely detailed classification of soils. However, with 10 orders, 29 common suborders, and countless great groups and subgroups, the number of possible combinations of terms is enormous; the system requires an intimate understanding of soil properties for its successful use. Table 3.10 is presented simply to show the approximate equivalents between the major soil categories in the S.C.S. classification and the older classification.

Table 3.9   Formative elements in names of suborders in the Soil Conservation Service classification.

Formative Element	Derivation of Formative Element	Mnemonicon	Connotation of Formative Element
acr	Gk. *akros*, highest.	acrobat	Most strongly weathered.
alb	L. *albus*, white.	albino	Presence of albic horizon (a bleached eluvial horizon).
alt	L. *altus*, high.	altitude	Cool, high altitudes or latitudes.
and	Modified from *Ando*.	Ando	Ando-like.
aqu	L. *aqua*, water.	aquarium	Characteristics associated with wetness.
arg	Modified from argillic horizon; L. *argilla*, white clay.	argillite	Presence of argillic horizon (a horizon with illuvial clay).
ferr	L. *ferrum*, iron.	ferruginous	Presence of iron.
hum	L. *humus*, earth.	humus	Presence of organic matter.
ochr	Gk. base of *ochros*, pale.	ocher	Presence of ochric epipedon (a light-colored surface).
orth	Gk. *orthos*, true.	orthophonic	The common ones.
psamm	Gk. *psammos*, sand.	psammite	Sand textures.
rend	Modified from Rendzina.	Rendzina	Rendzina-like.
ud	L. *udus*, humid.	udometer	Of humid climates.
umbr	L. *umbra*, shade.	umbrella	Presence of umbric epipedon (a dark-colored surface).
ust	L. *ustus*, burnt.	combustion	Of dry climates, usually hot in summer.

From Soil Survey Staff 1960.

Table 3.10   Soil orders of the Soil Conservation Service classification and approximate equivalents of the Great Soil Groups.

Orders in S.C.S. Classification	Approximate Equivalents
1. Entisols	Azonal soils, and some Low-Humic Gley soils.
2. Vertisols	Grumusols.
3. Inceptisols	Ando, Sol Brun Acide, some Brown Forest, Low-Humic Gley, and Humic Gley soils.
4. Aridisols	Desert, Reddish Desert, Sierozem, Solonchak, some Brown and Reddish Brown soils, and associated Solonetz.
5. Mollisols	Chestnut, Chernozem, Brunizem (Prairie), Rendzinas, some Brown, Brown Forest, and associated Solonetz and Humic Gley soils.
6. Spodosol	Podzols, Brown Podzolic soils, and Ground-Water Podzols.
7. Alfisols	Gray-Brown Podzolic, Gray Wooded soils, Noncalcic Brown soils, Degraded Chernozem, and associated Planosols and some Half-Bog soils.
8. Ultisols	Red-Yellow Podzolic soils, Reddish-Brown Lateritic soils of the U.S., and associated Planosols and Half-Bog soils.
9. Oxisols	Laterite soils, Latosols.
10. Histosols	Bog soils.

From Soil Survey Staff 1960.

Pedogenic Controls and Regimes

Soils can be grouped together systematically on the basis of profile similarities. In general, these result from a unique combination of certain factors that control the magnitude and types of soil-forming processes. The external factors of climate, biota, topography, parent material, and time have long been recognized as the prime controls, and they are commonly expressed as the soil equation (Jenny 1941):

$$S \text{ or } s = f\,(cl,\ o,\ r,\ p,\ t\ .\ .\ .)$$

where S = soil, s = any soil property, cl = climate, o = biota, r = topography, p = parent material, and t = time. Although the soil equation cannot be solved in quantitative terms (for discussion see Yaalon 1975), the relative effect of each factor has been determined, with some difficulty, by analyzing field sites where four factors are held essentially constant and one is allowed to vary (Jenny 1941).

The implication here is that soil-forming factors are independent variables, a perception that is probably invalid except for the factor of time. For example, numerous cases can be cited to demonstrate the interdependence of the various factors; climate directly controls vegetation and animal form, rock type (parent material) influences relief, and relief often affects climate. In addition, other difficulties arise because many soils are polygenetic; that is, they developed under more than one set of controlling factors. In such cases, the soil formed under the prior conditions becomes the parent material that will respond to the new controlling factors. Therefore, because of the inherent difficulties in isolating individual factors, it seems likely that solution of the soil-factor equation in quantitative terms that will satisfy all soil scientists may not be possible. Nonetheless, a qualitative understanding of which factors dominate particular environments may allow us to predict how soil properties will change from one region to another.

Parent material is usually considered to be rocks that are weathering in situ and unconsolidated sediment that was transported from its place of origin by various surficial processes and deposited at a different locality. It is generally conceded that parent material exerts its greatest control in the early stages of soil development or in very dry regions (Birkeland, 1974). However, highly impermeable materials or those having relatively stable minerals may maintain an influence on soil properties for longer time intervals or in other climatic zones. Perhaps the best demonstration of the effect is where textural and/or mineralogical variations of parent rock exist in the same climatic and topographic setting. For example, in the piedmont zone of North Carolina, soils developed from siliceous mica gneiss and granite gneiss are considerably different from those developed on diortes and gabbros. The greater content of ferromagnesian minerals in the mafic rocks led to formation of thick, clay-loam A horizons and reddish B horizons. These are notably absent in soils overlying the siliceous gneisses (see Buol et al. 1973 for discussion and references).

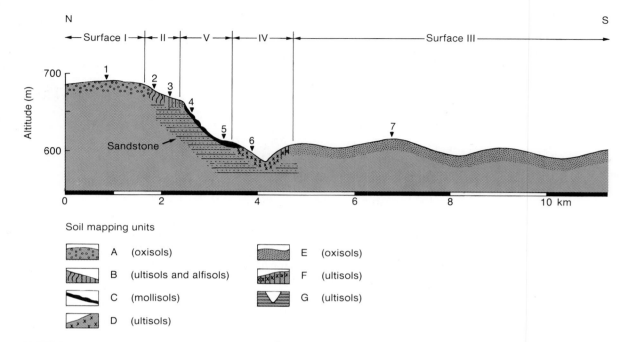

Figure 3.15.
Soil catena showing relationship between slope and soil type. (Reproduced from *Soil Science Society of America Journal*, Volume 41, 1977, p. 110, by permission of the Soil Science Society of America)

The relief factor essentially refers to the effect of local topography on soil development. Therefore, consideration of this factor is really an attempt to analyze soil changes brought about by the influence of slope. In this type of analysis, geomorphologists have found utility in the concept of a soil catena. A **soil catena** consists of a group of soil profiles whose characteristics change gradually beneath a sloping surface. The profile changes result because variations in soil-forming factors are produced by differences in geomorphic processes acting on the slope materials, especially drainage of groundwater, transport of surface sediment, and removal of mobile chemical elements. The steepness of the slope is important because it tends to control the magnitude of sediment transport by wash, creep, and mass movements, and it also allows underground water to move more rapidly through the soil. As a result, soils in the steep upper-slope location tend to be freely drained, oxidized to red-brown colors, and coarse-grained. Lower-slope soils, where gradients are less, commonly have higher moisture contents because the material is poorly drained. They tend to be clay-rich and blue-gray to gray in color. Midslope profiles are often transitional in color and texture between the upper- and lower-slope characteristics (fig. 3.15). Therefore, a catena manifests the interaction of soil processes and slope processes, and for this reason, the concept is significant in geomorphology.

The facility of a catena to reflect process is complicated because slopes are not always underlain by a single rock type. In addition, variations in climate, duration of soil development, and landscape evolution will make the meaning of catena profile changes more difficult to interpret. Nonetheless, the

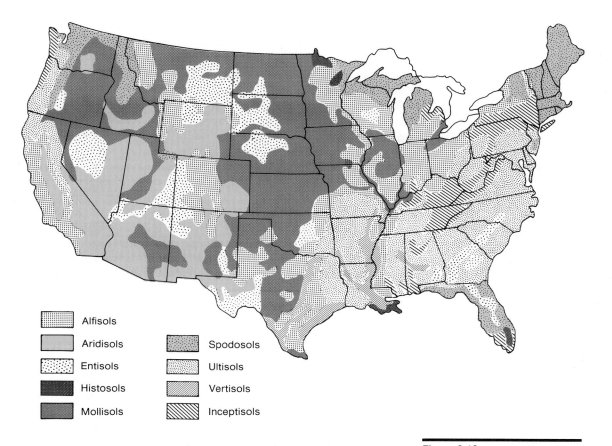

▦ Alfisols	
▨ Aridisols	▨ Spodosols
▨ Entisols	▨ Ultisols
■ Histosols	▨ Vertisols
▨ Mollisols	▨ Inceptisols

Figure 3.16.
The major occurrences of the ten soil orders in the United States. The boundaries are approximate and tentative.

catena concept can be successfully applied in any area regardless of geologic complexity or climate if sufficient analytical care is exercised (for more complete discussions, see Young 1972; Ollier 1976; Gerrard 1981).

In addition to steepness, slopes also affect soil development because their orientation often promotes differences in microclimate and vegetation even where the regional climate is the same. Due primarily to lower direct insolation and temperature, north-facing slopes tend to have greater moisture contents and vegetal cover than south-facing slopes. This leads to thicker A horizons with a higher organic content.

The effect of climate and its related control on vegetation have been recognized historically as the most dominant soil forming factors (*cl, o* in the Jenny equation). In fact, most of the Great Soil Groups in earlier soil classifications (see table 3.10) were based on the genetic relationship between soil and climate. Similarly, the distribution of S.C.S. soil orders in the United States is clearly related to climatic zonation (fig. 3.16). Thus, climate and vegetation can be used in a general way to predict what soil types may be expected in any region. This follows because certain combinations of climate and vegetation (called pedogenic regimes) produce distinct trends in soil-forming processes. Several examples of regimes can be used to illustrate this point.

Podzolization   The processes that result in the removal of iron and/or aluminum from the A horizon and their accumulation in the B horizon are referred to as *podzolization*. Podzolization is common in humid-temperature climates, especially in soils that develop under forest vegetation. For example, most soils in the eastern half of the United States where precipitation exceeds 70 cm annually are podzolized to some extent. To the west where precipitation is less than 56 cm, another pedogenic regime (calcification) becomes dominant. Podzolization produces a continuous translocation of mobile substances as they are released from parent minerals. Because forest vegetation uses less alkalis and alkaline earths than do grasses or shrubs, the surface litter is normally quite acidic. As a result, most soluble ions (Ca, Na, Mg) in the upper soil are leached, and colloids of all types adsorb H^+. Therefore, Spodosols, Alfisols, and Ultisols, the soils affected by podzolization, are notably acidic.

Where podzolization is very active, an E horizon may be found overlying a spodic B horizon. This bleached zone represents the ultimate form of podzolization because almost all the Al^{+3} and Fe^{+3} is removed and the residuum is highly concentrated in silica. Although the soil in a true podzol is acidic, the E horizon can be developed where the pH of the zone is still within the insolubility range of Al^{+3} and Fe^{+3}. Thus, most of the translocation of these insoluble ions involves chelation rather than dissolution by colloidal acids.

Beneath the zone of maximum eluviation, Spodosols and Alfisols are usually brown or grey-grown, but as the temperature increases they grade progressively into Ultisols having yellow and red colors (see table 3.10). The greater humus content in cooler regions, combined with the translocation and redeposition of organic matter, accounts for the drabber color in the higher latitudes. However, as the production of humus declines with increased temperatures, the ferric content rather than the organic matter becomes the predominant factor in determining the color of the B horizon. The characteristic color of podzolic profiles, therefore, changes from north to south in the eastern United States.

The end products of podzolic weathering include resistant primary minerals, iron and aluminum sesquioxides or hydrates, and kaolinitic clays. Most of the mobile ions have been removed from the system in solution, but some cations may be retained by fixation in the clay mineral structures. Clays accumulate in the B horizon in any or all of three ways: (1) the component polymers of clays can be carried in solution from the A horizon and recombined in the B horizon; (2) clays can be leached as already formed minerals and physically accumulated; or (3) clays can form in situ by chemical alteration of parent minerals that were originally in the position of the B horizon.

Laterization   The processes resulting in Oxisols constitute a pedogenic regime called *laterization*. The conditions needed for these processes to function are high rainfall and temperature, intense leaching, and oxidation. These conditions commonly exist in tropical regions, especially where wet and dry seasons alternate and drainage is good. It is now clear, however, that laterization does not necessarily require a tropical climate since rapid leaching of the proper lithologies can form Oxisols in a wide variety of climates, including subarctic. Thus, to consider Oxisols as *the* zonal soil of the tropics may not be absolutely correct.

Laterization is marked by two essential differences from podzolization: (1) organic accumulation is inhibited; and (2) silica is leached from the system in addition to the common mobile ions, leaving abnormally high concentrations of hydrated oxides of iron and aluminum in the soil. In some cases silica combines with available alumina to form kaolinitic clays, but where silica is more extensively leached the excess alumina is present as gibbsite $Al(OH_3)$ or, more rarely, boehmite ($AlO(OH)$). The type of parent material and the organic content exert a direct influence on whether the clays will be kaolinitic or gibbsitic. Where granites are the parent rocks, less silica will be removed during laterization because it exists as quartz rather than as amorphous SiO_2. Oxisols developed on granites, therefore, tend to have more kaolinite in their profiles; quartz may be a notable end product. Where basalts are the parent rocks, gibbsite tends to be more dominant because the more soluble amorphous silica is thoroughly leached, leaving little available to form the kaolinite. By observation, when limestone is the parent material, the main clay mineral seems to be boehmite.

Laterization is perhaps the most poorly understood of the pedogenic regimes. We know that iron and aluminum can be dissolved in acidic fluids, yet the environments of their accumulation are very acidic. Precisely how these substances are concentrated is not clear, and much controversy has arisen over interpretations concerning the environment of laterization. In light of this it may be worthwhile to stress again that although laterization should function efficiently in warm, humid climates, Oxisols are not necessarily restricted to the tropics. Such a binding inference can lead to rather hapless paleoclimatic interpretations, since Oxisols can form in cool climates, and Spodsols, Alfisols, and Ultisols are not uncommon in the tropics.

Calcification   A third pedogenic regime, *calcification,* functions in subhumid to arid climates where precipitation is insufficient to drive the soil water downward to the water table. Ions mobilized in the A horizon are reprecipitated in the B horizon where zones of $CaCO_3$ are commonly developed. The depth of the carbonate zone is a function of the annual precipitation (fig. 3.17) and in general represents a first approximation of the vertical extent of leaching. Arkley (1963), however, showed that this relationship holds only in regions without pronounced seasonality of rainfall or orographic effects. Nonetheless, for our purposes it seems useful to visualize the depth of the carbonate zone as a gross index of the magnitude of calcification. Soils derived from those processes all possess a K horizon or a calcium carbonate zone, but its position within the soil profile rises proportionately with decreasing rainfall and/or the efficiency of the calcifying processes.

Figure 3.17.
The depth of carbonate zones in soils as related to mean annual rainfall.

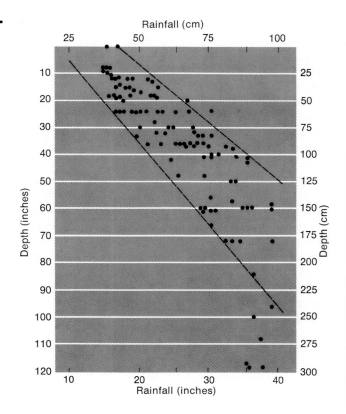

Calcification takes place most efficiently under a grass or brush vegetation because such plants utilize large quantities of alkali and alkaline earths in their life processes. When the plants die, the mobile ions are returned to the soil as the litter is decomposed. The amount escaping this recycling process depends on the precipitation, temperature, and related bacterial activity.

The two S.C.S. soil orders formed primarily by calcification are the Mollisols and Aridisols. Mollisols occur mainly in the temperate zones of the United States, especially in the northern Great Plains states (the Dakotas and Nebraska). In these soils the profile (fig. 3.18) usually has an undifferentiated, black A horizon ranging from 30 to 120 cm in thickness, which reflects the high accumulation of humus in cooler temperature zones. The B horizon is usually light-yellow to brown, and carbonate may exist there as disseminated material or in a distinct layer at the base of the horizon. In more arid areas the Aridisols become dominant. These soils are normally very immaturely developed; $CaCO_3$, if present as a secondary accumulation, will be near or at the surface. Generally, then, in cooler areas with less than 64 cm of annual precipitation, increasing aridity is shown in the profiles by a progressively less developed A horizon and a carbonate zone rising closer to the surface. Where the temperature is higher, humus production is less and a red hue becomes more prominent in the calcified soils. This gives rise to the reddish-colored Mollisols and Aridisols.

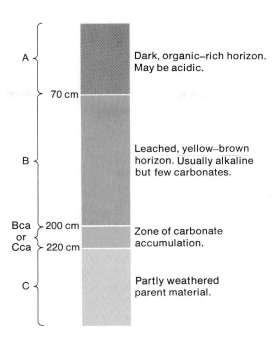

A

70 cm

B

Bca
or
Cca

200 cm

220 cm

C

Dark, organic–rich horizon.
May be acidic.

Leached, yellow–brown
horizon. Usually alkaline
but few carbonates.

Zone of carbonate
accumulation.

Partly weathered
parent material.

Figure 3.18.
Generalized profile of a Mollisol in
eastern Great Plains region of the
United States. Greater aridity
would decrease the size of the A
horizon and raise the carbonate
zone to a higher level. Depth
values will vary with local
conditions. Horizons indicated by
ca may now be designated with a
lowercase k.

Because of the immobility of the soluble ions, most clay minerals formed in the calcification regime are montmorillonite or illite types. In some cases kaolinite is abundant near the top of the profile where slightly acidic soil water is present, especially where humus is abundant. The retention of the alkaline earths, however, is not favorable for the development of kaolinite in the deeper zones.

Finally, as stated earlier, the factor of time in soil development is probably the only truly independent variable in the soil equation. There is no question that the development of diagnostic soil properties is a time-dependent phenomenon. Therefore, horizons, clay content, clay mineralogy, organic content, and other parameters can be expected to change during the tenure of soil formation. Recognize, however, that the value of each property changes at a different rate and that, given enough time, each will eventually reach a condition where the property no longer changes or its rate of change becomes negligible. At that time, the property has attained a steady-state relationship with the soil-forming environment. When all properties of a soil profile reach this condition, the soil itself is said to be in a steady state. Birkeland (1984) presents an excellent discussion about the amount of time needed for various diagnostic properties to culminate in a steady state. As shown in figure 3.19, an A horizon with pronounced organic accumulation will generally reach a steady-state condition before a B horizon can be recognized or while it is patently immature. Diagnostic B horizons (Bt, Bca) require more time. A true oxic B horizon presumably attains steady state only when all the weatherable minerals have been altered to stable forms and, therefore, is the diagnostic horizon requiring the most time for its complete development.

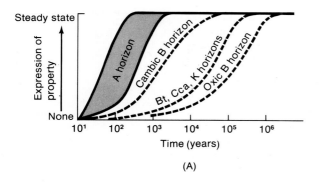

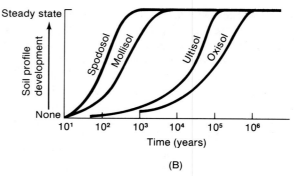

(A)

(B)

Figure 3.19.
Diagram showing the variations in time to attain the steady state for (A) various soil properties and (B) various soil orders.

Because diagnostic components of soil profiles form over variable amounts of time, the major soil orders that are based on those properties should also be time-dependent (fig. 3.19). Thus, the suggestion by Birkeland (1984) that the distribution of major soil orders should correlate with the ages of deposits or landscapes on which they develop seems reasonable. In fact, that suggestion is generally supported by the relationships between soil orders and deposit ages in the United States.

Geomorphic Significance of Soils

Our discussion thus far has been a very brief treatment of the basic processes involved in weathering and soil development. How is such information important and relevant to the study of geomorphology? There is no question that soil characteristics affect geomorphic processes in a number of significant ways; in fact, a unique soil property may dictate the mechanics of a surficial system. Once such a property is identified, it provides critical information needed for environmental control or for regional and local planning (McComas et al. 1969). For example, in certain situations soil properties are known to control the stability of building foundations (Baker 1975), hydrologic response to precipitation (Cooley et al. 1973), and the permanence of road construction (Weinert 1961, 1965). Thus, the importance of soils in physical geomorphic systems should not be underestimated. However, because soil properties are altered by time and climate change, soil formation has even greater importance in deciphering the sequence of events in Quaternary history. In fact, one of the greatest uses of weathering and soils is to establish relative ages of glacial deposits and, by inference, the sequence of glaciations. The same principles are employed outside of glaciated regions to suggest relative ages of deposits and often as evidence of climate change. A variety of weathering and soil characteristics have been used in this manner (see Burke and Birkeland 1979 for a detailed discussion).

The use of soils in Quaternary geomorphology is a dual-edged sword. The fact that soil-forming processes change with time is a basic ingredient of historical interpretation. However, the fact that soil-forming factors are not constant makes the record preserved in profiles more difficult to ascertain. This is especially true in areas that have been affected by the pronounced climatic fluctuations of alternating glacial and interglacial episodes. As the climate changes, the dominant factors of soil formation in any given area must also change accordingly. Thus, many soils preserve in their profiles characteristics that reflect more than one set of soil-forming factors. These *polygenetic soils* (sometimes called *complex soils*) complicate the record because thickness and maturity of soils are reliable indices of age only when the conditions developing the soils have been maintained continuously. When conditions change, the properties of the initial soil are supplanted by new characteristics; because such alterations vary in the degree of completeness, complex soils are very difficult to correlate with soils in other areas. A pre-Wisconsinan soil, for example, will accrue properties related to the controlling factors at the time of its formation. A change in those factors at some later time, perhaps post-Wisconsinan, will obviously cause the original soil properties to be out of phase with the new soil-forming environment. For all practical purposes, the pre-Wisconsinan soil becomes the parent material on which the post-Wisconsinan climate and biota are working, and a younger soil profile is superimposed on the older one. Separating the older soil from the younger one is a demanding field problem that, in this example, would require lateral tracing of the complex soil to a locale where Wisconsinan deposits intervene.

Soils that form on a landscape of the past are *paleosols*. They can be of three types: (1) *buried* soils are developed on a former landscape and subsequently covered by younger alluvium or rock; (2) *relict* soils were not subsequently buried but still exist at the surface; and (3) *exhumed* soils were at one time buried but have been reexposed when their cover was stripped by erosion (Ruhe 1965). Buried soils are immediately recognized as paleosols. Relict and exhumed soils are much more difficult to identify, because it must be proved that their properties are inconsistent with the modern environment or that their age is the same as the paleosurface on which they rest. Some additional geomorphic or stratigraphic data are usually needed to substantiate those requirements.

The most persuasive use of soils in Quaternary geomorphology arises when soil characteristics are painstakingly integrated with sequences of landform development and associated sedimentary deposits. An excellent example of this approach is documented in detailed studies of the Rio Grande valley and the adjacent slopes and intermontane basins near Las Cruces, New Mexico. These studies, sponsored by the U.S. Soil Conservation Service and known as the Desert Soil-Geomorphology Project, began with the work of R. V. Ruhe in the 1950s (Ruhe 1964, 1967) and culminated approximately 20 years later in thorough syntheses of the soil geomorphology (Gile and Grossman 1979; Gile et al. 1981). A detailed discussion of this enormous project cannot be entertained here. However, a brief abstract of its procedures and results can demonstrate how soils can be used in the analysis of Quaternary history.

The Desert Project area is physiographically divisible into distinct subareas (fig. 3.20). The east and northwest margins consist of semiconnected mountains that rise up to 2750 meters in elevation. The valley border zone is located in the valley of the Rio Grande river and is characterized by deposits and surfaces formed by the river and tributaries during and after the Pleistocene epoch. The flood plain of the modern river stands at approximately

Figure 3.20.
Block diagram showing major
landforms of the Desert Project.

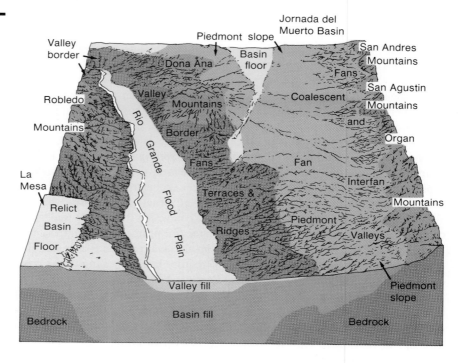

1200 meters. Between the mountains and the valley (the piedmont slope) exist
a number of intermontane basins that have been affected by tributary and
slope processes that function on the valley sides. The basin or piedmont areas
are composed mainly of alluvial fan deposits that emerge from the mountains
and coalesce into smooth alluvial plains sloping gently toward the valley bottom.
In some cases the piedmont deposits are graded to the valley-bottom deposits,
but more often they and their related surfaces are not physically connected to
the valley-border sequence.

The regional history was controlled by repeated climatic changes that
produced alternating periods of deposition and erosion. In the valley zone, these
alternations resulted in a series of river terraces, the surfaces of which are
underlain by alluvium of the Rio Grande river (fig. 3.21). Episodes of downcutting by the Rio Grande initiated trenching in the tributary arroyos of the
piedmont and valley-border areas. This isolated many of the geomorphic surfaces that formed as the upper level of a depositional event. Since the middle
Pleistocene, five depositional episodes, separated by intervening downcutting,
have created a sequence of surfaces that document their relative ages. Soil
formation on the various deposits and surfaces began at different times.

Knowing the geomorphic setting and realizing that climate change was
synchronous in the valley and piedmont areas, it might be easy to assume that
soil development was similar throughout the region. Actually, the meaning of
the preserved soils is very confusing because their developmental histories vary
on a local basis. In some zones, erosion has truncated diagnostic horizons of

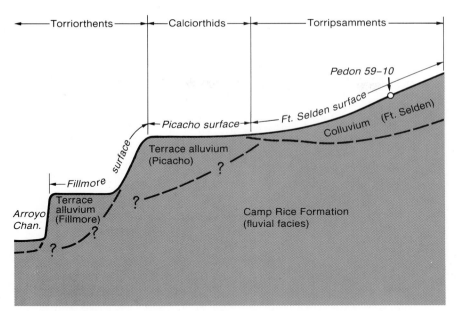

Figure 3.21.
Terraces and river alluvium in the valley border subarea near Las Cruces, N.M.

a profile. At other localities, soil profiles of one age were buried by material of a younger depositional event. For example, arroyo trenching initiated in the Rio Grande valley did not proceed entirely to the mountain front prior to the incidence of the next depositional phase. In such cases, deposits and soils in the upper piedmont zone were buried by subsequent fan development (fig. 3.22). Additionally, alluvial plain deposits may not extend all the way to the axial valley. Furthermore, precise dating by ^{14}C provided only a broad age framework for events in the area because the amount of datable material was limited. What results from these complications is that the history recognized in deposits and surfaces of the valley border cannot be directly correlated with those in the valley-side sequence, and soils developed in the two areas are different because of vagaries in local history, parent material, and microclimate.

The crux of this is that Quaternary history in the Desert Project area is unintelligible without the integration of stratigraphy, soils, and geomorphology. The history of the valley-border and piedmont areas was finally linked by relating the *degree* of soil development in each subarea to the relative age of the deposits and geomorphic surfaces within that subarea (Gile et al. 1981). This could be accomplished because at stable sites (no erosive disruption of profiles) age becomes the key soil-forming factor. This was especially revealed in the carbonate horizons, which become thicker and more indurated with age (fig. 3.23). Thus, even though soils of the two subareas have different characteristics and classifications, the relative degree of their development allowed them to be correlated and the surfaces associated with the soils could be placed into a developmental sequence (table 3.11).

Figure 3.22.
Cross section showing burial of deposits and surfaces by younger fan and terrace alluvium near Las Cruces, N.M.

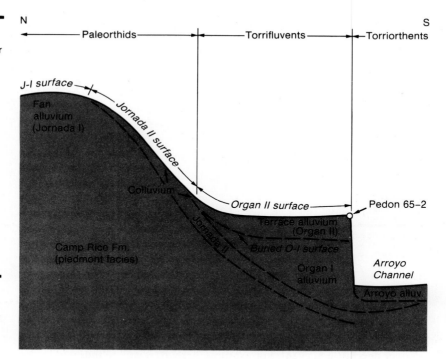

Figure 3.23.
Schematic diagram of the diagnostic morphology for the stages of carbonate horizon formation in the two morphogenetic sequences. Carbonate accumulations are indicated in black for clarity.

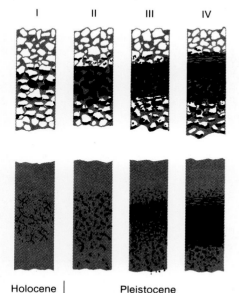

Holocene | Pleistocene

Stage	Gravelly sequence	Nongravelly sequence
I	Thin, partial or complete carbonate coatings.	Carbonate filaments and/or faint coatings on grains.
II	Carbonate coatings are thicker and there are some fillings in interstices.	Carbonate nodules separated by low-carbonate material.
III	Carbonate occurs essentially throughout the horizon, which is *plugged* in the last part of the stage.	
IV	A laminar horizon has formed on top of the plugged horizon.	

Table 3.11 General relationship between the degree of development in carbonate soils and geomorphic surfaces in subareas of the Desert Soils Geomorphology Project area near Las Cruces, New Mexico.

Age, yr B.P. or epoch	Geomorphic Surface		Stage of Carbonate Accumulation (cf. fig. 3.23)	
	Valley border	Piedmont slope	Nongravelly soils	Gravelly soils
Holocene				
0–100	(Dune)[a]	(Dune)[a]	—	—
100–7500	Fillmore[b]	Organ	I	I
> 7500 (latest Pleistocene)	Leasburg[b]	Isaacks' Ranch	II	II, III
Late Pleistocene	Picacho	Jornada II	III	III, IV
Late mid-Pleistocene	Jornada I	Jornada I	III	IV (multiple
Early mid-Pleistocene	La Mesa		IV	laminar zones)

From Gile 1975. Used with permission of Quaternary Research.
[a]The dunes are not formally designated by a geomorphic surface name.
[b]Where Fillmore and Leasburg cannot be distinguished they are grouped into the Fort Selden surface.

Summary

The processes of chemical weathering (hydrolysis, oxidation-reduction, solution, ion exchange) alter the exposed portion of the geologic framework and combined with organic processes, produce soils. The degree of chemical change depends on how mobile the ions of the parent minerals are under the external and internal controls on the weathering mechanics. In regions with abundant precipitation, highly mobile ions are usually removed from the weathered zone unless the original mineral structure is incompletely broken down and elements such as potassium are fixed in the system. In contrast, where leaching is minimal, mobile ions (Ca, Na, Mg) are concentrated in the weathered zone. Immobile ions (Fe^{+3}, Al^{+3}) may be transposed in very acidic ground waters or by special organic processes such as chelation. The mobility of most substances is also dependent on the pH and Eh of the weathering environment. The type of clay mineral formed in the weathering zone is usually a good indicator of the intensity of decomposition.

Soils are described and classified according to the soil profile. The character of the soil profile varies with parent material, climate, biota, topography, and the length of time involved in its formation. Three major pedogenic regimes—podzolization, laterization, and calcification—produce the dominant soil groups; however, changes in the controlling pedogenic factors may result in complex soils that show evidence of forming under more than one pedogenic regime. Soils are important elements in reconstructing geomorphic history, and they directly influence other surficial processes.

Suggested Readings

The following references provide greater detail concerning the concepts discussed in this chapter.

Birkeland, P. 1984. *Soils and geomorphology.* New York: Oxford Univ. Press.

Gile, L. H.; Hawley, J. W.; and Grossman, R. B. 1981. Soils and geomorphology in the Basin and Range area of Southern New Mexico—Guidebook to the Desert Project. New Mex. Bur. Mines Min. Resources, Mem. 39.

Hunt, C. B. 1972. *Geology of soils.* San Francisco: W. H. Freeman.

Loughnan, F. 1969. Chemical weathering of the silicate minerals. New York: American Elsevier.

Soil Survey Staff. 1960. *Soil classification, a comprehensive system—7th approximation.* Washington, D.C.: U.S. Dept. of Agriculture, Soil Conserv. Service.

———. 1975. *Soil taxonomy.* Washington D.C.: U.S. Dept. of Agriculture, Soil Conserv. Service, Agri. Hndbk. 436.

Physical Weathering, Mass Movement, and Slopes

4

Introduction

The transformation of rocks into unconsolidated debris is the prime geomorphic contribution of weathering and soil-forming processes. Whether the debris produced by weathering will resist erosion depends on the balance between the internal resistance of the materials and the magnitude of the external forces acting on them. The relative resistance of any natural substance is partly reflected in the character of the slope that develops on it. Extremely steep slopes, for example, can be maintained for long periods only if the underlying rock or soil is so tightly bound together that the forces and agents of erosion cannot lower the slope angle. On the other hand, gentle slopes in regions of low relief and elevation may be stable for relatively long time spans even if the underlying material is very friable. Clearly, then, slope characteristics provide us with useful information only when we understand the erosive processes attacking them.

In a large sense, the evolution of landscapes is the history of regional slope development. The formation of these slopes encompasses a multitude of geomorphic processes, and the properties of slopes reflect in subtle ways the temporal effect of these processes on the resisting framework. Thus, interest in slopes and slope-forming processes crosses the entire range of geomorphic thinking, from the analysis of a modern stability problem to the abstraction of geologic history.

The mechanics of slope erosion are in many ways closely related to the processes of physical weathering because the forces that disintegrate rocks and minerals simultaneously lower the internal strength of the unconsolidated cover. We will begin, then, with a brief discussion of physical weathering.

Physical Weathering

Physical weathering culminates in the collapse of parent material and its diminution in size. The continuing breakdown of rock takes place when stress is exerted along zones of weakness within the original material. These zones may be planar structures such as bedding or fractures that, upon rupture, produce fragments whose size and shape are controlled by the spacing of the planes. In other cases, failure may occur along mineral boundaries, resulting in an accumulation of particles similar in size and shape to the original rock texture. Although stresses are generated in different ways, the common bond that unites all processes of disintegration is that in every case a force within the material itself is responsible for its destruction.

The stress field involved in disintegration results from either expansion of rocks or minerals themselves or pressure generated by growth of a foreign substance in voids within the lithologic fabric. In each method the direction of the principal stress may change according to the process involved, but the most pronounced disintegration invariably occurs where the adjacent rocks exert the least confining pressure. Intuitively, then, we should expect disintegration to be most pernicious near the surface, where static load from overburden is minimal and fractures are abundant and closely spaced. With increasing depth, confining pressure increases, fractures are less common, and the disintegrating processes become less effective.

Expansion of Rocks and Minerals

Thermal Expansion   Rocks and minerals expand in response to several phenomena that can rightfully be considered as agents of physical weathering. There seems to be little doubt that the application of intense heat can cause physical disruption of rocks. The low thermal conductivity of rocks prevents the inward transfer of heat, allowing the external fringe of a rock mass to expand significantly while little, if any, change occurs below the outer few centimeters. Differential stresses are produced by this thermal constraint, and the rock exterior spalls off in plates or wedges 1–5 cm thick. As figure 4.1 illustrates, this process functions during forest fires; in semiarid forested mountains of the western United States, it may be the dominant process of physical weathering (Blackwelder 1927). Whether or not insolation can drive the process has been debated for many years. Many geologists have gradually, if not grudgingly, accepted the premise that diurnal temperature fluctuations are not severe enough to produce thermal spalling (Twidale 1968) because experimental studies (Griggs 1936a, 1936b) suggested that the process is not viable. Gray (1965), however, demonstrated that thermal spalling is indeed possible, and geomorphologists have reaffirmed thermal expansion as a method of rock disintegration (Ollier 1963, 1969; Rice 1976).

Figure 4.1.
Spalling of granitic boulder caused by heat expansion during forest fire. Beartooth Mountains, Mont.

Figure 4.2.
Expansion joints produced by
pressure release during valley
entrenchment. Vaiont River valley,
Italy. (From Kiersch, *Civil
Engineer,* 34, no. 3, p. 35, 1964.
Used with permission by the
American Society of Civil
Engineers.)

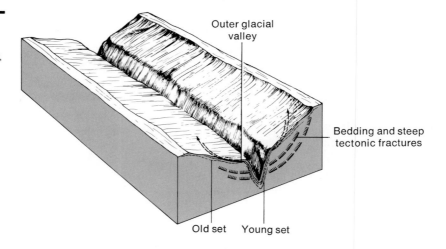

Outer glacial
valley

Bedding and steep
tectonic fractures

Old set   Young set

Unloading   Expansion of large segments of rock masses occurs when confining pressure is released by erosion. As denudation removes overburden, the
stress squeezing the underlying framework is lowered, and rocks tend to split
into widely spaced sheets, 1–10 m thick, that are oriented perpendicular to
the direction of pressure release. The sheeting tends to mirror the surface topography, and since outer sheets are relatively easy to erode, the process helps
perpetuate the surficial configuration because subsequent sheets develop with
a similar orientation. Although other processes aid in the removal of the sheets,
it can be readily documented that the original formation of the fractures is a
pressure-release phenomenon. Rock bursts in deep mines, for example, are
explicit proof that something as simple as excavation of tunnels can trigger a
rapid expansion of surrounding rocks. In the natural setting, postglacial entrenchment of the Vaiont River in Italy permitted valleyward expansion of
rocks and produced a joint system parallel to the valley sides (fig. 4.2). Hack
(1966) demonstrated that arcuate patterns of streams, ridges, and vegetal types
in the eastern United States are probably controlled by the position of curved
sheets that dilated during erosion of crystalline rocks.

Hydration and Swelling   Expansion also occurs when minerals are formed or
when they are altered by the addition of water to their structure. Although
the process begins as a chemical process called **hydration,** its physical side is
particularly obvious when clay minerals containing layers of OH or H_2O are
formed. The creation of the layered structure expands the minerals and propagates stress outward from the clay particle. Clays such as bentonite (Na-
montmorillonite), which do not have a fixed OH or water layer in their structure, have the capacity to absorb water into the mineral during periods of
wetting. The swelling produced by wetting exerts the same outward stress as
during clay formation. Most clays show the trait to some extent, but the percentage of expansion depends on the mineral type plus a myriad of other factors (table 4.1). Montmorillonite clays, for example, drastically lose their

Table 4.1 Expansion of common clay minerals by hydration.

Clay Mineral	% Expansion
Ca-Montmorillonite	
Forest, Miss.	145
Wilson Cr. Dam, Colo.	95
Davis Dam, Ariz.	45–85
Na-Montmorillonite	
Osage, Wyo.	1400–1600
Illite	
Fithian, Ill.	115–120
Morris, Ill.	60
Tazewell, Va.	15
Kaolinite	
Macon, Ga.	60
Langley, N.C.	20
Mesa Alta, N.M.	5

Adapted from Mielenz and King 1955, with permission of the California Division of Mines and Geology.

swelling capacity when sodium is replaced by some other cation (Mielenz and King 1955). Upon drying, the expanded clays lose part or all of the absorbed water, initiating an alternating swelling and shrinking sequence associated with episodes of wetting and drying. In contrast, the well-ordered hydrated clays have a stable structure, and destruction of the OH or water layer occurs only when the mineral is heated to at least 300°C. The disintegrating effect of these clays, therefore, occurs during their formation but, in contrast to the swelling clays, is exerted continuously until relieved. The use of hydrated clays as agents of disintegration is a one-shot affair, however, because once their expansive stress is released in a disintegrating event there is no way to reinstate the internal stress.

The effect of mineral expansion has been clearly demonstrated in the physical breakdown of granites in arid or semiarid regions (Wahrhaftig 1965; Eggler et al. 1969; Isherwood and Street 1976). In these settings, the major product of granite disintegration is a coarse, angular mass of rock and mineral fragments called *grus* (fig. 4.3), in which feldspars are often unaffected by decomposition. In the Laramie Range of Colorado and Wyoming, the sequence of grus development started during the Precambrian with formation of hematite by high-temperature oxidation along cleavage planes in the biotite (Eggler et al. 1969). Although this expanded the biotite in the direction of the c-axis, the stress was not sufficient to cause disintegration. It did, however, weaken the biotite's ability to resist further geomorphic attack. Subsequent near-surface weathering produced clays from the biotite with as much as 40 percent increase in volume, and the stress generated by this expansion shattered the granite into grus.

In other situations, the granite bedrock may be weakened in different ways prior to the grus development. For example, Folk and Patton (1982) show that the first stage of grus formation in granites of central Texas is the development of microsheet joints that parallel the weathering surface. Because these joints cut indiscriminately across mineral grains, they cannot be the result of biotite expansion. Instead, they precede and ultimately enhance the grus development.

It also seems certain that hydration of salts within pores of building stones and concrete develops sufficient stress to cause extensive spalling (Winkler and Wilhelm 1970). According to E. Winkler (1965), a similar process almost destroyed Cleopatra's Needle, an obelisk that was brought from Egypt to New York City in 1880. Salts trapped in spaces within the red-granite monument did not hydrate until they were placed in the humid climate of the eastern United States, but since then significant spalling has taken place.

Salt weathering is increasingly suggested as a significant component in the physical breakdown of rocks, especially as an explanation of minor weathering features such as tafoni. *Tafoni* are holes or depressions, usually less than

Figure 4.3.
Formation of grus by disintegration of granitic boulders. Upper part of photo shows grus matrix developed when boulders break apart. (Coin is silver dollar)

several meters in width and depth, that commonly form on the underside of rock masses or on steep rock faces. They often develop on granitic rocks set in arid climates. Salt weathering by crystallization of salt minerals or hydration expansion of salts has been suggested as the genetic cause of tafoni development (Evans 1969; Winkler 1975; Bradley et al. 1978). The precise origin of tafoni, however, remains a mystery (Evans 1969; Selby 1982) because the features form in a variety of climates other than arid (Calkin and Cailleux 1962; Martini 1978; Watts 1979) and on many rock types other than granite.

In most humid regions, the process of mineral expansion manifests itself in different end products. Rocks are peeled off to produce curved surfaces; the process on a large scale is called **exfoliation** and on a smaller scale **spheroidal weathering.** Even though the resulting large domelike masses or rounded boulders (shown in figs. 4.4 and 4.5) are probably in part a function of pressure release, it seems certain that water and mineral alteration are intimately involved (Gentilli 1968). Spheroidal boulders are formed because edges and corners of lithologic blocks are weathered more rapidly than flat surfaces, a phenomenon especially apparent where the parent rock has been fractured

Figure 4.4.
Northeast side of Half Dome taken from the subsidiary dome at the northeast end of the rock mass, revealing exfoliation on a gigantic scale. In the foreground is an old shell disintegrating into undecomposed granite sand. Yosemite National Park, Mariposa County, Cal.

into a blocky framework by perpendicular joint sets. The relatively fresh sphe-roidal cores are usually surrounded by a zone of disintegrated flakes and spalls that is enriched in secondary clay minerals. Simpson (1964), for example, found that the clay matrix in weathered graywacke increased by 5–10 percent in the spalled zone and also contained abundant vermiculite, an expandable clay not present in the fresh rock. Evidence such as this seems to indicate that outward expansion caused by the development of clay minerals peels off the fresh rock layer by layer, working progressively inward from the surrounding joint openings.

Growth in Voids

A second group of processes generate stress when some substance grows in spaces within the rock. The pressure gradient differs from that in the processes explained above because it is the openings that are expanded, not the parent minerals or rocks.

Microcracks in rocks can be produced by processes acting inside the earth (Simmons and Richter 1976; Whalley et al. 1982) and therefore may already be present before rocks are exposed at the surface. Because these spaces are

Figure 4.5.
Gabbro boulder showing spheroidal weathering. Himalaya Mine, San Diego County, Cal.

not expanded simultaneously or with equal magnitude or direction, the resultant pressures differ locally and the entire system is burdened with a differential stress field. Such a pressure distribution is conducive to fracturing or granulation; the processes responsible for its development are probably the dominant agents of disintegration.

Plants and organisms aid in the disintegrating processes, but their greatest effect usually occurs after the parent rock has already been converted into soil. Plant roots commonly grow in fractures of the parent rock and physically pry the solid material apart. Nonetheless, compared with other processes, rootlet growth is of minor consequence.

The most pervasive processes of physical weathering involve forces generated by crystallization of ice (frost action) or other minerals in rock spaces. In a perfectly closed system, water increases 9 percent in volume upon freezing and almost certainly produces hydrostatic pressures that exceed the tensile strength of all common rocks. **Frost action** is most effective when the rock is saturated prior to the freezing event. In fact, simple alternations of wetting and drying will sometimes fracture rocks, but the process is accelerated in combination with freezing (Mugridge and Young 1983).

If more than 20 percent of the available pore space is empty, the expansion pressure upon freezing may be less than rock strength, and shattering will not occur (Cooke and Doornkamp 1974). Some evidence exists to suggest that the intensity of frost action is related to the structure of the pores rather than simply the percentage of pore space. In a system containing a variety of pore sizes, ice crystals will preferentially grow in large pores rather than smaller ones (Everett 1961).

Minerals can also grow in rock spaces, with results similar to those of frost action, as figure 4.6 shows. Most commonly the process functions when percolating fluids evaporate within the pores, giving rise to supersaturated conditions and eventual precipitation of minerals. The pressures exerted in

Figure 4.6.
Pebble fractured by growth of calcite along planes of weakness. Near Roberts, Mont.

crystallization are probably greater than those produced by ice, but their absolute values depend on the concentration of the ionic constituents in the solution. The most common precipitates are sulfates, carbonates, and chlorides of very mobile cations (Ca, Na, Mg, K), and the process is therefore more prone to operate in arid and semiarid regions where the ions are rendered immobile by insufficient leaching.

The Significance of Water

Even a short review of physical weathering makes clear the importance of water in the disintegrating processes. Hydration, frost action, crystal growth, and swelling all require water as a basic component of the system. The amount of water need not be great. Many believe, for example, that even thin films of condensing dew in desert regions may be infinitely more destructive than insolation (Twidale 1968). Therefore, it seems fair to expect a direct relationship between climate and the prevalence of disintegration.

Peltier (1950) utilized mean annual temperature and precipitation to predict relative intensities of physical and chemical weathering; figure 4.7 shows these data. Physical weathering should be dominant where precipitation is readily available and the mean annual temperatures are near or below freezing. Presumably this analysis equates with the importance of frost action as a mechanical tool and with the fact that frost action is preeminent in those areas having the most freeze-thaw cycles during the year. The frequency of freeze-thaw events has been detailed for the United States by R. Russell (1943) and L. Williams (1964).

Where unusual local problems exist, the regional climatic characteristics may have little significance. In those cases it may be extremely important to understand in detail the climate-lithologic-weathering system, and a more sophisticated approach than those just reviewed will be necessary. An excellent example of this point was provided by Weinert (1961, 1965). In the eastern part of South Africa, the parent Karoo dolerite has been altered into a mature soil that is unsatisfactory for maintaining road foundations. In the western part of the region, this mature soil is not present. Instead, hydration of the micas has apparently disintegrated the dolerite into a grus that has considerable internal strength and is quite sound from an engineering viewpoint. Weinert found that the boundary between the sound and unsound surface materials (as defined by engineering properties discussed in the next section) could be mapped by the distribution of evaporation and precipitation in the area.

Physical Properties of Unconsolidated Debris

The resistance of unconsolidated debris to the forces of erosion is dependent on the physical properties of the material. In a real sense these properties determine whether slopes developed on any substance will be stable or, if they fail, the manner and rate of the resulting sediment movement. In addition, the physical properties help determine the shape of the slope profile when and if it attains an equilibrium condition. Clearly, then, the slope material itself directly influences the resulting process and landform. It is rather disconcerting to find that most geologists have only a vague knowledge of the basic physical

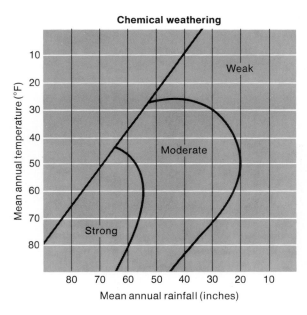

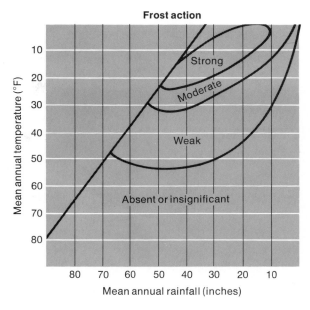

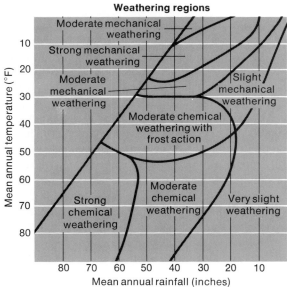

Figure 4.7.
Relative intensities of physical and chemical weathering under different temperature-precipitation conditions.

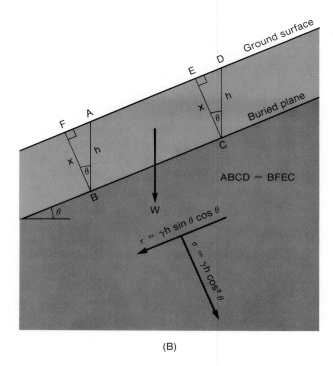

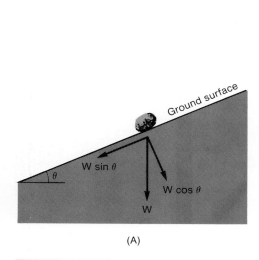

(A)

(B)

Figure 4.8.

Analyses of slope stability.
(A) Forces acting on a particle
resting on a slope surface.
(B) Stresses acting on a planar
surface covered by
unconsolidated material.

properties of soil, which have been identified through years of study by engineers, even though these characteristics directly control the mechanics of many geologic hazards.

Before we examine these properties, however, it may be helpful to look again at the concept of driving force, which was briefly discussed in chapter 1. Part (A) of figure 4.8 depicts a boulder resting on a sloping surface. The force of gravity acts vertically on the particle, and the force magnitude stems from the weight of the particle, *mg* (mass times the acceleration of gravity). Actually the weight may be resolved into two components, one acting perpendicular to the sloping surface and one acting parallel to it. The component acting parallel to the surface tends to promote downslope movement and is measured as $W \sin \theta$, where W is the weight in pounds or kilograms. The perpendicular component tends to keep the boulder in place by pushing it into the surface and thereby resisting the downslope motion; its magnitude is determined as $W \cos \theta$. Clearly, downslope movement of the particle is enhanced on steeper slopes because the sine value increases and the cosine value decreases as the angle θ is increased.

In the analysis of slope processes, however, engineers and scientists are usually concerned with the force acting on some plane existing below the ground surface along which movement of a block of overlying material takes place. See (B) of figure 4.8. In this case the exerted force derives primarily from the weight of the debris overlying the plane. Because the total weight of this material cannot be determined like that of a single, discrete boulder, it is calculated indirectly by multiplying the unit specific weight (γ) of the material (lb/ft³ or kg/m³) times the vertical distance (h) from the plane to the ground surface. The resolved components are now $\gamma h \sin \theta \cos \theta$ in the downslope direction and $\gamma h \cos^2\theta$ perpendicular to the plane. The reason for the change in equations is that the block of soil, represented by the parallelogram *ABCD*, is equal to the rectangle *BFEC*, and the angles *FBA* and *DCE* are equal to θ. Therefore,

$$\cos \theta = \frac{x}{h} \text{ and } x = \cos \theta h.$$

Since $W = \gamma x$, substituting from above gives us $W = \gamma h \cos \theta$. Thus, the pressure acting perpendicular to the plane is

$$\sigma = W \cos \theta = \gamma h \cos \theta \cos \theta = \gamma h \cos^2\theta$$

and the shear is

$$\tau = W \sin \theta = \gamma h \cos \theta \sin \theta.$$

Notice from the above that in the case of the buried plane, we are no longer talking about force because the value γh is given in units of *stress*, which by definition is the force acting on a specific area. This follows because

γ is in lb/ft³

h is in ft

$\gamma h = $ lb/ft³ $\times$ ft $=$ lb/ft².

When used in this sense stress and pressure are synonymous. The perpendicular component of the total stress is called *normal stress* and is usually indicated by the symbol σ. The downslope component is called *shear stress* and is denoted in mechanical analyses by the symbol τ.

Shear Strength

The properties of matter that resist the stresses generated by gravitational force are collectively known as the **shear strength.** The detailed analysis of internal strength is an extremely complex procedure, well beyond the scope of this book. For our purposes, we can safely say that shear strength of any material derives from three components: (1) its overall frictional characteristic, usually expressed as the angle of internal friction; (2) the effective normal stress; and (3) cohesion. These factors determine shear strength by the Coulomb equation,

$$S = c + \sigma' \tan \phi,$$

where S is shear strength (in units of stress), c is cohesion, σ' is effective normal stress, and ϕ is the angle of internal friction.

Internal Friction   Internal friction is composed of two separate types: *plane friction,* produced when one grain slides past another along a well-defined planar surface, and *interlocking friction,* which originates when particles are required to move upward and over one another. These are shown graphically in figure 4.9. The angle of plane friction is approximated by the angle at which

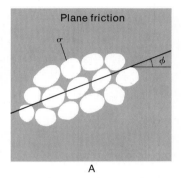

Plane friction

A

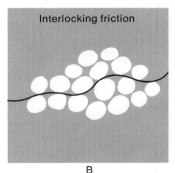

Interlocking friction

B

$\sigma = $ normal stress
$\phi = $ angle of internal friction

Figure 4.9.
Two types of internal friction.
(A) Plane friction, in which resistance occurs along a well-defined plane that may cut through individual grains. Angle of plane friction is approximated by angle at which sliding begins.
(B) Interlocking friction, in which resistance occurs around particle boundaries.

a block will begin to slide over another along the plane separating them. Once sliding begins, the frictional angle actually decreases slightly, requiring that a distinction be made between a static and a dynamic angle. In addition, the plane friction angle varies considerably with smoothness of the plane surface, moisture at the contact, mineralogy, and other factors.

Interlocking friction is usually greater than plane friction because extra energy must be used to move interlocked grains in an upward direction. The angle of interlocking friction also varies with mineralogy and moisture and, it would seem, must be affected by the density of packing within the mass. In loose particulate matter of any size, the angle of repose should approximate the angle of internal friction, but Carson and Kirkby (1972) point out that a platform holding a cone of debris at its angle of repose can be tilted up to 10° before the slope fails. This suggests that the angle of repose may be somewhat less than the maximum angle at which a slope can stand.

Effective Normal Stress   The importance of normal stress is its capacity to hold material together, thereby increasing the internal resistance to shear. In theory, normal stress acting perpendicular to a shear surface (fig. 4.9) is absorbed by the underlying slab at the point of contact between grains. In reality, some of the shear surface is occupied by openings filled with air or water. Since pore pressure exists in these interstitial spaces, it tends to support part of the normal stress. The total normal stress (σ) therefore includes two elements, effective normal stress (σ') and pore pressure (μ), such that

$$\sigma = \sigma' + \mu.$$

When considering internal friction, effective normal stress (stress exerted at the solid-to-solid contacts across the shear surface) is the critical parameter rather than total normal stress.

The value of pore pressure (μ) has a direct bearing on the effective normal stress because it can add to or detract from the total stress value. For example, figure 4.10 shows a distinct water table in unconsolidated material. At some level x below the water table, a hydrostatic pressure is exerted that is equal to the specific weight of the water (γ_w) times the vertical distance (h) between the water table and the level of x. Although the pore pressure beneath the water table is acting in all directions, Terzaghi (1936) points out that a portion of the pressure will be exerted in a direction opposite to the normal stress and therefore will provide some relief from the overburden weight. In contrast, in the unsaturated zone above the water table, some of the water will be prevented from moving downward because it is attached to particles by capillarity. Simply stated, this attached moisture increases the weight of the soil. Relating water content to the effective normal stress, three possible situations can be envisioned.

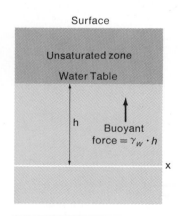

Figure 4.10.
Buoyant stress produced by hydrostatic pressure beneath a water table.

1. In a completely dry soil, the pore pressure is atmospheric and μ is zero. Therefore, the effective normal stress and the total normal stress are the same since

$$\sigma' = \sigma - 0.$$

2. Below the water table, pore pressure is positive ($>$ atmospheric pressure), causing the effective normal stress to be lower because

$$\sigma' = \sigma - \mu.$$

3. Above the water table, μ is negative and the effective normal stress is higher:

$$\sigma' = \sigma - (-\mu).$$

Because the effective normal stress directly influences internal friction, it is clear that dry or partially saturated soils, especially those with a high clay content, should have greater shear strength and stand at higher slopes than equivalent materials that are thoroughly saturated.

Cohesion   Figure 4.11 demonstrates the relationship between shear strength and effective normal stress. The graph indicates that as the effective normal stress increases, the values of shear strength also rise. The relationship between the two variables defines a straight line that passes through the origin of both axes. The angle between the line and the abscissa represents the angle of internal friction. The material represented in figure 4.11A has no discernible strength when the effective normal stress decreases to zero, a condition that is common in coarse, unconsolidated detritus. Solid rocks, however, possess shear strength even when σ' is removed (fig 4.11B) because the constituent particles are bonded or cemented together. The strength revealed here is *cohesion,* a factor that is unaffected by normal stress.

Clay-rich soils also have some cohesion, presumably because adsorption of ions and water by clay minerals creates a binding structure among the particles. The cohesive strength depends on the attractive force between the particles and the lubricating action of the interstitial liquid. As Grim (1962) points out, the molecules of the inner layers of water adsorbed to the surfaces of clays are oriented by the electrical charge of the minerals; because of this, fluidity and lubrication are not possible when moisture content is low, even though water is present. For clays to become plastic and exert a lubricating action, the adsorption of water layers must continue until the outermost ones can be held but are no longer fixed in a rigid, oriented position. Fortunately, simple tests can indicate how much water a soil can absorb before it begins to behave like a plastic substance; or, if more water is added, when the substance will lose all its cohesion and become a muddy fluid.

If water is gradually added to a dry, pulverized soil, the voids fill and the mixture becomes increasingly more plastic in its behavior. As more water is added, however, the cohesion decreases, and when all the pores are filled, any further input of water results in complete destruction of the internal fabric and the production of a fluid. Atterberg (1911) suggested two simple tests to indicate the transition from the solid to the plastic state and from the plastic to the liquid state; they are, respectively, the *plastic limit* and the *liquid limit.* The two limits are expressed as moisture contents determined by the weight of contained water divided by the weight of the dry soil. The range of water content between the two limits is the *plasticity index.*

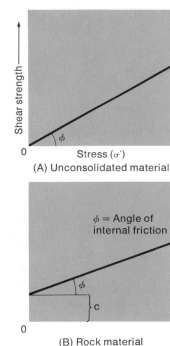

(A) Unconsolidated material

ϕ = Angle of internal friction

(B) Rock material

Figure 4.11.
Relationship between shear strength and effective normal stress. (A) Unconsolidated material has no shear strength when effective normal stress is zero. (B) Rock material has shear strength (c) from cohesion even where no effective normal stress is present.

Table 4.2 Average Atterberg limits in clays saturated with common cations.

	Montmorillonite[a]		Illite[b]		Kaolinite[c]		Halloysite[d]	
	Plastic Limit	Liquid Limit	Plastic Limit	Liquid Limit	Plastic Limit	Liquid Limit	Plastic Limit	Liquid Limit
Na^+	91	442	36	65	27	41	42	46
K^+	63	173	41	75	33	52	45	48
Mg^{++}	59	164	39	84	29	50	54	60
Ca^{++}	68	155	39	86	31	54	48	60

Adapted from Grim 1962. Used with permission of McGraw-Hill Book Company.
Data from White 1955.
[a] 4 samples
[b] 3 samples
[c] 2 samples
[d] 1 sample $2H_2O$, 1 sample $4H_2O$

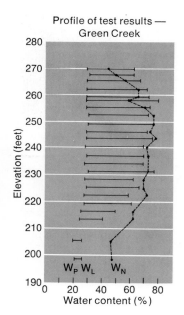

Profile of test results —
Green Creek

Figure 4.12.
Variation with depth of soil test results in the Leda clay, St. Lawrence River valley. Natural water content commonly is greater than the liquid limit in these sensitive soils. W_p is the plastic limit; W_L is the liquid limit; W_N is measured water content.

The values of Atterberg limits are affected by a number of auxiliary factors. First, they are related to the types of clay minerals included in the soil, although the precise limits for any particular clay species vary appreciably (see table 4.2). In general, plastic and liquid limits are higher for montmorillonite clays than for illites or kaolinites, mainly because the montmorillonites are able to disperse into very small particles with an enormous total water-absorbing area. The type of exchangeable cation in the montmorillonite, however, may engender wide variations in the limit values, especially in the case of Na^+ and Li^+. Plasticity indices vary in a similar manner. Second, limit values tend to increase when the particle size is smaller. The *activity*, defined as the ratio of the plasticity index to the abundance of clays (Skempton 1953), shows that plasticity increases proportionately with the percentage increase of clay-sized particles. As the activity increases, soils tend to have a lower resistance to shear, and so the parameter has some engineering significance. Finally, drying of a soil reduces its plasticity because shrinkage during the dehydration process brings particles closer together. The attraction between particles is so strengthened that penetration by water is difficult. Conversely, repeated wetting combined with only partial drying may increase plasticity (Grim 1962).

Although the tests for Atterberg limits are rather unsophisticated, the results are consistent when determined by more than one analyst, and they are geologically significant (Casagrande 1948; Seed et al. 1964). More importantly, the moisture content in fine-grained material is a guide to its internal strength. Some soils, called **sensitive soils,** exist with natural water contents above their liquid limits (fig. 4.12), an apparent inconsistency with the limit concept. The fact is, however, that some soils develop an open "honeycomb" structure (fig. 4.13) that is capable of holding water in excess of its liquid limit. Although the structure is unstable, in an undisturbed soil the material will be solid and possess some strength. The disruption of the internal structure by erosion, earthquake shocks, or other phenomena will cause the excess water to be released, and the solid material will become a fluid. A simple test for the liquid limit can reveal the presence of sensitivity and the potential for dangerous mass wasting.

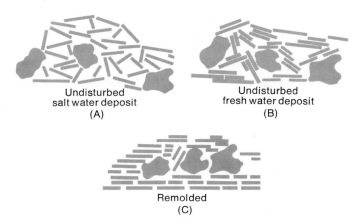

Figure 4.13.
Open soil structures in sensitive clays. Structures in (A) and (B) allow water content to exceed liquid limit. Disturbance of this structure causes temporary fluidity until substance is remolded (C). (From Lambe, 1953, pp. 315–18. Used with permission by the American Society of Civil Engineers.)

Measurement of Strength

Although some field tests have been developed, most detailed analyses of shear strength in soil are done in a laboratory by means of the direct shear test, the triaxial compression test, or the uniaxial (unconfined) compression test. (The details of these tests are provided in most textbooks of soil mechanics.)

Measurement of shear strength is a complicated task, made additionally difficult because some techniques do not yield certain critical data. In the direct shear test, for example, pore pressure cannot be determined. It should also be kept in mind that laboratory tests are designed to measure stress at the moment of failure, which is analogous to shear strength. Natural materials, however, especially soils, begin to deform well below the stress level that causes rupture (fig. 4.14) and may continue to deform at a rate that increases steadily with additional stress. Ultimately the material ruptures at a critical stress called the *breaking strength*.

Determination of strength in rocks presents an added complication because tests on small samples may not be reliable indicators of the overall strength of a large rock mass. This occurs because structures within the mass such as joints and other fractures may make the total mass relatively nonresistant to erosion. However, analyses of rock samples taken in zones between the fractures may indicate that the rock is quite strong. Obviously methods to estimate rock strength in the field are needed. One method devised by Selby (1980, 1982) is presented here as table 4.3. The Selby classification allows any rock mass to be placed into one of five categories representing overall rock mass strength. Placement in a specific group is based on numerical ratings (r) given for various strength parameters. All parameters are not considered to be of equal significance in the determination of strength, so each is assigned a percentage value representing its importance. The sum of the weighted values is an estimate of the rock mass strength.

The appeal of an approach such as Selby's is that a rapid estimate of rock mass strength can be made with normal geological field equipment except for the Schmidt hammer, which is a portable engineering tool designed to measure hardness (intact strength) of materials. In addition to stability recommendations, information gained from strength analyses may be important clues to understanding slope profiles and evolution.

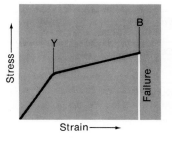

Figure 4.14.
Relationship between stress and strain. Permanent deformation begins at Y (yield stress) but rupture of the substance does not occur until the stress value reaches B (breaking strength). Material behaves like a plastic substance between Y and B.

Table 4.3 Field classification of rock strength.

Parameter	1 Very Strong	2 Strong	3 Moderate	4 Weak	5 Very Weak
Intact rock strength (N-type Schmidt hammer "R")	100–60 r : 20	60–50 r : 18	50–40 r : 14	40–35 r : 10	35–10 r : 5
Weathering	unweathered r : 10	slightly weathered r : 9	moderately weathered r : 7	highly weathered r : 5	completely weathered r : 3
Spacing of joints	>3 m r : 30	3–1 m r : 28	1–0.3 m r : 21	300–50 mm r : 15	<50 mm r : 8
Joint orientations	Very favorable. Steep dips into slope, cross joints interlock r : 20	Favorable. Moderate dips into slope r : 18	Fair. Horizontal dips, or nearly vertical (hard rocks only) r : 14	Unfavorable. Moderate dips out of slope r : 9	Very unfavorable. Steep dips out of slope r : 5
Width of joints	<0.1 mm r : 7	0.1–1 mm r : 6	1–5 mm r : 5	5–20 mm r : 4	>20 mm r : 2
Continuity of joints	none continuous r : 7	few continuous r : 6	continuous, no infill r : 5	continuous, thin infill r : 4	continuous, thick infill r : 1
Outflow of groundwater	none r : 6	trace r : 5	slight <25ℓ/min/10 m² r : 4	moderate 25–125ℓ/min/10 m² r : 3	great >125ℓ/min/10 m² r : 1
Total rating	100–91	90–71	70–51	50–26	<26

Mass Movements of Slope Material

Mass movements are of three primary types: slides, flows, and heaves, each having distinctive characteristics (Carson and Kirkby 1972). In *slides,* cohesive blocks of material move on a well-defined surface of sliding, and no internal shearing takes place concurrently within the sliding block. In contrast, *flows* move entirely by differential shearing within the transported mass, and no clear plane can be defined at the base of the moving debris. The velocity in flows tends to decrease from the surface downward. In *heave,* the disrupting forces act perpendicular to the ground surface by expansion of the material. This movement does not in itself provide a lateral component of transport, but it facilitates slow, downslope movement by gravity and serves as an important forerunner of more rapid mass movements such as rockfalls.

Although each of these movements could theoretically function alone, it seems certain that all are involved to some extent in most natural slope failures. Many processes can in fact only be explained by some combination of the primary types of movement. As figure 4.15 shows, mass movements can be thought of as multifarious events. The location of a particular process near

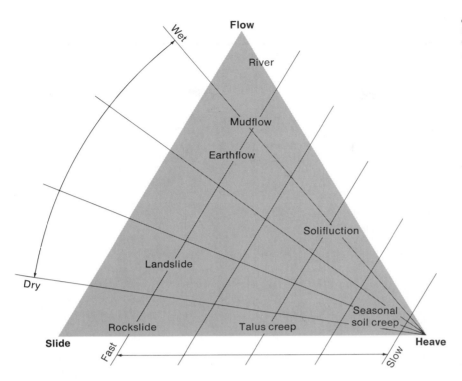

Figure 4.15.
Classification of mass movement processes.

a corner of the triangle in the figure indicates the dominance of the primary movement occupying that corner; the closer a term is to the corner, the more dominant that type of movement. Superimposed on the diagram are lines that represent the relative transport velocities of each process and the usual water content within the debris being moved.

Slope Stability

The various mass movements are alike in that all begin when the shear stress tending to displace material exceeds the resisting strength. As long as slope material maintains its internal resistance at a level greater than the driving impetus, no slope failure will occur. Stability, therefore, represents some balance between driving forces (shear stress) and resisting forces (shear strength), and can be expressed as a safety ratio:

$$F = \frac{\text{resisting force (shear strength)}}{\text{driving force (shear stress)}} .$$

F values greater than 1 connote slope stability, but as the ratio approaches unity a critical condition evolves and failure is imminent. Clearly, any factor that lowers the safety ratio (see table 4.4) can trigger mass movement, and this tendency can be produced by increasing the driving force, lowering the resistance, or both. Theoretically, failure occurs when $F = 1$; this value is an excellent example of a geomorphic threshold. In reality, however, because of

Table 4.4 Factors that influence stress and resistance in slope materials.

Factors That Increase Shear Stress

Removal of lateral support
    Erosion (rivers, ice, wave)
    Human activity (quarries, road cuts, etc.)
Addition of mass
    Natural (rain, talus, etc.)
    Human (fills, ore stockpiles, buildings, etc.)
Earthquakes
Regional tilting
Removal of underlying support
    Natural (undercutting, solution, weathering, etc.)
    Human activity (mining)
Lateral pressure
    Natural (swelling, expansion by freezing, water addition)

Factors That Decrease Shear Strength

Weathering and other physicochemical reactions
    Disintegration (lowers cohesion)
    Hydration (lowers cohesion)
    Base exchange
    Solution
    Drying
Pore water
    Buoyancy
    Capillary tension
Structural changes
    Remolding
    Fracturing

After Varnes 1958, with permission of the Transportation Research Board.

imprecise measuring techniques, some failures have occurred when the F value was slightly positive, and some slopes have been maintained temporarily with small negative values. Once failure does occur, the type of movement depends on precisely how the forces interact with one another.

The stability of slope material above a suspected plane of failure can be estimated if the components of stress and strength are known.

On shallow planar surfaces like that shown in figure 4.8(B), the driving stresses are numerically equal to $\gamma h \sin \theta \cos \theta$. Resistance is determined as shear strength, which is derived numerically as $s = c + (\gamma h \cos^2\theta - \mu) \tan \phi$. Therefore,

$$F = \frac{c + (\gamma h \cos^2\theta - \mu) \tan \phi}{\gamma h \sin \theta \cos \theta}.$$

In most analyses the vertical height of the water table above the slide plane is expressed as a fraction of the soil thickness above the plane (m), where $m = 1.0$ if the water table is at the surface, and $m = 0$ if it is at or below the sliding plane. Thus, the pore pressure can be noted as

$$\mu = \gamma_w mh \cos^2\theta$$

and

$$F = \frac{c + (\gamma - m\gamma_w) h \cos^2\theta \tan \phi}{\gamma h \sin \theta \cos \theta}.$$

The following hypothetical example will show how to determine whether the slope is stable or close to failure. If laboratory tests tell us that $\phi = 10°$, $c = 45$ lb/ft^2, $\gamma = 165$ lb/ft^3, $\theta = 8°$, $h = 12$ ft, $m = 0.6$, and $\gamma_w = 62.4$ lb/ft^3 then

$$F = \frac{45 + (165 - 0.6 \times 62.4)\, 12 \times .98 \times .18}{165 \times 12 \times .14 \times .99}$$

$$= \frac{45 + (165 - 37.44)\, 2.12}{274.43}$$

$$= \frac{45 + 127.56 \times 2.12}{274.43}$$

$$F = \frac{45 + 270.42}{274.43} = \frac{315.42}{274.43} = 1.15$$

The slope in our hypothetical case is stable, but changes in the controlling factors could easily lead to failure. For example, a rise in the water table to $m = 0.9$ would decrease the F value to 1.0 and presumably cause slope failure.

The analysis of slope stability for curved sliding surfaces such as those found in rotational slips is more complicated because it involves balancing moments (force $\times$ distance) rather than simply the forces acting on the surface. Rotational failures are usually analyzed by dividing the material above the slip plane into vertical slices and treating each slice by a modified form of the planar analysis discussed above. The values of all slices are then summed to determine the total resistance and shear stress and the estimate of the slope stability. (For an excellent discussion of slope stability, see Selby 1982.)

Heave, Creep, and Falls   Heave is instrumental in the process of **creep**, the almost imperceptibly slow movement of material in response to gravity. *Seasonal creep* or *soil creep* is the downslope movement of regolith that is aided periodically by the heave mechanism. No continuous external stress is placed on the mass; it moves under gravity when its cohesion and frictional resistance are spasmodically lowered. The process functions in the upper several feet of the soil, and its effect decreases rapidly with depth. The phenomenon of soil creep was first recognized in the latter part of the nineteenth century, and its ubiquity was gradually accepted as its effects were observed in the field. Historically, evidence that suggests a soil creep influence on slope materials has included downslope curvature of bedding (fig. 4.16), stone lines, downslope growth of trees or tilting of structures, and accumulations of soil upslope from a fixed obstruction (see Young 1972). Recently, however, observations and measurements of seasonal creep have become more sophisticated. Precise surveying methods and trenches such as Young pits (Young 1960) are fairly standard techniques in current studies of the process. A complete review of the common methods employed to measure soil creep can be found in Selby (1966) and *Revue de Géomorphologie Dynamique* (1967), and new techniques are continually being developed (Finlayson 1981).

Figure 4.16.
Creep in vertical Romney shale.
Western Maryland Railroad cut
one mile west of Great Cacapon.
Washington County, Md.

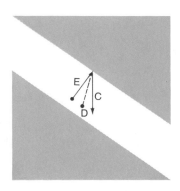

Figure 4.17.
Movement of near-surface
material by heaving. During
expansion (E), particle is
displaced perpendicular to the
surface. During contraction (C),
particle settles in a vertically
downward direction under
influence of gravity. Actual
movement is shown by line D.

Although burrowing animals and vegetation may cause random disturbances in soils, their effects are minor compared with the heave produced by
swelling or freezing and thawing. In the heave mechanism, expansion disturbs
soil particles perpendicular to the ground surfaces; when the soil contracts,
the vertical attraction of gravity acts on the particles. The expansion-con-
traction cycle, therefore, adds a lateral component to particle motion in any
soil having an inclined surface. Because gravity is reasonably uniform over
the Earth's surface, the distance of transport in each heave event, and presumably the rate of creep in any climate, should vary with the slope angle and
should decrease with depth beneath the surface. Schumm (1967a) has demonstrated a significant correlation between the rate of surficial rock creep and
the sine of the slope angle, but documentation of this relationship for fine soils
or below the surface is lacking. Actually, as figure 4.17 shows, the contraction
event is never perfectly downward but usually moves in a direction about
midway between the normal and the vertical. As the lateral distance traveled
in each heave is less than would be theoretically predicted, a clear relationship
between slope angle and creep rate may be difficult to demonstrate. In fact,
considerable evidence suggests that even the direction of creep movement is
often random on a short-term basis, complicating the presumed relationship
even more (Fleming and Johnson 1975; Finlayson 1981). In addition, detailed
measurements (Kirkby 1967) have shown the creep rate to decrease with depth
(fig. 4.18). Presumably this relates to the lower frequency of heaving at depth
and to the greater difficulty of expansion with increasing overburden. In any
case, below a depth of 20 cm movement ceases or becomes drastically smaller
(Young 1960; Kirkby 1967).

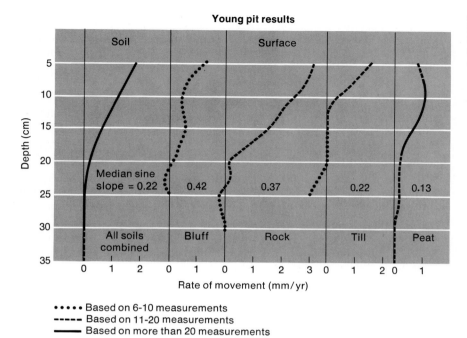

Young pit results

Figure 4.18.
Rate of creep as it relates to type and depth of material. Rate in all soils decreases with depth.

•••• Based on 6-10 measurements
‐‐‐‐ Based on 11-20 measurements
——— Based on more than 20 measurements

Downslope creep rates are extremely variable because of differences in slope angle, moisture content, and measuring technique (Caine 1981). Particle size also introduces variability in the rate because fine-grained clays tend to swell more upon wetting, and frost heaving is greatest in silty material. Nevertheless, available measurements indicate that downslope rates range from 0.1 to 15 mm/yr where the soil is vegetated. Rates may increase to 0.5m/yr or more on uncovered talus slopes where frost action is prevalent (Gardner 1979; Selby 1982). Volumetric rates can also be calculated as the volume of soil moved annually across a plane set perpendicular to the surface and parallel to the contour of the slope, for a unit horizontal distance along the plane. In semiarid regions the rate seems to be somewhat higher. In arctic climates a special kind of creep process called **solifluction** is extremely important in the geomorphic scheme; it is discussed in chapter 11.

A second type of creep, *continuous creep* (Terzaghi 1950), is fundamentally different from seasonal creep in that (1) it is driven by gravity alone, (2) it may affect consolidated rock, and (3) it can function at levels well below the surface. Continuous creep is the strain response to stress that is generated by the weight of overburden. It begins at the yield stress and continues even though no additional stress is placed on the material. Continuous creep is especially pronounced where rocks or semiconsolidated materials with low yield stress (fig. 4.14) are overlain by stronger substances. For example, a weak clay unit sandwiched between resistant strata is prone to deform by continuous creep. Excavations of any kind through that rock sequence will reduce the lateral confining pressure on the clay unit, and it will begin to flow. Creep of this type is not important in terms of the volume of material it moves or the distance of transport, but it is very significant as a precursor of rapid, sometimes catastrophic, mass movements. In many cases landslides are immediately preceded by accelerated creep; persons who intervene in the natural setting can trigger these events by not recognizing the potential for continuous creep (Kiersch 1964).

The heave mechanism is also an essential element in some rapid mass movements, especially falls. Falls in both rock and soils involve a single mass that travels as a freely falling body with little or no interaction with other solids (fig. 4.19). Movement usually is through the air, although occasional bouncing or rolling may be considered as part of the motion. Rock falls are most common where the parent material is well jointed and a steep slope is developed on the rock face. The fractures are enlarged progressively by heaving, mainly in the form of freezing and thawing, until the gravitational force exceeds the internal resistance. Undercutting of the rock or soil face by erosive agents acting at the base of the material accelerates the process. The removal of the subjacent support tends to increase tension in the overhang and so helps to create and expand incipient cracks.

Topples (fig. 4.19) are similar to falls except that forward movement of a material block is produced by rotation around a fixed hinge located at the base of the block. Although the process is not commonly cited, in some cases toppling may be the most important factor in cliff retreat (Caine 1982).

Slides   As in all classification, rapid mass movements are grouped according to the classifier's opinion as to what aspects of the phenomena are most important. Because most rapid movements are not observed as they occur, their fundamental properties of motion must be interpreted after the event. This, combined with the fact that no clear-cut distinction can be made between the primary modes of transport, has resulted in a number of viable classifications of rapid mass movements (Sharpe 1938; Ward 1945; Varnes 1958; Hutchinson 1968). The classification prepared by Varnes (1958) is adopted here as figure 4.20 because it relates well with our process orientation. The classification is based primarily on the type of material being moved and the primary type of movement. Additions to the classification (Varnes 1978) are shown in table 4.5.

As defined earlier, slides are slope failures that are initiated by slippage along a well-defined planar surface. The sliding mass is essentially undeformed; however, it may partially disintegrate during the sliding motion, giving rise to flow movement in the latter phase of the event. The plane of sliding may be shallow and approximately parallel to the ground surface as in the

(A)

(B)

Figure 4.19.
Types of mass movements.
(A) Rock fall on Interstate 70 west
of Denver, May 5–6, 1973.
(B) Rock topple leading to a rock
fall in Morgan County, Ky.

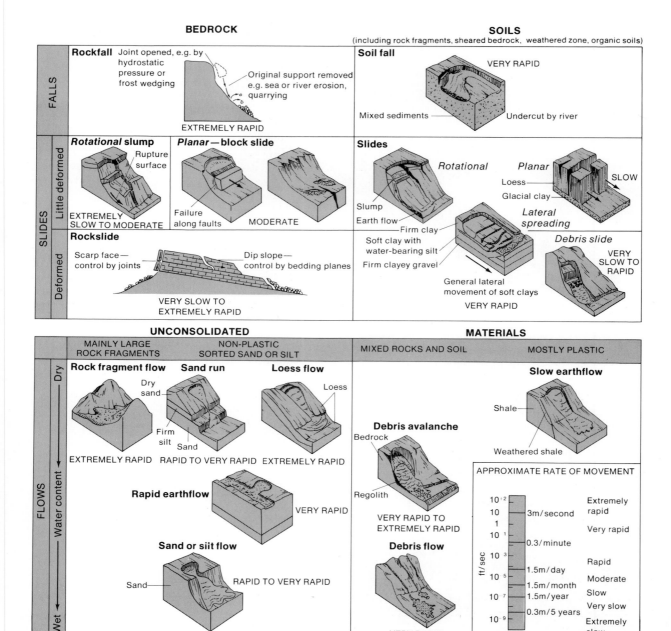

Figure 4.20.
Classification of landslides.

Table 4.5 Classification of mass movement types in different parent materials.

Type of Movement			Bedrock	Engineering Soils	
				Predominantly coarse	Predominantly fine
Falls			Rock fall	Debris fall	Earth fall
Topples			Rock topple	Debris topple	Earth topple
Slides	Rotational	Few Units	Rock slump	Debris slump	Earth slump
	Translational		Rock block slide	Debris block slide	Earth block slide
		Many Units	Rock slide	Debris slide	Earth slide
Lateral Spreads			Rock spread	Debris spread	Earth spread
Flows			Rock flow (deep creep)	Debris flow (soil creep)	Earth flow
Complex			Combination of two or more principal types of movement		

(Header: Type of Material spans Bedrock and Engineering Soils)

From Varnes 1978, *T. R. B. Special Report 176: Landslides.* Used with permission of the Transportation Research Board, National Academy of Sciences.

case of **rockslides** and **debris slides,** or it may penetrate to some depth as a concave surface along which **rotational slip** (fig. 4.21) or **slumping** may occur.

Slides on shallow planar surfaces, called **translational slides,** are the most common of sliding phenomena. Like all forms of mass movements, the initiation of translational slides occurs when the material's resisting strength is exceeded by the shear stress. As long as F is greater than 1 at the potential surface of sliding, the slope will be stable. If an increase in the driving force or a decrease in resistance brings the ratio to unity, sliding will ensue. Commonly the driving force is increased by an addition of mass to the sliding block, but other factors can produce the same effect. Earthquakes, for example, generate a horizontal mass force that passes through the center of gravity of the slope material and adds an extra driving element. Steepening of the slope also adds driving force because shear stress increases as the slope angle (θ) increases. Thus, human activity or erosion that increases slope may trigger mass movement. Actually this is a rather simplistic view since steepening also reduces the shear strength by complicated changes in cohesion, pore pressure, and effective normal stress (Terzaghi 1950; Carson and Kirkby 1972).

The sliding phenomenon can also be produced by a variety of events that reduce the internal resistance of the debris. From observation, sliding usually occurs after prolonged or exceptionally heavy rainfall, indicating that the lowering of resistance is predominantly a function of water. In the past, the water

Figure 4.21.
(A) Features of a rotational slide.
(B) Example of rotational slump in which toe area has disintegrated into an earthflow.

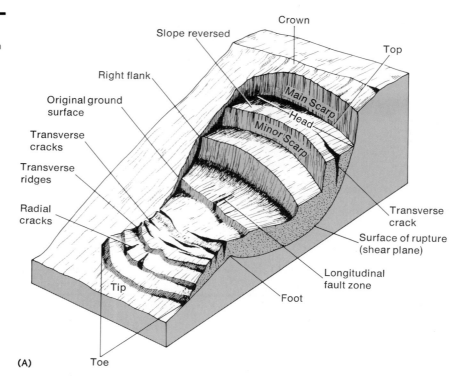

Crown

Slope reversed

Top

Right flank

Main Scarp

Head

Minor Scarp

Original ground surface

Transverse cracks

Transverse ridges

Radial cracks

Transverse crack

Surface of rupture (shear plane)

Longitudinal fault zone

Tip

Foot

(A)

Toe

(B)

effect was interpreted to be lubrication along the sliding surface. Terzaghi (1950), however, refutes this notion by pointing out that water applied to many common minerals such as quartz is actually an antilubricant. Furthermore, most soils in humid regions contain more than enough water to cause lubrication at all times, yet they also fail after rainstorms. Clearly water affects strength in other ways. You will recall that shear strength is a function of cohesion (c), effective normal stress (σ'), and friction (ϕ) such that

$$S = c + (\sigma') \tan \phi.$$

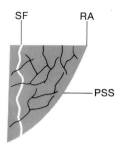

Figure 4.22.
Differences in rockslides caused by slab failure and by rock avalanche. Slab failure (SF) involves only the outer sheet of rock, which splits along one prominent fracture. A rock avalanche (RA) occurs when fracture system penetrates to the potential surface of sliding (PSS).

The response of these factors (c, σ', ϕ) to wetting is significantly more important in the initiation of slippage than is lubrication (Sidle and Swanston 1982). For example, the rise of the water table or the piezometric surface, which accompanies all prolonged rainfalls, may be the most common culprit in sliding. As the water table rises, the pore pressure (μ) at any point within the saturated mass increases, ultimately resulting in a decrease of effective normal stress (σ') and the concomitant reduction of shear strength.

Rockslides are usually associated with major structural features within the rock such as the stratigraphy of the rock sequence or joint patterns. Massive rock units normally have several prominent joint sets as well as a superimposed network of randomly spaced and oriented fractures. Prior to the creation of the joints these rocks have great shear strength associated with the high cohesion in lithified materials. Once joints begin to form, however, the process becomes self-generating because shear stresses are concentrated preferentially on the unfractured zones of solid rock. With time, more and more of the original rock is consumed by jointing, and eventually the near surface becomes a cohesionless mass of densely packed angular blocks (Terzaghi 1962). At this stage, except for particle size, the rock mass resembles an aggregate of dry sand in which the cohesion within the individual particles has little bearing on slope stability. The shear strength of the total mass stems entirely from internal friction because the cohesion across the joint openings is zero.

Rockslides can be divided into two types: **rock avalanches,** and **slab failure** (Carson and Kirkby 1972). Both obey the same mechanics, and they differ only in the amount of fracturing within the rock and the angle on the potential surface of sliding (fig. 4.22). In slab failure, cracks develop where a rock mass expands because the horizontal confining pressure is removed, allowing strain to proceed outward in the direction of the pressure release. In that sense, the process in rocks is similar to pressure-release sheeting. As lateral stress is removed, a tensional zone develops in the upper part of the mass, and cracks form within the zone. The tensional fractures penetrate to a depth that is controlled by the strength of the material. Because rocks usually have high tensile strength, fracturing does not penetrate the total depth of the tensional zone. This is not always the case in unconsolidated substances.

The stability of the outer slab depends on the depth of the fracture relative to the height of the unconfined surface that is undergoing expansion. Equations have been derived to predict the maximum height attainable before slab failure occurs (Terzaghi 1943), but predictive models are not always completely successful. For example, in one study (Lohnes and Handy 1968), tension cracks probably did not appear until slab failure was imminent. As a result, the unsupported face was considerably higher than would have been possible had the cracks developed at an earlier stage. In addition, Terzaghi (1962) considered the worst possible case—the weakest rocks—and found that most rock types could probably stand vertically up to heights of 1300 m. The observation that few vertical cliffs stand at this height even in stronger rocks indicates that fracturing drastically reduces the height that can be maintained on an unsupported rock face.

Rock avalanches occur when the joint network becomes essentially continuous down to the potential surface of sliding. An avalanche differs from slab failure in that it involves the entire mass above the sliding surface whereas slab failure includes only the material outside the outermost continuous joint; figure 4.22 shows both events. Both types of rockslide differ from rockfalls in that the rocks fail only when fractures intersect the potential surface of sliding. Falls, on the other hand, can occur above the potential slide plane. Both heaving and sliding processes may be involved in some natural mass movements (Schumm and Chorley 1964). In fact, the incidence of sliding is so complicated that Terzaghi (1950) was able to list 19 possible causes.

Flows   In true flows, the movement within the displaced mass resembles closely that of a viscous fluid, in which the velocity is greatest at the surface and decreases downward in the flowing mass. In many cases, flows are the final event in a movement begun as a slide, and the distinction between the two is usually unclear. For most types of flow, abundant water is a necessary component, but dry flows called *rock fragment* flows by Varnes (1958) do occur when rockslides or falls increase drastically in velocity and lose their identity as a unitized mass. When rocks slide or fall down a steep slope, the material disintegrates as it crashes into the relatively flat surface at the base of the slope. From there the rocks travel as masses of broken debris (called *sturzstroms* in German), moving with enormous velocities over the gentler slopes in the piedmont or valley bottom. For example, the wet, mud-soaked sturzstrom at Mount Huascarán, Peru, sped at approximately 400 km an hour over a distance of 14.5 km, wreaking destruction along its path and killing some 80,000 persons (Eriksen et al. 1970; Browning 1973).

The exact mechanism required to move these large bouldery masses (usually greater than 1 million m³) for such long distances is the subject of some controversy. Shreve (1966b, 1968) proposes that the material slides continuously on a layer of compressed air that was trapped beneath the debris as it came to the bottom of the steep slope. The air-lubrication hypothesis has been challenged by Hsu (1975) who suggests that sturzstrom movement is primarily a flow phenomenon, reviving a conclusion made earlier by Heim (1932). In most cases the highest strata involved in the fall are contained in the rear portion of the deposited debris. Shreve correctly indicates that this distribution negates viscous flow as the transporting mechanism, because in that process the uppermost layers are transported at higher velocities and so would be farther downstream in the blocky debris. The deposits, however, also have geometric features similar to those formed by lava flows and glaciers, and care must be used to distinguish between the different types of deposits (Porter and Orombelli 1980). Hsu therefore feels that the debris moves by flow, but that the mechanism differs from viscous transport in that individual particles are dispersed in a dust-laden cloud, and the kinetic energy driving the flow is transferred from grain to grain as they collide and push one another forward. The entire mass moves simultaneously until all the original energy is dissipated by friction from the particle collisions.

Flow by this process would explain the distribution of the source rocks within the deposits. It would also permit great distances of transport because the frictional resistance decreases when grains are immersed in a buoyant interstitial fluid that reduces the effective normal stress. Variations in flow distance and velocity in different events probably depend on the properties of the interstitial substance.

In soils, the transition from debris slides to debris flows also requires an increase in water content or another buoyant substance. In the intial phase, the mass breaks into progressively smaller parts as it moves downslope (fig. 4.20) even though the velocity of the advance may be slow. If the mass is wet, a **debris avalanche** may be generated as a long, narrow flow that extends well beyond the foot of the slope. The erosive nature of the movement commonly begins gulley formation on relatively undissected slopes, examples and descriptions of which can be found in Hack and Goodlett (1960) and Rapp (1960). **Debris flows** and **slides** usually result from heavy rains or the sudden melting of frozen soils. Torrential rain is especially effective in producing debris flows where vegetation has been stripped from a deep soil on moderate or steep slopes (Heller 1981; Renwick et al. 1982; Wasson and Hall 1982). For example, the annual southern California pattern of summer brush fires followed by torrential winter rains produces excessive runoff that converts the unbound soil into a high-density flow (Sharp and Noble 1953). These flows, following preexisting drainage, have tremendous erosive and transporting power, and may cause damage many kilometers downvalley while simultaneously eroding the sediment-yielding slopes in the headward reaches.

Earthflows and **mudflows** involve movement of fine-grained slope material and range from slow to rapid. In slow earthflows, the original failure of the slope is usually in the form of slump, often when the mass becomes saturated with groundwater. As explained earlier, a rising water table and pore pressure tend to lower shear resistance, and slippage results (Wells et al. 1980). If the slumped mass is relatively wet, it may slowly bulge forward at its front by viscous flow and take the form of tongues, superimposed piles of rolled mud, or bulbous toes (fig. 4.23). This movement may continue at a slow pace for many years until some stability is finally reached. Rapid earthflows or mudflows also begin as slides or debris avalanches (Pomeroy 1980; Cummans 1981); in fact, most can be traced back to bowl-shaped scars indicative of sliding. The primary slip, even if the motion is small, may remold the soil and cause it to lose much of its original undisturbed strength as it releases the water in excess of the liquid limit. The material changes instantaneously from a plastic solid to a viscous liquid and flows downslope away from the slipped zone.

To summarize, the distinction between slides and flows is often rather nebulous. Several generalizations can be proposed, however, to help put mass movements in some reasonable perspective:

1. Most flows rise naturally as the final stage of a movement that begins as a slide. For slope stability analyses, therefore, it is probably more important to understand the mechanics of sliding and the factors that might produce it.
2. The mobility or rate of mass movements depends to a large degree on the amount of water or other buoyant substances in the displaced material. This is particularly true in flow movements.

Morphology of Mass Movements

It is appropriate to ask how we can reconstruct the mode of mass transfer, especially since subtle transitions from one mechanism to another are common, and most interpretations of movement characteristics are made after the event is over. Unless some concrete relationship exists between the surface configuration of the displaced material and the genetic process, we are facing an insoluble dilemma. Fortunately some evidence has been presented to suggest that the desired morphologic relationship is real. In a study of 66 landslips in New Zealand, Crozier (1973) arranged the common types of mass movements into five primary process groups: *fluid flow* (mudflows, debris flows, debris avalanches), *viscous flow* (earthflow, bouldery earthflow), *slide-flow* (slump/ flow), *planar slides* (turf glide, debris slides, rockslides), and *rotational slides* (earth and rock slumps). Each of the 66 landslips studied was described quantitatively by seven morphometric indices, listed in table 4.6 and illustrated in figure 4.24. The relationship between each process group and the index values was tested statistically to ascertain whether the correlation was significant enough to warrant use of morphometry as a genetic determinant.

Crozier found that the classification index (D/L) was the best indicator of the process group, reaffirming Skempton's (1953) assertion about the importance of this parameter. As one would expect, the D/L value decreases

(A)

(B)

Figure 4.23.
Types of flows and slides.
(A) Lobate earthflow (center of
photo). Near Roberts, Mont.
(Photo by R. R. Dutcher)
(B) Hebgen Lake earthquake.
Upstream view of Madison slide,
showing the west edge of the
slide debris, blocked highway,
and dry river bed below the slide.
Madison County, Mont., August
1959. (Photo by J. B. Hadley. U.S.
Forest Service photo, fig. 54 in
U.S. Geol. Survey Prof. Paper
435-K 1964) (C) A slump-earthflow
caused by reactivation of an old
stabilized slump block. The
slump-earthflow follows the outline
of an old slump block apparently
subsidiary to the stabilized
landslide on the left side of the
active area. Charles Mix County,
S.D.

(C)

Table 4.6　Morphometric indices used to determine process of mass movement.[a]

Index	Description
Classification	D/L—maximum depth of displaced mass prior to its displacement over maximum length
Dilation	Wx/Wc—width of convex part of displaced mass to concave; indicates lateral spreading
Flowage	(Wx/Wc − 1) × Lm/Lc × 100—Lm is length of displaced mass; Lc is length of concave segment.
Displacement	Lr/Lc—Lr is length of the surface of rupture exposed in concave segment. Low value indicates instability.
Viscous Flow	Lf/Dc—Lf is length of bare surface on displaced material, Dc is the depth of the concave segment.
Tenuity	Lm/Lc—Indicates how dispersed or cohesive the material is during displacement.
Fluidity	Amount of flowage expected from particular type of material on distinct slope. Varies with water content.

After Crozier 1973. With permission of *Zeitschrift für Geomorphologie*.
[a]Compare figure 4.23.

markedly with greater flow (table 4.7) because the displaced material will extend farther downvalley than it would if moving as a sliding block. Some uncertainty will remain, however, unless the classification index is used in conjunction with other indices. Importantly, a definite inverse relationship was found between *D/L* and four other morphometric indices (flowage, tenuity, dilation, fluidity), each of which is presumably controlled by the water content of the material during its movement.

The relative mobility of sturzstroms may also be estimated from the geometry of the deposits. Heim (1932) observed that the distance traveled by a sturzstrom is a function of the height of the rockfall, the size of the mass, the characteristics of the mass, and the characteristics of the route followed. Later Shreve (1968) considered the *H/L* ratio to be an *equivalent coefficient of friction,* since in sliding *H* and *L* are related by

$$H = \tan \phi \, L,$$

where $\tan \phi$ is the coefficient of friction and *H* and *L* are respectively the height of fall and the horizontal distance of the movement. The *H/L* ratio should, therefore, indicate resistance and be inversely proportional to the mobility of the movement. Hsu (1975), however, concluded that the equivalent coefficient of friction is somewhat misleading, and he introduced another parameter called the *excessive travel distance* (*Le*) to reveal mobility characteristics. This factor is defined as the "horizontal displacement of the tip of a sturzstrom beyond the distance one expects from a frictional slide down an incline with a normal coefficient of friction of tan 32° (0.62)." It is expressed as

$$Le = L - H/\tan 32°.$$

In general, *Le* seems to correlate positively to the volume of the fallen mass.

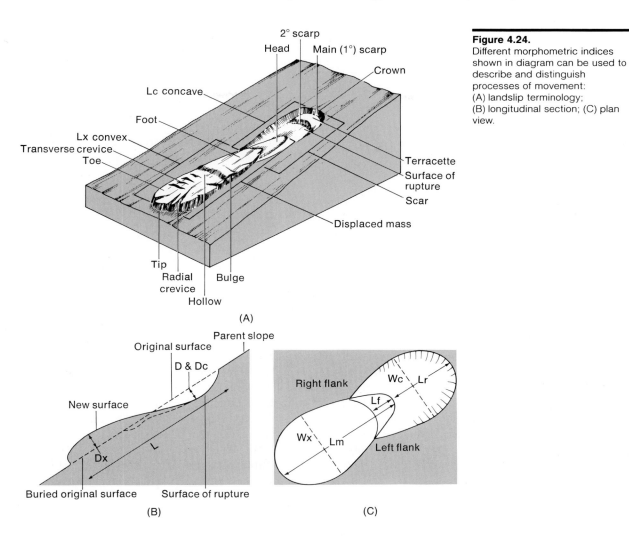

Figure 4.24.
Different morphometric indices shown in diagram can be used to describe and distinguish processes of movement: (A) landslip terminology; (B) longitudinal section; (C) plan view.

Table 4.7   Average values of the depth/length ratio in different types of landslips as calculated for different areas.

Type of Landslip	Average *D/L* Ratio
Flows	1.58
Planar slides	6.33
Rotational slides	20.84

After Crozier 1973. With permission of *Zeitschrift für Geomorphologie.*

Although more work is needed before a clear relationship between morphology and process can be defined, the morphometric approach exemplified by these studies holds real promise. Predictions of slope stability, possible modes of failure, and areas that might be affected are potential benefits if we can understand how previous movements occurred in any given region.

Slope Profiles

Profiles developed on the surface of natural slopes are regarded as reflections of the major geomorphic factors—climate, rock type and structure, time, and process. This rather innocuous statement leaves much to be desired because it may suggest to you that the relationship between slope form and geomorphic factors is straightforward. On the contrary, there is no simple method to decipher which factors will determine the precise characteristics of a slope profile. In fact, all of the factors are involved in some way to produce slopes. Therefore, many debates about the development of slopes revolve around *how much* a factor is involved in slope formation rather than *whether* a factor is involved. Invariably this leads to individual value judgments since very few, if any, studies can absolutely isolate the effect of one factor by keeping all others constant.

Geomorphologists have paid considerable attention to the geometry of slopes and the angles developed on different parts of the profile. Ideally, slope profiles can be divided into four, general components (fig. 4.25): a *convex* upper segment, a *cliff face* (or free face), a *straight* segment having a constant slope angle, and a *concave* segment at the hillslope base (Wood 1942; King 1953; Carson and Kirkby 1972). More detailed distinctions of components have been suggested. For example, the classification shown in figure 4.26 recognizes nine slope components and additionally suggests that each is associated with processes that probably dominate the zone. In any actual situation, all of the components shown may not be present in the profile, or they have negligible significance. The upper convexity, for instance, is usually more prevalent in humid-temperate regions than in semiarid or arid climatic zones because the soil creep process is known to be more important in the humid environment (fig. 4.26).

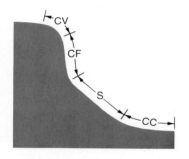

Figure 4.25.
Major components of slope profiles. CC = concave segment; S = straight segment; CF = cliff face; CV = convex segment.

In addition to the types of components found in profiles, measurements in a variety of climatic zones have revealed the interesting fact that slope angles appear to be concentrated in groups with rather small ranges of values (see Carson and Kirkby 1972; Young 1972). Most pronounced are those that cluster at 43°–45°, 30°–38°, 25°–29°, 19°–21°, 5°–11°, and 1°–4°. Although any slope angle is possible, the frequency of these recurring groups is tantalizing to geomorphologists because it probably reflects the underlying control of the great geomorphic variables.

Each of these groups has well-defined maximum and minimum values, which have been termed *limiting angles* by Young (1972) and *threshold angles* by Carson and Kirkby (1972). The general interpretation of the angular distribution is that angles within any group represent a stability regime for

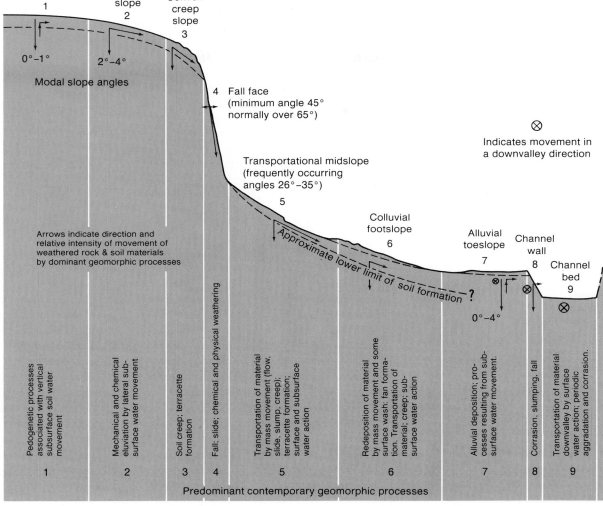

Figure 4.26.
Diagrammatic representation of the hypothetical nine unit landsurface model.

slopes formed in a particular climatic and lithologic setting. Under those conditions, threshold values can be exceeded if the intrinsic properties of the parent material are altered or if the climate changes. When threshold values are reached, any further change requires a fundamental response in the system that adjusts the slope angle and places it within a different stability group. Exactly why and how slopes adjust, however, is a debatable question. According to one hypothesis, groups and their limiting angles represent characteristic angles for the processes that are working on the slopes. Thus, it is clear that geomorphologists recognize process as an important ingredient in the development of slope components and slope angles.

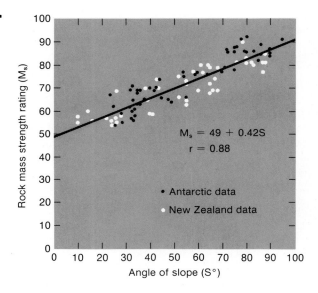

Figure 4.27.
The relationship between mass strength and profile angle for all rock units studied in Antarctica and New Zealand.

The significance of process was understood years ago by Gilbert (1877), who felt that the development of any slope was controlled by either weathering or sediment transport. Since then, the terms *weathering-limited* and *transport-limited* have been generally accepted to describe slopes formed under each process control. Weathering-limited slopes are created where the rate of soil or regolith production is lower than the rate of its removal by erosion. As a result, most of these profiles are determined by the character of the parent rock. Such profiles seem to prevail in dry climates or in mountainous terrain where erosion is rapid. In contrast, transport-limited slopes are formed where the rate of weathering is more rapid than erosion. Slopes produced under this regime normally develop on any unconsolidated parent material regardless of environment, but they are typically dominant in humid-temperate zones where vegetation cover is continuous. These profiles are less affected by parent rock and more dependent on the type and rate of slope processes.

Selby (1982) has made a cogent argument that weathering-limited slopes are directly dependent on the relative resistance of the underlying parent rocks. As evidence, he has demonstrated a high correlation between rock mass strength (see table 4.3) and the angle developed on various slope segments (fig. 4.27). A line drawn around the data points shown in figure 4.27 creates what Selby calls the *strength equilibrium envelope,* and the slopes represented by points within that envelope are referred to as *strength equilibrium slopes.* Presumably, as long as the mass strength is maintained, strength equilibrium slopes will keep a constant angle and the slope surface will retreat parallel to itself.

It is easy to visualize a cliff face controlled by a constant mass strength and retreating in a parallel manner. Whether we can expect rock strength to remain constant is debatable, however, because as discussed earlier once joints form in bedrock, the process becomes self-generating and more of the rock will become fractured. This, of course, decreases rock mass strength and requires an adjustment in the slope angle. In addition, during the process of slope retreat, material eroded from the cliff face by rockfalls accumulates lower in the profile as a talus. Technically, **talus** refers to the slope formed from the accumulation of debris eroded from a cliff face, although many geomorphologists have used the term in reference to the debris itself. In either sense, talus extends upslope with time and eventually may bury the original rock face (fig. 4.28). If and when this occurs, recognize that a transition has been made from a weathering-limited slope to a transport-limited slope and the strength of the parent rock is no longer significant. Instead, it is the relationship between resistance of the talus debris and the erosional forces that should determine the slope angles.

Theoretically it can be shown that in cohesionless material the angle of internal friction ϕ is equal to the slope angle θ (Carson and Kirkby 1972). Therefore, a strong correlation should exist between slope angles and the physical strength properties of unconsolidated material. In talus accumulates, values are uncommonly high (43°–45°) because the mass is densely packed and the rock fragments are interlocked (Carson and Kirkby 1972). If the void percentage is large, little pore pressure will develop and the stable angle of slope will correspond to the ϕ angle. Continuous breakdown of the talus deposits without much clay formation would produce a sandy matrix that should stand near the angle of repose for cohesionless sands, approximately 35°.

There is ample evidence that in a wide variety of climates a weathered mixture of rock rubble and soil underlies slopes between 25° and 28°. (Young 1961; Melton 1965b; Robinson 1966). As the original talus deposits are progressively broken down by weathering, the mass gradually loses its open-pore framework. During times of abundant water and high water tables, the material attains positive pore pressures that reduce the effective normal stress by buoyancy. This obviously lowers shear strength and changes the relationship between internal friction and the potential failure surface. Thus, the recurrence of slope angles at 25°–29° may lie in the mechanics associated with saturated soils. As summarized by Skempton (1964), cohesionless materials subjected to pore pressures are likely to experience shallow landsliding along failure planes that *approximate*

$$\tan \theta = \tfrac{1}{2} \tan \phi.$$

Assuming an original ϕ angle of 45° for coarse talus deposits and 35° for a sandy mantle, the stable slope developed on these materials when they become saturated would be about 26° and 19°, respectively. In clay-rich soils the ϕ angle is much lower, and stable slope angles are considerably less.

Carson (1969) proposed that instability in slopes requires the progressive replacement of steep slopes by gentler ones. In this model, many landscapes should go through more than one phase of instability, but the exact number depends on the characteristics of the rocks and how they ultimately break down. In the initial stage, a steep rock cliff is replaced by talus or slopes developed on thoroughly fractured rocks. This phase might be followed by a

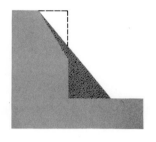

Figure 4.28.
Upslope extension of talus slopes.

change to lower slopes, and eventually to the gentle slopes formed on clay-rich soils. Each slope is only temporarily stable, for as weathering changes the mantle's properties and pore pressures vary, the mass reaches its slope threshold value. Further change causes the slope to adjust rapidly into a new stability range consistent with the revised properties of the mantle. Because of the variability of soil properties and pore pressures, any limiting angle values are possible, even though they apparently cluster in recurring groups. The types of material, the number of instability phases, and the threshold values combine in any area to control the progression of slope development. The net effect of the variables is eventually to form slopes that have long-term stability with respect to rapid mass movements; at that point, creep and surface water erosion become much more significant as slope processes.

The salient point of this discussion is that recurring angles measured on slopes may be easily explained by the relationship between erosive process and the different strengths of unconsolidated materials caused by textural variations. However, whether all slope materials experience an evolution in texture as envisioned by Carson is debatable, and perhaps unnecessary to explain slope angle and form.

Our discussion of slope profiles thus far has attempted to demonstrate that processes of weathering and erosion are intimately involved in slope development. Process, however, is not an independent variable because it is directly controlled by climate and geology. In fact, of the many variables cited as being responsible for hillslope form, only geology and climate can be considered as independent variables.

The Rock-Climate Influence

It was shown earlier that slopes in weathering-limited situations are controlled by the mass strength of the parent rock. This is especially significant in the maintenance of a cliff face. The lithologic influence on slopes is shown in both declivity and profile shape. Coherent rocks tend to support steeper slope angles, and with equal cohesion, the more massive the bedding, the steeper the slopes. Where strata contain alternately weak and resistant rocks, an irregular profile may develop, and resistant units will assume higher than normal angles where they overlie weaker rocks.

In regions where a cliff face is not present, lithology may still exert a control on slopes. It is an accepted fact that topography generally reflects lithology and that "resistant" rocks underlie hills and nonresistant rocks become the valleys. In this sense, however, resistance is not defined by intrinsic properties of a particular rock type but is a relative feature determined by how rapidly slopes developed on the rock retreat and whether the rock stands relatively high in the local topography (Young 1972). Therefore, it is not so much the rock itself that determines resistance, but whether the slopes formed over the rock are controlled by processes of weathering or processes of removal. If a slope is weathering-controlled, resistance is related to how rapidly the rock is weathered; it is a direct function of the rock properties. In transport-limited slopes, the resistance is attributable to the rate at which regolith can be eroded;

the properties of the weathered mass and the type and magnitude of the erosional processes become important in slope development. For these reasons, the resistance of a particular rock type and its influence on slopes can be reversed if the rock is located in different climates. For example, the characteristics of slopes formed on limestones in humid climates contrast markedly with those developed in arid climates.

With regard to climatic influence, geomorphologists have long recognized that the most common slope profile in humid-temperate regions is a distinct, convex upper slope and a concave lower slope. Contrary to some beliefs, straight slope segments do occur in regions with a humid-temperate climate, and some profiles do contain steep cliff faces. Most cliff faces, however, are ephemeral in the sense that as soon as undercutting ceases, a talus slope forms and will extend upslope until it covers the original cliff wall (fig. 4.28). If the lithology of the rock sequence underlying the slope is not uniform, cliff faces may persist because resistant units are maintained as caprocks when the weaker underlying strata retreat faster, essentially undercutting the stronger rocks.

Convex upper slopes are usually interpreted as a function of soil creep; the lower concavity probably results from soil wash, although not all slopes have this segment, particularly when there is active erosion at the slope base (Strahler 1950). The convexo-concave profile is most likely to be attained after mass movements have produced a long-term angular stability. At this stage, creep and wash become the dominant slope processes; the straight segment, representing stability of slope material, is gradually diminished in size. The processes of water erosion on slopes will be discussed in the next chapter. Recognize here, however, that water flowing over and through slope material combines with mass movement to mold slope profiles, and in some cases water erosion may be the dominant process involved.

Semiarid and arid climates tend to engender slope profiles that are more angular than those found in humid-temperate regions, even though the same convex, straight, and concave segments may be present (fig. 4.29). Steep cliffs usually are present above a straight, debris-covered segment that normally stands at angles between 25° and 35°. At the base of the straight segment a pronounced change in slope occurs, and angles decrease over a short distance to less than 5°, a normal slope for most desert plains. The limited vegetal cover and low precipitation associated with arid zones assure that mass movements occur at higher angles and that creep is subordinated to wash. As a result the upper slope convexity, so prominent in humid regions, is much less pronounced.

Straight segments are maintained by the wash process, which is accelerated on the sparsely vegetated surfaces. Unlike similar segments in humid climates, these usually have only a thin veneer of rock debris. They are not, then, slopes of accumulation such as talus slopes but instead probably represent true slopes of transportation, on which the amount of debris supplied to the straight segment from the cliff face or by weathering of the underlying rocks is removed in equal quantities to the desert plain. The angle of slope represents some balance between the processes that break debris down and

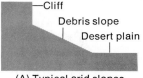

(A) Typical arid slopes

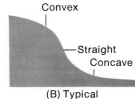

(B) Typical
humid-temperate slopes

Figure 4.29.
Typical slope profiles in (A) arid regions and (B) humid-temperate regions.

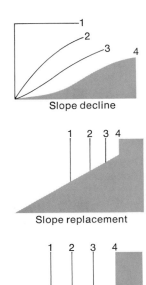

Figure 4.30.
Three hypotheses of slope evolution. Higher numbers indicate increasing age of the slope. (Adapted from Young, 1972, fig. 14, in *Slopes*. With permission of A. Young.)

the actual transporting mechanism (Schumm and Chorley 1966). Most students feel that a general relationship between particle size and slope angle can be demonstrated.

Although other climatic regimes have characteristic slope forms, in most cases they are produced by the same mechanics that operates in the humid-temperate or arid zones. In the periglacial environment a special influence is exerted by magnified frost activity; a more extensive treatment of that environment is presented in chapter 11.

Very little research has attempted to determine what aspects of hillslope profiles are most closely related to climate. An example of this approach is a study by Toy (1977). Toy utilized a rigorous statistical analysis to compare slope properties within two extended traverses in the United States (Kentucky to Nevada and Montana to New Mexico) along which considerable climatic variation is experienced. The selection of sampling localities was stringent. Parent rock at each measuring site was restricted to shales dipping at less than 5°. Each slope analyzed was south-facing, within 5 miles of a weather station having records for the same 21-year period used as the climatic base, and had no effects of human activity. Toy found that climate could account for 59 percent of the variability in the upper convex segments and 43 percent of the variability in the slope of the straight segments. Arid slopes in this study were shorter, had steeper straight segments, and had shorter radii of curvatures developed at the convex crests than slopes in humid regions. In addition, of the climatic variables used in the study, those most closely associated with slope variations were (1) spring and summer precipitation, (2) potential evapotranspiration, and (3) water availability (total precipitation minus total potential evapotranspiration during the 21-year period).

Toy's findings cannot be used to make sweeping generalizations about the climatic effect on slope profiles because they apply only to one type of parent rock. However, the study is a good demonstration of the research design needed to estimate the influence of one geomorphic factor by reducing or eliminating the effects of others.

Slope Evolution

In addition to geology and climate, the factor of time can also be considered as an independent variable. Its effect, however, is difficult to determine, especially when the time interval involved is very long. As we saw in chapter 1, some of the great debates in geomorphology revolve around the question of how slopes respond to continued erosion. Do slopes progressively flatten through time, providing landscapes with evolutionary steps or stages? Or do slopes reach an equilibrium between form and geomorphic factors that is maintained through time because slopes retreat in a parallel manner? Unfortunately, these questions are more easily asked than answered.

Three main types of slope evolution have been suggested: slope decline, slope replacement, and parallel retreat (fig. 4.30) In *slope decline,* the steep upper slope erodes more rapidly than the basal zone, causing a flattening of the overall angle. It is usually accompanied by a developing convexity on the upper slope and concavity near the base. Slope decline alone cannot in fact explain a concave profile on the lower slope unless some deposition occurs at

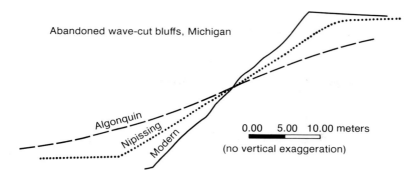

Abandoned wave-cut bluffs, Michigan

Algonquin
Nipissing
Modern

0.00   5.00   10.00 meters
(no vertical exaggeration)

Figure 4.31.
Profiles of bluffs of three different ages developed in the same material.

the base. In *slope replacement,* the steepest angle is progressively replaced by the upward expansion of a gentler slope developed near the base. This process tends to enlarge the overall concavity of the profile, which may be in either a segmented or a smoothly curved form. Slopes evolving by *parallel retreat* are characterized by the maintenance of constant angles on the steepest part of the slope. Absolute lengths of slope parts do not change except in the concave zone, which gets longer with time.

There is little doubt that all three types of slope evolution can be demonstrated in actual situations (Savigear 1952; Brunsden and Kesel 1973; Cunningham and Griba 1973; Haig 1979; Nash 1980; Selby 1980, 1982; Colman 1983). In general, slope decline is most notable in humid regions, and parallel retreat seems to be more prevalent in arid regions.

In most of the foregoing cases, demonstration of slope evolution is based on the *ergodic hypothesis* (Chorley and Kennedy 1971), which contends that, in the proper setting, spatial elements can be considered as equivalent to time elements, and space-time transformations are therefore acceptable. As an example, Nash (1980) suggests that profile variations in a series of lake bluffs in Michigan represent slope changes that have occurred during intervals between the development of the different bluffs. Because each bluff was formed in the same morainic material, it is assumed that the oldest bluff originally had a profile similar to that of the modern bluff. Therefore, the observed differences between the two profiles represent changes that have occurred on the older slope since the time of its formation (fig. 4.31).

The types of studies just mentioned have inherent value because they are based on real observations. However, they are probably valid only if significant changes in other geomorphic variables, such as climate, do not occur during the tenure of slope development. Thus, they are restricted to geologically short time spans. Because of this, the possibility remains that the observed changes are merely transitions toward an equilibrium form that, when attained after a greater period of time, would experience no further profile alteration.

Finally, it should be pointed out that numerous attempts have been made to characterize slope evolution by employing theoretical techniques such as numerical and simulation models (for references see Carson and Kirkby 1972; Young 1972; Selby 1982). Although modeling suggests routes that slope evolution may follow, the techniques suffer because the assumed character of the original slope profile is pure conjecture; i.e., there is no sure way to know the form of the initial profile. As a result, some geomorphologists believe they add little to our comprehension of slopes unless they are based on long-term and detailed field measurements (Dunkerley 1980; Selby 1982).

Summary

The processes of physical weathering tend to break rocks and unconsolidated debris into smaller particle sizes. The force needed to accomplish this disintegration is provided by expansion resulting from unloading, hydration of minerals, or growth of foreign substances in spaces within the parent material. Many important processes of disintegration require the presence of water. Physical weathering, combined with gravity, is instrumental in determining the type and rate of mass movements; ultimately it has a direct bearing on the slopes developed in any region.

Mass movements occur as slides, flows, and heaves, or by water-induced transport of surface debris. The magnitude and type of mass movement are partly dependent on the physical properties of the parent material. Shear strength (a function of internal friction, effective normal stress, and cohesion) determines how vigorously any substance will resist the force attempting to produce mass movement. Thus, slope failure or other mass movements can result from an increase in shear stress (driving force), a lowering of shear strength (resistance), or both. Physical weathering tends to decrease the shear strength of materials and thereby helps to initiate mass movements and control the form of the resulting slopes. Climate and lithology interact to influence slope profiles. The effect of time is shown by the manner in which slopes evolve.

Suggested Readings

The following references provide greater detail concerning the concepts discussed in this chapter.

Carson, M., and Kirkby, M. 1972. *Hillslope form and process.* London: Cambridge Univ. Press.

Crozier, M. J. 1973. Techniques for the morphometric analysis of landslips. *Zeit. f. Geomorph.* 17: 78–101.

Grim, R. 1962. *Applied clay mineralogy.* New York: McGraw-Hill.

Ollier, C. D. 1969. *Weathering.* Edinburgh: Oliver and Boyd.

Selby, M. J. 1982. *Hillslope materials and processes.* Oxford: Oxford Univ. Press.

Terzaghi, K. 1950. Mechanism of landslides. In *Application of geology to engineering practice,* edited by S. Paige, pp. 83–123. Geol. Soc. America Berkey Vol.

Varnes, D. J. 1978. Slope movement types and processes. In *Landslides—Analysis and control,* edited by R. Schuster and R. Krizek, pp. 12–33. Washington, D.C.: Transportation Research Board Spec. Rpt. 176.

Young, A. 1972. *Slopes.* Edinburgh: Oliver and Boyd.

The Drainage Basin—Development, Morphometry, and Hydrology

5

Introduction

Having examined the processes that break rocks apart and transport debris down slopes by mass movements, we are now ready to consider rivers—how they form and how sediment and water are delivered to the river system. Two basic truths about rivers were realized long before geomorphology emerged as an organized science: (1) streams form the valleys in which they flow, and (2) every river consists of a major trunk segment fed by a number of mutually adjusted branches that diminish in size away from the main stem. The many tributaries define a network of channels that drain a discernible, finite area recognized as the **drainage basin** or **watershed** of the trunk river.

Each basin is separated from its neighbor by a "divide," and so basins serve as excellent fundamental units of geomorphic systems. Any feature, fluvial or otherwise, within a basin can be reasonably considered as an individual subsystem of the basin, having its own set of processes, geology, and energy gains and losses. Furthermore, because it is possible to measure the amount of water entering a basin as precipitation and the volume leaving the basin as stream discharge, hydrologic events can be readily analyzed on a basinal scale. Similarly, most of the sediment produced within the basin limits is ultimately transported from the basin via the trunk river. Thus, considered on a long temporal scale, the rate of lowering of the basin surface can be estimated. The mechanics of fluvial processes usually reflect some balance between the sediment to be transported and the water available to accomplish the task.

Most earth scientists are introduced to watersheds by learning that drainage patterns or individual stream patterns often mirror certain traits of the underlying geology, as figure 5.1 and table 5.1 show. The gross character of patterns is a useful tool in structural interpretation (Howard 1967) and as a first approximation of lithology; a search of topographic maps and aerial photographs for a distinctive network arrangement is a logical first step in the study of regional geology.

In a hydrologic sense, however, prior to World War II, most basins were described in qualitative terms such as well-drained or poorly drained, or they were connoted descriptively in the Davisian scheme as being youthful, mature, or old. The mechanics of how river channels or networks actually form or how water gets into a channel was understood (or misunderstood) only in vague terms by both geologists and hydrologists. Realizing this early twentieth-century view of streams and drainages, it is startling to examine the avant-garde approach presented by R. E. Horton during this period. His attempt to explain stream origins in mathematical terms and to describe basin hydrology as a function of statistical laws can be cited as the birth of quantitative geomorphology. We now know that many of Horton's original ideas are only partially correct. However, there can be little doubt that modern geomorphic analysis of drainage basins has its roots in Horton's original work, and his thinking was instrumental in the rise of a new breed of geomorphologist (Horton 1933, 1945).

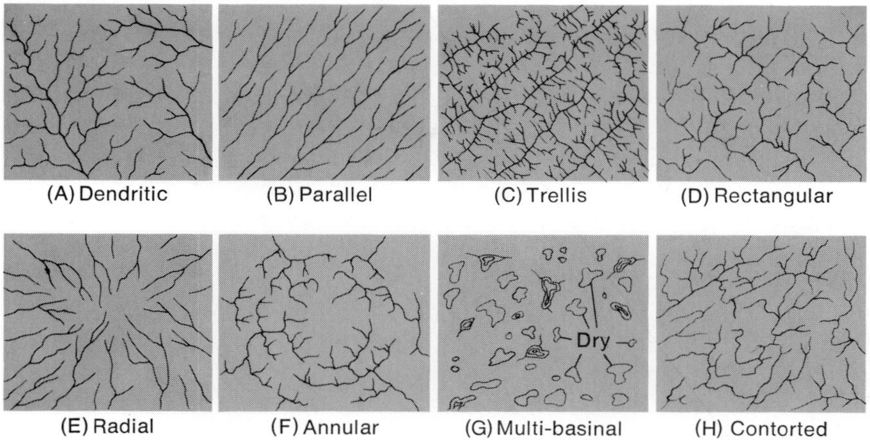

(A) Dendritic   (B) Parallel   (C) Trellis   (D) Rectangular

(E) Radial   (F) Annular   (G) Multi-basinal   (H) Contorted

Figure 5.1.
Basic drainage patterns.
Descriptions are given in
table 5.1.

Table 5.1   Descriptions and characteristics of basic drainage patterns illustrated in figure 5.1.

Basic	Significance
Dendritic	Horizontal sediments or beveled, uniformly resistant, erosion crystalline rocks. Gentle regional slope at present or at time of drainage inception. Type pattern resembles spreading oak or chestnut tree.
Parallel	Generally indicates moderate to steep slopes but also found in areas of parallel, elongate landforms. All transitions possible between this pattern and type dendritic and trellis.
Trellis	Dipping or folded sedimentary, volcanic, or low-grade metasedimentary rocks; areas of parallel fractures; exposed lake or sea floors ribbed by beach ridges. All transitions to parallel pattern. Type pattern is regarded here as one in which small tributaries are essentially same size on opposite sides of long parallel subsequent streams.
Rectangular	CAused by Joints and/or faults at right angles. Lacks orderly repetitive quality of trellis pattern; streams and divides lack regional continuity.
Radial	hills Volcanoes, domes, and erosion residuals. A complex of radial patterns in a volcanic field might be called multiradial.
Annular	Structural domes and basins, diatremes, and possibly stocks.
Multibasinal	Hummocky surficial deposits; differentially scoured or deflated bedrock; areas of recent volcanism, limestone solution, and permafrost. This descriptive term is suggested for all multiple-depression patterns whose exact origins are unknown.
Contorted	Contorted, coarsely layered metamorphic rocks. Dikes, veins, and migmatized bands provide the resistant layers in some areas. Pattern differs from recurved trellis in lack of regional orderliness, discontinuity of ridges and valleys, and generally smaller scale.

(handwritten annotations: "steep angle"; "Automate Resistant / Non Resistant folded"; "Right Angle"; "outwArd, steep slope"; "glacier deposit"; "omit" beside Contorted)

After Howard 1967. Used with permission of American Association of Petroleum Geologists.

Figure 5.2.
The slope hydrologic cycle. Some
of the precipitation (P) is
intercepted by vegetation (I) or
lost by evapotranspiration (ET).
Upon reaching the ground
surface, it becomes part of stream
discharge (Q) by direct runoff (R),
interflow (IF), or groundwater flow
(GW) after it reaches the water
table (∇).

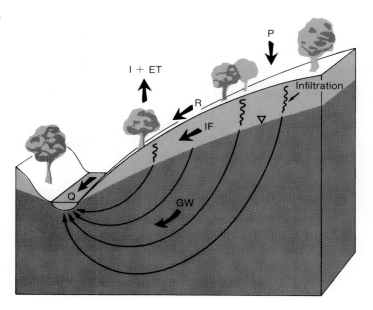

Slope Hydrology and Runoff Generation

The ultimate source of river flow is, of course, precipitation, which represents
the major influx of water to any drainage basin. Precisely how much of that
precipitation actually becomes part of stream flow and what route a particular
drop of water follows to reach a channel are topics of great concern to hydrologists. In fact, they constitute the components of what is commonly referred to as the *slope hydrological cycle* (fig. 5.2).

Rainfall does not usually strike the ground surface directly because its
fall is impeded by leaves or trunks of the vegetal cover. This process, called
interception, may significantly reduce the amount of rainfall striking the surface. Interception losses are quite variable because they depend on numerous
meteorological factors, vegetation types, land use, and seasonality. In addition,
the loss is dependent on storm duration, being high initially and decreasing
as the storm continues. During long storms, the vegetation may eventually
become saturated and all additional rainfall is passed to the surface. Nonetheless, interception may remove 10 to 20 percent of precipitation where grasses
and crops are the dominant vegetation and up to 50 percent under a forest
canopy (Selby 1982). Rainwater that does reach the ground surface may still
be prevented from becoming part of stream flow because it is susceptible to
further loss by **evapotranspiration.**

Infiltration

Once on the ground surface, precipitated water may follow different routes to
a stream channel. The water flowing into a channel in direct response to a
precipitation event is called **storm runoff** or **direct runoff.** Hydrologists have
always been concerned about how much runoff will issue from any precipitation event and how quickly the runoff will enter the stream channels. These

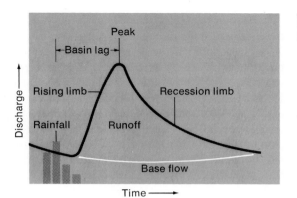

Figure 5.3.
Flood hydrograph.

factors determine if and when a flood will occur and how high the river level will be at peak flow. The response of a stream to a storm is determined by the use of a flood (or storm) hydrograph, which depicts river flow as a function of time (fig. 5.3). The volume of water (discharge) appears on the hydrograph in the form of direct runoff, or as **base flow** (water infiltrated to the water table and released more slowly from the underground system). Although discharge measured at a station may contain both of these components, it should be clear that direct runoff occurs only in response to a storm, while in dry periods all the discharge is base flow. The maximum or peak flow usually develops soon after precipitation ends, separating the hydrograph into two distinct segments, the rising limb and the recession limb. The rising limb in general reflects the input of direct runoff, and the increase in discharge with time is relatively rapid. The recessional phase, however, is controlled more by the gradual depletion of water temporarily stored in the system, and so the decrease in discharge with time is less pronounced. Also the shape of the recession limb is not influenced by the properties of the storm causing the increased discharge but is more closely related to the physical character of the basin. Since a hydrograph represents the sum of hydrographs from all subareas of a basin, if geology and topography are fairly constant throughout, then rainfalls having similar properties should generate hydrographs with the same shape. On this premise, a type hydrograph for a basin, called a *unit hydrograph,* has been developed, in which the runoff volume is adjusted to the same unit value (one for the entire area). The unit hydrograph has been used as a connecting link in many studies attempting to relate basin morphometry to hydrology.

In 1933 Horton suggested that the primary control of storm runoff is how easily rainwater can sink into the material on a sloping surface (a process known as infiltration). In general terms the infiltration theory of runoff can be described as follows. Imagine a hypothetical slope that is part of an undulating surface topography developed by weathering and mass movements. As precipitation falls on the slope surface, the water is absorbed into the ground at a rate called the **infiltration capacity,** which is a function of a number of factors such as soil texture and structure, vegetation, and the condition of the

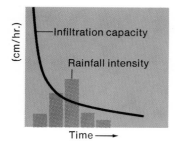

Figure 5.4.
Infiltration capacity and rainfall intensity plotted against time. Infiltration capacity decreases with duration of storm. Runoff occurs only when rainfall intensity is greater than infiltration capacity.

surface prior to precipitation. Thus, infiltration capacity varies on a large scale depending on regional geology, but it can also change on a local basis (even on a given slope) if the controlling factors vary. In addition, infiltration capacity usually changes markedly during any precipitation event. As figure 5.4 shows, it usually starts with a high value that decreases rapidly in the first several hours of the storm and then more slowly as rainfall continues, until it finally attains a reasonably constant minimum value. The infiltration capacity changes because surface conditions are being changed, especially as aggregated soil clumps are broken apart and surface entry by the water becomes more difficult as pores become clogged with the clay particles. Also, the infiltration capacity can be only as great as the lowest transmission rate in subsurface horizons. In prolonged storms, for example, a clay pan or a caliche zone in the B horizon may determine the ultimate infiltration rate. In the interval between rains, the infiltration capacity rises again as the surface dries and regains its aggregated structure, and the storage space within the soil increases as water gradually drains downward. The frequency of rainfall therefore becomes a significant factor, because a rapid succession of rains without enough time intervening will prevent the infiltration capacity from returning to its original high value. Rains of relatively small intensity can then trigger disastrous floods.

In Horton's model, as long as the infiltration capacity exceeds the rate at which rainfall strikes the surface (known as *rainfall intensity*), all incoming water will be infiltrated and none will run off. However, in those periods when rainfall intensity exceeds the infiltration capacity, runoff will occur as water moving down the slope surface, a process now known as **Hortonian overland flow.** The velocity of this flow usually falls in a range between 10 and 500m/hr, meaning that flow initiated at the top of a 100-meter slope would enter the channel within minutes or hours of its development (Dunne 1978). Thus, Hortonian overland flow was considered to be the primary determinant of peak flow and total direct runoff (fig. 5.3) that appears in stream channels in response to a storm. The basic assumption here is that all infiltrated water is delayed in its procession to stream channels because it percolates downward to the groundwater table, and from there moves riverward at the slow velocity of groundwater flow (fig. 5.2). Such base flow would eventually be part of river flow but would enter the channel after the direct runoff accruing from the initiating storm was moved downstream. This interpretation seemed to be supported by other hydrologic analyses, especially the unit hydrograph concept being developed at the same time (see discussion by Chorley 1978), and was readily accepted as the basic model of storm-generated runoff.

Overland flow traverses a surface as a broad shallow sheet or in small, linear depressions called *rills*. The flow is most common on slopes having limited soil development and a sparse vegetal cover such as those in arid or semiarid regions or in mountainous terrains. We now know, however, that Hortonian runoff is minor or nonexistent in humid-temperate regions that are densely vegetated and have well-developed soils (Kirkby and Chorley 1967; Rawitz et al 1970; Dunne and Black 1970a, 1970b). In those environments, even steady-state infiltration capacities normally exceed rainfall intensities, and only rare precipitation events can produce Hortonian overland flow.

In light of the above, we must ask the logical question as to where the water comes from to produce the rapid peak flows observed in small basins of humid regions if all rainwater reaching the surface is being infiltrated.

Subsurface Stormflow and Saturated Overland Flow

A major flaw in Horton's original model was the belief that infiltrated water moves directly downward under the influence of gravity. Research subsequent to Horton's analyses has demonstrated clearly that a lateral component of subsurface flow also exists in most slope materials. As water moves downward from the surface as a wetting front to the water table, it often encounters a soil horizon of low permeability that not only influences the equilibrium infiltration capacity but also tends to divert water downslope along its surface. This happens because a saturated condition builds up above the low permeability zone, and lateral flow is initiated parallel to the barrier. This type of water movement, known as **interflow** or **throughflow** (Kirkby and Chorley 1967), can occur above the water table and allow water to follow a more direct route to the channel than normal groundwater. Throughflow is probably most common in the more permeable A horizon.

Where no barrier exists, infiltrated water will reach the water table and elevate its position. The water table rises rapidly adjacent to the channel, where antecedent soil moisture is greatest, and slowly in the upper-slope zone. As a result, the water table steepens immediately next to the channel and generates accelerated groundwater flow in that area. The combination of throughflow and accelerated groundwater movement is called **subsurface stormflow.** It was first recognized and defined at the same time as Horton's infiltration theory (Hursh 1936; Hursh and Brater 1941), but the significance of the concept was not fully realized until much later.

Subsurface stormflow from the lower parts of slopes may actually produce runoff in the form of bank seepage in the early part of a given storm. Therefore, it is possible for some of the subsurface stormflow to contribute to the peak discharge. It is also possible, however, that in some situations much of the storm runoff is generated in the subsurface through macropores such as root channels or "pipes," which are linear openings having a much greater permeability than the normal soil matrix (Mosley 1979). These linear pathways provide a rapid delivery of subsurface flow to the channel and under the proper conditions may be significant contributors to the rising limb of the storm hydrograph. Nonetheless, most studies consider subsurface flow to be of minor consequence except during the recessional phase of runoff following peak flow. This occurs because measured rates of subsurface stormflow range between

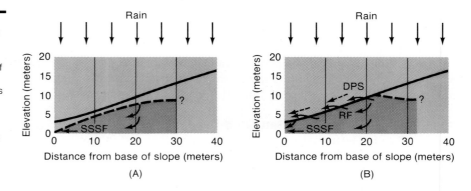

Figure 5.5.
Runoff from a steep, well-drained slope in northern Vermont. (A) Early in storm saturated zone (shaded) yields a small amount of subsurface stormflow (*SSSF*). (B) Late in storm water table rises to surface. Subsurface water returns to the surface as return flow (*RF*). Precipitation on the saturated area (*DPS*) adds to the return flow. (From *Water in Environmental Planning* by T. Dunne and L. B. Leopold. Copyright © 1978 by W. H. Freeman and Company. All rights reserved)

0.003 and 1.0 cm/hr (Dunne 1978), which is much too slow to become part of the peak discharge in most small basin flows. This slow response requires that another source of runoff be present to produce the rapid peak flow conditions.

The results of detailed studies conducted in Vermont (Ragan 1968; Dunne and Black 1970a, 1970b) showed that in many storms the level of the water table rises until it actually intersects the ground surface near the stream channel (fig. 5.5). In addition, zones of throughflow partially fed by laterally moving, infiltrated water may become saturated to the surface (Kirkby and Chorley 1967). This creates a situation where some of the infiltrated water is returned once again to the surface and begins to flow downslope toward the channel. The flow is aptly called return flow. **Return flow** is obviously a form of overland flow, but its origin and location is distinctly different from the Hortonian overland flow discussed earlier. The velocity of return flow is much greater than subsurface stormflow, attaining speeds of from 3 to 15 cm/sec (Dunne 1978). In addition, direct precipitation on the saturated area marked by return flow or zones of saturated throughflow adds significant amounts of water to the volume of direct runoff.

The combination of return flow and direct precipitation on saturated areas is called **saturated overland flow.** It is now documented that in many humid areas saturated overland flow is the major contributor of direct runoff to stream channels. For example, Ragan (1968) estimated that on average it supplied 55 to 62 percent of the total storm runoff in a small watershed near Burlington, Vermont, and was predominant in determining peak discharge. Subsurface storm flow provided 36 to 43 percent of the total flow and exerted its greatest influence during the recessional phase of the runoff.

It now seems clear that there are many sources of direct runoff other than Hortonian overland flow. It also is becoming apparent that the area of a watershed that actually provides runoff during any storm is not constant. This perception, known as the **variable source concept** (also called the **partial area**

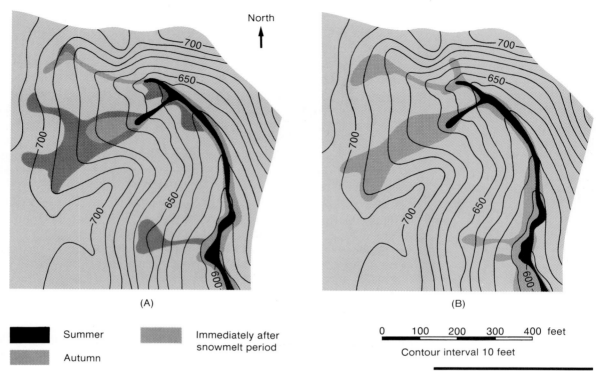

North

(A)   (B)

■ Summer   ▨ Immediately after
▨ Autumn   snowmelt period

0   100   200   300   400 feet

Contour interval 10 feet

Figure 5.6.
Variations in saturated areas on
well-drained hillslopes near
Danville, Vt. (A) Seasonal changes
of pre-storm saturated area.
(B) Expansion of saturated area
during a single, 46 mm rainstorm.
Solid black is at beginning of
storm. Light shade represents
saturated area at end of storm
where water table has risen to the
surface. (From *Water in
Environmental Planning* by
T. Dunne and L. B. Leopold.
Copyright © 1978 by
W. H. Freeman and Company. All
rights reserved)

concept), essentially indicates that the area over which quick runoff occurs
varies seasonally and during a given storm (fig. 5.6). This change is fundamentally controlled by topography, soil characteristics, antecedent moisture,
and rainfall properties. Areas with moderate to poorly drained soils, gentle
slopes, and concave recessions along the valley walls are prone to have the
greatest expansion of contributing areas during storms and on a seasonal basis.

Although the generalizations discussed above seem reasonable, they are
based on a limited number of studies and much more work is needed to refine
the model (Selby 1982). Nonetheless, the observations about runoff generation have an important bearing on the geomorphic processes operating within
a drainage basin, and further research in this field should be vigorously pur-
sued (Freeze 1980).

In the Horton model when rainfall intensity exceeds the infiltration capacity,
overland flow occurs and only then does erosion become possible (fig. 5.4). For
a stream channel to develop, the erosive force (F) of the overland flow must
surpass the resistance (R) of the surface to being eroded. According to Horton

**Initiation of Channels
and the Drainage
Network**

(1945), as overland flow begins to traverse the slope, the force it exerts on a soil particle depends on the slope angle, the depth of the water, and the specific weight of the water such that

$$F = \gamma \frac{d}{12} \sin \theta,$$

where γ is the specific weight, d the depth, and $\sin \theta$ the slope angle. Actually the force (F) represents a shear stress exerted parallel to the surface by the water. The stress progressively increases downslope because the depth rises as more and more water is added to the volume of the overland flow. Depending on the size of the slope material, a threshold stress is eventually reached at some point on the slope where $F > R$, and particles are dislodged or entrained. The processes and magnitude of slope erosion by water will be discussed later in the chapter when we examine sediment yield from drainage basins. Our concern here is the initiation of channels on sloping surfaces.

Surface resistance (R) is affected by the type and density of the vegetal cover. As we saw earlier, vegetation intercepts raindrops before they can strike the surface, thereby preserving cohesion in the aggregated-clay soil structures. In addition, rootlets tend to bind soil particles, and litter often serves as a protective mat above the surface material. Vegetation also inhibits the free flow of water and retards its velocity. In areas devoid of vegetal cover, the surface commonly develops a hard crust as it dries in the direct sunlight, which provides a high initial resistance to erosion. This may be destroyed during a storm, indicating that resistance, like the force of overland flow, may vary as a rainfall event progresses. It should also be noted that the factors that determine resistance are essentially the same ones that control infiltration capacity.

Erosion by overland flow begins when F exceeds R, taking the form of a series of shallow, parallel rills that are oriented perpendicular to the slope contours. Slight variations of the surface topography produce a greater depth of flow and more erosive force in the low spots. Erosion is accelerated at those points, and a rilled surface results rather than a slope that discharges water as an unconfined sheet. The actual point where rill formation begins depends on how efficiently the force of overland flow increases as the water moves down the surface. This ultimately relates to the infiltration capacity, the rainfall intensity, and the resulting rate of runoff (the *runoff intensity*).

Assuming a constant slope and runoff intensity, the distance between the watershed divide and the upper position of rills is a measurable segment called the *critical length* (X_c); the surface between X_c and the divide is recognized as a "belt of no erosion" (figs. 5.7 and 5.8). Horton considered the critical length to be the most important single factor in the development of stream networks, but its significance can perhaps be appreciated more fully on a local scale because it is highly sensitive to changes in the factors that control erosion. Assume, for example, that a farmer wishes to plant additional row crops in an area within the belt of no erosion. As the land is cleared of its original

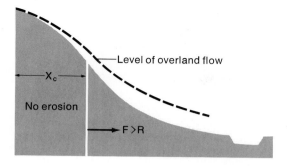

Figure 5.7.
Hypothetical slope showing overland flow. No erosion occurs until the force of overland flow (F) exceeds the resistance of the surface material (R). Upslope from that point no erosion occurs. X_c is the distance from the divide to point where erosion begins. (After Horton 1945. Courtesy of The Geological Society of America)

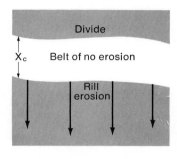

Figure 5.8.
Map view of critical length (X_c) and belt of no erosion. Rill erosion starts downslope from belt of no erosion when $F > R$. (Compare fig. 5.7)

vegetal cover and replaced by the crops, the resistance and infiltration capacity are drastically lowered. Under prevailing precipitation more runoff occurs, the force of overland flow increases, and X_c is shortened. The narrowing of the belt of no erosion allows rills and gulleys to form in the newly cropped region. Depending on their depth, these may make the area impassable for the heavy equipment needed for plowing and harvesting.

The analysis just presented is probably applicable in areas that are prone to the development of Hortonian overland flow. In humid-temperate regions, however, overland flow emanating from the upper slope areas is rare, and therefore the concept of a critical length has no particular significance. In cases where saturated overland flow dominates, the erosive action probably begins closer to the slope base (Kirkby and Chorley 1967; Kirkby 1969), and rills are gradually extended upslope. The headward growth most likely occurs because the headcut in any rill exposes a substratum that has a lower resistance to erosion than the slope surface. The formation of rills does not explain stream channels because a rilled surface is still part of a slope system. Their development, however, is the first necessary step toward a true river. It is now generally recognized that rills initiated on a slope cannot long remain as parallel unconnected channels. Deeper and wider rills develop where the length over which erosion can occur is the greatest. These master rills carry more water, and because of the greater depth, they undergo downcutting until all the flow is contained within the channel and the rill becomes a tiny stream. Because they become slightly entrenched, master rills capture adjacent rills when bank caving or overtopping during high flow destroys the narrow divides between them. The repeated diversion of rills, a process called *micropiracy,* tends to obliterate the original rill distribution, and gradually the initial slope parallel to the master channel is replaced by slopes on each side that slant toward the main drainage line.

The development of new slope direction in accordance with the master channel was called *cross-grading* by Horton (1945). In the final stage, only one stream, confined in the master rill channel, crosses the slope. The side slopes presumably develop a new rill system sloping to the position of the initial stream, and the process repeats itself, culminating in a secondary master rill serving as an incipient tributary. Each smaller tributary evolves in a similar way until the network of streams takes form.

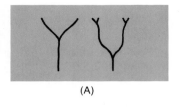 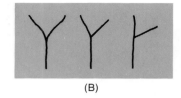 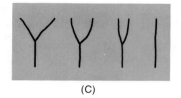

(A) (B) (C)

Figure 5.9.
Development of bifurcation angles (A) The original angle is preserved. (B) One branch becomes dominant. (C) Angle decreases and branches merge into one channel; occurs on steep slopes. (From Schumm 1956. Courtesy of The Geological Society of America)

The network pattern develops by repeated division of single channel segments into two branches, a process known as *bifurcation.* Schumm (1956) suggests that the angle between the limbs of a bifurcated channel probably evolves in one of three possible ways (fig. 5.9): (1) both limbs grow headward while preserving the original angle at their juncture; (2) one branch straightens its course and becomes dominant; or (3) the angle on steep slopes progressively decreases until the branches reunite into a single channel. Any or all of these procedures might be found in the evolution of a network, constrained only by the fundamental erosive controls and the geologic framework. Divides between adjacent basins are predetermined by the extent to which streams can expand headward. Because critical length varies with resistance, the areal extent of the uneroded uplands partially reflects the geology and its history. Within the basin itself, smaller interfluves may be present where cross-grading has not operated. These small areas parallel the main stream and preserve the slope of the original surface.

Basin Morphometry
One of Horton's greatest contributions was to demonstrate that stream networks have a distinct fabric, called the *drainage composition,* in which the relationship between streams of different magnitude can be expressed in mathematical terms. Each stream within a basin is assigned to a particular order indicating its relative importance in the network, the lowest order streams being the most minor tributaries and the highest order, the main trunk river.

Figure 5.10 shows several methods of ordering streams. Horton's cumbersome method was refined by Strahler (1952a) so that stream segments rather than entire streams become the ordered units. As the figure shows, a segment with no tributaries is designated as a first-order stream. Where two first-order segments join they form a second-order segment; two second-order segments a third-order segment, and so forth. Any segment may be joined by a channel of a lower order without an increase in its order; i.e., third-order segments may have an infinite number of second- or first-order tributaries. Only where two segments of equal magnitude join is an increase in order required. The apparent inconsistency in Strahler's method of not accounting for all tributaries is removed in the network analysis conceived by Shreve (1966a, 1967). He considers streams as links within the network, with the magnitude of each link representing the sum of the link numbers of all tributaries that feed it. That is, networks in which the downstream segments are of the same magnitude have equal numbers of links within their basins. Shreve's link system gives a number that at any point within the basin is equal to the number of first-order streams upstream from that point.

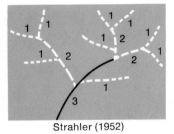
Horton (1945)

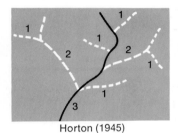
Strahler (1952)

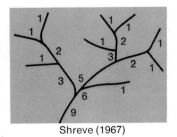
Shreve (1967)

Figure 5.10.
Methods of ordering streams within a drainage basin.

Table 5.2 Common morphometric relationships.

Linear Morphometry

Stream number in each order (N_o)	$N_o = R_b^{s-o}$
Total stream numbers in basin (N)	$N = \dfrac{R_b^s - 1}{R_b - 1}$
Average stream length	$\bar{L}_o = \bar{L}_1 R_L^{o-1}$
Total stream length	$L_o = \bar{L}_1 R_b^{s-1} \left(\dfrac{u^s - 1}{u - 1} \right) u = R_L/R_B$ where
Bifurcation ratio	$R_b = N_o/N_{o+1}$
Length ratio	$R_L = \bar{L}_o/\bar{L}_{o+1}$
Length of overland flow	$\ell_o = \dfrac{1}{2D}$

Areal Morphometry

Stream areas in each order	$\bar{A}_o = \bar{A}_1 R_a^{o-1}$
Length-area	$L = 1.4A^{0.6}$
Basin shape	$R_F = \dfrac{A_o}{L_b^2}$
Drainage density	$D = \dfrac{\Sigma L}{A}$
Stream frequency	$F_s = \dfrac{N}{A}$
Constant of channel maintenance	$C = \dfrac{1}{D}$

Relief Morphometry

Relief ratio	$R_h = H/L_o$
Relative relief	$R_{hp} = H/p$
Relative basin height	$y = h/H$
Relative basin area	$x = a/A$
Ruggedness number (Melton 1957)	$R = DH$

Adapted from Strahler 1958. Courtesy of the Geological Society of America.
s = order of master stream
o = any given stream order
H = basin relief
P = basin perimeter

Although the Strahler system of numbering was most commonly employed in earlier studies of basin morphometry, many researchers now find Shreve's link ordering conducive to highly sophisticated analyses that are well beyond the purpose of our treatment here (for example, see Smart and Wallis 1971; Abrahams 1980; Abrahams and Miller 1982). It is now generally recognized that in every basin a group of measurable properties exist that define the linear, areal, and relief characteristics of the watershed (table 5.2). These variables seem to correlate with stream orders, and various combinations of the parameters obey statistical relationships that hold for a large number of basins. Two general types of numbers have been used to describe basin morphometry or network characteristics (Strahler 1957, 1964, 1968). *Linear scale* measurements allow size comparison of topographic units.

The parameters may include the length of streams of any order, the relief, the length of basin perimeter, and many other measures. The second type of measurement consists of *dimensionless numbers,* often derived as ratios of length parameters, that permit comparisons of basins or networks. Length ratios, bifurcation ratios, and relief-length ratios are common examples. Table 5.2 gives the basic and most commonly used linear, areal, and relief equations, but many more relationships not shown have been derived from these.

Linear Morphometric Relationships   The establishment of stream ordering led Horton to realize that certain linear parameters of the basin are proportionately related to the stream order and that these could be expressed as basic relationships of the drainage composition. Much of linear morphometry is a function of the *bifurcation ratio* (R_b), which is defined as the ratio of the number of streams of a given order to the number of the next higher order. The primary use of the bifurcation ratio is to allow rapid estimates of the number of streams of any given order and the total number of streams within the basin. Although the ratio value will not be constant between each set of adjacent orders, its variation from order to order will be small, and a mean value can be used. Also, as Horton pointed out, the number of streams in the second highest order is a good approximation of R_b. When geology is reasonably homogeneous throughout a basin, R_b values usually range from 3.0 to 5.0.

The *length ratio* (R_L) is similar in context to the bifurcation ratio; it is the ratio of the average length of streams of a given order to those of the next higher order. The length ratio can be used to determine the average length of streams in an unmeasured given order ($\overline{L}_o$) and their total length (L_o). The combined length of all the streams in a given basin is simply the sum of the lengths in each order. For most basin networks, stream lengths of different orders plot as a straight line on semilogarithmic paper (fig. 5.11), as stream numbers also do.

The relationships between stream order and the number and length of segments in that order have been repeatedly verified and are now firmly established (Schumm 1956; Chorley 1957; Morisawa 1962; Chose et al. 1967; Selby 1967; and many others).

Areal Morphometric Relationships   The equity among linear elements within a drainage system suggests that areal components should also possess a consistent morphometry, since dimensionally area is simply the product of linear factors. The fundamental unit of areal elements is the area contained within the basin of any given order (A_o). It encompasses all the area that provides runoff to streams of the given order, including all the areas of tributary basins of a lower order as well as interfluve regions. Schumm (1956) demonstrated (fig. 5.12) that basin areas, like stream numbers and lengths, are related to stream order in a geometric series.

Although area by itself is an important independent variable (Murphey et al. 1977), it has also been employed to manifest a variety of other parameters (shown in table 5.2), each of which has a particular significance in basin geomorphology. One of the more important factors involving area is the

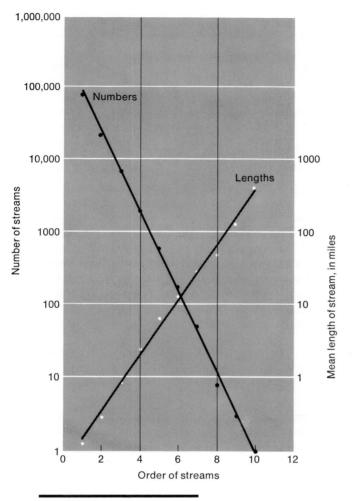

Figure 5.11.
Relation of stream order to the
number and mean lengths of
streams in the Susquehanna River
basin. (After Brush 1961)

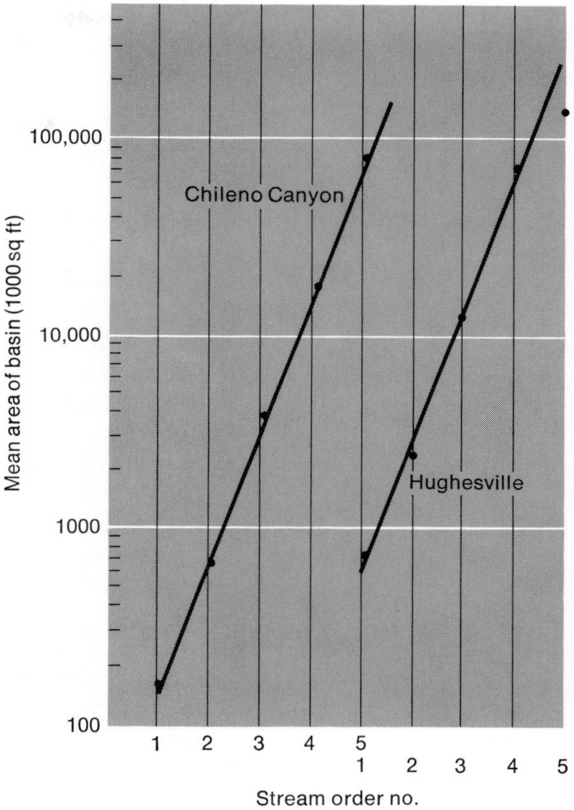

Figure 5.12.
Relationship between stream
order and mean basin area in two
drainage basins. (After Schumm
1956. Courtesy of The Geological
Society of America)

drainage density (D), which is essentially the average length of streams per
unit area and as such reflects the spacing of the drainageways. The drainage
density is of interest because it is directly controlled by the interaction between geology and climate. As these two factors differ from region to region,
wide variations in D can be expected (table 5.3). In general, resistant surface
materials or those with high infiltration capacities have widely spaced streams
and, consequently, low drainage densities. As resistance or surface permeability decreases, runoff is usually removed in a greater number of closely
spaced channels, and D tends to be much higher. As a rule of thumb, where
geology and slope angles are the same, humid regions develop a thick vegetal

Table 5.3 Drainage density in regions with different geology and climate.

Drainage Density[a]	Climate	Geology	Area
3–4	Humid-temperate	Resistant sandstone, flat-lying	Appalachian Plateau
8–16	Humid-temperate	Nonresistant, flat-lying rocks	Central-Eastern U.S.
20–30	Dry summers— subtropical, seasonal	Fractured and weathered igneous and metamorphics	Southern California
50–100	Semiarid	Variable lithology and structure	Rocky Mountains
200–400	Arid-semiarid	Flat-lying, non-resistant sedimentary rocks	Badlands, S. Dakota
1100–1300	Humid-temperate	Weak clays	Northern New Jersey

Data from Strahler 1968 and Schumm 1956.
[a]In miles/sq. mi.

mat that increases resistance and infiltration and thereby perpetuates a lower drainage density than would be expected in more arid basins. Thus, drainage density not only reflects the geologic framework, it may serve as a useful parameter in climatic geomorphology (Daniel 1981). Methods for rapid estimation of drainage density have been devised (McCoy 1971; Mark 1974; Richards 1979; Bauer 1980).

The density factor is also related to a parameter known as the *texture ratio* (T):

$$T = N/P,$$

where N is the number of crenulations on the most irregular contour and P is the length of the basin perimeter. Ratios determined for many areas show the mean texture ratio to be a useful descriptive number. A region with a coarse texture has a mean value < 4, medium texture $4-10$, and fine texture > 10; figure 5.13 shows a fine-textured region where drainage density is high. Drainage density is related to the texture ratio as a simple power function

$$D = aT^b,$$

where a and b are constants. The texture ratio can therefore be used as a substitute to indicate relative density values.

The drainage density not only reflects geology and climate, but it also has been used as an independent variable in the framing of other morphometric parameters. For example, the *constant of channel maintenance* and the *length of overland flow* (table 5.2) both utilize a reciprocal relationship with density to demonstrate the link between factors that control surface erosion and those that describe the drainage net (Schumm 1956). The constant of channel maintenance indicates the minimum area required for the development and maintenance of a channel; that is, the ratio represents the amount of basin area needed to maintain one linear unit of channel length. As Schumm points out (1956, p. 607) this relationship requires that drainage networks develop in an

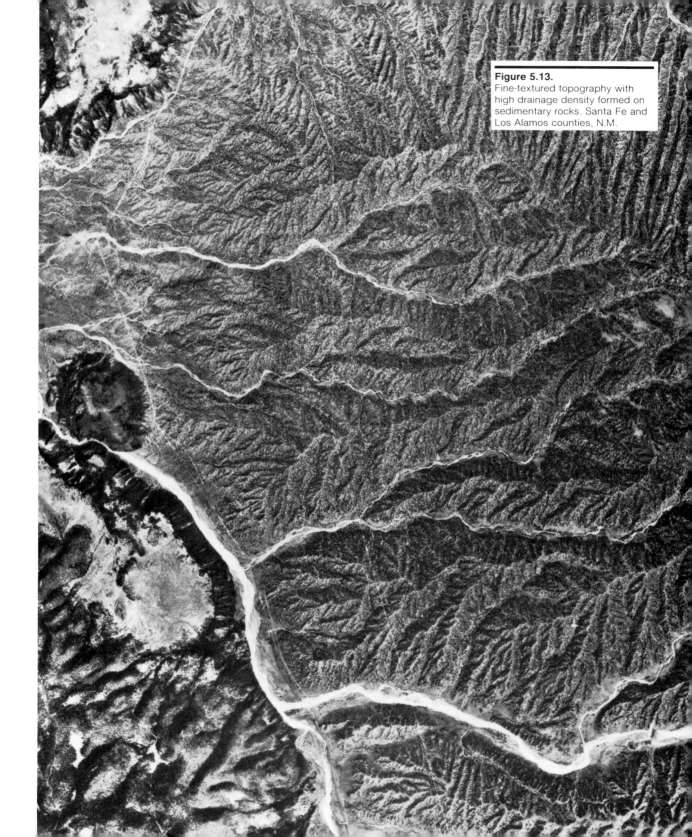

Figure 5.13.
Fine-textured topography with high drainage density formed on sedimentary rocks. Santa Fe and Los Alamos counties, N.M.

orderly way because the meter by meter growth of a drainage system is possible only if sufficient area is available to maintain the expanding channels. The *ruggedness number* (drainage density × basin relief) is another parameter that employs drainage density; it is useful in relating morphometry to flood peak discharge (Patton and Baker 1976).

Relief Morphometric Relationships   A third group of parameters shown in table 5.2 is used to indicate the vertical dimensions of a drainage basin; it includes factors of gradient and elevation. Like stream numbers, length, and area, the average slope of stream segments in any order approximates a geometric series in which the first term is the mean slope of first-order streams. This relationship is reasonably valid as long as the geologic framework is homogeneous. Channel slopes and surface slopes are closely akin to the parameters for length. Horton suggested, for example, that the length of overland flow as a function of only the drainage density is at best an approximation because overland flow also depends on slope parameters.

As relief refers to elevation differences between two points, slopes that connect the points are integral factors. The choice of reference points differs, but the most useful relief parameters are the *maximum relief* (highest elevation in the basin–lowest elevation in the basin) or the *maximum basin relief* (highest elevation on the basin perimeter–the elevation at the mouth of the trunk river). The *relief ratio* (Schumm 1956), the maximum basin relief divided by the longest horizontal distance of the basin measured parallel to the major stream, indicates the overall steepness of the basin.

A different relief study is the *hypsometric analysis* (Strahler 1952b), which relates elevation and basin area. As figure 5.14 shows, the basin is assumed to have vertical sides rising from a horizontal plane passing through the basin mouth and under the entire basin. Essentially, a hypsometric analysis reveals how much of the basin is found within cross-sectional segments bounded by specified elevations. The relative height (y) is the ratio of the height (h) of a given contour above the horizontal datum plane to the total relief (H). The relative area (x) equals the ratio a/A, where a is the area of the basin above the given contour and A is the total basin area. The hypsometric curve (fig. 5.14B) represents the plot of the relationship between y and x and simply indicates the distribution of mass above the datum. The form of the curve is produced by the *hypsometric integral* (HI), which expresses, as a percentage, the volume of the original basin that remains. In natural basins most HI values range from 20 to 80 percent, the higher value indicating that large areas of the original basin have not been altered into slopes. Low integral values simply mean that much of the basin stands at low elevation relative to the area of the original upland surface. One objection to the hypsometric analysis is the tedium involved in determining the integral, but several alternative methods of calculation (Chorley and Morley 1959; Haan and Johnson 1966; Pike and Wilson 1971) have removed much of this difficulty.

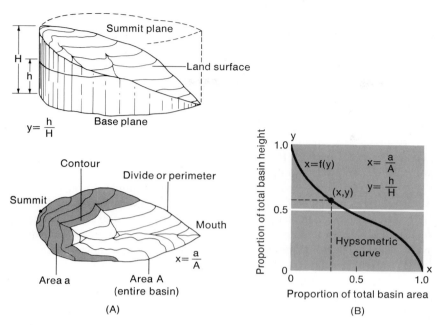

Figure 5.14.
Ingredients of a hypsometric analysis. (A) Diagram showing how dimensionless parameters used in analysis are derived. (B) Plot of the parameters to produce the hypsometric curve. (From Strahler 1952. Courtesy of The Geological Society of America)

Basin Evolution

Although morphometric values differ from basin to basin, each network still obeys the same statistical relationships discussed above. Many authors have suggested that morphometry reflects an adjustment of geomorphic variables that is established under the constraints of the prevailing climate and geology (Chorley 1962; Leopold and Langbein 1962; Strahler 1964; Doornkamp and King 1971; Woldenberg 1969; and many others). Essentially this means that once a network is established, the basinal characteristics can be defined by the same quantitative terms at any time during the drainage growth. As the basins and networks evolve, an equilibrium is eventually produced by the interplay of climate and geology and maintained as a time-independent phenomenon. Once the components within a basin become balanced, any changes of climate or geology will be compensated for by adjustments of the basin parameters in such a way that the relationships of drainage composition will be preserved. As originally conceived, however, these relationships issued from well-developed stream systems, and the measurements needed to derive the equations were made on topographic maps of these basins. Such an approach provides no insight as to how quickly morphometric balance is attained or what changes in its character occur as the basin ages. It seems appropriate, therefore, to consider the influence of time on the morphometry of a basin.

Some studies have provided a glimpse into the question of how rapidly morphometry is established and what changes occur in its nature as the basin evolves. There seems to be little doubt that a quantitatively balanced drainage net forms rapidly in erodible material. This was clearly demonstrated by Schumm (1956) in the Perth Amboy (N.J.) badlands, by Morisawa (1964) on the uplifted floor of Hebgen Lake (Montana), and by Kirkby and Kirkby (1969) on the raised beach and harbor floor around Montague Island (Alaska). These studies showed that a balanced drainage composition had evolved in a period no longer than a few years (in the Alaska case, only a few days). Certainly, the time needed for drainage development to occur in erodible material is insignificant in terms of geological time. However, the amount of time needed in absolute terms probably varies according to the resistance of the material, the climate, and the initial slope angles. Unfortunately, examples of drainage establishment are documented only in areas where the least resistant materials underlie the system. Precisely how long it takes to form a balanced network in regions underlain by resistant crystalline rocks is rather conjectural.

Drainage evolution has also been examined experimentally. Parker (1976, discussed in Schumm 1977), using the Rainfall-Erosion Facility at Colorado State University, found that patterns grow by headward extension until dissection reaches the watershed margin. Drainage density increases during this phase, especially along the outside of the expanding drainage. Further evolution, however, revealed that density achieves an equilibrium value maintained because tributaries near the center of the basin are lost while new tributaries form and extend near the watershed margin. The total drainage density and the mode of growth depended on the slope of the basin. The basin with a lower slope had a higher density because tributaries formed inside the basin during the early extension growth, while in the steeper basin, streams extended to the margins before interior links were added. In general, Parker's work showed that the network will develop as many streams and as much length as is needed to efficiently drain the basin. However, during the period of growth, drainage density may change, and in different parts of the basin, the change may be in opposite directions.

Once basin elements attain a statistical balance, further changes in morphometry are usually revealed in the shape, drainage texture, or hypsometry of the basin. The area limits of any basin presumably are determined by the hydrophysical controls denoted by Horton and by the competition for space between adjacent basins. During the period of expansion to its peripheral limits, however, each parameter should change in such a manner as to maintain the original quantitative relationships among the factors. Hack (1957) showed that for a large number of basins, the stream length and basin area are related by the simple power function

$$L = 1.4\, A^{0.6},$$

where L is the distance from any locality on a stream to the divide at the head of the longest segment above the given locality, and A is the basin area above the given locality. Hack noted that to preserve the original geometric balance, each variable must change at the same rate and the exponent in the equation should be 0.5. The larger exponent he observed requires that basins become more narrow and elongate as they grow. Elongation is very striking in the growth of parallel drainage patterns, as exemplified in the postglacial history of the Ontonagon region of northern Michigan (Hack 1965). There the network formed as the Ontonagon Plain was exposed during the destruction of

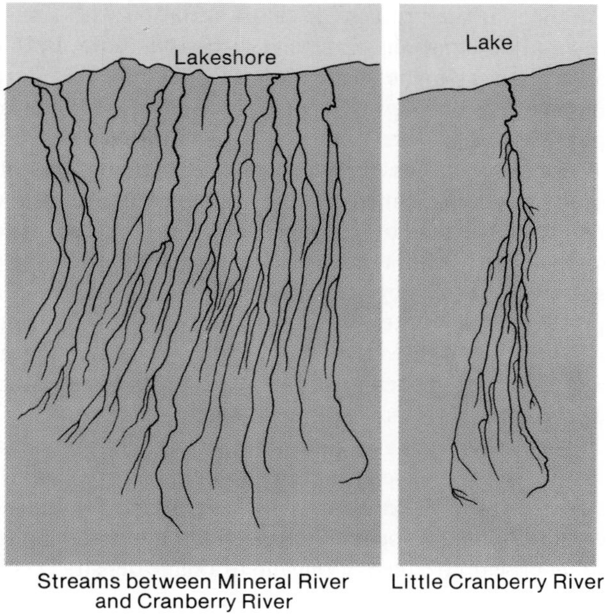

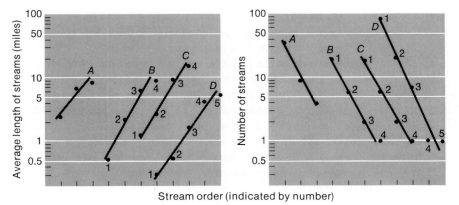

Figure 5.15.
Growth of two parallel drainage patterns in the Ontonagon region, northern Michigan. (After Hack 1965)

Lakeshore

Lake

Streams between Mineral River and Cranberry River

Little Cranberry River

Figure 5.16.
Relation of stream order to stream length and number in drainage basins of the Ontonagon Plain, Mich. *A,* Streams between Mineral and Cranberry rivers; data from maps; stream orders not known. *B,* Little Cranberry River. *C,* Weigel Creek. *D,* Mill Creek. (From Hack 1965)

Stream order (indicated by number)

glacial Lake Duluth. Initial stream development followed parallel grooves that extended down the plain to the lake margin (fig. 5.15), and although this accentuated the basin elongation, figure 5.16 demonstrates that the lengths and areas of every order follow the 0.6 power function recognized elsewhere.

The same tendency has been noted in arid basins (Miller 1958). Thus, the trend toward elongation has probably been noted in enough widely divergent settings to suggest that it is an inherent property in the growth of drainage basins (Jarvis and Sham 1981). Some studies, however, suggest that as small basins evolve into large ones, they reach a stage where they widen as fast as they elongate. The exponent then will reach a value that approximates 0.5 (Mueller 1972; Shreve 1974; Moon 1980).

Part of the problem of shape variation with time is similar to the experimental observations of density discussed above. Both of these factors may be associated with the suggestion by Schumm (1956) that the region of most intense erosion migrates with time, being near the mouth during the early phase of basin development and near the headward margin in later stages. These observations demonstrate that various parts of a large basin may be evolving at different rates. The drainage texture also seems to vary with time. As demonstrated by Ruhe (1952), both drainage density and stream frequency increase systematically in areas underlain by glacial deposits of progressively increasing age. The rate of the textural change is not constant; it probably was greatest during the first 20,000 years and then decelerated. The marked transition in the rate of textural evolution perhaps represents the time at which complete equilibrium was established and the basins were filled with as many streams as possible (Leopold et al. 1964).

It is interesting to note that throughout the period of growth in Ruhe's study, the channel lengths and numbers seem to obey Horton's geometric laws, suggesting that texture may be a surrogate for time in the analysis of basin evolution. This proposition was given added credence when Melton (1958) found stream frequency (F) and drainage density (D) to be related by a simple equation

$$F = 0.694\ D^2.$$

In deriving this equation, Melton used as his sample 156 basins with widely divergent geology, erosional histories, and presumably, ages. Since all stages of drainage development were thrown into the statistical pot, the significant empirical relationship between the variables most likely reflects a general trend followed by basins as they grow. In fact, it was suggested that the dimensionless ratio F/D^2, called the *relative density,* should indicate how completely the stream network fills the basin (Melton 1958). High relative density values suggest that stream lengths are short and the basin outline is not yet completely filled with the stream network. As the drainage evolves and expands into each basin niche, the ratio decreases until equilibrium is established.

It has been commonly accepted that Melton's growth law is an example in which spatial parameters can be substituted for time. However, the universal applicability of the growth equation has been questioned. Abrahams (1972) suggested that space and time are interchangeable in morphometry only if the basins analyzed are environmentally similar and of the same order. Further studies, however, did not completely substantiate this premise (Wilcock 1975), and a reasonable argument could be made that all factors—linear, areal, and relief—are probably involved in the statistical relationships affected by time.

Even though it may be necessary to include all types of factors in the derivation of growth laws, it is clear that alterations of basin morphometry do occur with time. These changes do not violate the steady-state concept because parameters vary with one another in a systematic way. It is tempting to assume, therefore, that a well-balanced network with discernible morphometry

Figure 5.17.
Map of terrace deposits in a small piedmont area of the Beartooth Mountains, Mont. Major stream piracy occurred near   A between deposition of Burnett Ranch gravels and deposition of Luther gravels. (From Ritter 1972)

evolved in an orderly manner from infancy to its present state. Realistically, such an assumption may be totally erroneous because basin morphometry may tell us little, if anything, about basin *history.*

A cogent example of this is the drainage system of Volney Creek, a small basin in southern Montana within the watershed of the Yellowstone River. Volney Creek heads in the piedmont region of the Beartooth Mountains, rising approximately 3 km from the mountain front. Its basin, covering an area of 60 km², is underlain entirely by Mesozoic and Tertiary clastic sedimentary rocks. The basin exhibits all the normal morphometric characteristics of nearby piedmont basins, and it is presumed to be in equilibrium because its morphometry is balanced statistically as discussed above. The geomorphic history of the basin, however, shows that its evolution was anything but orderly (Ritter 1972). Two terraces standing well above the present level of Volney Creek are capped by gravel containing crystalline clasts that had to be derived from the Beartooth Mountains; in fact, upstream tracing of the terraces demonstrates that they cross through the present basin divide and continue mountainward in the neighboring valley of West Red Lodge Creek (see map, fig. 5.17). The significance is that prior to early Wisconsinan glaciation (Burnett Ranch time in fig. 5.17), the valley now occupied by Volney Creek was the drainage avenue for the master stream of the area. The crystalline-rich gravel capping the terraces indicates that the paleoriver drained from the mountains, and its basinal area must have been vastly greater than that of the modern Volney

Creek. Immediately preceding the glaciation, the river was diverted into its present position in the valley of West Red Lodge Creek, leaving the Volney segment abandoned until it was occupied by the very small modern river.

Diversions of the Volney Creek type are common in piedmont regions and may be important phenomena in the expansion of any drainage system, especially in the early stages (Howard 1971). Such changes, nonetheless, are catastrophic events in basin development because they drastically alter basin properties such as area, relief, and stream length. The internal adjustment to changes spurred by piracy must occur rapidly, for most basins possess a balanced morphometry even though such spasmodic events must be commonplace.

Basin Hydrology

It is important for planning purposes to estimate how much water exists in a drainage basin and whether it is available for use. Hydrologists usually employ a concept known as the **water balance** or **hydrologic budget** to make such estimates. The water balance simply refers to the balance that must exist between all water entering the basin (input), all water leaving the basin (output), and changes in the amount of water being stored. The major types of input are rainfall and snow, and the major outputs are in the form of evapotranspiration and streamflow. Water is stored as soil moisture and groundwater, and changes in these values actually represent losses or gains of available water. For example, a positive change in groundwater storage indicates that the underground reservoir is being recharged, but in the budget it represents a loss because the availability of the water is being lowered. On the other hand, groundwater runoff (base flow) represents an input because water availability is increased as it is released from storage.

Underground Water

The increased demand for water that accompanied population growth, industrial expansion, and extended irrigation in the United States has brought with it a marked increase in the utilization of groundwater. The volume of water contained in cracks and pores in the Earth's underground reservoir is enormous, being estimated at almost 8 million km^3 in the outer 5 km of the crust (Todd 1970). Unfortunately, this vast resource is not evenly distributed, and some regions are blessed with abundant groundwater while others are seriously deficient. Ironically, many of the most rapidly expanding areas of the United States are in regions with low reserves of surface and groundwater. The lack of available water, coupled with a growing demand, poses a real challenge to geologists and requires that we continue to expand our knowledge concerning the distribution, movement, and utilization of groundwater. Groundwater geology is a separate discipline in itself; however, because it is an important part of basin hydrology, we must briefly examine how groundwater systems work.

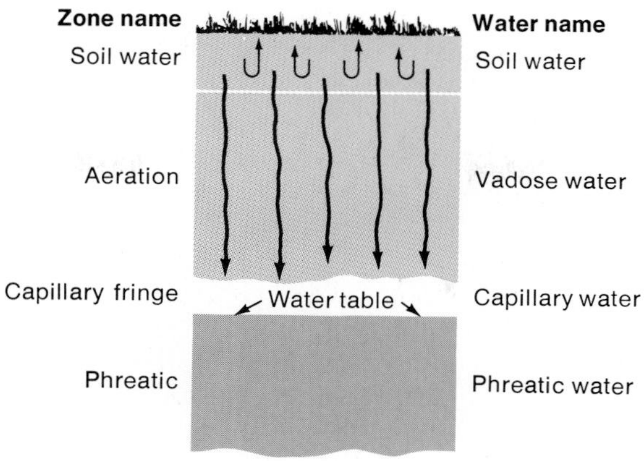

Figure 5.18.
Schematic diagram of zones and water types in the groundwater profile. (Not to scale)

The Groundwater Profile   Groundwater in porous and permeable rocks or unconsolidated debris usually has a rather distinct distribution that can be visualized as a vertical zonation known as the *groundwater profile,* shown in figure 5.18. Below the *zone of soil moisture,* water moves downward under the influence of gravity into and through the *zone of aeration.* In this zone, pore spaces are partly occupied by water, called *vadose water,* and partly by air that is physically connected to the atmosphere. At some lower depth all the pore spaces are occupied by water, marking the beginning of a thin, saturated zone, the *capillary fringe,* where water, held in tension, will not drain freely into a well. Beneath the capillary fringe the spaces are also filled with water in a zone known as the *phreatic zone.* The distinction between these two saturated zones is that phreatic water will drain freely into a well because the hydrostatic pressure in the phreatic zone is greater than atmospheric pressure. The *water table* marks the top of the phreatic zone and represents the level at which the hydrostatic pressure is equal to the atmospheric pressure.

Movement of Groundwater   After descending to the phreatic zone, groundwater does not stagnate but is capable of further movement. Unlike flow in the zone of aeration, however, movement in the phreatic zone is not entirely controlled by gravity. Each unit volume of water contained in the phreatic zone possesses a certain amount of potential energy, called its *potential* or *head.* The amount of potential varies from droplet to droplet, being dependent on each drop's pressure and elevation above some datum. When pressure and elevation are known, the potential for each unit volume of water can be calculated, and water particles having the same potential can be contoured along surfaces known as *equipotential surfaces* (fig. 5.19). Although some diffusion occurs, groundwater particles move along paths that are perpendicular to the equipotential surfaces.

Figure 5.19.
Movement of groundwater according to distribution of potential in the underground system. Water moves from high to low potential and perpendicular to the equipotential surfaces.

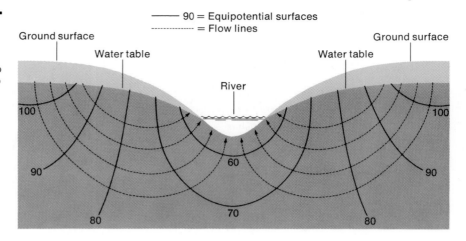

Figure 5.20.
Water table and loss in head as water moves from well 1 to well 2 in unconfined aquifer.

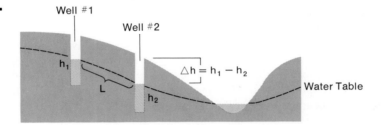

The movement of groundwater is a mechanical process whereby some of the initial potential energy of the water is lost to friction generated as the water moves. It follows that water moving from one point to another must have more potential energy at the beginning of the transport route than at the end. Thus, combining the above, groundwater always moves according to the following rules: (1) it moves from zones of higher potential towards zones of lower potential, and (2) it flows perpendicular to the equipotential surfaces (Hubbert 1940).

The velocity and discharge of groundwater flow is directly proportional to the loss of potential (head) that occurs as water moves from one point to another (fig. 5.20). This was demonstrated by Darcy in his experiments on flow through permeable material and led to his law of groundwater flow expressed as

$$V = K \frac{h_1 - h_2}{L},$$

where V is velocity, h is the head, L is the distance between points 1 and 2, and K is a constant of proportionality representing the permeability (or the *hydraulic conductivity*) of the medium. In figure 5.20, the velocity of flow can be calculated from the difference in hydrostatic level between two wells

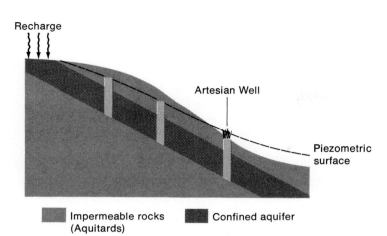

Figure 5.21.
Confined aquifer and piezometric surface.

$(h_1 - h_2)$ if the permeability of the material is known. Discharge can also be calculated by adding the cross-sectional area to the equation so that

$$Q = KIA,$$

where Q is discharge, A is the cross-section area of an aquifer ($w \times d$), and I, called the *hydraulic gradient*, equals $\dfrac{h_1 - h_2}{L}$.

As mentioned above, the top of the phreatic zone is an important hydrologic feature known as the water table (fig. 5.18). Because hydrostatic pressure everywhere along the surface of the water table is equal to atmospheric pressure, the potential of water there is completely a function of elevation, and water at that surface will always move from higher to lower elevations. This partially explains why the water table is generally a mirror image of surface topography. Where surface water and groundwater systems are physically connected, the levels of rivers, lakes, swamps, etc., are merely surface extensions of the underground water table (fig. 5.19).

Aquifers, Wells, and Utilization Problems   *Aquifers* are lithologic bodies that store and transmit water in economic amounts. The most common aquifer is called an *unconfined aquifer* because it is open to the atmosphere and its hydrostatic level (the level at which water stands in an open hole) is within the water-bearing unit itself. The hydrostatic level in an unconfined aquifer is the water table. In some other aquifers the water is held in a porous and permeable unit that is not connected vertically to the atmosphere but is overlain and underlain by impermeable layers called *aquitards*. Water contained in these *confined aquifers* (fig. 5.21) will rise above the top of the aquifer when it is penetrated by an open hole. The level to which water will rise is called the *piezometric surface*, and its height above the aquifer itself depends on the difference in potential at the point where precipitation enters the aquifer (recharge zone) and the position of the hole (fig. 5.21). Sometimes the piezometric level is above the elevation of the ground surface, and water will flow freely out of the hole without pumping as *artesian flow.*

Figure 5.22.
Successive cones of depression caused by drawdown of water table during 72-hour pumping of unconfined aquifer.

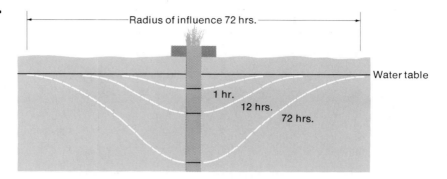

The development of an aquifer for water supply requires wells; the larger the demand, the larger and more numerous the wells. As a well is pumped, the hydrostatic level (water table or piezometric surface) surrounding the well is molded into an inverted cone known as a *cone of depression* (fig. 5.22). The cone develops because the release of water (or of pressure in a confined aquifer) is greatest near the well, causing pronounced lowering of the hydrostatic level, called *drawdown,* adjacent to the well. The effect of pumping decreases away from the well, and so the drawdown is less at the perimeter of the cone. In the initial phase of pumping, the rate of drawdown is high, but as pumping continues the rate gradually decreases until the cone attains a nearly constant form. The dimensions of the quasi-equilibrium cone depend on the rate at which the well is pumped and the hydrologic properties of the aquifer.

Utilizing groundwater can result in a number of environmental problems if the system is not carefully studied before development. For example, excessive drawdown can occur locally if wells are placed too close to one another and the radius of influence (maximum diameter of the cone of depression) of adjacent wells overlaps. This produces abnormally high drawdown in the zone of overlapping because the actual lowering of the water level is the sum total of drawdown produced by all the interfering wells. On a regional scale, most problems arise when more water is pumped from the aquifer over a period of years than is returned to the aquifer by natural or artificial recharge, a practice known as *overdraft.* In such a case the aquifer is actually being mined of its water, and on a long-term basis the normal hydrostatic level may be drastically lowered.

The effects of continued overdraft differ, but two types of responses will demonstrate the problems that can result from misuse of the groundwater system. First, in confined aquifers, the drawdown of the piezometric surface reflects the decrease of pressure within the aquifer. Because the pressure is lowered, water from the overlying aquitard seeps downward into the aquifer. As the aquitard drains, the normal load exerted by the weight of the overlying rocks compacts the aquitard and decreases its thickness. Ultimately, the process culminates in measurable subsidence of the ground surface. Some areas (Mexico City, Las Vegas, Central Valley, California, Houston–Galveston area) have experienced 1 to 5 meters of overdraft subsidence, creating a variety of annoying and hazardous conditions such as cracking of buildings, strain on buried pipelines, and destruction of well casings.

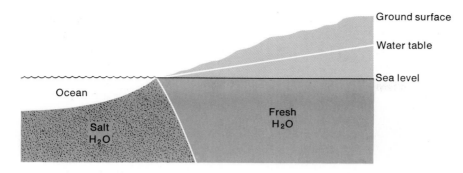

Figure 5.23.
Relationship of salty and fresh groundwater in an aquifer of a coastal region. Lowering of water table by overdraft requires a 40-fold rise in the boundary between the fresh and salt water and increases the possibility of pollution of the aquifer by salt water.

A second major effect of overdraft usually occurs in coastal regions where the aquifer is physically connected to the ocean. There two fluids (ocean water and fresh water) having different densities are separated by a sharp boundary, as diagrammed in figure 5.23. The location of the interface depends on the hydrodynamic balance between fresh water (density = 1.000 g/cm³) and salt water (density = 1.025 g/cm³). In general, the depth below sea level to the saltwater boundary is about 40 times the height of the water table above sea level. In such a situation, continued overdraft can cause pollution of the aquifer because minor lowering of the water table necessitates a much greater rise of the saltwater–freshwater interface, a phenomenon known as *saltwater intrusion*. For example, a 2 m drawdown of the water table requires a concomitant 80 m rise of the salt water, and any well extending to a depth greater than the new interface level will be polluted with nonpotable water.

Many cities along the coasts of California, Texas, Florida, New York, and New Jersey have been affected by saltwater intrusion. The phenomenon, however, can function wherever two fluids of different density exist within the same aquifer. For example, the water supply of Las Vegas, Nevada, is in jeopardy from pollution by high-magnesium groundwater located about 24 km south of the city. The impending intrusion is due to the large overdraft from the aquifer beneath Las Vegas.

Surface Water

The major loss of water from most drainage basins occurs as stream discharge, that is, the volume of water passing a given channel cross section during a specified time interval, or

$$Q = wdv,$$

where Q is discharge in ft³/sec(cfs) or m³/sec(cms), w is width, d is depth, and v is velocity. Measurement of discharge is a relatively simple procedure whereby the total width of the channel is divided into evenly spaced segments. The depth and velocity are measured in each compartment, and total discharge is determined by summing the discharges of all the subsections. The difficult measurement to obtain is velocity. Surface velocity is not a good estimate of the mean velocity, and so measurements must be made within the current. Many devices have been developed to measure velocity, but probably

Figure 5.24.
Photo of Price Current Meter.

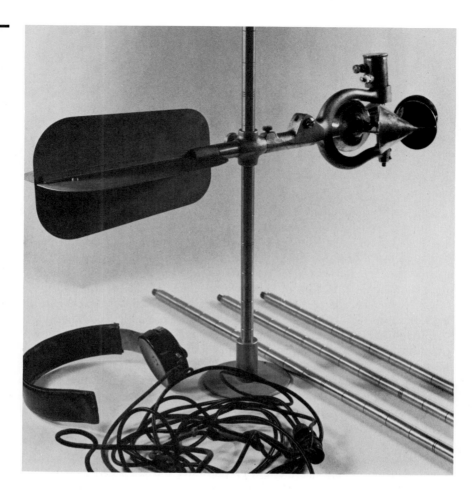

the most widely used is the Price current meter. This instrument, designed by W. G. Price in 1882, has a group of conical cups mounted on a vertical rod that is rotated by the force of the water striking the cups (fig. 5.24). As the shaft spins, it periodically closes an electrical circuit, the moment of closure being recognized by the meter operator as a clicking sound in earphones connected to the circuitry. The number of axial revolutions per unit time is easily converted into velocity values.

To be of any scientific value, discharge must be measured repeatedly at the same locality, giving hydrologists a better understanding of how flow varies with time. In the United States these sampling localities, called *gaging stations,* have been in operation since the late 1800s; therefore, a wealth of flow data for a large number of streams is available from the U.S. Geological Survey and many state water surveys. At most stations, discharge is measured every 15 minutes and the average of these values published as a *mean daily discharge.* The *mean annual discharge* is the average of the daily values over the entire period of record, presuming that the station has been maintained for

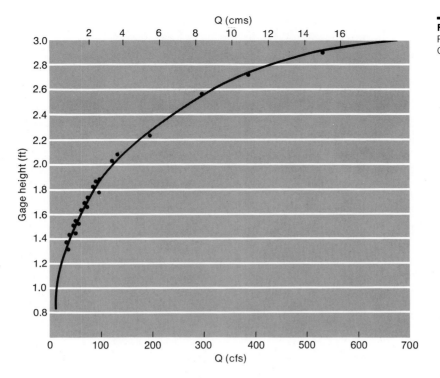

Figure 5.25.
Rating curve for low flow, Rock Creek near Red Lodge, Mont.

longer than one year. Actually, discharge is not measured as described above because the procedure is too time-consuming. Instead it is estimated from a **rating curve** (fig. 5.25), which relates a wide spectrum of discharge values to the elevation of the river above some datum. Once the rating curve is constructed, the height of the river above the datum (called the *stage* or *gage height*) is the only variable observed directly, and its value is used to predict the discharge.

Geomorphologists are interested in the frequency and magnitude of flow events because each has a decided bearing on how watershed systems work. The frequency of a given discharge is compiled into a **flow duration curve** (fig. 5.26), which relates any discharge value to the percentage of time that it is equaled or exceeded. At any station, then, the lowest daily discharge in the period of record will be equaled or exceeded 100 percent of the time. The largest flow will be equaled only once out of the entire number of days in the sample, giving it a percentage value slightly greater than zero.

Another common approach to finding the frequency-magnitude relationship is to consider only the peak discharge during each year (*annual series*) or only the discharges above some predetermined value (*partial duration series*). The statistical samples in the series approaches are much smaller than in analyses utilizing daily records; however, they are useful in studies of major

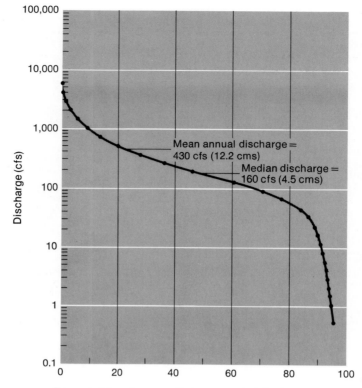

Figure 5.26.
Flow duration curve for the Powder River near Arvada, Wyo., 1917–1950. (From Leopold and Maddock 1953)

flow events. The annual discharges shown in table 5.4 have been ranked according to their magnitude during the years of record, and a recurrence interval for each flow is shown. The **recurrence interval,** simply the average time between two flow events of equal or larger magnitude, is calculated as

$$R = \frac{n + 1}{m},$$

where R is the recurrence interval in years, n is the total number of discharge values in the sample, and m is the rank of a given flow.

A plot of discharge and recurrence intervals on probability graph paper (fig. 5.27) allows hydrologists to estimate the magnitude of a flood *to be expected* within a specified interval of time. Various types of probability graphs have been used in these analyses (lognormal, Gumbel Type I, Gumbel Type III, Pearson Type III), but no one type is inherently better than the others (see discussion in Dunne and Leopold 1978 and Benson 1971).

On Rock Creek (Mont.), as figure 5.27 shows, the flow should equal or exceed 81 cms (2860 cfs) once during each 25-year interval. As rivers unfortunately do not understand statistical theory, there is no reason to believe that 25-year floods will be evenly distributed over time. In the next 50 years there may be two 25-year floods in successive years, or floods may occur in the first and last years of the period, or a flow equaling or exceeding the discharge of

Table 5.4   Annual flood series data, 1932–1963, for Rock Creek near Red Lodge, Mont. Each maximum annual flood is ranked, the highest being ranked 1. Recurrence intervals have been calculated.

Year	Maximum Flood		Rank Magnitude (m)	Recurrence Interval (R)
	cfs	cms		
1932	935	26.5	22	1.45
1934	533	15.1	31	1.03
1935	1490	42.2	9	3.56
1936	1240	35.1	15	2.13
1937	1930	54.6	5	6.4
1938	991	28.0	21	1.52
1939	661	18.7	29	1.10
1940	673	19.0	28	1.14
1941	780	22.1	24	1.33
1942	1840	52.1	6	5.33
1943	2010	56.9	3	10.6
1944	1630	45.1	7	4.57
1945	1990	56.3	4	8
1946	774	21.9	25	1.28
1947	846	23.9	23	1.39
1948	1070	30.3	19	1.68
1949	1190	33.7	17	1.88
1950	1100	31.1	18	1.78
1951	1460	41.3	10	3.2
1952	2590	73.3	2	16.0
1953	1300	36.8	13	2.46
1954	1570	44.4	8	4.0
1955	578	16.4	30	1.06
1956	1430	40.5	11	2.90
1957	3110	88.0	1	32
1958	1250	35.4	14	2.29
1959	1200	34.0	16	2.0
1960	680	19.2	27	1.19
1961	751	21.3	26	1.23
1962	1030	29.1	20	1.60
1963	1350	38.2	12	2.67

(Handwritten annotation to the right of the table:)

$m = 7 \quad n = 31$

$R = \dfrac{31 + 1}{7} = \dfrac{32}{7} = 4.57$

a 25-year flood may not happen at all. However, the probability that a flow of a particular magnitude will occur in any year is the reciprocal of the recurrence interval

$$P = \frac{1}{R},$$

where P is the probability of flow being equaled or exceeded in any one year and R is the recurrence interval. Thus, a 25-year flood has a 4 percent chance of occurring in any given year; a 10-year flood has a 10 percent chance, and so on. Exactly which year it will happen is simply unpredictable.

Even though it is impossible to predict when a given flow will occur, magnitude-frequency analyses have practical value in river management, especially for lower frequency floods such as the 20-year (q_{20}), 10-year (q_{10}), or 5-year (q_5) floods. Using an annual series, the *mean annual flood,* which is

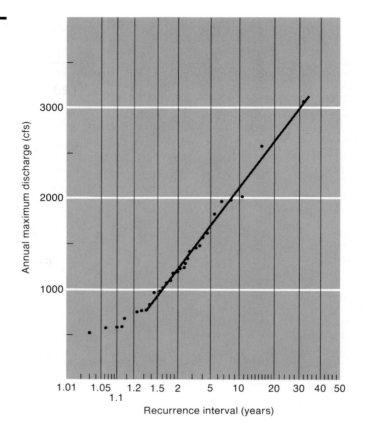

Figure 5.27.
Flood frequency curve for Rock Creek near Red Lodge, Mont.

the arithmetic mean of all the maximum yearly discharges in the sample, is the flow that should recur once every 2.33 years ($q_{2.33}$). In our Rock Creek example the mean annual flood has a discharge of 1290 cfs and plots at a recurrence interval of approximately 2.33 on our probability curve (fig. 5.27).

When longer time intervals are involved in watershed management, it may be important to know the probability of a particular flood's occurring during the projected life of a design structure. This is normally estimated by the probability equation

$$q = 1 - \left(1 - \frac{1}{T}\right)^n,$$

where q is the probability of a flood with a recurrence interval (T) occurring in the specified number of years (n). Thus, the chance of a 50-year flood's occurring in the next 50 years is 63 percent (Costa and Baker 1981).

The U.S. Geological Survey publishes extensive flood-frequency information on a state and regional basis. Techniques for deriving flood frequency curves from gage data or for estimating maximum probable floods at any site have been described in detail (Dalrymple 1960).

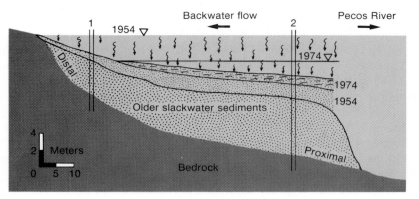

Figure 5.28.
Schematic of on- and off-lap sequences and peak flood stage in a tributary valley for the 1954 and 1974 floods on the Pecos River, Texas. Sections in the proximal region (area 2) contain both floods, while distal regions (area 1) farther up the tributary record only the larger 1954 flood. Paleostage reconstructions are based on the elevation of the most distal sediments of each flood unit.

Paleofloods   Perhaps the single most difficult problem faced by hydrologic planners is to estimate the magnitude of the largest flood that can be expected to occur in any given basin. Such estimates require extension of the flood-frequency curve beyond the limit provided by measured flow events. This is very risky business because critical decisions about the type and cost of flood protection may be based on these estimates. Simple extension of the frequency curve along the trend of the line requires questionable assumptions, and most hydrologists agree that line extensions past twice the period of record are invalid (see Costa and Baker 1981). In some cases historical observations about stages attained in floods that occurred prior to installation of gaging stations may be used to extend the curve (Benson 1950). In the United States, however, such observations would only extend the curve several hundred years at best. In addition, the accuracy of observations found in historical records are suspect, and the method may not be any more valid than simple line extension.

One of the more promising techniques to extend flood-frequency curves involves the use of Holocene stratigraphy. Essentially, the method utilizes the positions of overbank deposits in the sequence of floodplain sediments as indicators of rare flow events. Techniques used to make the analyses are thoroughly presented in Costa (1978), Patton et al. (1979), and Baker et al. (1983).

The most common stratigraphic approach is to utilize *slack water* sediment deposited by rivers confined in relatively narrow bedrock valleys. In these settings, large stage increases result from small increases in discharge, and the valley cross sections are not subject to major change during floods. In high-flow conditions, water is backflooded into tributary mouths, shallow caves, or protected areas downstream from bedrock spurs jutting into the valley. The low-velocity slack water entering these areas deposits fine-grained silts and sands which contain organic matter that can be dated by radiocarbon analyses, thereby providing the recurrence interval for the flood. Projection of the highest level of the deposit into the main valley gives an estimate of the peak flow depth (fig. 5.28); the discharge can then be estimated by using the slope-area hydrologic method (for details, see Dalrymple and Benson 1967).

Slack water deposits have been used to extend frequency curves over a period of 2000 to 10,000 years; in addition, the technique has been applied in widely diverse climatic zones (Moss and Kochel 1978; Patton et al. 1979; Patton and Dibble 1982; Kochel et al. 1982; Baker et al. 1983). The method, however, is subject to certain assumptions that may lead to error. Nonetheless, comparison of results obtained by this method with those from conventional techniques of curve extension shows the promise of the slack water analysis. For example, the 1954 flood of the Pecos River in Texas was estimated by conventional methods to have a recurrence interval ranging from 81 years to 10 million years. The slack water method estimate is 2000 years, a much more reasonable value (Kochel et al. 1982).

Other stream deposits and approaches may be used to determine the age and/or magnitude of past floods (Costa 1978, 1983). For example, overbank gravel deposits associated with soils or organic-rich layers can also be used in a manner similar to that employed in the slack water method (Costa 1974a; Patton and Baker 1977).

Effect of Physical Basin Characteristics   Physical characteristics of a basin play a significant role in determining the magnitude of the flood peak. The **basin lag,** for example, is the time needed for a unit mass of rain falling on the basin to be discharged from the basin as streamflow. It is usually estimated as the time interval between the centroid of rainfall and the peak of the hydrograph (fig. 5.3) and probably consists of two separate parts: (1) the time involved in overland flow and (2) the channel-transit time. Lag time for any basin is fairly consistent, with only minor variations in the parameter caused by the position of the storm center relative to the gaging site. Obviously lag should increase with the size of the basin, but comparisons of basins of equal size show that lag time may be as much as three times greater in *sluggish* streams than in those with short lag times, called *flashy* streams. Lag, therefore, must be determined by more than drainage area alone, and since it is presumed to be influenced by the geomorphic framework, it should be related to morphometry in some discernible way.

There is no question that the time elements of basin hydrology have a monumental influence on the magnitude of peak discharge during floods. Take, for example, the progression of the flood crest recorded at several gage stations in the Susquehanna River basin during the flood in June 1972 (fig. 5.29). Most of the precipitation spawned by hurricane Agnes entered the basin during the period between June 19 and June 22. In the minor tributary Bald Eagle Creek, the flow peaked at 143 cms (5050 cfs) on the night of June 22, indicating a relatively quick response to the storm. In the larger Juniata River, the flood crested about 12 hours later with a considerably higher discharge value of 3538 cms (125,000 cfs). Far downstream on the main Susquehanna River, the flood peak did not occur until 12 hours after the Juniata peak, when discharge rose to 30,564 cms (1,080,000 cfs). The fact that discharge peaked later on the main stem, and at a significantly higher magnitude than in the tributaries, shows the importance of timing in basin hydrology.

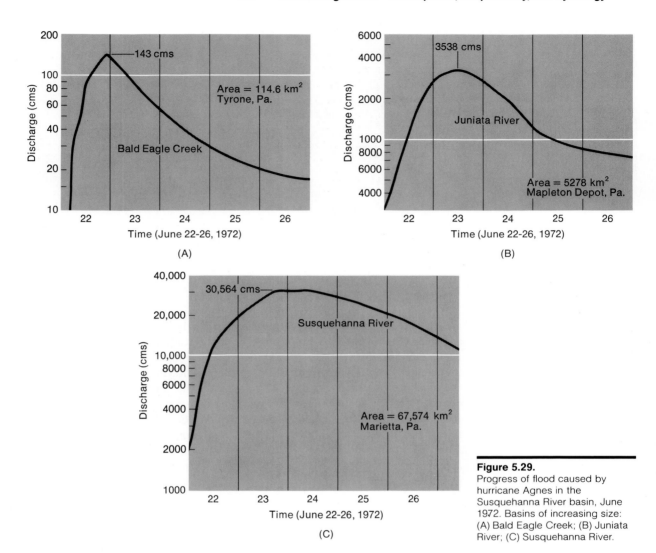

Figure 5.29.
Progress of flood caused by hurricane Agnes in the Susquehanna River basin, June 1972. Basins of increasing size: (A) Bald Eagle Creek; (B) Juniata River; (C) Susquehanna River.

As suggested above, lag time and peak discharge are positively related to basin size. More important, however, is the fact that simple expansion in size cannot totally explain the observed flow characteristics during a flood. Using the same data as before and replotting the peak discharge as Q/km^2, figure 5.30 reveals the interesting hydrologic property that discharge per unit area is much higher for the smallest tributary than for the massive basin of the main stem. Had the increase in discharge during the 1972 Susquehanna flood been only a function of increased basin area, the peak flow at the Marietta station should have been a simple product of the Q/km^2 value of Bald Eagle Creek times the drainage area above Marietta, or 1.25 cms/km² × 67,574 km² = 84,468 cms (2,978,454 cfs); but this is almost three times the actual

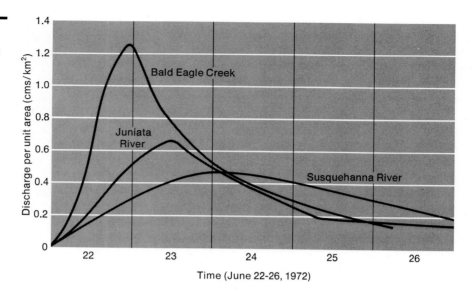

Figure 5.30.
Discharge per unit area during flood caused by hurricane Agnes in the Susquehanna River basin, June 1972. Note that the main river has considerably less discharge/area than tributaries. Main stem also peaks later and discharges flood water over a longer time period, showing the effect of storing water on floodplains.

measured value. Somewhere within the basin, a built-in flood control mechanism exists that not only holds down the magnitude of peak Q but simultaneously maintains abnormally high flow over a longer period of time. Most lowering of the peak flow is due to the ability of floodplains to store or retard the movement of larger volumes of water until the flood crest passes a given channel locality. In addition, differences in flood-crest travel time on the numerous tributaries in the basin and vagaries introduced by variable sources of runoff may help to retard the downstream peak flow. Thus, a portion of runoff never adds to the increasing peak flow downstream but shows up in the record after the crest has passed as part of the recessional limb of the flood hydrograph.

Morphometric Relationships   We now know that stream hydrology, as defined by the discharge hydrograph and by time elements such as flood frequency and lag, is significantly related to many components of basin and network morphometry. The interdependence of morphometry and hydrology is statistically real but does not necessarily indicate a cause-and-effect relationship—one factor is not the cause of changes in the other. The high correlation probably exists because both factors vary in a consistent way with the same underlying climatic and geologic controls. In general, area and relief factors are closely related to flow magnitude, and length elements to the timing of hydrologic events. All morphometric types, however, are themselves so complexly woven together that no one factor can be isolated as a completely independent variable. (Murphey et al. 1977).

Since basin area and peak discharge are highly correlative, we should expect that many other areal parameters would be similarly related to discharge. In fact, every factor involving area differs in its success as a predictor of discharge, but one parameter, drainage density, seems to have considerable value as a gage of peak flow. In a study of 15 small basins in the southern and central Appalachians and the Interior Low Plateau region, Carlston (1963) demonstrated a very close relationship between drainage density and the mean annual flood (fig. 5.31). Notably, the basins in his sample have wide variations in relief, valley-side and channel slopes, and precipitation characteristics; yet none of these factors seems to disrupt the flood magnitude-drainage density relationship. Carlston suggests that the general capacity of a terrain to infiltrate precipitated water and transmit it through the underground system is the prime controlling factor of the density-mean annual flood relationship in basins up to 260 km² in area. In larger basins, channel transit time plays the dominant role in the flow character. The rate of base flow, found to be inversely related to drainage density, is also dependent on the terrain transmissibility. Thus, as Horton suspected earlier, high transmissibility (as evidenced by infiltration capacity) spawns low drainage density, high base flow, and a resultant low-magnitude peak flood. In contrast, an impermeable surface will generate high drainage density in order to efficiently carry away the abundant runoff; base flow will be low and peak discharge high.

Patton and Baker (1976) found that basin relief and drainage density are the two variables that best distinguish areas of different flash-flood potential. Since the ruggedness number is the dimensionless product of relief and drainage density, this number should be a suitable guide to the potential for flash floods. Indeed, Patton and Baker found that high-potential basins tend to have greater ruggedness numbers than low-potential watersheds. It appears, therefore, that drainage density, by itself or in combination with other morphometric variables, may be an important guide to how a basin will function hydrologically. Dingman (1978), however, cautions that the relationship between drainage density and flow can be overriden by other effects in the basin such as floodplain or channel storage. In addition, where saturated overland flow is the major source of runoff, drainage density may tend not to be related to the efficiency at which a basin is drained.

In spite of the apparent difficulties, the possible interrelationships between hydrology and morphometry are seemingly infinite, and even though the parameters are so complexly related that equations will not explain all the variability, the approach has some validity and should not be abandoned in future research. The hydrogeomorphic approach is especially applicable in determining regional flood hazards (Baker 1976).

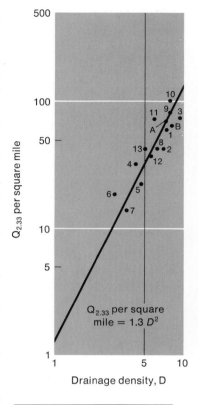

Figure 5.31.
Discharge (mean annual flood, $Q_{2.33}$) controlled by drainage density in 13 basins. (From Carlston 1963)

Basin Denudation

In addition to being hydrologic entities, basins are also geographic compartments where sediment is manufactured, eroded, and deposited, and from which, given sufficient time, the debris will ultimately be removed. The amount of sediment leaving a basin can be readily converted into an estimate of lowering of the basin surface, called **denudation,** which is usually expressed as a time parameter or rate. Denudation seems to have no rigorous definition, but because it implies removal of basin material, it is commonly used as a synonym for erosion. The two differ, however, because denudation considers only those eroded products that are removed completely from the basin, assuming in that consideration that the sediment is derived in equal portions from all subareas of the watershed. It therefore presupposes equal surface lowering over the entire basin. Denudation rates tell us little about those erosive processes that simply redistribute sediment *within* the basin; nor do they indicate that at any given time some parts of the basin are probably aggrading rather than eroding. Denudation, then, is the long-term sum of the overall erosive process, and even though it is analogous to erosion it is not precisely the same. Furthermore, recognize that the basin surface may not actually lower even with a high denudation rate if active uplift of the basin is proceeding at a greater rate.

Estimates of modern denudation are usually based on measurements of stream load made at gaging stations or on the volume of reservoir space lost when sediment accumulates behind a dam. All types of load (suspended, bed, and dissolved) are included in the analyses at gage stations, which require that the weight values of load be converted into volumetric terms. Once the volume of sediment leaving the basin is determined, it is divided by the area of the watershed above the gage station to provide the third (or vertical) dimension, which represents the magnitude of surface lowering (for details see Ritter 1967). Rates are commonly expressed in inches or centimeters per 1000 years. In most basins the amount of solid load is the predominant type of material lost. Additionally, as we will see in the next chapter, the type and volume of solid load has a monumental influence on river behavior. Thus, we must examine the mechanisms by which slopes are eroded and what factors affect the mode and rate of the erosive processes.

Slope Erosion and Sediment Yield

A raindrop possesses a considerable amount of kinetic energy, derived from its mass and the velocity it attains during its fall. Under the influence of gravity, a raindrop accelerates until its force is equal to the frictional resistance of the air, the speed at that point being the *terminal velocity*. As the distance needed to attain this condition is very short, most rain strikes the surface at its terminal velocity, although the absolute speed varies with wind, turbulence, drop size, etc. In high-intensity rains, drops usually reach a maximum size of approximately 6 mm and a terminal velocity of about 9 m/sec. The impact of such rain can directly displace into the air particles as large as 10 mm in diameter and, by undermining downslope support, can indirectly set even larger pebbles in motion. The amount of soil moved by splash depends on several interrelated factors. First, the kinetic energy of raindrops is directly related to splash movement (Kneale 1982). However, it is interesting to note that the kinetic energy of raindrops sometimes varies in unexpected ways. For example, Mosley (1982) found that rain passing through a forest canopy had greater total kinetic energy than normal rainfall and that rainsplash was three times greater under the canopy than in open areas. Second, the type of soil being struck is extremely important in determining the magnitude of splash movement. Free (1960), for example, found that splash loss varied as $E^{0.9}$ for a silt loam soil and $E^{1.46}$ for a sand, where E is the kinetic energy. Over a five-year period the total splash loss from the sandy surface was calculated at 1600 tons/acre, an amount three times greater than the loss from the loam, probably because the fine-grained soil had greater cohesion. Actually the manner in which the soil particles aggregate (Luk 1979) and the dispersive properties of the surface material (Yair et al. 1980; Rendell 1982) are more significant controls than simple textural composition. Third, the rate and amount of splash transport appear to be a function of slope angle (Savat 1981; Reeve 1982), but the precise relationship is quite variable and not easily determined (Bryan 1979).

In addition to direct transportation, splash has several other erosion-inducing effects on the soil. By detaching particles, it destroys the structure of the soil and breaks apart resistant aggregates of clays. These physical processes make the soil much more susceptible to erosion by surface flow. Furthermore, as splash disperses the clays, they tend to form a fine-grained crust as they settle back on the surface. This crust forms a semipermeable barrier that reduces infiltration and promotes runoff, thereby increasing soil loss by overland flow.

Wash   Most natural slopes are too irregular to permit a uniform flow of water over the entire surface; flow is deeper over depressions and shallower over flat reaches or high spots. The variable depth of flow produces differences in the eroding and transporting capabilities of the water so that, in detail, *wash* does not imply that a regular sheet of debris is being carried continuously down the slope surface. In areas where sheet flow might be possible, only fine-grained particles can actually be moved efficiently, and those only if the surface has been prepared for erosion by rainsplash or weathering processes that reduce cohesion. In areas of concentrated flow, larger sediment can be moved, but the ability to erode depends more on the hydraulic force of the water and less on the condition of the surface.

When rainfall and flow become intense, small shallow channels may be formed, channels that periodically shift their position so that in the long run erosion is more or less even across the slope. In fine-grained soils, a set of well-defined subparallel rills is usually formed. Rills vary in size with the erodibility of the soil, but normally they are only several centimeters wide and deep. Heaving and other processes can obliterate these tiny channels in periods between high rains, especially in highly seasonal climates where rain may be lacking for months at a time. The periodic destruction of rills allows new channels to form in an entirely different location and ensures less than equal lowering of the entire slope surface. Some rills escape this spasmodic destruction by entrenching to greater depths, a difficult task that only a few of the largest rills accomplish. These "master" rills become relatively permanent and eventually evolve into true rivers.

In soils that are sandy or coarser, the channels are usually braided because the material, although easily eroded, is transported with difficulty. Sediment commonly accumulates as temporary bars within the channel, and these subdivide the channel and the flow into a multitude of small passageways. Most braided channels are wider (up to 5 m) and deeper (1–10 cm) than rills, but they also change their position regularly because the bar deposits require a continuously shifting channel environment.

Sediment Yield (Soil Loss)   Assuming that all factors can be assessed within reasonable limits of uncertainty, the soil loss by water erosion should follow the **Universal Soil Loss Equation**

$$A = RK(LS)CP,$$

where A is average annual soil loss, R is the rainfall factor, K is the erodibility factor, LS the slope length-steepness factor, C the cropping and management factor, and P the conservation factor (for details see Smith and Wischmeier 1962). The precision of such an equation is questionable because in reality it predicts total soil *movement* rather than total soil loss. Along a typical slope surface, some zones may be experiencing active deposition while others are being eroded. In addition, the equation sometimes underpredicts soil loss. Haigh and Wallace (1982), for example, showed that ground losses from strip-mine dumps in Illinois were twice as great as those estimated by the U.S.L.E. Nonetheless, the analysis is widely used, and it probably can provide useful ballpark data for predicting sediment loss from slopes.

Factors Affecting Sediment Yield   A number of interrelated geologic, hydrologic, and topographic factors cause the magnitude of sediment yield to vary widely from region to region. The most important of these are (1) precipitation and vegetation, (2) basin size, (3) elevation and relief, (4) rock type, and (5) human activity.

Precipitation and Vegetation   Intuitively we would expect the amount of sediment yielded from any basin to be related in some systematic fashion to the amount of incoming precipitation. Actually the correlation between the two is not so direct as we might hope, because precipitation is complexly interrelated with other factors that influence its erosive capability. The amount of runoff from any given precipitation, for example, varies with temperature. Vegetation, a function of both precipitation and temperature, serves as a protective screen against the erosion of surface material. Precipitation, then, cannot be considered as a completely independent variable in the realm of denudation, even though it may be a dominant factor.

In an important paper, Langbein and Schumm (1958) documented the relationship between sediment yielded from basins averaging 3900 km² in area and *effective precipitation,* a parameter derived by adjusting the magnitude of precipitation to values expected at a mean annual temperature of 10°C (50°F). Under those conditions, they were able to show that as precipitation rises from zero, the sediment yield increases rapidly to a maximum yield value at about 30 cm of effective precipitation (see fig. 2.17). Any increase in precipitation above 30 cm promotes a decline in sediment yield because the density and type of vegetation begin to play an active role in protecting the slopes from erosion. Vegetation generally begins to exert a control on erosion when the ground cover is between 8 percent and 60 percent, values typical in semiarid and subhumid climates. Thus, as L. Wilson (1973) suggests, the Langbein-Schumm curve may be valid for regions with a continental climate but may not be applicable in other climatic regimes, especially nonseasonal types.

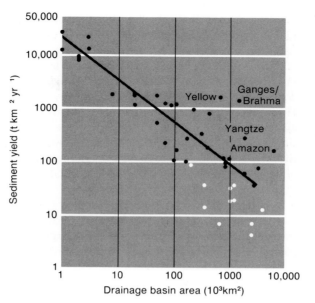

Figure 5.32.
Comparison of sediment yields and drainage basin areas for all major sediment-discharging rivers (greater than 10×10^6 t yr^{-1}). Open circles represent low-yield rivers draining Africa and the Eurasian Arctic. Smaller basins have larger yields, although the largest rivers (Amazon, Yangtze, Ganges/Brahmaputra, and Yellow) all have greater loads than their basin areas would predict.

In any case, the demonstration by Langbein and Schumm that the relationship between precipitation and sediment yield is nonlinear and very complex seems to be a valid geomorphic observation. Precisely at what precipitation the maximum values of sediment yield will occur depends on the specific climatic setting (for example, see Fournier 1960), and total values may relate more to the seasonality of the climate than to the mean annual precipitation (L. Wilson 1973). Some studies tend to support that contention (Corbel 1959; L. Wilson 1972; Jansen and Painter 1974). It should also be noted that variations in sediment yield under different climates may be partially offset by an increase or decrease in dissolved load. Normally, dissolved load will increase regularly with precipitation, but maximum solution is probably reached at about 63 cm of annual precipitation, with little additional dissolution resulting from higher precipitation (Leopold et al. 1964). Regardless, the amount of dissolved load usually does not exceed solid load even in regions of high precipitation, where chemical loss should be great (Li 1976; Leigh 1982). This probably results because the magnitude of solution loss is dominated by rock type rather than climate or vegetation (Garrels and Mackenzie 1971).

Basin Size   A number of studies have suggested that sediment yield decreases markedly as the size of the drainage basin increases. This hypothesis is supported by remarkably high yields in very small basins of the midwestern United States (Schumm 1963c) and also by data from the major sediment-discharging rivers of the world (Milliman and Meade 1983; fig. 5.32) The explanation for this phenomenon seems to lie in several topographic realities: (1) small basins generally have steep valley-side slopes and high-gradient stream channels that efficiently transport sediment; (2) in basins filled to capacity with streams, the drainage density always remains high near the basin divide, but it may decrease with time in the central part of the basin; (3) floodplain area increases as the basin expands, especially in the central and lower reaches of the basin (Hadley and Schumm 1961).

The integration of these factors leads us to realize that in natural basins most sediment is produced in the small headward subareas, but during its downstream transit a significant portion may be stored in the floodplain system. How long it will remain in storage depends on the rigor of the geomorphic and hydrologic processes. It has been assumed that sampling over a long enough time would show that sediment stored within the basin is flushed rapidly from the system during episodes of rejuvenation. Inclusion of such spasmodic bursts of erosion in a long-term sample would tend to temper the variations in yield that appear to be related to basin size.

There are reasons to believe, however, that the relationship between sediment yield and basin size is less real than it appears because we tend to emphasize the extreme erosion in small marginal basins, and we do not fully comprehend the equilibrium state of the master streams. For example, there is ample evidence that the downstream decrease in sediment yield occurs during episodes of accelerated erosion in small basins because storage in stream channels and valleys is increased (Trimble 1977). This occurs because the main channel is unable to transport the additional load. Thus, sediment yielded from the basin mouth decreases during a time of high erosion in the marginal basins, indicating that rivers are in disequilibrium conditions throughout the basin. The reverse can also happen; i.e., a reduction of erosion in the headland basins by sediment control measures starves the main channel load and creates erosion of the previously stored sediment, leading to increased yield measured at the basin mouth (Trimble 1977). Thus, the apparent basin size-sediment yield relationship may simply stem from the possibility that different parts of drainage basins are responding in opposite directions to external erosional stimuli in the small marginal zones. Modern sediment yield values, therefore, may be related to recent changes in land use practice rather than an inherent areal control.

Elevation and Relief   Mountainous terrains with excessive elevation and relief are known to produce abnormally high sediment yields, particularly where rocks are nonresistant (Corbel 1959; Hadley and Schumm 1961; Schumm 1963c; Ahnert 1970) or affected by recent or current tectonism (Li 1976). For basins at least 3900 km² in area, the greatest denudation rates average about 0.9 m/1000 years and probably occur in mountain belts where relief and elevation are greatest and rocks are erodible (Schumm 1963c).

In the arid climate of the western United States, sediment yields are a function of the relief-length ratio (fig. 5.33). Utilizing this fact and holding area constant, Schumm (1963c) showed that the relationship between relief and denudation is definable in quantitative terms. It has also been shown that

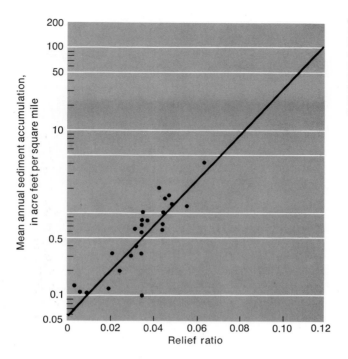

Figure 5.33.
Relation between mean annual sediment accumulation in reservoirs and the relief ratio for basins located on the Fort Union formation in the upper Cheyenne River basin. (From Hadley and Schumm 1961, fig. 31)

elevation alone produces disparate denudation rates since low-lying areas in any climatic regime yield less sediment than higher basins of comparable size (Corbel 1959).

Significantly, relief and elevation analyses demonstrate clearly that denudation rates are not constant through time. As relief and elevation of a basin are gradually diminished during its evolution, the rate of surface lowering decreases proportionately, and each successive interval of stripping requires a longer period of time (fig. 5.34).

Rock Type   With similar climate and topography, basins underlain by clastic sedimentary rocks and low-rank metamorphics usually produce abundant suspended loads and so are characterized by higher rates of denudation than regions of crystalline rocks or highly soluble sedimentary rocks (Corbel 1964). The relationship of lithology and denudation is poorly understood in a quantitative sense and may be obscured by other rock characteristics, such as fracturing, which cause different lithologic units to behave similarly with respect to denudation. Even so, properties that are commonly a reflection of lithology, such as the infiltration capacity, seem to be systematically related to sediment yield (fig. 5.35), indicating that the lithologic influence is real. At this time, however, geomorphologists have not been able adequately to reveal its fundamental character.

Figure 5.34.
Relation of denudation rates to relief-length ratio and drainage-basin relief. Denudation rates are adjusted to drainage areas of 1500 square miles. Curve is based on the average maximum denudation rate of 3 feet per 1000 years when relief-length ratio is 0.05. (From Schumm 1963)

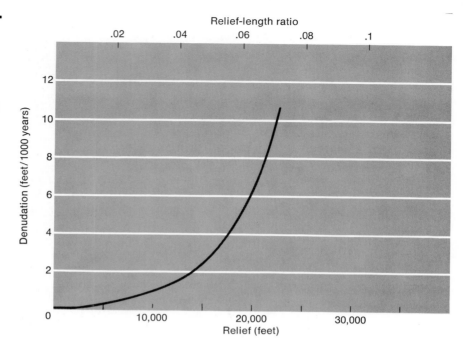

Figure 5.35.
Relation between infiltration capacity and sediment yield, indicating lithologic influence. (Data from Hadley and Schumm 1961)

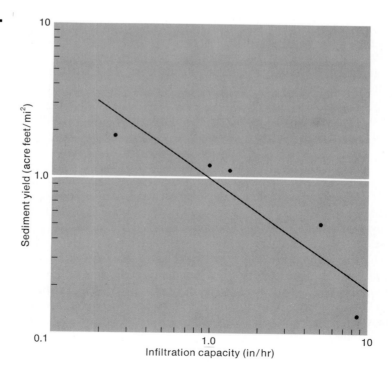

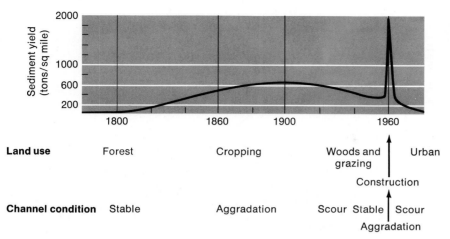

Figure 5.36.
Changes in sediment yield and
channel behavior in one area
under various types of land use.

The Human Factor   Most estimates of denudation are significantly inflated
where human activity disturbs the natural setting. Exactly how much humans
accelerate erosion varies with the type of land use and the particular environment, but it seems reasonable to propose that human interference has the potential to alter drastically the natural sediment yield (Moore 1979; Dunne
1979; Toy 1982).

Evidence indicates that human activities may increase detrital loads by
at least an order of magnitude (Judson 1968a; Meade 1969); chemical loads
are expanded by pollutants introduced into streams or the atmosphere (Meade
1969). One important contributor to accelerated erosion is the replacement
of mature forest cover by intensely cultivated land (Toy 1982). Studies in the
United States show an increase in sediment yield of 1 to 3 orders of magnitude
when cropping is substituted for the natural vegetation (Ursic and Dendy 1965;
Wolman 1967). With proper soil conservation techniques, however, the effect
can be reversed and sediment yield values will decrease dramatically (Trimble
and Lund 1982).

Construction associated with urbanization causes an even more dramatic
rise in the sediment yield (fig. 5.36), but after construction is completed the
values decrease rapidly because much of the surface is protected from erosion
by our concrete citadels (Wolman 1967). Urbanization also affects the runoff
characteristics within a drainage basin (McPherson 1974); consequently, the
concentration of sediment in streams, as influenced by humans, depends on
variations in both hydrology and sediment yield.

Several workers have suggested that the abnormalities induced by human
activities may be great enough to invalidate the use of sediment yields to calculate rates of denudation in drainage basins (Douglas 1967; Trimble 1977).
Trimble (1977) was able to show that in large drainage basins in the southeastern United States, lowering of upland surfaces was proceeding at a rate
of 95 mm/100 yr; in contrast, the denudation rate calculated by sediment in
the streams was only 5.3 mm/100 yr. Thus, the *delivery ratio* (sediment yield

Figure 5.37.
General linkages between sediment storage sites and erosional processes. Boxes indicate storage elements; listed below each box are erosional processes mainly responsible for mobilizing sediment in that element. Arrows show transfers between elements. Labels on arrows qualify or restrict location of transfers. (From Lehre 1982)

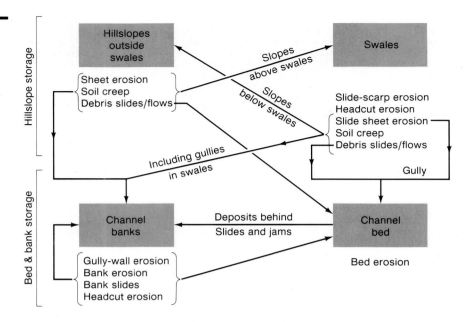

as a proportion of upland erosion) was only 6 percent. Therefore, sediment yield may be a poor indicator of how rapidly the upland area of a basin is being lowered. The major portion of sediment eroded from the uplands is, of course, being stored within the basin by deposition.

In addition, Meade (1982) has suggested that the accelerated erosion produced by early settlers in the eastern United States while clearing the land for cultivation has been largely arrested by soil conservation and reduced farming. However, sediment stored in the basins in response to the earlier settlement is now being eroded and is augmenting the present river loads and will continue to do so for decades or centuries. The point is that modern river loads may be reflecting erosional events that occurred in the distant past.

Sediment Budgets

It is clear from the above that storage of sediment within a basin is a significant geomorphic phenomenon. Its importance is further realized with the concept of a sediment budget (Dietrich and Dunne 1978; Kelsey 1980; Lehre 1982). **A sediment budget** is a quantitative analysis of a drainage basin that shows the relationship among erosion of basin materials, discharge of sediment from the basin, and the associated changes in sediment storage. In essence, it is an accounting sheet of ins and outs of sediment and as such is similar to the hydrologic budget discussed earlier. Of greater importance, however, is its significance in planning and land management because it distinguishes the dominant erosional processes in the basin and the circumstances under which storage of sediment may change. Additionally, the sediment budget demonstrates the process linkage that we discussed in chapter 1 (fig. 5.37).

Table 5.5 Sediment budget for a small drainage in northern California for the years 1971–1974.

Year	Rainfall	Recurrence Interval of Peak Flow	Mobilization on Slopes (1)	Production to Channels (2)	Redistribution on Slopes (3)	Yield: Bed + Susp. Load (4)	Bed + Bank Storage (5 = 2 − 4)
	Millimeters	*Years*			*Metric tons km²*		
1971–72	602	1.5	86	148	−71	24	+124
1972–73	1184	15–20	1219	985	+317	1420	−435
1973–74	1046	3–5	1960	1575	+389	630	+945
1971–74	2832	—	3265	2708	+635	2074	+634

From Lehre 1982.

To make a complete sediment budget analysis one must identify and quantify *sediment mobilization* (processes that initiate motion and move sediment any distance), *sediment production* (sediment reaching or given access to a channel), and *sediment yield* (sediment actually discharged from the basin). Having made such measurements, a balance sheet is constructed to reveal the progression of sediment through the basin. For example, table 5.5 shows the sediment budget for the years 1971–1974 in a small drainage basin northwest of San Francisco (Lehre 1982). In dry years or years without extreme flow events (1971–1972, 1973–1974), most mobilized and produced sediment was stored within the basin. In contrast, during the year having a flow event with a 15- to 20-year recurrence interval (1972–1973), the amount of sediment mobilized and produced was less than the amount discharged from the basin. This indicates that sediment was taken out of storage during that year, presumably by erosion of channel banks and beds. By continuous observation, Lehre (1982) was able to demonstrate that the variations in annual sediment budgets shown in table 5.5 were accompanied by different erosive processes. In the years of low rainfall and/or peak flow, sediment was mobilized by spalling, rainbeat, and minor sliding and was moved toward the channels by sheetwash. The lack of water, however, assured a minimum transport distance and most sediment went into storage. In the year of high peak flow (1972–1973), sediment was mobilized and produced mainly by debris slides and flows. It was quickly delivered to the channels and because of the coincidence of high channel discharge was removed from the basin.

Sediment budget analysis is a relatively new way of looking at the inner workings of the drainage basin, and we do not yet know what its applications will be. However, it seems to have considerable promise for land management, especially in small basins that are unstable geomorphically and subject to a variety of interrelated processes.

Rates of Denudation

Regardless of the problems inherent in denudation-rate analyses, various estimates have been made. Table 5.6 presents a random sample of modern denudation rates for basins of varying size in the United States. The values are imprecise and probably high because in most cases they do not include an

Table 5.6 The influence of geology and climate on suspended-load denudation in basins of different size in the United States.

Basin	Location	Area (mi²)	Average Annual Suspended Load (tons × 10³)	Denudation (in/1000 yr)
Mississippi	Baton Rouge, La.	1,243,500	305,000	1.3
Colorado	Grand Canyon, Ariz.	137,800	149,000	5.6
Columbia	Pasco, Wash.	102,600	10,300	0.5
Rio Grande	San Acacia, N.M.	26,770	9,420	1.8
Sacramento	Sacramento, Cal.	27,500	2,580	0.5
Alabama	Claiborne, Ala.	22,000	2,130	0.5
Delaware	Trenton, N.J.	6,780	998	0.8
Yadkin	Yadkin College, N.C.	2,280	808	1.8
Eel	Scotia, Cal.	3,113	18,200	30.4
Rio Hondo	Roswell, N.M.	947	545	3.0
Green	Palmer, Wash.	230	71	1.6
Alameda	Niles, Cal.	633	221	1.8
Scantic	Broad Brook, Conn.	98	7	0.4
Napa	St. Helena, Cal.	81	63	4.1

Data from Judson and Ritter 1964.

adjustment for human impact. Nonetheless, they do indicate the general tendency for denudation rates to fall between 2.5 and 15 cm per 1000 years when considered on a regional scale. Judson (1968b) recalculated the denudation rate for the entire continental United States by subtracting the effect of human occupancy from earlier estimates. His figure of 3 cm/1000 years agrees rather well with the denudation in very large drainage basins that are mostly unaffected by humans (Gibbs 1967) and perhaps represents a reasonable approximation for denudation on a continental scale.

Although methods other than analyses of sediment wedges have been employed (Eardly 1967; Ruxton and McDougall 1967; Clark and Jager 1969), most past rates of denudation are estimated from measurements of sediment accumulation in depositional basins. A valid estimate can be made only if (1) the volume of sediment derived by erosive processes can be accurately determined, (2) the boundaries of the source area are definable, and (3) the time interval of sediment accumulation can be ascertained within reasonable limits. It is very difficult to meet all these requirements: noneroded matter such as pelagic and volcanic rocks add to the depositional volume; material eroded from the basin as dissolved load may not be returned to the deposit by chemical precipitation; and absolute dates that bracket the time of deposition are necessarily imprecise. Nonetheless, some estimates of past rates have been made for large portions of North America (Gilluly 1949, 1955, 1964; Menard

Table 5.7 Denudation estimates based on solid and dissolved loads delivered to the oceans by major rivers of the continents.

Continents	Solid Load (t/km^2/yr)[a]	Dissolved Load (t/km^2/yr)[b]	Total Load	Denudation Rate (cm/1000 yr)
North and Central America	84	33	117	4.00
South America	97	28	125	4.28
Europe	50	42	92	3.15
Asia	380	32	412	14.10
Africa	35	24	59	2.02
Australia	28	2	30	1.03

[a]Solid load from Milliman and Meade (1983).
[b]Dissolved load from Garrels and Mackenzie (1971).

1961). It is interesting to note that the rates found in these studies of large regions are within the same order of magnitude as those based on modern stream data. Such similarity prompted the hypothesis (Ritter 1967) that when viewed on a large enough area or over a long time interval, denudation rates will probably be about the same. Based on current evidence, the average value will probably fall somewhere between 2.5 cm and 15 cm per 1000 years. That range is generally supported by estimates of solid and dissolved loads being delivered to the oceans from the world's continents (table 5.7).

The use of denudation rates calculated for continent-size areas (table 5.7) can lead to terribly incorrect conclusions. For example, a 3 cm/1000 yr rate suggests that 300 m of surface lowering will be accomplished over an entire continent in a 10-million-year period. Such a rate might be used as evidence to support the generally accepted canon of geomorphology that most of the earth's topography is no older than Pleistocene, i.e., almost all landscapes formed in the last 2 million years. Therein lies the fallacy of denudation rates because we know that large regions of Tertiary and older landforms do exist, especially in the Southern Hemisphere. For example, radiometric dates and geologic evidence in southeastern Australia show that much of that landscape was in its present form by mid-Miocene, and some upland surfaces originated in the Mesozoic (Young 1983). The incompatibility of these observations and denudation analyses from river sediment arises because denudation rates are unrealistically spread evenly over entire continents. Actually, the interiors of continental plates probably experience extremely slow denudation and may easily preserve old landscapes (Young 1983). In contrast, continental plate margins where active tectonism is occurring probably have enormously high denudation rates. Combining the two subareas provides an average rate for the continent that is indicative of neither. The point here is that a denudation rate calculated for a large basin or region tells us nothing about the tenor of erosion occurring in any component part of that basin or region, and the overall rate should never be used in that sense.

Summary

In this chapter we examined a remarkable statistical balance among the spatial characteristics of river networks and the watersheds that contain them. Because the parameters of this morphometry also relate in a significant way to the hydrologic and erosional properties of most watersheds, drainage basins serve as primary units for systematic analyses of geomorphology. Drainage basins and their river networks probably evolve according to fundamental hydrophysical laws, but their ultimate character is conditioned by the geological framework and the external constraints of climate. An equilibrium condition, defined in terms of mathematical balance, is probably attained early in the growth history of most basins. This does not indicate, however, that basins evolve in an orderly way with time. Geologic catastrophes that upset equilibrium tend to be filtered out in a morphometric sense because basinal parameters apparently adjust to changes rather quickly.

Water flowing in basin rivers is derived from variable sources, and sediment reaching stream channels is produced and delivered by numerous erosive processes operating on basin slopes. Both water and sediment are amenable to budget analyses, which provide basic data for watershed planning. The amounts of water and sediment entering stream channels are functions of the physical properties of the basin and the effect produced by human activities. Estimates of basin denudation can be made on the basis of total sediment yield, but difficulties are created by sediment storage within the basin.

Suggested Readings

The following references provide greater detail concerning the concepts discussed in this chapter.

Chorley, R. J. 1969. *Water, earth and man.* London: Methuen and Co.

Costa, J. E., and Baker, V. R. 1981. *Surficial geology—Building with the Earth.* New York: John Wiley & Sons.

Doornkamp, J. C., and King, C. A. M. 1971. *Numerical analysis in geomorphology: An introduction.* London: Edward Arnold Ltd.

Dunne, T. 1978. Field studies of hillslope flow processes. In *Hillslope hydrology,* edited by M. J. Kirkby, pp. 227–94. New York: John Wiley & Sons.

Dunne, T., and Leopold, L. B. 1978. *Water in environmental planning.* San Francisco: W. H. Freeman.

Horton, R. E. 1945. Erosional development of streams and their drainage basins: Hydrophysical approach to quantitative morphology. *Geol. Soc. America Bull.* 56: 275–370.

Judson, S. 1968. Erosion of the land. *Am. Scientist* 56: 356–74.

Patton, P. C.; Baker, V. R.; and Kochel, R. C. 1979. Slack water deposits: A geomorphic technique for the interpretation of fluvial paleohydrology. In *Adjustments of the fluvial system,* edited by D. D. Rhodes and G. P. Williams, pp. 225–53. Dubuque, Iowa: Kendall/Hunt.

Schumm, S. A. 1963. Disparity between present rates of denudation and orogeny. U.S. Geol. Survey Prof. Paper 454-H.

Fluvial Processes

6

Introduction

It seems fair to say that fluvial action is the single most important geomorphic agent. Although other surficial processes are significant, streams are so ubiquitous that their influence in geomorphology can hardly be overestimated. As discussed in chapter 1, most geomorphic analyses of stream processes in the early twentieth century were based primarily on logic and qualitative observation. The few quantitative studies of fluvial mechanics were somehow lost in the wave of geomorphology that used fluvially produced landforms to reconstruct history. Our understanding of river mechanics is hampered by the difficulty of obtaining measurements in a natural channel. Much of our knowledge of fluvial mechanics has therefore been derived from studies in flumes, where slope and velocity can be varied and any variable can be held constant. Unfortunately, flumes have a limited range of discharge and depth, and although the application of flume results to real systems gives an excellent first approximation, it cannot take into account the complexities of the many everchanging and interdependent variables in natural streams (see Maddock 1969). We still need much more data about natural rivers before we can hope to utilize our knowledge in a predictive way. Nonetheless, there has been a rebirth of interest in the details of fluvial mechanics, and data are now accumulating faster than one can assimilate their meaning. This explosion of information is undoubtedly the key to the advancement of geomorphic thinking, for without it our basic tenets will never be critically evaluated. In this chapter we will examine briefly much of the current thinking about streams; those interested in greater detail than space allows here are referred to a number of excellent books about the fluvial realm: Leopold et al. 1964; Leliavsky 1966; Morisawa 1968; Raudkivi 1967; Chorley 1969a; Graf 1971; Schumm 1972, 1977; Gregory and Walling 1973; Richards 1982; Hey et al. 1982.

The River Channel

Basic Mechanics

The ability of a river to do geomorphic work represents a balance between driving and resisting forces. It depends on how much potential energy is provided to the flow and how much of that energy is consumed in the system by the various resisting elements. The resistance generated along the channel perimeter or within the flow is controlled by the shape of the channel, the size and concentration of sediment, and the total volume and physical properties of the water. To illustrate, let us examine what happens to a constant discharge of water flowing in a long, steeply inclined channel that has no tributaries. Gravity tends to accelerate the flow downstream continuously unless the increase in velocity is moderated by friction within the water and turbulence generated along the channel perimeter. If these resisting elements were constant and less than the gravitational force, acceleration would continue along the entire length of the channel. In natural rivers, however, the intensity of resistance is not constant but increases with the flow velocity. At any point in a river, therefore, the velocity represents the balance between the energy causing flow and the energy consumed by resistance to flow.

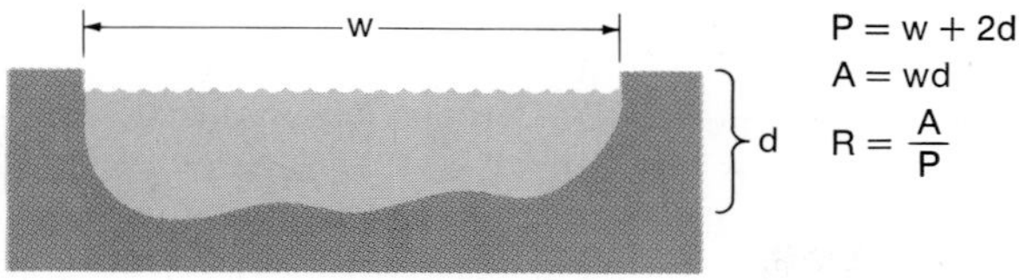

$$P = w + 2d$$
$$A = wd$$
$$R = \frac{A}{P}$$

Figure 6.1.
Cross-sectional measurements of a stream channel: w = width; d = depth; A = area; R = hydraulic radius; P = distance along wetted perimeter.

Resistance is spread through the water by viscous or turbulent processes and is proportional to the velocity or its square. In **laminar flow,** particles of water move in straight paths that are not disrupted by the movement of neighboring particles. In this type of flow regime, internal friction is the dominant resisting force, and its intensity is proportional to the velocity of flow. Because the change in velocity with depth is linear, resistance increases uniformly from the water surface to the bed, the rate of increase being determined by the molecular viscosity of the fluid. Most resistance in laminar flow results from intermolecular viscous forces as layers or particles of the fluid slide smoothly past one another (Leopold et al. 1964). The controls on viscosity are internal characteristics of the fluid such as temperature and suspended sediment concentration.

In **turbulent flow,** the water does not move in parallel layers; its velocity fluctuates continuously in all directions within the fluid. Water repeatedly interchanges between neighboring zones of flow, and shear stress is transmitted across layer boundaries in another form of viscosity, called *eddy viscosity*. Eddy viscosity greatly increases the dissipation of energy and the flow resistance. In turbulent flow resistance is proportional to the square of the velocity. Resistance and velocity, therefore, do not change uniformly with depth. In addition, because turbulence is generated along the channel boundaries, most resistance in this type of flow is due to external factors such as the channel configuration and the size of the bed material.

As depth and velocity increase, the conditions at which laminar flow changes to turbulent can be predicted by a dimensionless parameter called the **Reynolds number** *(Re):*

$$Re = \frac{VR\rho}{\mu},$$

where V is the mean velocity, R the hydraulic radius, ρ the density, and μ the molecular viscosity. The hydraulic radius R is determined by the relationship

$$R = \frac{A}{P},$$

where A is the cross-sectional area of the channel and P the wetted perimeter (fig. 6.1). In wide, shallow channels the hydraulic radius closely approximates the mean depth.

Since the factor $\frac{\mu}{\rho}$ defines a fluid property called *kinematic viscosity* (ν), the Reynolds number represents a ratio between driving and resisting forces because

$$Re = \frac{VR\rho}{\mu} = \frac{VR}{\nu} = \frac{\text{driving force}}{\text{resisting force}}.$$

In normal situations true laminar flow occurs where Re values are less than 500, and well-defined turbulent flow when Re is greater than 750.

Another dimensionless number used to describe the conditions of flow is the **Froude number** (F_r):

$$F_r = \frac{V}{\sqrt{dg}},$$

where d is depth and g is gravity. The Froude number is important because it can be used to distinguish subtypes of turbulent flow called *tranquil flow* ($F_r < 1$), *critical flow* ($F_r = 1$), and *rapid flow* ($F_r > 1$).

Flow Equations and Resisting Factors

Flow and resistance have been the concern of hydraulic engineers for centuries, and a number of equations have been derived to express the relationships between the factors. Two equations of great importance to students of rivers are the **Chezy equation** and the **Manning equation.** Both derive from equating driving and resisting forces in nonaccelerating flow, and both have been employed in a variety of fluvial investigations. The Chezy equation

$$V = C\sqrt{RS},$$

derived in 1769, shows that velocity is directly proportional to the square root of the RS product, where S is the slope of the channel. The Chezy coefficient (C) is a constant of proportionality that relates to resisting factors in the system.

The Manning equation originated in 1889 from an attempt by Manning to systematize the existing data into a useful form. The equation

$$V = \frac{1.49}{n} R^{2/3}S^{1/2}$$

is similar to the Chezy formula in that velocity is proportional to R and S. In addition, the factor n, called the Manning roughness coefficient, is also a resisting element that is closely related to the Chezy coefficient because as

$$C(RS)^{1/2} = \left(\frac{1.49}{n}\right) R^{2/3}S^{1/2}$$

$$\text{then } C = 1.49 \left(\frac{R^{1/6}}{n}\right).$$

Manning's n is presumed to be a constant for any particular channel framework; consequently, it has been used extensively in analyses of river mechanics

Table 6.1 Manning roughness coefficients (*n*) for different boundary types.

Boundary	Manning Roughness $n(\text{ft}^{1/6})$
Very smooth surfaces such as glass, plastic, or brass	0.010
Very smooth concrete and planed timber	0.011
Smooth concrete	0.012
Ordinary concrete lining	0.013
Good wood	0.014
Vitrified clay	0.015
Shot concrete, untroweled, and earth channels in best condition	0.017
Straight unlined earth canals in good condition	0.020
Rivers and earth canals in fair condition—some growth	0.025
Winding natural streams and canals in poor condition—considerable moss growth	0.035
Mountain streams with rocky beds and rivers with variable sections and some vegetation along banks	0.041–0.050

(table 6.1). The U.S. Geological Survey, for example, has developed a visual guide for rapid estimation of Manning's *n* (Barnes 1968). It should be noted that substituting depth for hydraulic radius in calculations using the Manning equation will introduce considerable error (Tinkler 1982). Therefore, *R* rather than *d* should always be used to calculate velocity or resistance coefficients.

Although resistance coefficients are defined by hydraulic characteristics (*S, R, V*), they are not independent of other factors because in alluvial channels their values vary with particle size, sediment concentration, and bottom configuration. For example, in the same reach of a river, different flow conditions (as determined by the Froude number) may mold bottom sediment into a variety of bed forms (fig. 6.2), which in turn have a decided influence on the value of Manning's *n* (table 6.2). Irregularity of the channel bottom due to other factors such as bars, riffles, and bends may have an equally important influence on the roughness (Leopold et al. 1964; Simons and Richardson 1966). In fact, Prestegaard (1983a) was able to demonstrate that in gravel-bed reaches of 12 western U.S. rivers, bar resistance accounted for 50 to 75 percent of the total flow resistance. Particle size of bed material also causes variation of roughness values. Wolman (1955) demonstrated this last fact in the Brandywine Creek (Pennsylvania), where the resistance-particle size relationship is defined by the equation

$$\frac{1}{\sqrt{F}} = 2 \log \frac{d}{D_{84}} + 1.0,$$

where *F* is another resistance parameter called the *Darcy-Weisbach resistance coefficient, d* is water depth, and D_{84} is the particle diameter, which is equal to or larger than 84 percent of the clasts on the channel bottom. Increased particle size should produce a correlative increase in flow resistance.

Figure 6.2.
Bed forms in alluvial channels and their relation to flow conditions.
F = Froude number, *d* = depth.

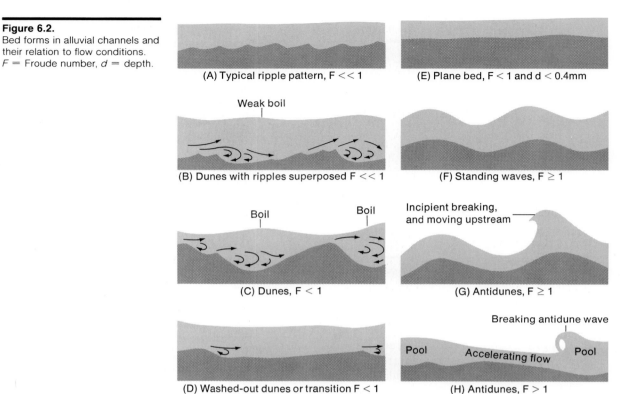

(A) Typical ripple pattern, F << 1

(E) Plane bed, F < 1 and d < 0.4mm

(B) Dunes with ripples superposed F << 1

(F) Standing waves, F ≥ 1

(C) Dunes, F < 1

(G) Antidunes, F ≥ 1

(D) Washed-out dunes or transition F < 1

(H) Antidunes, F > 1

Table 6.2 Variation of Manning *n* values with changes in bed form occurring under different flow conditions.

Bed Form	Manning *n* $(ft^{1/6})$
Lower Regime	
Ripples	0.017–0.028
Dunes	0.018–0.035
Washed-out Dunes or Transition	0.014–0.024
Upper Regime	
Plane bed	0.011–0.015
Standing waves	0.012–0.016
Antidunes	0.012–0.020

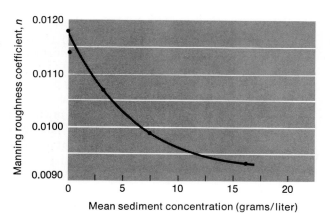

Figure 6.3.
Effect of suspended load concentration on the Manning roughness coefficient *n*.

Clearly, external boundary conditions such as channel shape and particle size generate a large amount of resistance. Some of these factors produce turbulence in the form of eddies and secondary circulation. In contrast, sediment concentration (the amount of sediment per unit volume of water) affects resistance internally. This modification was first detailed by Vanoni (1941, 1946), who showed that an increase in the concentration of suspended sediment tends to lower the resistance (fig. 6.3). As the concentration increases, the turbulent effect presumably is reduced because the mixing process within the fluid is dampened. All other factors being equal, sediment-laden water should flow at a higher velocity than clear water.

Sediment in Channels

Although most energy in a stream is dissipated by turbulence, a small part is used in the important task of eroding and transporting sediment. These processes, often taken for granted, are extremely complex and poorly understood, yet they underlie some of our most basic geological concepts. We will briefly review the more significant ideas about the relationship between river flow and sediment.

Transportation

In general, fine-grained sediment (silt and clay) is transported suspended in the water by the supporting action of turbulence. **Suspended load** usually moves at a velocity slightly lower than that of the water and may travel directly from place of erosion to points far downstream without intermittent stages of deposition. Coarse particles may also travel in true suspension (Francis 1973), but they are likely to be deposited more quickly and stored temporarily or semipermanently within the channel. Except for short spasms of suspension, coarse sediment usually travels as bedload. **Bedload** refers to sediment transported close to or at the channel bottom by rolling, sliding, or bouncing. How long coarse debris remains stationary within a channel depends on the size of the debris and the flow characteristics of the river; such debris probably is immobile more than it is in motion. Bradley (1970) showed that gravel can be stored in channel bars long enough for weathering to drastically weaken its resistance to abrasion.

Because of fluctuating discharge, at any given time a single particle may be part of either the bedload or the suspended load. As this makes the distinction between the two load types unclear, other terms have been devised to relate sediment more appropriately to river flow. **Wash load** consists of particles so small that they are essentially absent on the stream bed. In contrast, **bed material load** is composed of particle sizes that are found in abundant amounts on the stream bed (Colby 1963). While most, if not all, bedload is bed material load, most bed material load is transported as suspended load.

The relationship between wash load and discharge is poorly defined because most streams at any flow can carry more fine-grained sediment than they actually do. In fact, the concentration of fines is a function of supply rather than transporting power; therefore, it is relatively independent of flow characteristics. Coarse sediment, on the other hand, is usually available in amounts greater than a stream can carry, and so its concentration should correlate more significantly with the parameters of flow such as depth and velocity. The problem, however, is that direct measurement of bedload is extremely difficult because hand-held instruments can sample for only short periods, and when they are placed on the channel bottom the flow regime is disrupted. In addition, where bedload has been continuously measured, the amount of sediment passing a given channel cross section varies significantly on a time interval as short as one minute (Leopold and Emmett 1977). Furthermore, the amount of bedload varies drastically in different subwidths of the channel cross section.

Because of the difficulties surrounding direct measurement, most estimates of bedload discharge are made by means of empirical equations relating flow parameters and bedload transport rates (Meyer-Peter and Muller 1948; Einstein 1950). These equations, however, are themselves problematical because small errors in measurement of the flow parameters produce large variations in the computed load. In addition, the equations are valid only when hydrologic and sedimentologic conditions are the same as those that prevailed at the time the equation was derived (Andrews 1981).

The problems inherent in deciphering the relationship between parameters of flow and bedload transport are difficult but not insurmountable. Some very good bedload measurements have been made on the East Fork River, Wyoming (Leopold and Emmett 1976, 1977), where the U.S. Geological Survey has installed a concrete trough across the channel floor. Sediment moving along the river bottom falls onto conveyor belts rotating within the trough and is carried to the channel side where it is weighed (fig. 6.4). In addition, other measuring techniques and instruments have been used successfully (Helley and Smith 1971; Emmett 1980; Reid et al. 1980).

(A)

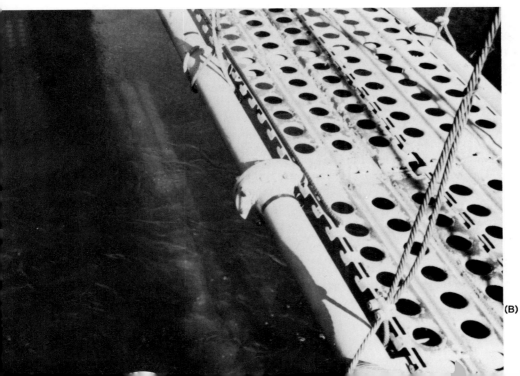

(B)

Figure 6.4.
U.S.G.S. bedload sampling
station, East Fork River, Wyo.
(A) View across river showing
suspension bridge and drive
mechanism of conveyor belt
bedload sampler. (B) Bedload trap
on streambed visible below
suspension bridge.

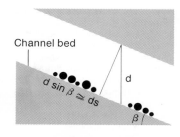

Channel bed

Figure 6.5.
Component of flow weight exerted
as shear stress on a channel
bottom. The critical shear stress
is the depth-slope product (*dS*)
multiplied by the specific weight
of the water and *β* is the angle of
slope.

Entrainment and Bank Erosion

The temporal and spatial variations in bedload transport rate noted at the East
Fork River station are related to the mechanics involved in moving coarse-
grained sediment. Most large particles do not move great distances during any
transporting event. Instead, their downstream migration is characterized by
spasmodic bursts of short-distance movement separated by periods during
which they come to and remain at rest. The processes that initiate the bursts
of motion experienced by any particle are collectively known as **entrainment.**
The amount of sediment entrained depends directly both on the erosive power
of the flow and on the size of particles on the bed surface that are in the proper
position to be eroded. Two streams with identical flow conditions may have
different bedload or bed material discharge if one flows across a fine-sand
bottom and the other over a cobble bed.

The term *competence* refers to the size of the largest particle a stream
can entrain under any given set of hydraulic conditions. It should be evident
that the value of competence to geomorphology depends on how we measure
the sediment being moved and, more importantly, how accurately we can determine the flow conditions. Although this may seem simple enough, in practice it is an excruciating problem for several reasons: (1) particles are entrained
by a combination of fluvial forces including direct impact of the water, drag,
and hydraulic lift, and each of these may be best defined by a different parameter of flow; (2) flow velocity is neither constant nor easily measured, especially during high discharge; and (3) sediment of the same size may be
packed together differently or have shape properties that cause abnormal responses to the same flow conditions. Thus, any investigation into the mechanics of competence must settle for only partial success until we can eliminate
or inhibit some of the inherent variability.

Historically, two hydraulic factors have been utilized to represent the flow
condition in the competence relationship. The first, *critical bed velocity,* is
used to demonstrate the relationship between velocity and entrainment. It has
been long known (Rubey 1938) that the volume or weight of the largest particle moved in a stream varies as the sixth power of the velocity. The *sixth-
power law* provides a sound theoretical basis for competence studies, but it is
less satisfying in practice because accurate measurement of bed velocity is
exasperating—if not impossible—in high-energy streams. Mean velocity, an
easily determined substitute in competence work, may not be reliable unless
the depth is also considered in the analysis (Vanoni et al. 1966).

The second factor, *critical shear stress* (sometimes called *critical tractive
force*), signifies the downslope component of the fluid weight exerted on a bed
particle (fig. 6.5). It is proportional to the depth-slope product and can be
expressed by the DuBoys equation for boundary shear:

$$\tau_c = \gamma RS,$$

where τ_c is the critical shear stress, γ the specific weight of the water, R the
hydraulic radius, and S the slope. In most streams transporting coarse bedload, R is closely approximated by mean depth, but as indicated earlier, care

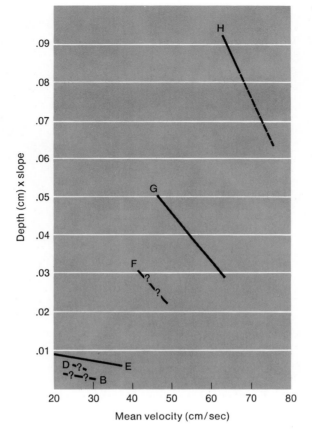

Figure 6.6.
Sediment particles of different sizes begin to move on the streambed at different values of mean velocity and depth-slope. Smallest particles (*B*) move mainly as a function of *dS* while largest particles (*H*) move primarily as a function of the mean velocity. (From Rubey 1938)

should be taken before a substitution of parameters is made (Tinkler 1982). The use of critical shear stress in competence studies has been criticized (Yang 1973), but the simple reality that depth and slope in a river are easier to measure than bed velocity makes it an appealing parameter.

Precisely how the two methods correlate with each other is not completely understood, but Rubey (1938) presented evidence to suggest that in the size range between fine sand and pebbles, critical bed velocity becomes more important in the entrainment process as particle size increases (fig. 6.6). Smaller sizes move as a function of the *dS* product and seem to be relatively independent of velocity. Thus, the shear stress approach may be completely valid only for smaller sizes or low-velocity flows, and very fine-grained sediment requires higher velocities for its entrainment than the sixth-power law would predict. Wolman and Brush (1961) found similar trends in a later flume study of the movement of sand-sized debris.

Figure 6.7.
Mean velocity at which uniformly sorted particles of various size are eroded, transported, and deposited.

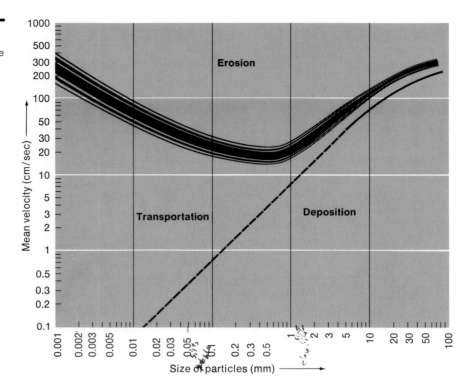

The above observations fit rather nicely the curves produced by Hjulström (1939) and shown in figure 6.7, which relate current velocity, particle size, and process; as figure 6.8 shows, Rubey's size classes seem to fall along the trend of the boundary between erosion and transportation, that is, the threshold velocity needed to initiate motion. The velocity that produces erosion of clay-sized particles is in some cases as great as that needed to entrain larger material. This explains the commonly observed phenomenon of coarse particles being transported across stationary material of a smaller size.

Unfortunately, flumes are not useful in the study of competence when particles are larger than pebble size. Most competence investigations of coarser sediment have therefore been made in natural rivers or canals and, for reasons explained earlier, employ shear stress as the diagnostic hydraulic variable (Lane 1955; Fahnestock, 1963; Kellerhals 1967; Scott and Gravlee 1968; Church 1972, 1978; Baker 1973b; Baker and Ritter 1975). Shear stress analyses such as these have produced widely divergent results concerning competence. Andrews (1983) suggests that much of this variation comes from our failure to consider sediment characteristics in the competence analyses. He was able to show that the size distribution of bed material has a significant effect on the shear stress required to entrain a particle of any given size. The variation in estimates of competence also prompted Bradley and Mears (1980) to combine a number of earlier techniques to gage the velocity and depth associated with entrainment.

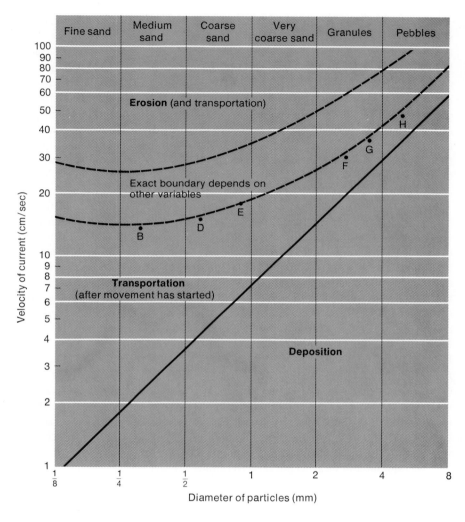

Figure 6.8.
Portion of Hjulström curve with size grades *B* to *H* placed on diagram. Compare figures 6.6 and 6.7. (Adapted from Rubey 1938)

In a related type of study, Bagnold (1973, 1977) proposed that entrainment and transportation of bedload may be analyzed in terms of stream power. **Stream power** is defined as

$$\omega = \gamma QS,$$

where ω is stream power and γ, Q, and S are specific weight, discharge, and slope, respectively. If power is considered per unit area of the stream bed, it essentially becomes a combination of shear stress and velocity because

$$\omega = \frac{\gamma QS}{\text{width}} = \gamma dS\bar{v} = \tau \overline{V},$$

where $\overline{V}$ is mean velocity. Where available stream power is greater than that needed to transport load, scour of bed alluvium (entrainment) will occur. Bull (1979) suggests that stream power analyses can be used to determine the likelihood of a threshold response in rivers.

The processes of entrainment determine the type and magnitude of erosion that occurs on the channel floor. It is incorrect, though, to assume that the only significant erosion is vertically directed. Bank erosion, which proceeds laterally, not only contributes to the sedimentary load but also, through its control on channel width, exerts a direct influence on other channel processes. A large number of studies have identified the major processes involved in bank erosion (see Thorne, 1982). Invariably researchers of this phenomenon arrive at the striking conclusion that bank erosion is rarely, if ever, accomplished by a single process but instead involves some combination of processes unique to the individual setting. In general terms, two major categories of erosion have been recognized: *fluvial entrainment* and processes of *weakening and weathering* (Thorne 1982).

Fluvial entrainment promotes bank erosion in two ways. First, sediment may be entrained directly from the bank surface by the shear stress generated in river flow, a process usually referred to as *corrasion.* Second, differential corrasion often produces an overhanging ledge of cohesive sediment because a noncohesive layer in the bank material has been more rapidly eroded (Thorne and Tovey 1981). The overhanging bank sections, called *cantilevers,* are mobilized when the undercut cohesive material finally fails and drops to the surface below. A similar process operates where vertical fractures, called *tension cracks,* exist in the floodplain sediment. In these cases, lateral trimming of the bank at the water level intersects the tension crack and produces downward failure along the fracture plane. This type of movement has been called *soil fall* (Brunsden and Kesel 1973), *earth fall* (Twidale 1964), *slab failure* (Hagerty 1980), and *shallow slip* (Thorne 1982). The failure is usually associated with release of river pressure against the banks and constant removal of debris from the bases of the banks. Normally, slab failure does not involve as much bank material as other mass movements discussed below, but the process is very significant because it occurs frequently (Thorne and Tovey 1981).

Weakening and weathering tend to reduce the strength of bank materials and thereby promote instability and failure. The mechanics of failure depend on many variables such as geometry and structure of the bank and the physical-chemical properties of the bank material. The most important control on weakening of bank material is the soil moisture condition. This condition depends on both climate and bank properties, and it is transformed into bank erosion by processes that (1) reduce strength within the bank and (2) act on the bank surface to loosen and detach particles and their aggregates. For example, many researchers note a distinct seasonality associated with bank erosion rates, winter and spring rates being considerably greater than those in the summer. Presumably this results from high moisture contents and the effect of frost action in the winter and spring months (Wolman 1959; Twidale 1964; Hill 1973; Thorne and Lewin 1979).

Where saturated banks are found in poorly drained, cohesive sediment, positive pore pressure can decrease the strength of bank material. This is especially true in high, steep banks after prolonged precipitation or rapid drawdown of the river level. Under these conditions bank failure may occur by rotational sliding (see figure 4.21). In some cases, the stratigraphy of the floodplain sediment plays an important role in bank failures, especially where cohesive layers rest above and below a noncohesive layer. Usually the noncohesive layer consists of permeable sands or silts in contrast to the cohesive material, which is normally richer in clay. In these cases, the cohesionless zone often serves as an avenue of pronounced seepage of underground water. Sapping at the base of the overlying unit may occur, or the seepage may actually transport material from the noncohesive unit in a process known as *piping* (Deere and Peck 1959; Hagerty 1980). In either case, internal strength is lost and failure of the overlying mass usually follows. In addition, the top horizon of the clay-rich cohesive unit beneath the zone of seepage may become a lubricated surface that serves as a sliding plane for overlying material. This type of failure, referred to as a *planar slide* (see figure 4.20), usually functions on sloping surfaces, but where shear resistance on the plane of sliding is very low, the movement can occur on a horizontal surface. Planar slides (fig. 6.9) have been recognized as important factors in the erosion of bluffs along the Mississippi River (Brunsden and Kesel 1973) and, in combination with other types of movement, are probably quite common (Varnes 1958). They require, however, that vertical fractures (tension cracks) exist in the bank sediment.

Figure 6.9.
Bank erosion along the Osage River in central Missouri by lateral spreading and planar sliding.

The important message in the above is that many times riverbank erosion has nothing to do with rivers. Often it is a mass movement phenomenon controlled by the texture and stratigraphy of floodplain sediment and triggered by the movement of groundwater. In light of this, the rate of bank erosion in alluvial channels can be enormous or minuscule, depending mainly on the character of the bank materials. In general, banks that are composed of fine-grained sediment or are densely vegetated (Hadley 1961; D. G. Smith 1976) have more resistance to corrasion than channels with sandy or gravelly banks. The actual process of erosion, however, may differ; clay-rich banks usually retreat by undercutting and failure of large blocks of the bank (Stanley et al. 1966; Laury 1971), while more coarse-grained banks erode by dislodgement and sloughing of individual particles. Even highly cohesive banks may therefore erode rapidly if the dominant process is undercutting and/or mass failure.

Deposition

If the entrainment of sediment logically represents a threshold of erosion, a similar threshold must exist when sediment in transport is deposited. Suspended rock and mineral fragments tend to settle to the bottom at a rate that depends on the density of the water, the fluid viscosity, and the size, shape (Komar and Reimers 1978), and density of the sediment. The distance any suspended particle will be transported in one event depends on its fall velocity and on whether its downward settling is offset by turbulent forces in the water. Coarse particles tend to be deposited during minor fluctuations in velocity (see figure 6.7). Variations in any of the fluvial properties make the channel floor a dynamic interface where some particles are being entrained while others simultaneously are being deposited. The net balance of this activity depends on local conditions rather than on the average cross-sectional hydraulics; in fact, local effects may be different from those that occur over a long reach of the channel (Colby 1964). It is known that scour and fill may be happening in the same channel reach at any given time, and a number of alternating events may coincide with fluctuating discharges. A long episode in which less sediment leaves the bed than is returned results in a distinct period of aggradation, and the converse situation leads to an episode of degradation. Long-term events are caused by changes external to the channel (climatic or tectonic) that introduce into any stream reach more sediment than the available discharge can transport. These events will be considered in more detail later when we examine the relationship of time in channel mechanics.

Fluvial deposition is important to geomorphology in several ways. On a long-term basis, continued deposition results in landforms that reflect distinct periods of geomorphic history. The stratigraphy of the associated deposits indicates the types of rivers involved in the aggradational phase (Schumm 1977) and therefore provides clues to environmental reconstructions. On a short-term basis, deposition creates bottom forms such as dunes, bars, and riffle-pool sequences that are closely interrelated with channel pattern and the character and distribution of flow within the channel (for example, see Schumm et al. 1982). These relationships are important parts of river mechanics and will be discussed again in later parts of this chapter. Finally, you should recognize that the short- and long-term mechanics of deposition have implications beyond the boundaries of geomorphology. They are clearly basic to sedimentology and stratigraphy and, interestingly, may be key factors in subdisciplines of economic geology such as exploration for valuable placer deposits (Schumm 1977).

The Frequency and Magnitude of River Work

At this juncture we can logically ask when and how fluvial work is done. Is it the superevent of very high discharge that happens once in a millennium that causes rivers to do what they do, or is it the normal flow that is repeated time and time again? The answer probably depends on one's perception of geomorphic work.

Geomorphic work is usually estimated in one of two ways. First, Wolman and Miller (1960) suggest that the work done by a river can be estimated by the amount of sediment it transports during any given flow. They concluded that in most basins 90 percent of the total sediment load (i.e., 90 percent of the work) is removed from the watershed by the sum of rather ordinary discharges that recur at least once every five or ten years. There seems to be little argument concerning that conclusion. It can be justified theoretically and by measurement; for example, loads transported during some very large floods (Stewart and LaMarche 1967; Scott and Gravlee 1968) represent a smaller percentage of the total load removed from their basins than the normal flows of high frequency and low magnitude.

The second way to estimate geomorphic work, with perhaps greater implication, is to assess the conditions under which rivers make adjustments that control or maintain their channel morphologies. Primarily on the basis of their sediment analyses, Wolman and Miller (1960) also suggested that river channels form and reform within a range of flows between a lower limit set by the demands of competence and an upper limit where flow exceeds bankfull and is no longer confined to the channel. That is, channel configuration itself is probably related to high-frequency discharge, and its precise character is presumably an indication of river work. This hypothesis also received considerable support and, indeed, was reinforced in studies which suggested that channel morphology is adjusted during flows having a recurrence interval of between 1.1 and 2 years, i.e., bankfull discharge (Kilpatrick and Barnes 1964; Dury 1973). Therefore, the discharge that determines the characteristics and dimensions of a channel is known as the **dominant discharge,** and its frequency and magnitude was implicitly accepted to be bankfull discharge.

It seems justified to say that river channel morphology is maintained in all environmental settings by geomorphic work done during a dominant discharge or within a distinct range of flows. The real question, however, is whether bankfull discharge is the dominant discharge for all rivers. For example, Harvey and his colleagues (1979) found that river flows in northwest England redistributed bed material between 14 and 30 times a year and changed overall channel form from 0.5 to 4 times a year. In addition, the effect of major floods on channel and floodplain configuration seems to vary with the environmental setting (Costa 1974; Gupta and Fox 1974; Baker 1977; Moss and Kochel 1978).

This prompted the suggestion (Wolman and Gerson 1978) that the Wolman-Miller principle should be modified to include factors that control the work of floods in different environments.

One of the major factors controlling flood effect is **recovery time,** essentially the time needed for a river to recover its equilibrium form after a major flow event has disrupted the channel configuration. In humid climates the recovery time is relatively short, perhaps 1–20 years. In semiarid to arid climates, however, recovery periods tend to be much longer. The significance here is that the dominant discharge cannot have a recurrence interval shorter than the recovery time. This follows because the effects of one flood would not be removed (and the channel returned to its presumed equilibrium form) before another channel-modifying flood occurred. Thus, river channels in humid climates may be controlled by flows of intermediate frequency and magnitude (bankfull to 10-year flood), but it seems likely that larger discharges are more important in the work of arid or semiarid rivers.

The Quasi-Equilibrium Condition

Every river strives to establish an equilibrium relationship between the dominant discharge and load by adjusting its hydraulic variables (e.g., channel width and depth, velocity, roughness, water slope, etc.). This normal fluvial condition has been aptly referred to as a "quasi-equilibrium" state (Leopold and Maddock 1953; Wolman 1955) because the flow variables are mutually interdependent, meaning that a change in any single parameter requires a response in one or more of the others. The difficulty involved in understanding rivers becomes evident when you consider that discharge and load are in continuous flux, and so all the hydraulic variables must always be adjusting. Obviously equilibrium as a steady-state condition cannot be attained in a river—thus the term quasi-equilibrium.

Hydraulic Geometry

The quasi-equilibrium condition was first demonstrated in a landmark study by Leopold and Maddock (1953). Using abundant flow records compiled at gaging stations throughout the western United States, they set out to determine the statistical relationship—the **hydraulic geometry**—between discharge and other variables of open channel flow. Because every river has wide fluctuations in discharge, any given channel cross section must transport a range of flows that come to it from the adjacent upstream reach. Discharge, therefore, serves as an independent variable at any station, and the changes in width, depth, velocity, or other variables can be observed over a wide spectrum of discharge conditions (fig. 6.10). At a station each of the factors (w, d, v) increases as a power function such that

$$w = aQ^b$$
$$d = cQ^f$$
$$v = kQ^m,$$

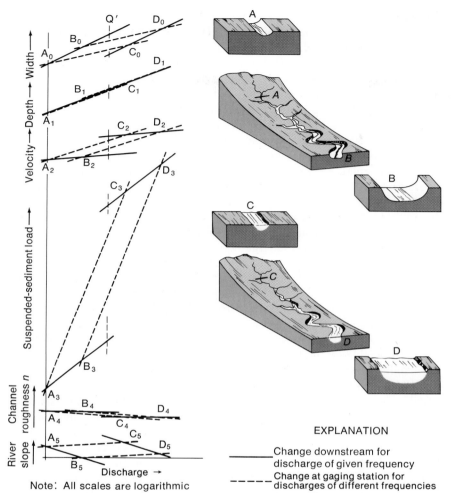

Figure 6.10.
Hydraulic geometry relationships of river channels comparing variations of width, depth, velocity, suspended load, roughness, and slope to discharge at a station and downstream. (From Leopold and Maddock 1953)

EXPLANATION

——————— Change downstream for discharge of given frequency

- - - - - - - Change at gaging station for discharges of different frequencies

Note: All scales are logarithmic

where *a, c, k, b, f,* and *m* are constants. The exponents *b, f,* and *m* indicate the rate of increase in the hydraulic variable (*w, d, v*) with increasing discharge. Because discharge (*Q*) equals the product of width, depth, and velocity, the relationship can be expressed as

$$Q = aQ^b \times cQ^f \times kQ^m$$

or

$$Q = ackQ^{b + f + m}$$

and it follows that ($a \cdot c \cdot k$) and ($b + f + m$) must each equal 1. Leopold and Maddock found that the average at-a-station values of *b, f,* and *m* for a large number of midwestern and western streams were .26, .40, and .34, respectively. The exponent values, however, will differ with climate and geology, and average values will not fit any particular stream. Essentially the at-a-station exponents tell us what portion of the increase in discharge will be caused by an increase in each of the component variables.

As discussed earlier, discharge also increases with the expansion of drainage area, and so on most rivers, it must increase downstream. The question is how much of the downstream increase in volume is due to each of the variables of width, depth, and velocity. To make this analysis, care must be taken to ensure that the variables are measured during the same flow conditions. On a given day, for example, a disastrous flood with high w, d, and v values may be occurring in an upstream reach while flow conditions far downstream are normal. A comparison of the hydraulic variables in these two widely divergent frequencies of flow would be misleading. Obviously the frequency of the discharge must be considered for any observations to be valid.

Leopold and Maddock (1953) found that the mean annual discharge is equaled or exceeded about 20 to 25 percent of the time on a large number of rivers. The mean annual discharge may not be the flow that determines the character of the channel; its absolute value is relatively low, being equaled or exceeded one out of every four days, and the water level is well below the bankfull stage. Nonetheless, its availability in published reports makes it a useful parameter to determine downstream hydraulic geometry. In sum, at-a-station and downstream hydraulic geometry differ in that one (at-a-station) compares flows of vastly different frequencies while the other (downstream) analyzes variables at the same frequency of Q even though the absolute values in cfs units are different.

Width, depth, and velocity increase downstream with increasing mean annual discharge (fig. 6.10). The average values of b, f, and m for western streams are .5, .4, and .1, respectively, but they vary from region to region and for any particular stream within a region. In general, the rate of change in depth (f) is relatively consistent in both downstream or at-a-station geometry; values seem to range from .3 to .45. Width usually increases much more rapidly and with more consistent values downstream than at a station, probably because the b value for any cross section depends more on bank cohesiveness than on discharge (Wolman and Brush 1961; Knighton 1974; Williams 1978). Velocity increases more rapidly at a station than it does downstream.

The nonchalance of the statement that velocity increases downstream indicates how much fluvial geomorphology has changed since the Leopold and Maddock study. The suggestion that mean velocity, and probably bed velocity (Leopold 1953), increase downstream came as a shock to most geologists, who intuitively "knew" that small tributaries flowing on steep slopes must be traveling faster than the low-gradient trunk rivers. The surprise at this new interpretation of velocity was probably due mainly to geologists' inclination to consider slope as the major, if not the overriding, control of velocity. The importance of slope seems to be entrenched in geologic thinking and emerges in our basic concepts. Further, slope has always been involved in the interpretation of sediment transportation and deposition, and it usually is the easiest parameter to reconstruct for an ancient fluvial regime. For example, the surface of coarse-grained deposits such as terrace alluvium or fan debris is in many cases the original stream bottom, and its gradient represents the channel slope at the time of deposition. Slope, then, is a valuable tool in geologic interpretation and a prime factor of fluvial processes, but as we will see later, it cannot be singled out as the only adjustable variable in a stream channel or even the dominant one.

The possibility of a downstream increase in velocity should have been suspected because Manning's equation tells us that depth plays a greater role than slope in determining velocity. In a stream with a constant roughness, if depth increases downstream at the same rate that slope decreases, an overall increase in velocity will occur. Increased depth simply overcompensates for the loss of velocity due to a gradually decreasing slope.

All geomorphologists, however, do not accept the idea of a downstream increase in velocity. Carlston (1969), for example, suggested that on large rivers downstream velocity is probably constant, but on smaller streams it may increase or decrease according to local controls. Mackin (1963) sounded a more serious objection when he cautioned against the injudicious use of empirical methods as the basis for sweeping geomorphic conclusions, citing the downstream velocity interpretation as a specific example. Indeed, some streams do decrease in velocity downstream (Brush 1961; Carlston 1969), but as Mackin points out, these individual cases may be obscured when they are placed on the same scatter diagram with measurements taken on other streams with different fundamental controls. The increase in velocity, therefore, may be the result of a particular methodology; as such it represents the exposition of a general trend rather than a conclusive rule. It is difficult to find fault with Mackin's argument, but in defense of empiricism we must recognize that streams do not lend themselves to a purely rational or scientific approach. The complex interaction in a river cannot be adequately expressed in mathematical terms, and the multitude of variables that define the system are constantly adjusting and readjusting to minor variations in flow. There may always be an element of indeterminancy in river mechanics that is simply beyond rational comprehension (Leopold and Langbein 1963). It may be better to strive for general trends and admit their fallibility in specific cases than to wait endlessly for a detailed scientific explanation.

Hydraulic geometry analyses have become standard in fluvial geomorphic studies for making comparisons of rivers of different climates, physical settings, and size (Wolman 1955; Leopold and Miller 1956; Myrick and Leopold 1963; Fahnestock 1963), even though the values are probably controlled by local factors (Miller 1958; Hadley 1961; Wolman and Brush 1961; Thornes 1970; Knighton 1974; Park 1977). In fact, the case has been made that the most probable channel form is one in which adjustments of the hydraulic variables to changing discharge are minimal (Langbein 1964, 1965; Knighton 1977; Williams 1978). This concept, known as the *minimum variance theory,* suggests that the shape of channels is predetermined by local geomorphic constraints that allow adjustments to be made with the least variation in the hydraulic variables. If no constraints are present, all variables have equal variance and the exponent values will each be 0.33. Minimum variance is an oversimplification of reality because it only predicts the most probable exponent values. Thus, any individual cross section may have values different from the predicted values.

Figure 6.11.
Relation of suspended load to discharge in Powder River at Arvada, Wyo. (From Leopold and Maddock 1953)

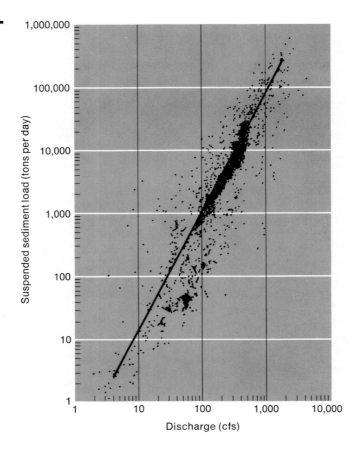

The utility of hydraulic geometry in geomorphic studies has yet to be satisfactorily documented. Park (1977) found that variations in sets of *b-f-m* values do not even distinguish rivers in diverse climates. On the other hand, Rhodes (1977) feels that all rivers can be categorized on the basis of various ratios of exponent values and/or other river properties (Froude number, roughness) that are controlled by the exponential values. The groups proposed by Rhodes reflect basic fluvial mechanics; this suggests that hydraulic geometry should be useful in predicting how any particular river will work. For example, one group includes rivers in which the rate of increase in velocity (*m*) exceeds the combined changes in width and depth (*b* + *f*). Such rivers should experience a rapid increase in competence with rising discharge, a condition that is probably needed to entrain coarse bedload (Wilcock 1971).

In most streams the amount of suspended sediment at a station increases rapidly with increasing discharge (fig. 6.11), varying as the simple power function

$$L = pQ^j,$$

where *L* is suspended load and *p* and *j* are constants. The at-a-station value of *j* is normally > 1, indicating that the influx of sediment to the river is

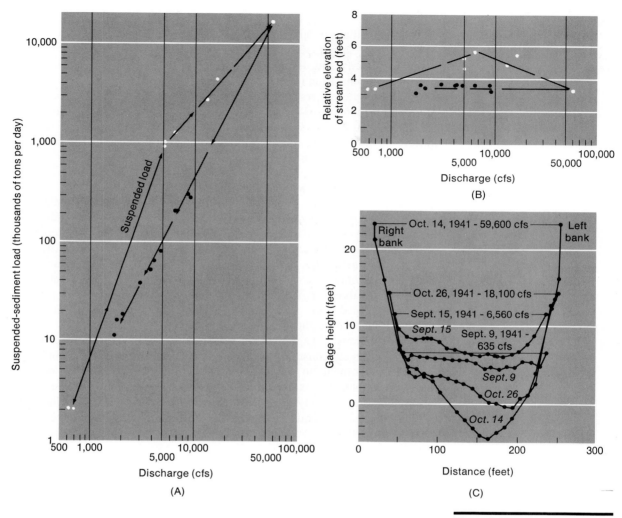

(A)

(B)

(C)

Figure 6.12.
Changes in (A) suspended load,
(B) stream-bed elevation, and
(C) water-surface elevation with
discharge during flood of
September–December 1941, San
Juan River near Bluff, Utah. (From
Leopold and Maddock 1953)

greater than the addition of water. Interestingly, the dramatic increase in sediment content is not necessarily caused by scouring of the channel floor. Several studies have shown that scouring can occur at different times during the
passage of a flood. It may occur at peak flow, during rising flow, or in the
waning part of the flood (Leopold and Maddock 1953; Foley 1978; Andrews
1979). It has also been shown that scouring and filling can occur simultaneously in close proximity within the same channel reach (Foley 1978; Andrews 1979).

In those situations where scouring occurs at peak flow or during the recession phase, deposition may take place in the rising stages of the flood, precisely
when the suspended load is increasing rapidly (fig. 6.12). Scouring takes place
later in the flood when velocities and sediment loads are lower. This observation would mean that the bulk of sediment added to a river during floods
derives from the valley-side slopes of the watershed, channel bank erosion, or

tributary input. The observation also helps explain the notable variation in suspended load at any given discharge (evident in fig. 6.11) because the sediment acts as an independent variable that demands compensatory adjustments in velocity and depth. Since the value of *j* is greater than unity, the sediment concentration (suspended load/unit volume H_2O) increases, thereby lowering the internal resistance to flow, as discussed earlier. The decreased *n* prompts an increase in velocity and enhances the river's ability to carry a greater load. In this way, a high rate of increase in suspended load seems to initiate the adjustments in hydraulic geometry that are needed to carry that load through the section.

The Influence of Slope

It was suggested earlier that channel slope has always been recognized as a prime adjustable property of rivers. Geologists and geographers traditionally have carefully studied the river gradient and generally have accepted the proposition that a concave-up longitudinal profile (change in elevation with increasing length) is the channel form assumed by rivers in equilibrium. There is abundant evidence to substantiate the importance of slope in a river striving to maintain balance, but whether the gradient adjustment operates to the exclusion or subservience of other variable changes is a question of considerable debate in modern geomorphology.

Observations at gage stations show that the slope of the water surface remains relatively constant during flows of different magnitudes, indicating that the adjustments to increasing discharge must be made by the other hydraulic variables. For example, we cannot call on a dramatic increase in slope to produce the relatively high rate of increase in velocity (*m*). The increasing velocity must be generated by an increase in depth, a decrease in roughness, or both. Downstream the channel gradient does exert an influence, because in most rivers there is a notable decrease in slope. Roughness, however, usually remains fairly constant (Leopold and Maddock 1953) because of the offsetting effects of a decrease in particle size (decreases *n*) and a decrease in sediment concentration (increases *n*). As a result, any increase in velocity downstream can best be justified by the increase in depth, explaining the low *m* values in that direction.

The relationships between slope and other hydraulic parameters reveal the complexities of quasi-equilibrium, but they do not explain why slope usually decreases downstream or what external factors may control the form of the longitudinal profile. As early as 1877, G. K. Gilbert concluded that slope was inversely related to discharge, and since *Q* increases with basin area and stream length, it is axiomatic that slope should decrease downstream. However, in most rivers particle size generally diminishes downstream, prompting many observers to suggest that channel gradient adjusts to the bed material. Actually both factors are probably involved. Rubey (1952) demonstrated that if channel shape is constant the slope will decrease with (1) a decrease in particle size, (2) a decrease in total load, or (3) an increase in discharge. Rubey concludes that the channel gradient at any point along the river is a function

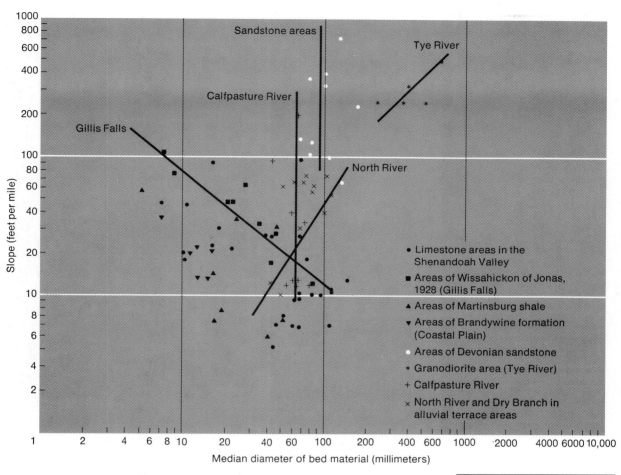

Figure 6.13.
Scatter diagram showing relation
between channel slope and
median size of bed material in
selected rivers. Maryland and
Virginia. (From Hack 1957)

of both sediment and discharge. If Rubey is correct, then slope is dependent,
or partially so, on all hydraulic variables because they are also related to discharge.

Many studies subsequently have shown the correctness of Rubey's analysis. In one of these studies, comparing stream profiles in areas of differing
geology, Hack (1957) found no consistent correlation between slope and bed-
material size when all sample localities from a geologically divergent region
were considered together. His plot, shown in figure 6.13, reveals a tremendous
scatter; for example, streams with a median particle size of 60 mm have slopes
ranging from 1.1 m/km (6 ft/mi) to 37 m/km (200 ft/mi). It was only after
Hack added a third variable, drainage area, to the analysis that a significant
relationship became apparent (fig. 6.14), and slope could be defined by the
equation

$$S = 18 \left(\frac{M}{A} \right)^{0.6},$$

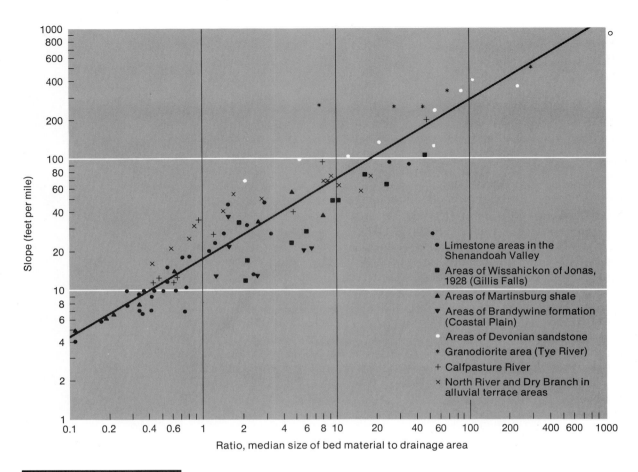

Figure 6.14.
Relation between slope and the ratio of median size of bed material to drainage area in selected streams, Maryland and Virginia. See figure 6.13 for comparison. (From Hack 1957)

where M is the median size of the bed material in millimeters, A is area in mi², and S is slope in ft/mi. Since basin area can normally be used as an index of discharge (Leopold et al. 1964), Hack's study reinforces Rubey's contention that both Q and sediment are determinants of slope. It does not indicate which factor is the principal determinant; indeed, one would expect the relationship to be defined by different mathematical equations in different physical settings. Nonetheless, Hack demonstrated that streams flowing within a single geological unit and having the same drainage area should have similar channel slopes and particle size.

On a local scale, slope is less dependent on discharge and more closely related to sediment and bed characteristics. For example, Prestegaard (1983b) found that components of bed roughness (particle size and bottom configuration) are the major determinants of water-surface slopes in a number of gravel-bed streams. In rivers such as those, particle size has an important influence on slope over a range of distances from local (one to three times the channel width) to an entire reach (100 m–300 m). Bed configuration exerts an influence only on a scale relating to the entire reach length.

Before assuming that slope seems to be understandable and possibly even predictable, note once again that Rubey's conclusions were based on the assumption of a constant channel shape. Such an assumption is probably more than nature will allow because in most alluvial rivers the channel shape is subject to change. Thus, although slope depends on both sediment and discharge, fluctuations in those variables do not *demand* an adjustment in the channel gradient. It is equally possible that changes may be counterbalanced by variations in the cross-sectional shape of the channel.

Channel Shape

Logic tells us that, unless velocity is completely unrestrained, rivers with a large mean annual discharge have greater cross-sectional areas than streams with smaller average flows. This fact has been verified repeatedly by casual observation and documented by the relationships exposed in hydraulic geometry. It is significant, however, that many rivers with the same mean annual discharge have different cross-sectional areas; even when the total area is the same, the width-depth ratio may vary considerably. Obviously factors other than discharge alone influence the shape of river channels, and so we look to the channel sediment for an explanation.

Schumm (1960) presented cogent arguments to suggest that channel shape, as defined by W/D, is determined primarily by the nature of the sediment in the channel perimeter. Where perimeters have a high percentage of silt and clay (particles $<$ 0.074 mm), channels tend to be narrow and deep. In contrast, wide, shallow channels seem to be characteristic of rivers having coarse-grained perimeters. Schumm's data, collected from semiarid and arid climate streams, show that W/D is related to the percent of silt-clay by the equation

$$F = 255M^{-1.08},$$

where F is the width-depth ratio and M is the percent of silt and clay in the channel perimeter. The magnitudes of the mean annual discharge or the mean annual flood do not seem to affect this relationship; in fact, only 40 percent of the variability in channel shapes can be accounted for by discharge alone (Schumm 1971).

In streams that are not aggrading or degrading, the channel shape is also related to the load being transported. Bedload is more efficiently transported in a wide-shallow channel because higher velocities are nearer the channel floor (Lane 1937). Suspended loads are carried best in channels having lower width-depth ratios (narrow, deep channels). This prompted Schumm (1963a) to classify rivers on the basis of their load types; more important, it demonstrates another viable method by which a river can be adjusted. A channel reach suddenly burdened with a different type of load than the one it previously carried may as easily alter its channel shape to accommodate the new load as change its gradient by deposition or erosion. The repeated changes in the shape of the Cimarron River in Kansas since 1880 are an excellent example of this phenomenon (Schumm and Lichty 1963).

The responses envisioned by Schumm are based on relationships determined in semiarid to arid climates. In other climates the sediment control on channel characteristics may be different from those noted above (Baker 1978). Therefore, widespread application of the equations without consideration of climates is not encouraged.

Factors other than the character of sediments can affect the cross-sectional shape of channels. It has been shown that root systems of riparian vegetation exert a drastic decrease in bank erosion and therefore control the channel width (Smith 1976). In contrast, large trees that fall into a channel may increase bank erosion by diverting flow around a tree jam into an unprotected bank (Keller and Swanson 1979; Keller and Tally 1979). In arid regions sediment by precipitation of $CaCO_3$ may increase bank cohesion and thereby prevent expansion of channel width (Van Arsdale 1982).

Channel Patterns

Our discussion of adjusting mechanics in rivers has thus far centered on the balancing factors that function in a channel cross section or a series of cross sections considered in a downstream direction. Rivers also have characteristic forms extending over long stretches of their total length which, when observed in plain view, display a distinct geometric pattern. The pattern that a river adopts is now recognized as another manifestation of channel adjustment to the prevailing discharge and load. In that sense, channel patterns are another important aspect of river mechanics.

Patterns are usually classified as straight, meandering, or braided, even though the boundaries between types are sometimes arbitrary and indistinct. The distinction between straight and meandering, for example, is based on a property called **sinuosity.** Although there is no complete agreement on how sinuosity should be determined (Leopold and Wolman 1957; Schumm 1963b; Brice 1964), we will follow Schumm's definition that it is the ratio of stream length (measured along the center of the channel) to valley length (measured along the axis of the valley). The transition between a straight and a meandering stream is usually placed at a sinuosity value of 1.5, but again this value probably has no particular mechanical significance.

The braided pattern, characterized by the division of the river into more than one channel, is more easily discerned. The designation becomes vague, however, when only part of the river's total length is multichanneled. How much of the river must consist of divided reaches to constitute a braided system is an individual decision. Another complication is that some single-channeled rivers become distinctly braided in times of high flow, requiring that stage be considered when deciding on the pattern classification. In addition, there is no reason to expect that a river will display the same pattern for its entire length. In fact, because patterns are probably a function of discharge and load, minor variations in those factors downstream may easily generate different patterns, especially in very large watersheds.

Straight Channels

Most streams do not have straight banks for any significant distance, making the straight pattern a rather uncommon one. It may seem strange, then, that straight streams display many of the same channel features as the more common meandering pattern. As figure 6.15 illustrates, straight reaches often contain accumulations of bed material called **alternate bars** that are positioned successively downriver on opposite sides of the channel. A line connecting the deepest parts of the channel, called the **thalweg,** migrates back and forth across the bottom. In rivers with a poorly sorted load, the channel floor undulates into alternating shallow zones called **riffles** and deeps called **pools.** The pools are directly opposite the alternate bars, and the riffles are about midway between two successive pools. Clearly, a straight channel implies neither a uniform stream bed nor a straight thalweg, and the spacing of bars, riffles, and pools is closely analogous to that in a meandering channel (Leopold et al. 1964; Dury 1969).

Sequences of pools and riffles are very important manifestations of how bedforms, flow, and sediment transport are interrelated in rivers to maintain quasi-equilibrium. This is true regardless of channel pattern, indicating that the tendency to develop bars and/or pool and riffle sequences must result from some fundamental property of moving water. Pool-riffle sequences have distinct spatial characteristics that are largely independent of the material type forming the channel perimeter, even if the channel is cut in bedrock (Keller and Melhorn 1978). Successive riffles in straight channels are usually spaced about five to seven channel widths from one another and, in alluvial channels, are composed of more coarse-grained sediment than the intervening pools. Riffles are also wider and shallower than pools at all discharges (Richards 1976a).

At low discharge, riffles are characterized by rapid flow and steep water surfaces; in contrast pool velocities are low and surface gradients are gentle. Assuming that coarse particles cannot be moved under the low flow conditions existing in the pool, logic tells us that the pools should be destroyed by aggradation when sediment moves downstream from adjacent riffles. Consequently, the question arises as to how the riffle-pool sequence is continuously maintained. Actually, it has been demonstrated (Keller 1971; Richards 1976b; Andrews 1979) that as discharge increases, the *rate* of increase in bottom velocity and mean velocity is greater in the pools than in the riffles. Thus, a velocity reversal occurs at some particular discharge, above which the pool velocity is greater than the riffle velocity (fig. 6.16). As a result, coarse particles entrained from the riffles during rising discharge can be moved through the pool when discharge exceeds the point of velocity reversal. In the recessional phase, the largest particles are deposited on the riffles while discharge is still above the velocity reversal. Ultimately, fine sediment is eroded from the riffles and trapped in the adjacent pools, leaving a coarse-grained lag on the riffle and a thin, fine-grained deposit in the pool. Thus, pools and riffles seem to be formed and maintained by scouring (pools) and deposition (riffles)

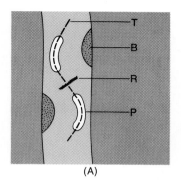

(A)

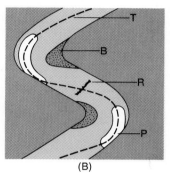

(B)

Figure 6.15.
Features associated with (A) straight and (B) meandering rivers. T = thalweg; B = bar; R = riffle; P = pool.

Figure 6.16.
Rate of increase in mean velocity over a pool and riffle with increasing discharge. East Fork River, Wyoming. (After Andrews 1979)

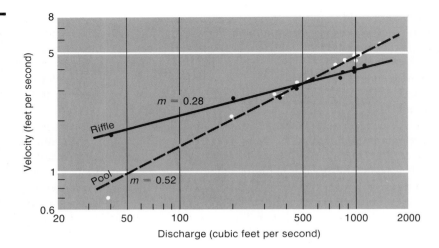

Figure 6.17.
Schematic diagram of convergent flow and secondary circulation over a pool (A) and divergent flow and secondary circulation over a riffle (B) in a straight channel.

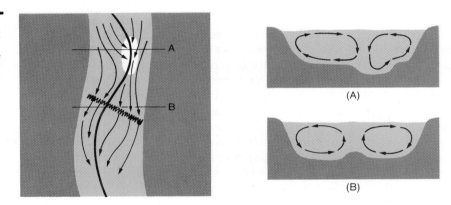

under conditions of relatively high discharge. Presumably this discharge occurs with moderate frequency and, therefore, is probably not associated with flooding.

Pool-riffle distributions are somehow related to the fundamental mechanics of open-channel flow that produce secondary water movements. The secondary motion occurs as cells circulating in planes that are transverse to the normal downstream component of flow (Leliavsky 1966; Leopold 1982). The direction of cell circulation over the pools is the opposite of that operating over the riffles (Keller and Melhorn 1973; Richards 1982), and the surface portion of the cells can be noted as zones of converging or diverging flow. In **convergent flow,** surface water is accelerated, and the maximum velocity occurs near the bottom because isovels in the water are depressed. This facilitates scour (fig. 6.17). In **divergent flow,** surface water tends to spread outward toward the banks. In this case bottom velocities are retarded and deposition is common (fig. 6.17). Significantly, convergent flow is common over pools and divergent flow is normally located over riffles (fig. 6.17).

Meandering Channels

The most common river form by far is the meandering pattern, also shown in figure 6.15. Meandering reaches contain the same physical components observed in straight channels (pools, riffles, bars), distributed in a similar way. The thalweg also migrates back and forth across the channel, impinging against the outer bank of the meander bends and crossing to the opposite side near the riffles.

Secondary circulation is a prime property of flow in a meandering system. Presumably, as water is forced against the outer bank of a meander, its slightly elevated level at that point gives the flow a circulating motion. The water moves along the surface toward the undercut bank and along the bottom toward the **point bar** (fig. 6.18). This corkscrew motion, called **helical flow** or **helicoidal flow,** has traditionally been thought of as a single rotating cell that reaches its greatest velocity slightly downstream from the axes of the meander bends at the position of the pools; the velocity decreases gradually downstream until the flow approaches or reaches the next riffle and the lateral circulation disappears. From this transition zone to the next downstream meander, the spiral velocity increases again, but the cellular pattern is opposite to its original direction because the orientation of the meander itself has been reversed (fig. 6.18). The magnitude of the lateral velocity varies but it can be great enough to influence the transport of sediment. Particles eroded from an undercut bank, for example, may be dragged into the center of the channel by the bottom limb of the helical cell. When the motion reverses at the transition, some of the sediment is returned to its original channel side and deposited on the next downstream point bar.

We now know that this model of helical flow is probably oversimplified. For example, several researchers (Hey and Thorne 1975; Bathurst et al. 1979) have observed two helicoidal cells at the meander apex and at the points of inflection (riffles) in several British rivers. At the meander apex, the two rotating cells meet at the surface in a zone of convergence that promotes scouring close to the undercut bank (fig. 6.19). At the transition, the flow diverges over the position of the riffle (fig. 6.19) and tends to induce deposition in that zone. This suggests that the polarity of the rotating cells must change between the meander axis and the next downstream inflection point. Such a distribution requires that an intermediate zone of no secondary circulation must be present where one type of helicoidal movement is dissipated and another begins (fig. 6.19). The precise relationships between helical flow and sediment transport are, therefore, still to be understood. The data needed to make such an analysis are very difficult to collect and require detailed field measurements (Bhowmik 1982).

Meandering rivers shift their positions across the valley bottom by eroding on the outer banks of meander bends and simultaneously depositing point bars on the inside of the bends. Even though the location of the river varies with time, there is no compelling reason to suggest that the shape or hydraulic properties of the river stray far from average values as long as the prevailing controls of climate and tectonics remain unchanged. In fact, meanders in rivers of all sizes are dimensionally similar, with consistent geometric and hydraulic relationships as in table 6.3 and figure 6.20.

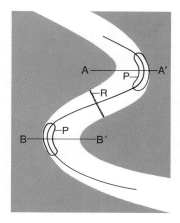

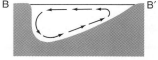

Figure 6.18.
Traditional model of helicoidal flow at successive bends in a meandering river. *P* represents a pool and *R* represents a riffle.

Figure 6.19.
Model showing two secondary
cells and surface flowlines in a
meandering river.

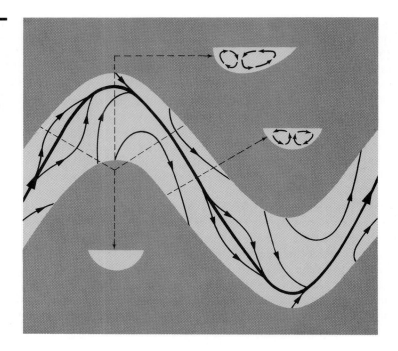

Figure 6.20.
Geometric parameters of a
meander: λ = wavelength;
A = amplitude; r_m = radius of
curvature.

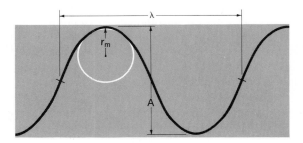

Table 6.3 Empirical relationships between parameters that define meander
geometry.

Dependent Relationship	Source
Wavelength	
$\lambda = 6.6\, w^{0.99}$	Inglis (1949)
$\lambda = 10.9\, w^{1.01}$	Leopold and Wolman (1957)
$\lambda = 4.7\, r_m^{0.98}$	Leopold and Wolman (1957)
$\lambda = 30\, Q_{bf}^{0.5}$	Dury (1965)
Amplitude	
$A = 18.6\, w^{0.99}$	Inglis (1949)
$A = 10.9\, w^{1.04}$	Inglis (1949)
$A = 2.7\, w^{1.1}$	Leopold and Wolman (1957)

λ = wavelength (ft)   *A* = amplitude (ft)
w = width (ft) at bankfull stage Q_{bf} = bankfull discharge (cfs)
r_m = radius of curvature (ft)

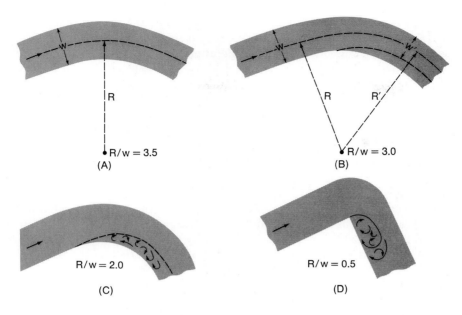

Figure 6.21.
Relation between radius of
curvature, width, and flow
properties. Decreasing ratio r_m/w
in (B) causes flow to break away
from inside of bend and create
eddying zone (C and D). R = r_m.
(After Bagnold 1960)

The most revealing geometric property is the meander wavelength, which relates to many other variables including discharge, width, and the radius of curvature (r_m). The relationships are in many cases almost linear and undoubtedly reflect basic mechanical principles. For example, using the equations in table 6.3 and considering them as linear, we find that where all parameters are measured in feet,

$$\frac{r_m}{w} = \frac{\dfrac{\lambda}{4.7}}{\dfrac{\lambda}{10.9}} = \frac{10.9}{4.7} = 2.3.$$

Actual measurement of this ratio shows most rivers to have values between 2 and 3 (Leopold and Wolman 1960), suggesting that the relationship probably is a function of channel curvature exerting an influence on flow. Bagnold (1960) showed that as the radius of curvature decreases $\left(\text{decreasing } \dfrac{r_m}{w}\right)$, the main filament of flow tends to shift toward the outer bank, causing a concomitant decrease in resistance on the inside of the bend. Greater curvature will continue to decrease resistance until a critical value of $\dfrac{r_m}{w}$ is attained, when flow along the inner bed becomes unstable and breaks away from the boundary (fig. 6.21). This creates eddy currents along the inside boundary, increasing the energy dissipation and so effectively establishing a minimum resistance for the flow. In most fluid systems, eddying begins when the curvature ratio is between 2 and 3, suggesting that the large number of real meanders having these values probably represents a quasi-equilibrium between flow and geometry. Hickin (1974) found the critical $\dfrac{r_m}{w}$ value to average 2.11 on the

Beatton River in British Columbia. He demonstrated that once a developing meander attains the critical curvature, it exerts significant control on the subsequent rate and direction of lateral migration.

Before the recent developments in fluvial geomorphology, most geologists accepted the premise that random diversions of flow by slumped boulders or fallen trees were the ultimate causes of meanders. Such aberrations undoubtedly can start meanders, and once an initial bend develops, the sinuous nature is transmitted downstream and causes more bends to form (Friedkin 1945). Helical flow is also propagated from the initial curve. Callander (1969) showed that any deflection of downstream flowlines on the river surface creates hydraulic instability because secondary flow is initiated. Sediment on the channel floor begins to move toward the inner bank. This results in the formation of an incipient bar, thalweg migration, and scouring of the bottom. In addition, two circulating cells are present at this time (Leopold 1982). Thus, it is clear that the ingredients of meandering are already present in straight streams.

Although it may be tempting to write off meandering as the result of random perturbations of flow direction, the change from straight to meandering requires that other factors operate within the system. First, Parker (1976) showed that secondary flow alone is not the cause of meandering. The instability associated with the beginning of helicoidal cells will not culminate in a meandering pattern unless a second process triggers the action. In alluvial rivers, the necessary process is bedload transport. Second, the meandering pattern will not develop without bank erosion. The bank erosion must be a local phenomenon rather than the widespread retreat of both banks, a process that would lead to creation of a wide-shallow channel.

Third, the development of a meandering pattern requires certain energy adjustments within the flow. Bank erosion and flow around curves dissipate available energy, and therefore the transition from a straight to meandering pattern implies that energy must be consumed in a different way. Most analyses of river energy indicate that meandering streams are probably closer to an equilibrium condition than straight streams because (1) meandering tends to dissipate energy in equal amounts along the length of the channel and (2) under the constraints of (1), meandering tends to minimize the total energy expenditure (to do the least work) or the rate of energy expenditure. This is accomplished by adjusting the curvature geometry or the channel gradient.

If a straight stream is prevented from making any of the above adjustments, it is doubtful that it can change into a stable, meandering pattern. For example, Schumm (1963b, 1967b) showed that sinuosity decreases and meander wavelength increases when a river transports sand and gravel rather than a fine-grained suspended load. Presumably this results because the available river energy must be used to entrain and transport bedload, and therefore the energy adjustments coincident to meandering (such as lowering slope) cannot be made.

Braided Channels

A basic part of the formation of the braided pattern is the division of a single trunk channel into a network of branches (fig. 6.22) and the growth and stabilization of intervening islands. Detailed studies of braided systems in both flumes and rivers (Leopold and Wolman 1957; Fahnestock 1963; Church 1972; Rust 1972; N. D. Smith 1970, 1974) show that divided reaches have different channel properties than adjacent undivided segments. Braided zones are usually steeper (fig. 6.23) and shallower; total width is greater although each channel may be narrower than the undivided trunk; and changes in channel positions and the total number of channels are likely to be extremely rapid (Fahnestock 1963; Church 1972; N. D. Smith 1974). Fahnestock, for example, documented lateral shifting of the channels up to 122 meters (400 ft) in eight days in the braided segment of the White River in Washington.

The Origin of Braids   Any explanation of the origin of braids is necessarily oversimplified because, like all fluvial processes, it involves the simultaneous interaction of a number of factors. According to Fahnestock (1963), the most important of these are the following:

1. *Erodible banks.* Most investigators of channel patterns feel that bank erosion is perhaps the most necessary factor in creating a braided system. If bank erosion is prohibited by material cohesiveness or vegetation, it is unlikely that a braided pattern will develop.

Figure 6.22.
Photograph of typical braided stream pattern. McKinley River, Alaska.

2. *Sediment transport and abundant load.* Almost every braided river transports large volumes of bedload, and much of the channel shifting is prompted by temporary deposition of bars across entrances to branches of the network. It is incorrect, however, to assume that braided rivers are overloaded, since many times braids form when the channel is actively being eroded (Leopold and Wolman 1957), and at the same time, undivided segments immediately downstream from an evolving braided reach are being actively aggraded. Even though the load exceeds the transporting capacity of the channel, no braiding occurs.

3. *Rapid and frequent variations in Q.* Fluctuations in discharge tend to produce the alternating erosion and deposition that seem to be a necessary part of braiding mechanics. However, some of the channel shifting and much of the increase in the number of channels may simply be a matter of reoccupying abandoned channels during high river stages. It is also significant that laboratory studies have produced braids under constant discharge, indicating that discharge may sustain the pattern, not cause it.

Although slope and other channel properties of braided streams are different from those of meanders, they probably are not factors in the origin of braids. More likely they represent the geometric modifications brought about by particular sediment and discharge requirements. For example, Parker (1976) emphasized that slope and the width-depth ratio are important manifestations of the controls leading to braiding versus meandering. If slope and w/d are high under conditions of dominant discharge, the pattern will probably be braided. In addition, braids result from a river's tendency to form bars (Parker 1976), but the type of bars developed may be a significant factor in maintaining a braided pattern (Church and Jones 1982). Therefore, it seems safe to say that braids do not necessarily connote instability. The pattern simply represents another condition a river may establish in response to external controls. It may be maintained for a long period of time and possibly is as close to true equilibrium as the meandering pattern.

Experimental studies suggest that braid development follows a distinct sequence of events, illustrated in figure 6.23 (Leopold and Wolman 1957). During high flow a portion of the coarse load being transported is deposited because of some local channel condition. This initial accumulation becomes the locus of an incipient longitudinal bar because reentrainment of the particles requires a greater velocity than did their transportation and deposition (fig. 6.7). Continued deposition here allows the bar to grow both upward and in the downstream direction. As particles move across the reach, they are deposited on the lower end of the bar where depth suddenly increases and velocity decreases. These changes occur because the width and discharge remain constant over that segment of the cross section, and since $q = wdv,$ an increase in depth requires a decrease in velocity. Most smaller particles move easily over the growing bar, but some may be trapped in the interstices between the larger grains.

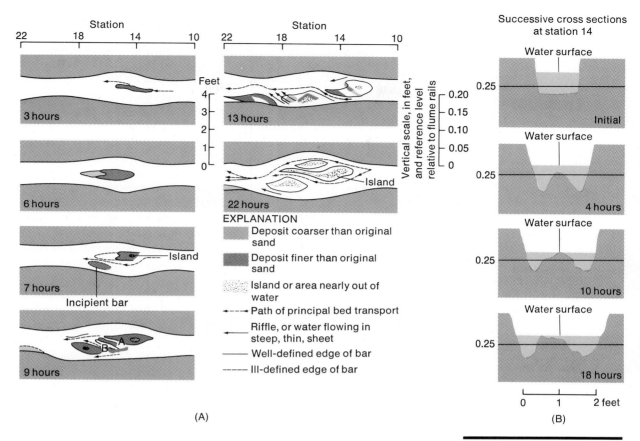

Figure 6.23.
Stages in the development of a braid in a flume channel. (A) Sequential development of the pattern at various times. (B) Cross sections at one station along the flume. (From Leopold and Wolman 1957)

As the expanding bar begins to occupy a significant portion of the channel area, the channel is no longer wide enough to contain the total flow, and as flow is deflected around the bar, the banks are eroded. Simultaneously, the bar itself may be trimmed and the channel somewhat deepened. These processes combine effectively to enlarge the channel on both sides of the bar, allowing the water level to be lower at any equivalent discharge. Eventually the bar emerges as an island flanked by two distinct channel branches. Bars and islands do not necessarily remain fixed in position or shape since they are also susceptible to lateral erosion (N. D. Smith 1974). In documented cases, however, vegetation may spread rapidly on the islands, especially if overbank flows or wind provide silt as a capping layer; the rooting of vegetation tends to increase the resistance to erosion. The whole process from initiation to stabilization may take as little as two years. Bar growth can also begin when a river reoccupies an abandoned channel. In such an event, an erosional remnant may become a core upon which new bar growth will develop (Eynon and Walker 1974).

Other bar types also are common in braided systems. N. D. Smith (1974) recognizes transverse, point, and diagonal bars in addition to the longitudinal type. Transverse bars are tabular bodies that grow by downstream migration of foreset beds developed more or less perpendicular to the current. These bars form when sand moving along the bottom encounters a shallow depression where velocity is lowered (Jopling 1966). In response, the sand is deposited as a "delta" that builds upward until the flow velocity increases to its former level, and the top of the bar becomes the channel floor across which sand can once again be transported. At low flow, transverse bars may be exposed and dissected into a series of small channels.

Smith (1971) noted significant differences between longitudinal and transverse bars in the Platte River in Nebraska. Transverse bars normally are composed of sand and are better sorted and more fine-grained than longitudinal bars. In addition, they show well-developed planar cross-beds in contrast to the crude horizontal stratification found in the coarse sediment of the longitudinal bars (Rust 1972). A more recent analysis of the macroforms (bars) and the meaning of their sedimentary structures is given in Crowley (1983).

Anastomosed Channels   Another type of multichanneled river is an **anastomosed stream.** Historically, the terms "anastomosed" and "braided" have been used interchangeably, but this practice should be discontinued because the normal characteristics of the two patterns and their formative mechanics are considerably different. Smith and Smith (1976) define an anastomosed river as one having an interconnected network of low-gradient, relatively deep and narrow channels that have variable sinuosities and stable banks composed of fine-grained sediment and vegetation. The significant point here is that these rivers have some characteristics of both meandering and braided streams but overall are distinct from each of those patterns. For example, the pattern is multichanneled, indicating a relationship with braiding; however, the banks are not erodible, which as indicated above, is a prime requisite in the formation of braids.

The essential ingredients needed to form anastomosed rivers are (1) rapid aggradation caused by downstream base level rise and (2) stable, cohesive banks that retard channel migration. Under these controls, floodplain surfaces are rapidly elevated by overbank deposition. It is also possible that gravel may simultaneously fill the channel (Smith and Smith 1976).

The Continuity of Channel Patterns

If channel patterns reflect specific adjustments to fluvial variables, it follows that boundaries between the patterns should be definable in terms of those variables. It also follows that each pattern must be stable within certain threshold limits of the controlling factors. When the limits are exceeded, a viable fluvial response would be a pattern change. Patterns, then, may be ephemeral fluvial properties, especially in segments where the values of the controlling factors are critically close to the threshold condition.

As we saw earlier, straight reaches in suspended-load rivers are rare enough to be considered as unstable and probably transitional to the meandering form. This observation is supported by energy analyses and by the fact that the physical components of straight reaches are analogous to those in meanders. For example, in straight reaches the spacing of successive riffles is about five to seven times the channel width. In meanders the relationship

$$\lambda = 10.9 w^{1.01},$$

where λ and w are measured in feet, suggests that the two successive riffles found in one complete wavelength are spaced about the same as those in straight reaches. Straight reaches, therefore, will probably not remain long in that form unless the banks are unerodable or a coarse-grained load requires the river to use all its energy for transportation, leaving none to be dissipated in meander bends. Even so, special cases may exist. Distributaries in deltas, for example, are straight and perhaps remain so through a combination of very low gradient and very low bedload values. In addition, there is no compelling reason to expect that meandering, once achieved, is forever stable. The mechanics that produce meandering also cause chutes and cutoffs that return the reach to straightness (Keller 1972; Lewin 1976), or they proceed to multilooping (Brice 1974) or other complexly woven forms (fig. 6.24).

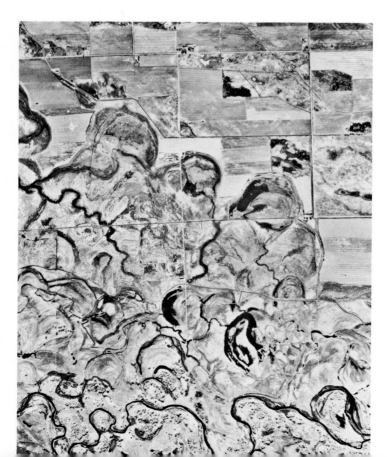

Figure 6.24.
Complex of loops and abandoned channels in strongly meandering river pattern. Rio Grande in Alamosa County, Colo.

Figure 6.25.
Relation of slope and discharge.
Lines represent threshold slopes
at various discharges as
determined in different studies.

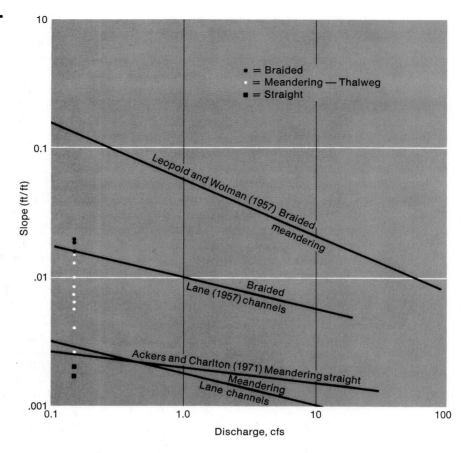

The threshold marking the transition between meandering and braided is perhaps more precisely, although empirically, defined. Several studies have suggested a threshold boundary between the two forms based on the relationship between slope and discharge (Lane 1957; Leopold and Wolman 1957; Ackers and Charlton 1971). Although the limiting values are not consistently the same (fig. 6.25), it seems obvious that at the threshold an increase in slope at any given discharge (or an increase in discharge at any given slope) will change a meandering pattern to a braided pattern. It has also been demonstrated (Schumm and Khan 1972) that the threshold values between patterns can be defined as well, if not better, by the slope-sediment load relationship. This suggestion was reinforced by Osterkamp (1978), who found that in many Kansas streams the lowest gradient of the braided pattern is best defined when bed material characteristics are combined with mean discharge to serve as the independent variable.

In sum, we can say with some assurance that the combined effects of discharge and sediment delimit the stability range for any channel pattern. Slope is probably not the inducing agent in pattern change but more likely adjusts as a dependent variable, along with the pattern, to changes in the sediment-discharge regime.

The physical operations within rivers are driven by their attempts to establish and maintain the most efficient conditions for transporting water and sediment. Because every river has a unique combination of these two factors, the parameters that define the equilibrium state must differ from river to river and even in various segments of the same river. Every student of rivers understands that discharge and load are not constant, and thus the element of time becomes a significant factor in any consideration of fluvial mechanics. Remember that discharge and load are not in themselves independent variables because they are ultimately a function of climate, geology, and tectonics. Furthermore, it is unreasonable to expect either discharge or load to change independently of each other since both factors are related to the same, more basic, controls. A change in climate, for example, will surely alter discharge, but it may also prompt a simultaneous change in the character of the load because vegetation and weathering will likewise adjust to the new climatic regime.

The importance of time is that the type of fluvial variable most likely to act to maintain equilibrium may depend on the time span being considered. Figure 6.26 shows a gradual increase in mean annual discharge provided to a river over a period of several hundred years. Presumably the load characteristics are also changing for the reasons stated above. Within the period, normal variations in discharge and load are accommodated by instantaneous adjustments of hydraulic geometry. Major floods may occasionally alter the valley topography or divert the river to a new position, but the channel itself will reorganize according to the average flow and load conditions. At some point, however, the gradually changing mean values of load and discharge can no longer be balanced by hydraulic variables under the prevailing channel configuration or pattern. One flow event, perhaps not even a major flood, will eventually exceed the stability limits of the original channel morphology; a fluvial threshold is passed and a major rearrangement of the channel pattern or its configuration must take place.

During this long-time interval, the river was in a perpetual quasi-equilibrium condition if one considers only the instantaneous responses of hydraulic geometry (*w, d, v,* sediment concentration, roughness) to short-lived events such as floods. It is clear, however, that during the same interval, the river was *approaching* a different equilibrium condition that required a particular channel morphology or pattern to balance the new mean values of discharge and load.

As discussed in chapter 1, we see, then, that fluvial equilibrium depends on the time scale one uses to define it. Schumm and Lichty (1965) addressed this problem by recognizing the importance of different geomorphic time spans and indicating that some variables of fluvial geomorphology are dependent or independent according to the time span being considered. Channel morphology, for example, is an independent variable during steady time and exerts a direct control on the hydraulics of flow. During graded time, however, channel morphology is a dependent variable.

Rivers, Equilibrium, and Time

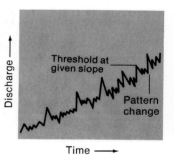

Figure 6.26.
Diagram showing gradually increasing discharge with time. Discharge variations are accommodated by variables of hydraulic geometry until threshold is reached. At threshold, a major change in the river character, such as a pattern change, is required to carry the increased average discharge.

Our interest here is in the adjustments a river might make to counterbalance changes in discharge and load that occur over a period of tens or hundreds of years, the time interval known as *graded time* (Schumm and Lichty 1965). Channel morphology is the main dependent variable on this temporal scale, largely determined by mean values of the controlling factors. Rivers during this episode may appear to be quite stable, if stability is judged by hydraulic geometry. Even the morphology may show little change since its adjustment may be imperceptibly slow.

Adjustment of Gradient

A common response of channel morphology to changes extending over a graded time span is the alteration of slope. As discussed earlier, in alluvial rivers the normal downstream decrease in gradient promotes a concave-up longitudinal profile. The concavity, however, is usually not perfectly smooth in detail but is commonly interrupted by perturbations. These can be caused by reaches where the channel is floored by bedrock or by local zones of erosion or deposition. Local filling may be initiated by an influx of bed material load that is too coarse or too great in volume to be transported on the preexisting gradient. For example, incision by upstream tributaries might provide so much additional load that deposition would occur in the downstream reaches. A change in particle size of the load may also be impressed on a trunk river if tributary basins are experiencing a change in fundamental controls. This is especially common where tributaries drain mountains that are subject to spasms of glaciation or uplift.

If a coarser load is produced, deposition should occur at the confluence of the two rivers until the local gradient is increased sufficiently to allow the bedload to be transported. As the channel floor is raised by deposition, the slope of the river upstream is effectively lowered, and a wave of filling may spread through the channel network. In contrast, a change that produces finer or less load may induce a river to entrench its channel. The most apparent cause of this response is the construction of large dams (Williams and Wolman 1984). These effectively starve the river of bedload in the reaches immediately downstream from the construction, and the river uses its excess energy to lower its gradient by downcutting. The entrenchment, however, may be stopped if the channel floor becomes armored by particles in the alluvium that are too large to be entrained on a lower slope (Livesey 1965; Hammad 1972).

Presumably, then, gradients will adjust to counteract changes in the load-discharge relationship. Recognize, however, that even if a change in slope occurs, there is no requirement that it will be propagated throughout the length of the system. Other responses involving the channel cross section are equally able to provide the necessary counteraction away from the point of equilibrium disruption (Leopold et al. 1964; Leopold and Bull 1979).

A clear indication of a river gradient undergoing active readjustment is a short, oversteepened segment of the longitudinal profile known as a knickpoint. **Knickpoints** are created by any process that results in the lowering of base level. Such an event causes pronounced channel incision immediately upstream from the site of base level decline because the river is attempting to

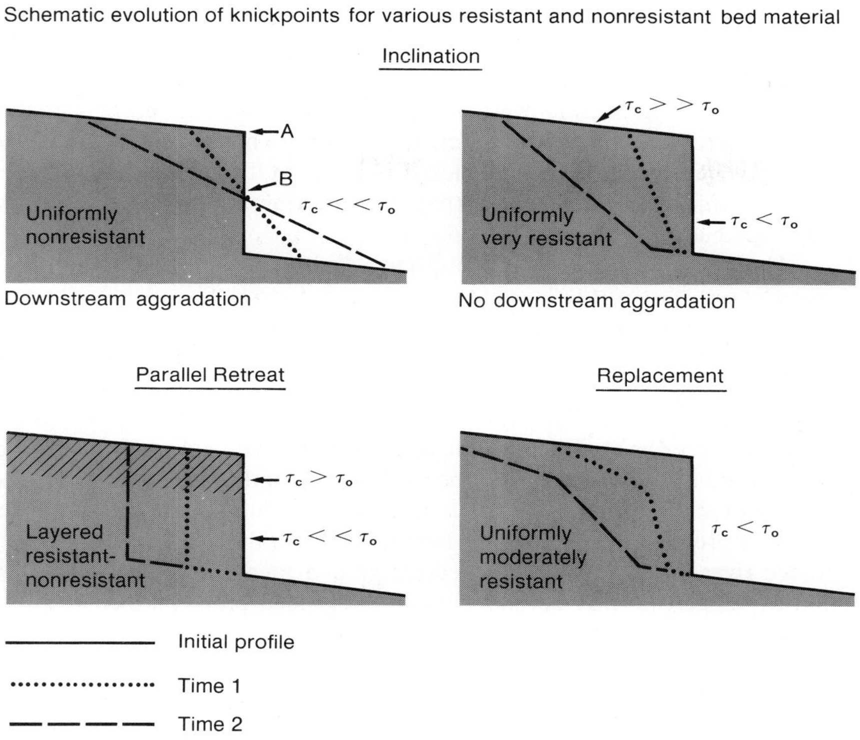

Schematic evolution of knickpoints for various resistant and nonresistant bed material

Inclination

Uniformly nonresistant

$\tau_c << \tau_o$

Downstream aggradation

$\tau_c >> \tau_o$

Uniformly very resistant

$\tau_c < \tau_o$

No downstream aggradation

Parallel Retreat

Layered resistant-nonresistant

$\tau_c > \tau_o$

$\tau_c << \tau_o$

Replacement

Uniformly moderately resistant

$\tau_c < \tau_o$

——————— Initial profile

················· Time 1

— — — — Time 2

Figure 6.27.
Models of knickpoint evolution for various types of bed material. τ_c is critical bottom shear stress needed to initiate erosion. τ_o is actual bottom shear stress. The knickpoint lip, shown at A, is the break in slope where the channel becomes oversteepened. The knickpoint face at B extends from the lip to the base of the knickpoint.

establish a new equilibrium condition. It is generally accepted that knickpoints will migrate upstream with time and, in doing so, may even initiate a wave of erosion throughout a river basin. In fact, rates and distance of the headward erosion are amenable to modeling (Pickup 1977) and have been expressed in mathematical terms (Begin et al. 1981). Knickpoint behavior can also be studied experimentally (Brush and Wolman 1960; Begin et al. 1980; Gardner 1983). Nonetheless, it seems clear that the details of the headward erosion depend on the character of the geologic and hydrologic setting.

Both flume and field data suggest that in channels formed of cohesionless material, pronounced knickpoints will be smoothed out after only a short distance of upstream migration (Brush and Wolman 1960; Morisawa 1964). Where the channel is composed of bedrock or cohesive sediments, the knickpoint may retreat for considerable distances and still preserve a vertical headcut, although Gardner (1983) showed that this does not always occur.

Gardner (1983) suggests that knickpoint evolution can occur in any of three different modes, depending on the balance between the shear resistance of the bed material and the shear stress produced in the river flow (fig. 6.27). In *knickpoint inclination* there is a uniform change in slope of the knickpoint face, the details of which depend on the material resistance and whether the river can transport the sediment away from the front of the knickpoint. Where the shear stress needed to erode this debris (τ_c) is less than the actual shear stress in the river (τ_o), aggradation will occur and the fill becomes part of the readjusted gradient (fig. 6.27). *Parallel retreat* is characterized by retreat of the near-vertical knickpoint face without a change in its inclination. This mode of migration is produced best if layering exists in the parent material such that a more resistant zone overlies a less resistant zone (fig. 6.27). The third type of migration, *knickpoint replacement,* occurs when erosion of the bottom takes place upstream from the knickpoint lip as well as along the face. It results in a knickpoint profile consisting of two distinct zones in which the original slope has been modified (fig. 6.27).

The changes in channel slope discussed above are responses familiar to geologists as part of the concept of the *graded river.* The idea of a graded condition in rivers finds its roots, once again, in the writings of G. K. Gilbert, who indicated that "equilibrium of action" in streams consists of a mutual adjustment between velocity, discharge, slope, and load. Later W. M. Davis (1902) refined these ideas by introducing the terms "grade" and "graded slopes" to describe the balanced fluvial condition. Although terminology left wide latitude for interpretation, the early conceptual models clearly stressed the importance of slope in the adjusting process, paving the way for the widely accepted premise that a concave longitudinal profile is the trademark of a graded river. In addition, Davis tied the graded condition to his cycle of erosion by suggesting that development of the graded profile, which is the optimum form to transport sediment, marks the beginning of the mature stage of the cycle.

The idea of a graded river did not go unchallenged, and in fact it was in answer to these challenges that Mackin (1948) put the concept into perspective and provided for the first time a clear definition of a graded river as

> . . . one in which, over a period of years, slope is delicately adjusted to provide, with available discharge and with prevailing channel characteristics, just the velocity required for the transportation of the load supplied from the drainage basin. The graded stream is a system in equilibrium; its diagnostic characteristic is that any change in any of the controlling factors will cause a displacement of the equilibrium in a direction that will tend to absorb the effect of the change.

The concept of grade as an equilibrium condition is valuable in understanding fluvial mechanics even though it probably overstates the role of slope. The construction of Hoover Dam, for example, caused a radical decrease of load in the Colorado River downstream from the dam. The river did not adjust to the altered load by a slope change; instead, an increase in roughness brought about the expected decrease in velocity (Leopold and Maddock 1953). Other studies have shown dramatic changes in the channel shape to be the prime factor involved in the balancing process (Schumm and Lichty 1963; Knox 1972). Thus, every altered load condition does not have to be countered with a modification of declivity alone. This has tempted geomorphologists to consider the possibility that all rivers flowing in alluvial channels are graded, except that they adjust their slope and/or other channel characteristics to transport their loads. In that sense, grade becomes analogous to quasi-equi-librium, and the graded condition represents the most probable state for the channel configuration and flow properties (Langbein and Leopold 1964).

In spite of the advantages in considering grade and quasi-equilibrium as equivalent, it is important to remember Mackin's words "over a period of years," since they may state the true distinction of a graded river (Knox 1975). It is known, for example, that actively downcutting streams are still in quasi-equilibrium as defined by their hydraulic geometry; that is, the hydraulic vari-ables are perfectly adjusted to flow and their measurement would not indicate that any fluvial response is occurring or, for that matter, that any change re-quiring a response has occurred. It is also true, however, that the very fact of progressive entrenchment indicates the river is approaching a different equi-librium, one established over a period of years. It is not, perhaps, necessary to require every graded-time adjustment to be made by a change in slope, but it is important to recognize the time distinction. Graded-time adjustments seem to be made by the variables of channel configuration and by pattern changes as alluded to by Schumm and Lichty (1965). Significantly, the initial response may appear in the form of a hydraulic variable (such as n in the Hoover Dam situation), but this may not be the ultimate response. If the inducing change is minor, the variables of hydraulic geometry may absorb it, just as they do during floods. But if the change is major, and especially if it occurs gradually, the final response may be one involving the channel configuration or pattern, and changes in those factors take time.

Adjustment of Shape and Pattern

As explained above, one of the more significant advances in recent years is the growing awareness of fluvial geomorphologists that rivers can respond to altered discharge and/or load in ways other than cutting or filling of the channel (for example, see Dury 1964). In a series of papers Schumm (1965, 1968, 1969) pieced together many of the empirical equations we have examined into a comprehensive, though qualitative, model of possible river adjustments to altered hydrology and load. The following equations are the basis for what Schumm (1969) refers to as *river metamorphosis:*

$$Q_w^+ Q_t^+ \simeq \frac{w^+ L^+ F^+}{P^-} S^{\pm} d^{\pm}$$

$$Q_w^- Q_t^- \simeq \frac{w^- L^- F^-}{P^+} S^{\pm} d^{\pm}$$

$$Q_w^+ Q_t^- \simeq \frac{d^+ P^+}{S^- F^-} w^{\pm} L^{\pm}$$

$$Q_w^- Q_t^+ \simeq \frac{d^- P^-}{S^+ F^+} w^{\pm} L^{\pm}.$$

In these equations Q_t is the percentage of the total load transported as bed material load (sand-sized or larger), and Q_w can be either the mean annual discharge or the mean annual flood. The other variables are width (w), depth

Figure 6.28.
Adjustments of the Murrumbidgee River, Australia, to changes in climate and sedimentology. PC_2 (paleochannel 2) functioned at time of greater aridity; PC_1 (paleochannel 1) was sinuous and much larger than present river. Modern river is highly sinuous and flows within the channel limits of PC_1. (After Schumm 1968)

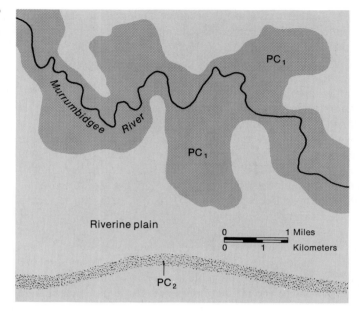

(d), slope (S), meander wavelength (L), width-depth ratio (F), and sinuosity (P). The plus or minus exponents indicate whether the variables are increasing or decreasing.

To exemplify the use of these equations, let us assume that a large area is clear-cut of its natural forest cover. We can expect an increase in Q_w, because infiltration rates will be lowered and direct runoff will increase, as well as an increase in Q_t, because coarse sediment normally stabilized on slopes by rooting now makes its way to the channel; the coarse sediment also will be moved more frequently because of the increased peak discharge. With both Q_w and Q_t increasing, we can expect increases in width, wavelength, and W/D and a decrease in sinuosity. Depth and slope may vary in either direction. Slope will probably increase, however, because the channel becomes straighter, and depth will probably be constant or decrease since both w and W/D increase.

An excellent geological example of river metamorphosis is found in the history of the Murrumbidgee River which flows across a large alluvial plain in New South Wales, Australia (Schumm 1968). As figure 6.28 illustrates, the present highly sinuous river flows within a floodplain containing large oxbow lakes and other features that preserve an older and larger channel of the Murrumbidgee (paleochannel 1). Evidence of a still older, low-sinuosity channel (paleochannel 2) is also present on the plain. The morphologic, sedimentologic, and hydrologic characteristics for the three channels are presented in table 6.4. Pedogenic and geomorphic data confirm that during the tenure of paleochannel 2 the climate was more arid than at present, and at the time of paleochannel 1, more humid than now. Using the present river as a norm,

Table 6.4   Fluvial, geometric, and sedimentological data comparing the modern river with paleochannel 1 and paleochannel 2, Murrumbidgee River, Australia.

	Median Grain Size (mm)	Channel Silt-Clay Percent (M)	Width (ft)	Depth (ft)	Width-Depth Ratio	Sinuosity	Gradient (ft per mi)	Bankfull Velocity (ft per sec)	Bankfull Discharge (cfs)	Sand Discharge (tons per ft. per day)	Sand Discharge (tons per day)	Meander Wavelength (ft)
Murrumbidgee River near Darlington Point	0.57	25	220	21	10	2.0	0.7	3.0	10,000	9	2,000	2,800
Paleochannel 1		16	460	35	13	1.7	0.8	4.2	51,000	45	21,000	7,000
Paleochannel 2 Northern (Kearbury pit):												
Small	0.55	1.6	600	9	67	1.1	2.0	5.2	23,000	90	54,000	18,000
Medium			1,000	12	83			6.3	73,000	140	140,000	
Large			1,700	20	90			8.8	290,000	300	510,000	
Central (Kulki pit):												
Small	0.60	3.4	500	8	63	1.1	2.0	4.8	19,000	70	35,000	15,000
Medium			800	9	90			5.2	35,000	80	64,000	
Large			800	14	57			7.0	77,000	210	178,000	

After Schumm 1968.

Schumm (1968) compared the adjustments in channel parameters that occurred in the Murrumbidgee River under changing climates. In a change toward aridity (present $\rightarrow$ paleochannel 2), the decrease in precipitation over the entire basin should decrease Q_w and increase Q_t. According to the equations of river metamorphosis, to transport the changed load with less water, the channel should become shallower and probably wider (d^-, F^+). Wavelength should probably increase (L^+) because sinuosity should decrease (P^-) and slope increase (S^+). Comparison of the Murrumbidgee variables in the table shows that the actual changes agree with those predicted by Schumm's equations.

An increase in precipitation (present $\rightarrow$ paleochannel 1) would increase Q_w but probably cause little change in sediment yield or size because the density of vegetation would also increase (as discussed in chapter 5). In response, the slope and W/D would show little change, but the channel would become wider and deeper, and the dimensions of meandering (wavelength and amplitude) increase (table 6.4).

One of the more significant observations made by Schumm is that changes in the Murrumbidgee gradient in response to altered controls were made without cutting or filling the channel. Most of the slope adjustment was accomplished by changes in the length of the river due to variations in sinuosity.

To summarize, our present understanding of rivers requires that we rethink the meaning of equilibrium and its relation to fluvial mechanics. The following emerge as prime points for future consideration and should be of special interest to engineers and geologists:

1. The adjustment of slope in rivers can be made by a major change in channel pattern rather than by vertical filling or trenching.
2. The initial response of a river may not be the same as its end response. In addition, regulation of a river may generate responses that do not remain local but alter channel morphology over great lengths of the river system.
3. Both discharge and sediment must be considered in the prediction of fluvial mechanics. The greater width associated with channels of low discharge (such as paleochannel 2 in the Murrumbidgee, table 6.4) is contrary to many regime equations relating width and discharge. The tendency of the river's response to a change in discharge, therefore, can be counterbalanced by a simultaneous change in the character of the load.

Summary

River action, like all geomorphic processes, behaves according to the driving and resisting forces built into the system. For example, a river will entrain, transport, or deposit sediment depending on the driving energy given to the water by velocity, depth, and slope, and on the amount of that energy consumed by the resistance to flow offered by elements such as channel configuration, particle size, and sediment concentration. The work demanded of a river—the amount of load it must handle under prevailing discharges—is determined by the geological and climatic character of the drainage basin. Each river develops a particular combination of shape, gradient, and hydraulic variables (called the hydraulic geometry) that allows it to accomplish its work most efficiently. The river will attempt to maintain its high efficiency by adjusting the above properties whenever discharge or load vary. Because discharge and load fluctuate continuously, equilibrium as a steady-state condition can never be attained, and the river variables must perpetually be adjusting. Nontheless, the normal variations of discharge and load are accommodated by hydraulic geometry.

The type of channel pattern (straight, meandering, braided) a river displays and the longitudinal profile are other fluvial characteristics controlled by the basin environment. Each pattern originates in a specific manner, and its geometric form is designed to facilitate the work of a river, measured as the prevailing values of discharge and load. Once established, the pattern will be maintained as long as the normal variations in load and discharge can be absorbed by the mechanics of hydraulic geometry.

Major long-term changes in climate or basin tectonics may alter the average discharge and/or load to a point where adjustments of hydraulic geometry can no longer maintain the most efficient system. When those threshold values of discharge or load are reached, major fluvial responses in the form of pattern changes, degradation or aggradation, or dramatic revisions of the width-depth ratio will occur to reestablish the greatest fluvial efficiency. An excellent example of these major reactions has been detailed for the Murrumbidgee River in Australia. It is important to recognize, however, that our present knowledge does not allow us to predict which of the possible adjustments will occur in response to major changes in the fundamental controls.

The following references provide greater detail concerning the concepts discussed in this chapter. **Suggested Readings**

Hey, R. D.; Bathurst, J. C.; and Thorne, C. R. 1982. *Gravel-bed rivers.* New York: Wiley-Interscience.

Leopold, L. B., and Maddock, T., Jr. 1953. The hydraulic geometry of stream channels and some physiographic implications. U.S. Geol. Survey Prof. Paper 252.

Leopold, L. B., and Wolman, M. G. 1957. River channel patterns; braided, meandering, and straight. U.S. Geol. Survey Prof. Paper 282-B.

Leopold, L. B.; Wolman, M. G.; and Miller, J. P. 1964. *Fluvial processes in geomorphology.* San Francisco: W. H. Freeman.

Richards, K. 1982. *Rivers.* London: Methuen and Co.

Schumm, S. A. 1968. River adjustment to altered hydrologic regimen, Murrumbidgee River and paleochannels, Australia. U.S. Geol. Survey Prof. Paper 598.

————. 1969. River metamorphosis. *Am. Soc. Civil Engrs. Proc., Jour. Hyd. Div.,* HY 1:255–73.

————. 1977. *The fluvial system.* New York: Wiley-Interscience.

Fluvial Landforms

7

Introduction

In chapter 6 we examined the basic mechanics of fluvial processes and found that the activity within a stream channel is generally related to the energy possessed by the river and to the ways that energy is utilized to carry water and sediment most efficiently. Rivers, however, are more than natural sluices; they also mold the geologic setting into discernible topographic forms. They accomplish this primarily through the erosional capability inherent in the movement of sediment-laden water, and through the deposition of debris that occurs when the transporting energy is less than the demands being made on it. Some fluvial features are purely erosional; the topographic form is clearly one of sculptured rock, and little, if any, sediment is associated with the feature. Others may be entirely depositional, and the exposed topography is formed by the burial of an underlying surface that existed before the covering sediment was introduced. In these cases, the bedrock framework may have no influence on the surface configuration. Many features spring from some combination of both erosion and deposition; the pure cases are probably end members of a continuum of possible forms.

If rivers establish or nearly establish some form of equilibrium, it seems reasonable to expect that fluvial features—the tangible results of river work—will somehow reflect the balanced processes that created them. Here we will investigate these end products of fluvial action and, wherever possible, document how the properties of features may reveal the processes involved in their origin.

Floodplains

Floodplains are perhaps the most ubiquitous of fluvial features, found in the valley of every major river and in most tributary valleys. However, a precise definition of a floodplain is more difficult than one might expect. Topographically and geologically speaking, a floodplain is the relatively flat surface occupying much of a valley bottom and is normally underlain by unconsolidated sediment. The sediments of most valley bottoms are not necessarily a function of the river occupying the valley but may be deposited there by a variety of geomorphic processes. Nonetheless, to be considered as part of the floodplain, the surface and the sediments must somehow relate to the activity of the present river. The definition must also have a hydrologic connotation, since the floodplain is a surface subject to periodic flooding. It can easily be defined in terms of hydrology as the water level attained in some particular stage of the river. Detailed analyses (Wolman and Leopold 1957) demonstrate that most topographic floodplains are subjected to flooding nearly every year or every other year. The recurrence interval of bankfull stage, for example, averages about 1.5, indicating that most rivers leave their channels two out of every three years. If the surface flanking the river has any relief, however, part of the *topographic* floodplain will not be inundated by the annual or biannual flood that marks the *hydrologic* floodplain.

Engineers may view floodplains in a different sense, being more concerned with how much of the total discharge moves across the floodplain surface during high flow events (Bhowmik and Demissie 1982). In addition, engineers and basin managers are primarily concerned with how much damage a flood will cause and therefore have little interest in yearly floods. In fact many engineers use the phrase *flood-damage stage*—the water level where overflow begins to cause damage—to mean the flood stage. In general, the damage stage is well above both the level of bankfull and the average elevation of the topographic floodplain.

Regardless of how it is defined, a floodplain plays a very necessary role in the overall adjustment of a river system. It not only exerts an influence on the hydrology of a basin (lag, etc.) but also serves as a temporary storage bin for sediment eroded from the watershed. Therefore, floodplains are features that are both the products of the river environment and important functional parts of that system.

Deposits and Topography

Floodplains are composed of a variety of sediments that are created by diverse processes and accumulate in distinct subenvironments within the valley bottom. Most floodplains can be differentiated into deposits of channel fill, channel lag, splays, colluvium, lateral accretion, and vertical accretion (Happ et al. 1940; Lattman 1960). Near the valley sides, *colluvium,* resulting from unconfined wash and mass wasting, may be prominent in the floodplain sequence; toward the axis of the valley, these deposits grade into alluvial-type deposits. Coarse debris from which the fines have been winnowed are interpreted as *channel lag* deposits, in contrast to *channel fill,* which consists of a poorly sorted admixture of silt, sand, and gravel. *Splay* deposits are composed of material spread onto the floodplain surface through breaks in natural levees and usually are more coarse-grained than the overbank sediments that they cover. The most important deposits in the floodplain framework are those of **lateral accretion** and **vertical accretion,** which in some cases can be separated on the basis of particle size, the laterally accreted sediment usually being sands and gravels, more coarse-grained than the vertically accreted silts and clays (shown in the photograph in fig. 7.1). Point bars, however, the most common deposit of lateral accretion, may sometimes have the same texture as the overbank sediments (Wolman and Leopold 1957). Therefore, particle size is not an infallible criterion for distinguishing between vertical and lateral accumulation.

Valleys of large rivers such as the lower Amazon and Mississippi have been actively aggrading for thousands of years. As a result, they may have incompletely formed floodplains (Tanner 1974) or special features caused by rapid elevation of the valley floor (Baker 1978). Nonetheless, most valleys of large rivers are occupied by well-defined floodplains that consist primarily of laterally and vertically accreted deposits. These are usually associated with specific depositional environments. In the Mississippi River valley, for example, they have been broadly categorized (Fisk 1944) as (1) channel types, including point bars, chutes and sloughs, and sand ridges of meander scrolls; and (2) overbank types, consisting of splays, natural levees, and backswamps (fig. 7.2). The surface of the floodplain may have considerable microrelief that reflects the fluvial mechanics in the various depositional environments.

Figure 7.1.
Normal sequence of floodplain stratigraphy. Lower two-thirds of deposit is composed of coarse-grained, laterally accreted, point bar gravel. Yardsticks show how gravel in lowest unit dips slightly to the left. Upper third of deposit is vertically accreted silt deposited during overbank flow. Some gravel at top of floodplain is also overbank debris. Sexton Creek, Shawnee National Forest, southern Illinois.

The floodplain surface is most irregular in a zone close to the river where point bars are molded into alternating ridges and swales that Leopold and his colleagues (1964) refer to as *meander scrolls*. The characteristic scroll topography may start as a longitudinal bar with a narrow trough to its rear (Sundborg 1956) or simply as a low ridge of sediment that accumulates on the inside of a meander bend during bankfull flow (Kolb and Van Lopik 1958; Hickin 1974). When the high discharge subsides, the ridge is exposed and rapidly vegetated, becoming, for all practical purposes, the new channel bank. The next high flow repeats the process. As the river shifts across the valley by undercutting on the outer bank, the successive ridges and intervening swales that characterize meander scroll topography develop simultaneously (fig. 7.3). Periodically, flow will break across the point bar surface, often occupying a particular swale and scouring the surface into more pronounced low channels called **chutes.** Wolman and Leopold (1957) measured velocities of up to 3 fps in chute channels, a flow that is capable of eroding the surface and transporting coarse sand. Chute erosion tends to accentuate the scroll topography,

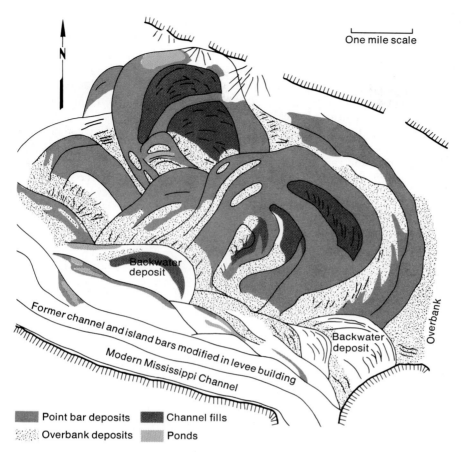

Figure 7.2.
Map showing complex distribution of various types of deposits on a portion of the Mississippi River floodplain near Grand Tower, Ill. (Courtesy of S. E. Harris, Jr.)

Backwater deposit

Former channel and island bars modified in levee building

Modern Mississippi Channel

Backwater deposit

Overbank

▨ Point bar deposits   ▨ Channel fills
░ Overbank deposits   ▨ Ponds

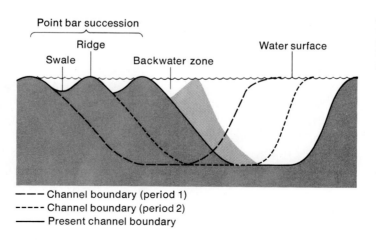

Figure 7.3.
Meander scroll topography formed by point bar deposition in a laterally migrating river.

Point bar succession

Ridge

Swale   Backwater zone   Water surface

– – – Channel boundary (period 1)
----- Channel boundary (period 2)
——— Present channel boundary

and even though the slough areas may be silted in by overbank deposition, the scalloped profile of scroll topography (fig. 7.4) can be preserved for hundreds of years (Hickin and Nanson 1975). (For greater detail about point bar mechanics and deposits, see McGowen and Garner 1970.)

In addition to the irregularity caused by scrolls on the inside of meander bends, some topographic relief near the channel is due to the formation of **natural levees.** Natural levees stand along most major rivers as low ridges that commonly are broader than one might expect, often extending for hundreds of meters. Levees are usually highest near the active channel and slope gradually toward the valley sides. They owe their character to the retardation of flow velocity when rivers leave the channel, which results in the largest suspended particles being deposited adjacent to the bank. Levee deposits are

Figure 7.4.
(A) Diagram of two profiles of meander scroll topography that has been preserved for considerable periods of time. Beatton River, Canada.
(B) Meander scroll topography on abandoned point bar of the Mississippi River in Pike County, Ill. Local relief between ridges and swales is about 3 meters.

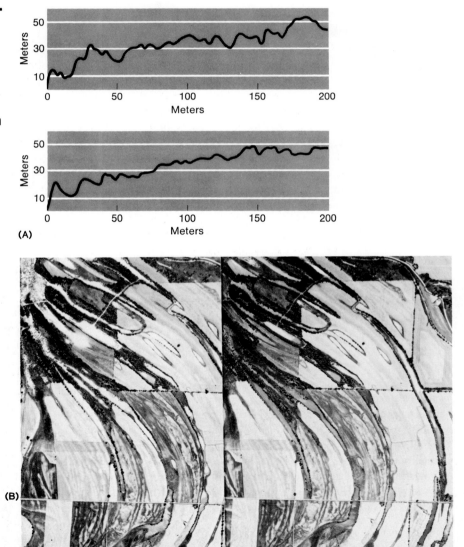

therefore more coarse-grained than most other overbank sediment and are accreted more rapidly. For example, Kesel and his colleagues (1974) found a
net addition of 53 cm on natural levees during the two-month overflow of the
Mississippi River in 1973, while only 1.1 cm accumulated in the backswamp
area during the same period.

As one proceeds away from the river, the floodplain topography becomes
much more regular, and its flatness is interrupted only by oxbow channels or
by splay deposits that have been debouched onto the backswamp surface. The
zone away from the active river is characterized by gradual accumulation of
overbank sediment that tends to subdue any existing relief. Abandoned channels represented by oxbows or oxbow lakes gradually fill with silts and clays,
leaving *clay plugs* as the only evidence of the former channel. Old sloughs,
chutes, or cutoffs can also fill with overbank sediment, adding to the general
reduction of relief away from the channel. It is incorrect, however, to assume
that all overbank sediment is fine-grained. Where banks are cohesive, coarse
gravel can be transported onto the surface of the floodplain (McPherson and
Rannie 1967; Costa 1974a; Ritter 1975; see fig. 7.5).

Figure 7.5.
Large lobe of overbank sand and
gravel deposited on the floodplain
surface in December 1982 flood.
Lobe sediment is 1.0 to 2.0 m
thick and 100 m wide. Gasconade
River near Mt. Sterling, Mo.

Table 7.1   Rates of lateral migration of rivers in valleys.

River and Location	Approximate Size of Drainage Area (square miles)	Amount of Movement (feet)	Period of Measurement	Rate of Movement (feet per year)
Tidal creeks in Massachusetts		0	60–75 yr	0
Normal Brook near Terre Haute, Ind.	±1	30	1897–1910	2.3
Watts Branch near Rockville, Md.	4	0–10	1915–55	0–0.25
	4	6	1953–56	2
Rock Creek near Washington, D.C.	7–60	0–20	1915–55	0–0.50
Middle River near Bethlehem Church, near Staunton, Va.	18	25	10–15 yr	2.5
Tributary to Minnesota River near New Ulm, Minn.	10–15	250	1910–38	9
North River, Parnassus quadrangle, Va.	50	410	1834–84	8
Seneca Creek at Dawsonville, Md.	101	0–10	50–100 yr	0–0.20
Laramie River near Ft. Laramie, Wyo.	4,600	100	1851–1954	1
Minnesota River near New Ulm, Minn.	10,000	0	1910–38	0
Ramganga River near Shahabad, India	100,000	2,900	1795–1806	264
	100,000	1,050	1806–1883	14
	100,000	790	1883–1945	13
Colorado River near Needles, Calif.	170,600	20,000	1858–83	800
	170,600	3,000	1883–1903	150
	170,600	4,000	1903–1952	82
	170,600	100	1942–52	10
	170,600	3,800	1903–42	98
Yukon River at Kayukuk River, Alaska	320,000	5,500	170 yr	32
Yukon River at Holy Cross, Alaska	320,000	2,400	1896–1916	120
Kosi River, North Bihar, India		369,000	150 yr	2,460
Missouri River near Peru, Nebr.	350,000	5,000	1883–1903	250
Mississippi River near Rosedale, Miss.	1,100,000	2,380	1930–45	158
	1,100,000	9,500	1881–1913	630

From Wolman and Leopold 1957. See this work for data sources for individual rivers.

In general, the relationship between floodplain deposits and river processes would be straightforward if rivers would stay in the same place for extended periods of time. Actually, as the rates in table 7.1 indicate, most rivers migrate laterally across the valley bottom quite rapidly, forcing the depositional environments also to shift their location with time. A backswamp region, for instance, may include remnant deposits of a channel. The displacement of one environment by another adds to the complex maze of floodplain deposits, emphasizing the point that floodplains are dynamic rather than static fluvial features.

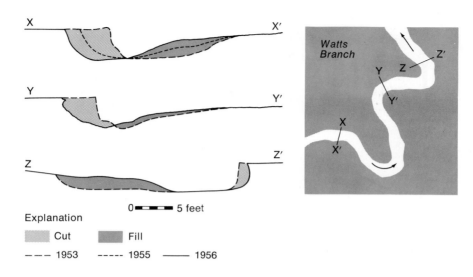

Figure 7.6.
Progressive lateral erosion and point bar deposition (cross sections) in Watts Branch near Rockville, Md. (From Wolman and Leopold 1957)

Explanation

Cut	Fill

――― 1953   ----- 1955   ―――― 1956

The Origin of Floodplains

It is now generally accepted that two dominant fluvial processes act simultaneously to develop most floodplains. As described earlier, maximum erosion in meandering rivers takes place on the outer bank just downstream from the axis of curvature. At the same time, sediment accumulates in point bars that build up along the inside of the meander bend. Detailed study of these processes has documented that bank erosion and point bar accumulation are volumetrically equal during any given period of lateral and downvalley migration of the meander bends (Wolman and Leopold 1957). In addition, data from the same study show that the point bars tend to increase in height until they reach the level of the older part of the floodplain. It seems clear that a meandering river can shift its position laterally without changing the channel shape or dimensions.

In channels where coarse sediment is an important part of the load, the point bars tend to collect sediment that is easily distinguished from that of overbank origin. During low flow, sediment of all sizes may be temporarily trapped in the channel floor, but at the peak of bankfull discharges this material will be removed along with any debris eroded from the undercut banks. The coarse sediment is deposited on the point bars, now submerged. As figure 7.1 shows, these deposits commonly display cross-beds dipping into the channel. Over a period of years, the point bars expand laterally, being progressively spread across the valley bottom as a thin sheet of sand or gravel (Mackin 1937; Leopold et al. 1964). If the load is not characterized by coarse-grained particles, the spreading of lateral accretion deposits proceeds in exactly the same way (fig. 7.6), but the point bar sediment may be more difficult to distinguish from the overbank materials.

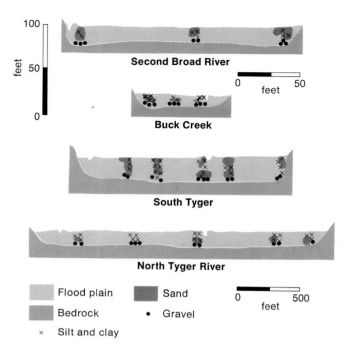

Figure 7.7.
Cross sections of floodplains in North Carolina and South Carolina. (From Wolman and Leopold 1957)

The maximum thickness of laterally accreted deposits is determined by the depth to which a river can scour during recurring floods. Natural channels are probably scoured to a depth 1.75 to 2 times the depth of flow attained during a flood (Wolman and Leopold 1957; Palmquist 1975). This rule of thumb generally fits the scouring observed in a variety of perennial rivers (Leopold et al. 1964, p. 229), but where the channel width is prevented from expanding by bridge supports or extremely resistant banks, the scour depth may reach 3 or 4 times the water depth. The thickness of lateral accretion deposits should, therefore, increase gradually downvalley along with the normal increase in discharge and depth noted on most rivers.

The second dominant process in the origin of floodplains is overbank flow. As rivers wander across their valley bottoms, they normally leave their channel confines during periodic flooding and deposit fine-grained sediment on top of the floodplain surface. Because of this, the floodplain is vertically accreted, and in most floodplain stratigraphy, a thin layer of overbank silt and clay rests on the laterally accreted point bar deposits described above (figs. 7.1, 7.6).

Ideally, then, the entire floodplain sequence consists of a relatively thin accumulation of laterally and vertically accreted sediments that have been spread more or less evenly across the valley bottom (fig. 7.7). The two types of deposits can differ in their textural properties (although the process mechanics does not require it), but they will be essentially the same age.

An analogous development is associated with floodplains of braided rivers except that the systems are more dynamic and less regular. Bars and bank erosion, for example, are not restricted to one particular side of the channel,

Table 7.2 Increment rates of overbank depositon in major floods.

River Basin	Flood	Average Thickness of Depositon (feet)
Ohio River	Jan.–Feb., 1937	0.008
Connecticut River	March, 1936	.114
Connecticut River	Sept., 1938	.073
Kansas River	July, 1951	.098

From Wolman and Leopold 1957. See this work for references on individual data sources.

and the river can shift its position without laterally eroding the intervening material. Abandoned channels and islands gradually coalesce into a continuous floodplain surface. Normally, floodplain sediments in a braided stream system might reasonably be expected to be less thick and more irregular. Recognition of the true floodplain sequence, however, may be complicated by the fact that braided systems are commonly associated with long-term valley aggradation. In such a case, the floodplain sequence might appear to be enormously thick. However, floodplains relate to the hydraulics of the present river only, and the sedimentary pile it produces is probably just a thin skim on top of the fill.

The model of floodplain origin just described raises the question as to which process—lateral migration or overbank flow—plays the dominant role. There is probably no universal answer to this question, because each system obeys its own unique combination of controlling factors. Nonetheless, evidence suggests that most floodplains result primarily from the processes associated with lateral migration. Perhaps the most persuasive argument for that conclusion uses a comparison of the rates involved in the two competing processes. Table 7.2 is a random sampling of sediment increments measured during floods. Although incomplete, it shows that backswamp deposition during overbank flow tends to be limited, probably because the maximum sediment concentration in a flood often occurs before bankfull stage is reached. That is, most sediment is transported from the system before overbank conditions are attained (Wolman and Leopold 1957; Moss and Kochel 1978).

Assuming the same vertical increment with each flood, the level of a floodplain built entirely by overbank deposition should increase at a progressively decreasing rate, as shown by the curves in figure 7.8. The initial growth would be rapid because flooding would occur frequently, and perhaps 80 to 90 percent of the floodplain construction would take place in the first 50 years (Wolman and Leopold 1957; Everitt 1968). However, as the surface grows higher relative to the channel floor, the stage needed to overtop the banks is also increasing. The surface is inundated less frequently, and the rate of growth is drastically retarded. The fact that most floodplains are occupied with water nearly every year argues against the importance of overbank deposition in their construction.

Figure 7.8.
Increase in elevation of floodplain with time. Lower curve from empirical data collected on floodplain of the Little Missouri River; upper curve was derived theoretically for Brandywine Creek (Pa.) by Wolman and Leopold (1957). Note different vertical scales.

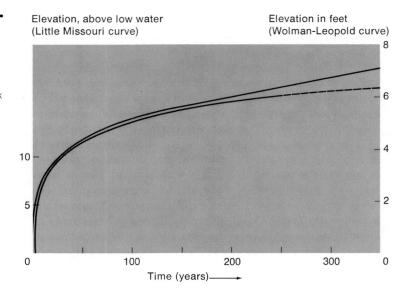

Elevation, above low water (Little Missouri curve)

Elevation in feet (Wolman-Leopold curve)

Time (years) ⟶

At reasonable increment rates for overbank deposition, a 3 m thick floodplain sequence would probably take several thousand years to accumulate. Assuming this is a valid estimate, it is instructive to note again the lateral migration rates given in table 7.1, which indicate that most large rivers migrate quite rapidly. This is especially true when their meander geometry is adjusted for efficient lateral shifting. Hickin and Nanson (1975) suggest that the rate of lateral migration in meander bends of the Beatton River (British Columbia, Can.) is greater when r_m/w is about 3 (see chapter 6). At values higher or lower than 3, the rate of channel migration decreases dramatically. This observation has been supported by theoretical analyses (Begin 1981).

It seems certain that the magnitude of vertical accumulation depends primarily on the rate at which the river migrates laterally. The total thickness will approximate the vertical accretion that can be accomplished in the time the river takes to migrate the entire width of the valley. For example, if a floodplain is a kilometer wide and the river shifts laterally at a rate of 2 m a year, it will take 500 years for the river to complete one swing across the valley. At any given locality, perhaps several meters of overbank sediment will accumulate in that time, but the entire deposit will be reworked by lateral erosion when the river reoccupies that position. The lateral migration rate thus becomes a controlling and limiting factor on the thickness of overbank deposition. The apparent preeminence of lateral processes in floodplain construction does not mean that overbank deposition is unimportant. Indeed, vertical accretion may be the dominant process involved during the initial stage of floodplain development (Schumm and Lichty 1963; Everitt 1968), even though lateral erosion may subsequently rework the sequence. Furthermore, in some cases rivers might lack the widespread lateral movement needed to rework the entire floodplain, and portions of the surface could continue unimpeded growth by overbank deposition (Ritter et al. 1973; Kesel et al. 1974; Smith and Smith 1976; Nanson and Young 1981).

In summary, floodplains appear not only to be formed by balanced fluvial systems but also to serve as integral parts of the system. They are constructed by simultaneous processes of lateral migration and overbank flooding. Point bars, the deposits of lateral accretion, are spread in a rather even sheet across the valley bottom, while overbank deposits accumulate over the entire floodplain surface away from the channel. The floodplain acts as a storage area for sediment that cannot be transported directly from the basin when it is eroded.

Floodplains are usually considered to be features associated with stable rivers, but there is no overriding reason why they cannot be present when a channel is undergoing long-term aggradation or degradation. In fact, the observed frequency of overbank flooding can continue during valley filling if the channel floor and the floodplain surface are raised at the same rate. Once the thickness of the valley deposits exceeds the limits of a reasonable scouring depth, however, the sediment below that depth can no longer be considered as part of the active floodplain. In a degrading channel, the floodplain becomes a terrace when channel incision prevents the river from inundating the surface annually or biannually.

Fluvial Terraces

Terraces are abandoned floodplains that were formed when the river flowed at a higher level than at present. The surface of the terrace is no longer related to the modern hydrology in that it is not inundated as frequently as an active floodplain. Topographically, a terrace consists of two parts: a **tread,** which is the flat surface representing the level of the former floodplain, and the **scarp,** which is the steep slope connecting the tread to any surface standing lower in the valley (fig. 7.9). The very presence of a terrace indicates an episode of downcutting: some change must occur between the conditions prevailing during formation of the tread and those producing the scarp. Usually the downcutting phase begins as a response to climatic or tectonic changes, but these are not always necessary. The tread surface normally is underlain by alluvium of variable thickness, but in a pure sense, these deposits are not part of the terrace. To avoid confusion, it is better to limit the term to the topographic form and refer to the deposits as fill, alluvium, gravel, etc.

Types and Classification

Howard and his coauthors (1968) categorize terraces as erosional or depositional. **Erosional terraces** are those in which the tread has been formed primarily by lateral erosion. If the lateral planation truncates bedrock, the terms *bench, strath,* or *rock-cut terrace* are commonly used. If the erosion crosses unconsolidated debris, the terms *fill-cut* or *fillstrath* (Howard 1959) have been suggested. **Depositional terraces,** the second major grouping, are those terraces where the tread represents the uneroded surface of a valley fill. Figure 7.10 illustrates both types.

(A)

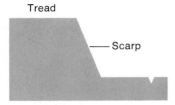

Figure 7.9.
(A) Parts of a fluvial terrace.
(B) Terraces and point bars along
the Snake River in western Idaho.

(B)

Erosional terraces, especially rock-cut types, are identifiable by the following, rather distinct, properties (Mackin 1937): (1) they are capped by a uniformly thin layer of alluvium in which the total thickness is controlled by the scouring depth of the river involved; and (2) the surface cut on the bedrock or older alluvium is a flat mirror image of the surface on top of the capping alluvium (fig. 7.10). In contrast, the alluvium beneath the tread of depositional terraces varies in thickness and commonly exceeds any reasonable scouring depth of the associated river. Although the tread surface may be flat, the surface beneath the fill can be very irregular (fig. 7.10.)

Another classification scheme is based on the topographic relationship between terrace levels within a given valley, as illustrated in figure 7.11. In this method, terrace treads that stand at the same elevation on both sides of the valley are called **paired** (matched) **terraces** and presumably are the same age. If the levels are staggered across the valley they are said to be **unpaired** (unmatched) **terraces.** Most investigators interpret unpaired terraces as erosional types, formed by a stream simultaneously cutting laterally and downcutting very slowly. Levels across the valley, therefore, are not exactly equivalent in age but differ by the amount of time needed for the river to traverse the valley bottom. Actually, unpaired terraces can also be depositional in origin if the entrenchment between two episodes of valley filling occurs at the valley sides rather than along the valley axis (Ritter 1967). Paired and unpaired terraces can be of any origin (erosional or depositional) and develop from either bedrock or alluvium. The terms, therefore, are entirely descriptive and carry no genetic connotation.

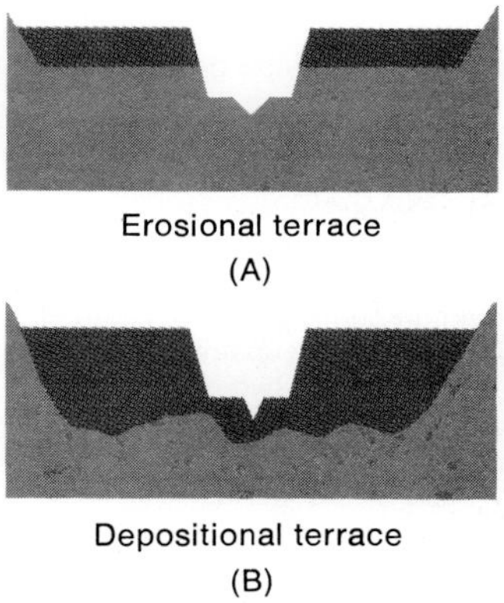

The Origin of Terraces

Depositional Terraces   The development of a depositional terrace always requires (1) a period of valley filling and (2) subsequent entrenchment into or adjacent to the fill. This cyclic pattern is necessary because the alluvium at the tread surface takes its form from purely depositional processes. The tread, in fact, represents the highest level attained by the valley floor as it rose during aggradation. The initial entrenchment that forms the terrace scarp is primarily vertical, and so the tread surface is virtually unaffected by subsequent lateral erosion at a lower level (see fig. 7.10).

Valley filling occurs when, over an extended period, the amount of sediment produced in a basin exceeds the amount that the river system can carry away. Prolonged aggradation is usually triggered by (1) glacial outwash, (2) climate change, or (3) changes in base level, slope, or load due to rising sea level, rising local or regional base level, or an influx of coarse load because of uplift in source areas. Where tectonics are ruled out, the balance between load and discharge is determined primarily by climatic processes, although it may be driven by glaciation and may be complexly interrelated with sea level changes, etc. (see discussion in chapter 2). Although entrenchment has been considered theoretically (Foley 1980a) and studied experimentally (Shepherd and Schumm 1974), details of the mechanical processes involved are still not understood. Nonetheless, incision, like filling, can be triggered by tectonic events or climate change.

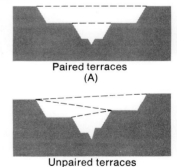

Erosional terrace
(A)

Depositional terrace
(B)

Figure 7.10.
(A) Erosional terrace. Thin alluvial cover with truncation of underlying bedrock along smooth, even surface. (B) Depositional terrace. Terrace scarp underlain by alluvium that is highest level of fill deposited in valley. Note thickness of alluvium and irregular bedrock surface beneath the fill.

Paired terraces
(A)

Unpaired terraces
(B)

Figure 7.11.
Terraces classified on basis of topographic relationships.
(A) Paired terraces have treads at same level on both sides of the valley. (B) Unpaired terraces stand at different elevations on either side of the valley.

Figure 7.12.
Physiographic and geologic controls of stream piracies in piedmont regions. Main river coming from mountain carries coarse-grained load on high gradient. Tributaries that head in piedmont region carry fine-grained load on gentle gradient. At an equal distance upstream from juncture of the two rivers, the tributaries stand at lower elevation and are in position to capture the poised master stream.

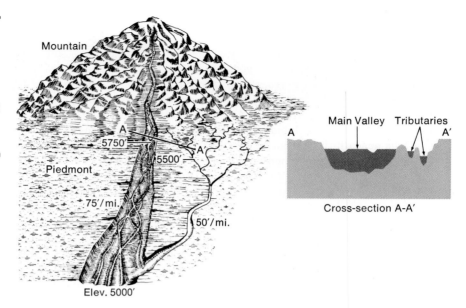

Spontaneous filling and cutting can also result from physical processes that have no relationship to tectonics or climate (see Foley 1980b). For example, small streams that rise in the plains surrounding high mountains often have gentler valley gradients than do the larger rivers that head in the mountains. This trait develops best where the piedmont area is underlain by easily eroded siltstones and shales. Streams originating there adjust their gradients to the fine-grained sediment released from the weakly resistant rocks. The mountain rivers, however, must transport coarse bedload derived from the resistant rocks in the mountain core, and they do so most efficiently by developing a steeper channel gradient. Because of this unique physical control, the main river stands at a higher elevation than its tributaries at an equal distance upstream from their confluence (fig. 7.12). It is well established that such a lithologic and drainage distribution leads to repeated stream captures when headwardly eroding tributaries intersect the position of the master stream (Rich 1935; Mackin 1936, 1937; Hunt et al. 1953; Hack 1960b; Denny 1965; Ritter 1967, 1972). The mountain stream is diverted into a lower tributary valley and is contained there until the process functions again.

The sudden influx of coarse load into the valley of the capturing tributary produces an untenable fluvial condition because the master stream cannot transport its oversized debris on the low valley gradient established by the tributary. The obvious result is filling of the valley until the gradient increases to an incline capable of transporting the mountain load under the prevailing discharge. Subsequent downcutting, often along the valley side, produces a depositional terrace (Ritter 1972). The enigma of depositional terraces formed in this manner is that the eroded surface beneath the gravel was formed by one river (the tributary) while the filling was caused by another (the mountain stream).

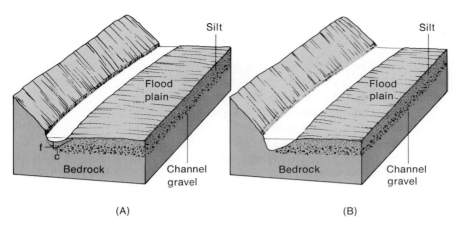

Figure 7.13.
Stages in the development of a rock-cut terrace. (A) During low water stages, fine sediment (*f*) is deposited during normal flow and coarse sediment (*c*) is deposited at the end of a high water event. (B) High water stage entrains all the channel sediment and scours the underlying bedrock before coarse detritus is deposited again on the channel floor. (Source: Mackin 1937. The Geological Society of America)

Erosional Terraces   One sometimes wonders if any aspect of fluvial processes escaped the genius of G. K. Gilbert. The following statement is contained in his remarkable discussion of the origin of floodplains:

> . . . The deposit is of nearly uniform depth, descending no lower than the bottom of the water-channel, and it rests on a tolerably even surface of the rock or other material which is corraded by the stream. The process of carving away the rock so as to produce an even surface, and at the same time covering it with an alluvial deposit, is the process of planation. (Gilbert 1877, pp. 126–27)

Clearly Gilbert presupposed the process of lateral erosion long before any detailed understanding of its mechanics existed. Certainly he provided a theoretical base for the early analyses of fluvial terraces, and his thinking probably represents a cornerstone in the classic model of rock-cut terraces developed later by Mackin (1937) and illustrated in figure 7.13.

As we have said, erosional terraces are those in which lateral erosion is the dominant process in constructing the tread. Mackin (1937) presented an excellent description of the terrace origin (see that work for details of the process; the mechanics were described in the previous section). Briefly, as rivers migrate across the valley bottom, they erode one bank while simultaneously depositing point bar debris near the other. The bar sediment later becomes the capping terrace alluvium. It is usually thin and of constant thickness, and it sits on a flat surface eroded across the underlying bedrock or sediment. The buried surface is carved during floods when scouring penetrates the debris lying on the channel floor. For this to occur, the scouring depth of the river must be great enough to remove the entire pile of channel alluvium and expose the suballuvial material to short-lived erosion (fig. 7.13). Continual shifting of the channel position back and forth across the valley, combined with the occasional scouring, creates beneath the alluvium the bevelled surface that is a mirror image of the plane surface on top of the deposit.

The sheet of alluvium is almost always present in an erosional terrace, but it is not a prerequisite and is certainly not the paramount characteristic of the feature. That role falls to the laterally eroded surface. Thus, it is probably acceptable to ignore the alluvium and consider the cut surface to be the terrace tread. Like any approach to terraces, this may or may not lead to difficulties in the field, depending on the particular situation.

Erosional terraces are normally thought of as the "equilibrium" model of the terrace line. It would seem that the development of a tread surface by lateral planation should require not only time but also a long period of stability during which base level and channel functions are constant, and no vertical disruptions by filling or cutting occur. Nonetheless, even this logical rule of thumb has exceptions. For example, near Pyramid Lake (Nevada), the Truckee River has formed six erosional terraces during a period when its base level, represented by the lake, was rapidly declining (Born and Ritter 1970). The highest and oldest terrace formed sometime between 1925 (when its level was beneath the lake), and 1938, when aerial photography showed it to be a well-developed landform. It now stands approximately 10 m above the Truckee River, which is rapidly downcutting to keep pace with the declining lake level. Apparently each terrace was formed during one major flood when the river, in high flow, was able to erode laterally at a dramatic rate into its unresistant banks. The terrace levels are carved into noncohesive lake sediments recently exposed because Pyramid Lake, in hydrologic imbalance, has dropped almost 25 m in this century. Thus, where banks are easily eroded, time and stability are not essential factors in the formation of erosional terraces. In fact, it is difficult to imagine any geomorphic setting in greater disequilibrium than the Truckee River system near Pyramid Lake.

Terrace Origin and the Field Problem

Understanding local terraces and establishing a regional pattern of terrace development are not only basic in historical geomorphology, they are also useful in providing information for regional planning, land management, water supply, and locating sand and gravel for building materials. Acquiring such knowledge, however, is a painfully slow procedure requiring field study and correlation of surfaces within a valley or between valleys. Determining terrace origins is not so easy as we would like to think. Although we postulate guidelines for recognizing terraces of different origins, terraces in the real world develop in such a variety of ways that the exceptions almost become the rules. Basically, terraces are terraces are terraces—and we should probably not generalize about features that defy generalization. Each terrace sequence must be examined according to its own geologic, climatic, and tectonic setting without preconceived ideas about its origin.

Using terraces to interpret geomorphic history is a monumental task for several fundamental reasons. First, terraces are rarely preserved intact along the length of a valley but instead are segmented into isolated and physically separated remnants, often kilometers apart. Reconstruction of the original longitudinal profile of the terrace surface requires correct correlation of the remnants, and every method used in that procedure is burdened with fundamental assumptions that may be invalid in certain situations (for details see D. W. Johnson 1944; Frye and Leonard 1954).

Second, more than one terrace can result during a period of downcutting. This indicates that entrenchment, representing the response to a threshold-exceeding change, is not a continuous, unidirectional erosional event that results in a lowered river gradient. Instead, as discussed in chapter 1, the response is complex (Schumm 1977). It involves pauses in a downcutting phase during which the river may form erosional terraces by lateral planation (Born and Ritter 1970; Ritter 1982) or depositional terraces by valley alluviation (Womack and Schumm 1977). The point here is that complex response results in multiple terraces formed during the adjustment to a single, equilibrium-disrupting event. This complicates the historical interpretation.

Third, it may seem that erosional terraces are really not that much different from depositional terraces. After all, both are usually covered with alluvium, and if one walked across that alluvial surface there would be nothing to indicate what type of terrace lay beneath. Nonetheless, a real and very important difference does exist—one that cannot be disregarded or minimized. When an erosional terrace forms, the capping alluvium is deposited *at the same time* that the underlying surface is eroded. In significant contrast, the surface beneath a depositional terrace was present before the influx of the alluvial fill; a finite time gap separates the deposition from the cutting of the underlying surface. Failure to recognize this subtle distinction between erosional and depositional terraces can lead to drastically different reconstructions of geomorphic history. For example, the differing interpretations of the terrace sequence in the Bighorn Basin of Wyoming show the problem well and demonstrate the difficulty of obtaining sufficient field data for interpretive purposes.

Mackin (1937) divided the Cenozoic history of the Bighorn Basin into two major phases: (1) a long period of basin filling throughout most of the Tertiary, followed by (2) rejuvenation and basin excavation that has continued to the present. Within the basin are a series of terrace levels standing at elevations of from 330 to 6 meters above the present rivers. Mackin felt that each level represented a rock-cut bench formed when downward excavation ceased and allowed lateral erosion to become the dominant fluvial process. The evidence supporting this interpretation seemed to be clear. Where later entrenchment exposed the terrace gravels, they were thin, constant in thickness, and resting on a flat, truncated bedrock surface (fig. 7.14A). All the ingredients of a rock-cut terrace were observed, and to interpret them as such was certainly reasonable.

In a later study, however, Moss and Bonini (1961) were able to gain additional information about the subsurface framework of several key terraces by running seismic profiles across the features perpendicular to the axis of the Shoshone River valley. Instead of the expected flatness, the bedrock surface beneath the alluvium showed considerable relief, and in places the gravel

Figure 7.14.
Interpretations of the Cody
Terrace near Cody, Wyo. (A) As
rock-cut terrace, based on
observed alluvial thickness. (B) As
depositional terrace, based on
seismic profiles across the terrace
(Moss and Bonini 1961)

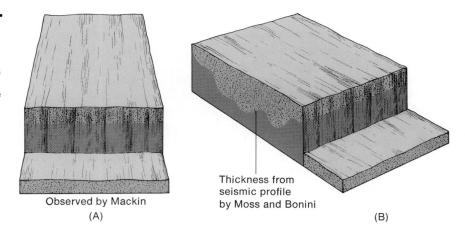

Observed by Mackin
(A)

Thickness from
seismic profile
by Moss and Bonini

(B)

Figure 7.15.
Difficulty in interpreting terrace
origin from field data: If
downcutting exposes only part of
fill, terrace may appear to be rock-
cut. Data across the terrace are
needed to determine true
thickness of the fill.

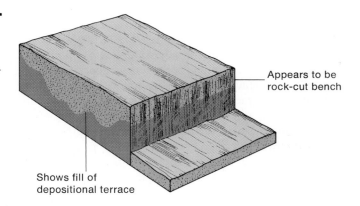

Appears to be
rock-cut bench

Shows fill of
depositional terrace

thickness was well beyond a reasonable scour depth for rivers of this type (fig. 7.14B). They interpreted these characteristics to mean that the surface beneath the gravel represented the valley topography that existed before it was buried by the influx of a later fill. They concluded that the terraces are depositional and that the fill was outwash from glaciers in the nearby Absaroka Range. If all the terraces had this origin, the history of the basin would change significantly. The general excavation phase, in this interpretation, was periodically interrupted by glaciofluvial filling of the valleys, not by valley widening. In addition, considerable time elapsed between erosion of the underlying bedrock surfaces and creation of the terrace treads.

It is instructive to note that the Moss and Bonini interpretation was made possible by techniques and data not available to Mackin in 1937 or, for that matter, to most of us today. What Mackin observed was the edge of a fill where it intersected an eroded valley (fig. 7.15). The much-needed third dimension across the terrace could be reconstructed only with the proper field equipment and approach.

The topography of almost every region reflects an adjustment between dominant surficial processes and lithology. When the rocks have diverse resistances, geomorphic processes tend to maximize the relief between regions of greatest and least resistance. Nowhere is this more apparent than in areas where mountains and plains adjoin, especially where the climate is arid or the region has undergone recent tectonism. Aridity serves to buffer the smoothing effects of vegetation; vertical tectonic activity accentuates relief by bringing more resistant basement rocks toward the surface, where they are commonly etched into the cores of topographic mountains.

The sloping surface that connects the mountain to the level of adjacent plains is the **piedmont.** It extends from the mountain front to a floodplain or playa, either of which can mark the base level for geomorphic processes that function on the piedmont surface (fig. 7.16). Piedmonts consist of a number of geomorphic landforms, but most commonly they are composed of eroded bedrock plains called **pediments** and depositional features called **alluvial fans.** The relative percentage of the total piedmont area occupied by either of these features probably depends on the unique combination of local geomorphic variables.

Alluvial Fans

Alluvial fans have been investigated most extensively and in great detail in regions of arid or semiarid climate. This does not mean that fans are absent in other climatic zones. On the contrary, humid climate fans and their deposits have been examined in such diverse settings as humid-glacial (Boothroyd and Ashley 1975), humid-periglacial (Ryder 1971a, 1971b; Wasson 1977), humid-tropical (Mukerji 1976; Wescott and Ethridge 1980) and humid-temperate (Hack and Goodlett 1960; Williams and Guy 1973; Kochel and Johnson 1984). Fans developed in every climatic setting are linked together by a similar planview geometry, but other aspects of morphology and depositional processes may vary considerably (table 7.3).

Schumm (1977) suggests that fans can be of two major types. **Dry fans** are those created by ephemeral flow, and **wet fans** are developed by perennial stream flow. Clearly, the mode of formation has a climatic connotation because ephemeral flow is normally associated with low groundwater tables and spasmodic rainfall, conditions typical in arid climates. Wet fans are obviously more dominant in humid climates, where perennial stream flow is the norm.

In spite of their recognition in all environments, fans so dominate the piedmont zone in arid climates that most of our detailed understanding of fan development derives from studies in that setting. Thus, unless specified, the treatment that follows refers to arid climate (dry) fans.

Alluvial fans are one end of an erosional-depositional system, linked by a river, in which rock debris is transferred from one portion of a watershed to another. Fans are largest and most well developed where erosion takes place in a mountain and the river builds the fan into an adjacent basin. Deposits tend to be fan-shaped in plan view and are best described morphologically as

Piedmont Environment: Fans and Pediments

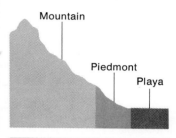

Figure 7.16.
Physiographic components of a mountain-basin geomorphic system.

Table 7.3 Generalized characteristics of alluvial fans formed in different environments.

Parameter	Arid Fans	Humid-Glacial Fans	Humid-Tropical Fans	Virginia Humid-Temperate Fans
Fan Morphology				
Plan View	Broad fanlike symmetrical	Broad fanlike symmetrical	Broad fanlike symmetrical	Broad fanlike to elongated
Axial Profile	Segmented (20–100m/km)	Smooth (1–20m/km)	Smooth	Segmented (40–100m/km)
Thickness	Up to 100's m	Up to 100's m	Up to 100's m	5 m to 20 m
Area	Small	Very large	Large	Small
Depositional Processes				
Major Processes	Debris flow Braided stream Sheet flood Sieve flood	Braided stream	Braided stream Debris flow	Debris flow (avalanche)
Return Interval	1–50 yr Discrete events	0–few days Seasonally constant	Seasonally constant to discrete	3000–6000 yr Discrete events
Fan Area Activated	10–50%	80–100%	30–70%	10–70%
Triggering Processes	Heavy rain Snow melt	Meltwater Outwash	Heavy rain Monsoon	Heavy rain Hurricane
Discharge	Flashy	Seasonal	Seasonal	Flashy

After Kochel and Johnson 1984. Used with permission of Canadian Society of Petroleum Geologists.

a segment of a cone radiating away from a single point source (fig. 7.17). The point source represents the spot where the master river of the watershed emerges from the confines of the mountain; it doubles as the apex of the conical shape. The point source can also shift away from the mountain front to a position well down the original fan surface if that surface has been entrenched at some time during its development. In those cases, the mountain stream, still occupying a confining channel, traverses a portion of the older fan material. The stream eventually emerges downfan as the point source for a still younger fan. Adjacent fans often merge at their lateral extremities; the individual cone shape is lost, and a rather nondescript deposit is formed covering the entire piedmont. These coalesced fans are commonly referred to as *bajadas, alluvial aprons,* or *alluvial slopes.*

The surfaces of fans can often be subdivided into major zones (Denny 1965, 1967) called *modern washes, abandoned washes,* and *desert pavements,* which reflect their participation in modern fan processes (fig. 7.18). On the Shadow Mountain fan in Death Valley, washes make up about two-thirds of the surface area, but only a few contain unweathered gravel and accommodate present-day floods. These modern washes are the primary areas of deposition on the fan surface. Most washes have scrub vegetation and gravel coated with desert varnish in their channels, indicating that they are abandoned channels

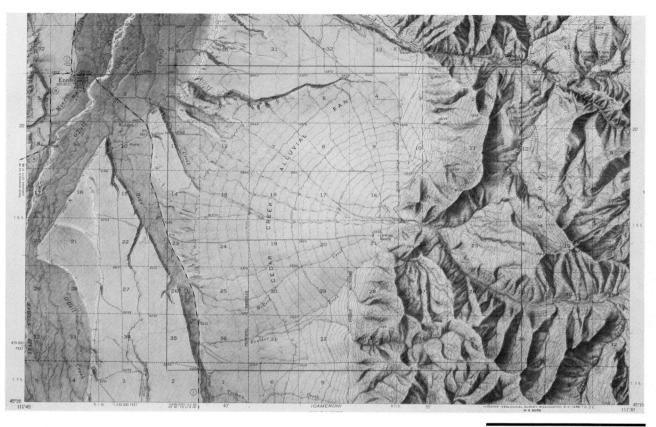

Figure 7.17.
Map view of topography on typical alluvial fan. Note downslope deflection of contours indicating convex cross-fan profile.

and have not been flooded for considerable time, perhaps several thousand years. Desert pavements are surfaces of tightly packed gravel that armor, as well as rest on, a thin layer of silt, presumably formed by weathering of the gravel. Pavements are the primary erosional areas of a fan. They have not received sediment for a long time, as evidenced by the thick varnish coating the pebbles, the pronounced weathering beneath the silt layer, and the striking smoothness of the surface, due to obliteration of the original relief by down-wasting into depressions. Pavements characteristically are cut by gullies that head within the pavement area and may be meters or even tens of meters deep. Because the gullies carry a locally derived fine-grained load, they often meander and, importantly, may stand at lower elevations than adjacent modern washes that head in the mountains.

The distribution of pavements and their gullies in relation to modern washes creates a geomorphic situation quite similar to that discussed earlier for the creation of terraces in a piedmont region (see fig. 7.12). Flow in the modern washes is periodically diverted into the gullies, changing part of the desert pavement area back into an active wash and shifting the position of the modern washes. Simultaneously, the lower segment of the captured wash is abandoned and, over a long period of time, imperceptibly converts to a desert pavement.

Figure 7.18.
(A) Large alluvial fans at the base of the Panamint Range in the north end of Death Valley, Inyo County, Cal. (B) Components of fan in Death Valley region. (Adapted from Denny 1965)

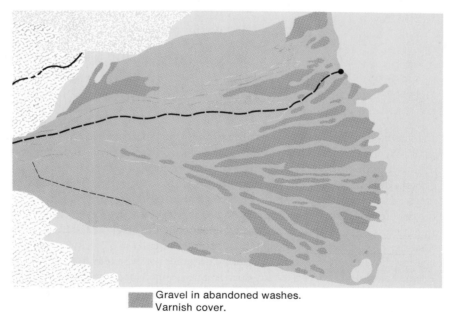

Gravel in abandoned washes. Varnish cover.

Gravel in desert pavement areas. Weathered and varnished.

Gravel in modern washes. Unweathered and no varnish.

Evaporites on floor of Death Valley.

Undifferentiated gravel.

Sedimentary and metasedimentary rocks of mountains.

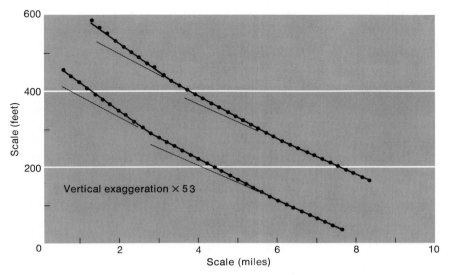

Figure 7.19.
Straight segments of several radial profiles on an alluvial fan. Concave profile stems from lower gradient on each basinward segment of the fan. (After Bull 1964)

Fan Morphology   The longitudinal slope of an alluvial fan generally decreases downfan even though its precise value at any point depends on the load-discharge characteristics of the fluvial system. Near the mountain front, slopes are commonly very steep, although they probably never exceed 10° (Cooke and Warren 1973). Fans gradually flatten to their lower extremity, called the *toe,* where gradients may be as low as 2 m per kilometer ($\approx$ 10 ft/ mi). The steepest gradients are usually associated with coarse-grained loads, low discharges, high sediment production in the source area, and transport processes other than normal streamflow (Blissenbach 1954; Bluck 1964; Bull 1964a; 1964b, 1968; Hooke 1967, 1968; Hooke and Rohrer 1979). These factors often conflict in the same region. In Fresno County, California, for example, fans derived from basins underlain by mudstones or shales are 33 to 75 percent steeper than fans of the same size related to sandstone basins (Bull 1964a, 1964b). The low gradient expected because of the small particle size is offset by a high rate of sediment production. Fan gradients may also be related to parameters of drainage basin morphometry (Melton 1965a); however, most statistical measurements probably reflect only the more basic controlling factors. Absolute slope values at any given point, then, may represent a myriad of controls in the erosional-depositional system and other factors such as its position relative to the fan axis (Hooke and Rohrer 1979).

Two slope characteristics deserve special attention. First, the gradient of most fans near the mountain front is approximately the same as that of the mountain river where it merges with the fan apex. Deposition on the upfan surface, therefore, is not initiated by a dramatic decrease in gradient as the master river passes from the mountain onto the fan. Second, although fans are concave-up from the apex to the toe, their longitudinal profiles are usually not a smooth exponential curve. Instead, on many fans the concavity stems from a junction of several relatively straight segments, each successive downfan link having a lower gradient (fig. 7.19).

Figure 7.20.
Relation of fan area to drainage-basin area for a number of fans in California and Nevada. See Bull 1968 for sources of data and equations of the regression lines.

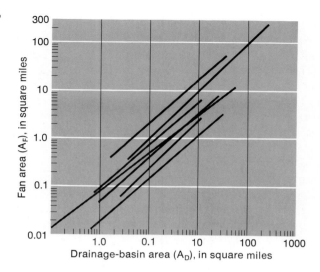

The changes in fan slope, represented by individual segments, are genetically related to changes in the channel of the trunk river upstream from the fan apex. For example, in many fans intermittent uplifts have increased the stream gradients, and in response to each event, a new fan segment has formed, gradually adjusting its slope until it approximates the newly formed steeper slope of the trunk river (Bull 1964a, 1964b; Hooke 1972). Under this particular control, each segment as one moves upfan is steeper and younger, and its deposits are graded to the level of the next lower segment. Segmentation may also result from climatically induced changes in the load-discharge balance, and the segments do not always become steeper toward the apex. Thus the overall longitudinal profile may be very sensitive to historical changes in the balance between the erosional and depositional parts of the system (Bull 1964, 1968).

It is now firmly established that the area of a fan is statistically related by a simple power function to the area of the basin supplying the sediment such that

$$A_f = cA_d^n,$$

where A_f is the area of the fan and A_d is the area of the drainage basin. The exponent n is the slope of the regression line in a full logarithmic plot of the two variables; it measures the rate of change in fan area with increasing drainage basin area. The coefficient c indicates how much the fan "spreads out."

Bull (1968) showed that the relationship is generally very similar for a group of fans representing a variety of environments in the western United States, and the mean value of n is approximately 0.9 when A_f and A_d are measured in square miles (fig. 7.20). The coefficient c, however, seems to vary widely, reflecting the effect on fan dimensions of geomorphic factors other than drainage basin size. Chief among these are climate, source rock lithology, tectonics, and the original space available for fan growth in the collecting basin. For example, Hooke and Rohrer (1977) suggest that c probably relates to the competition for space in the depositional zone. A particular fan that receives a large volume of sediment from its drainage basin would tend to thicken faster than its neighbor and spread outward at the expense of the neighboring fan area. Thus, even where drainage basins in the same region have equal areas, the areas of their fans may differ by as much as an order of magnitude if these other characteristics differ greatly.

Several specific examples demonstrate the effect of these factors and the significance of fan morphometry. Bull (1964) showed that fans derived from basins underlain by fine-grained sedimentary rocks are almost twice as large as those derived from sandstone basins of equal size. The regression lines have approximately the same slope ($n = 0.91$ and 0.98), but the effect of particle size shows up in the value of the coefficient c, which varies from 0.96 for sandstone to 2.1 in the mudstone drainage basins. Actually, a high density of joints and fractures is probably more important than the texture of the rocks. Highly fractured shales will weather into small chips and will be easily eroded. This produces the large loads and high values of c (Hooke and Rohrer 1977). The effect of tectonics is revealed in the fans of Death Valley, where eastward tilting of the valley permitted fans on the west side of the valley to grow larger while those on the east side were stunted by burial beneath the playa (Denny 1965). The c values are 1.05 for fans on the west side of the valley and 0.15 for those on the east side.

Processes, Deposits, and Origins Any model that proposes to explain the geomorphic meaning of alluvial fans must be based on discernible facts concerning both fan morphology and the processes that function in the fan system. Evidence that some of the pertinent facts are still missing can be found in the unresolved controversy about the stability or instability of modern fans. Although fans in the California-Nevada region have been studied in greater detail than anywhere else, little agreement exists about their equilibrium conditions. Some authors view fans as steady-state forms, neatly adjusted and in a continuing dynamic equilibrium (Denny 1965, 1967; Hooke 1968). Others see them as actively growing (Beaty 1970) or being dissected (Hunt and Mabey 1966), processes that presumably indicate that fans may be approaching an equilibrium condition but have not sensibly attained it. Bull (1975b) cites fans as features that do not attain a steady-state condition but instead develop under the control of allometric change (see Bull 1975a, Bull 1975b). Still another interpretation is that fan characteristics can only be explained by cyclic changes in processes initiated by climatic fluctuations (Lustig 1965), and so every fan has vestigial properties unrelated to the present conditions. Facing these widely divergent viewpoints, it seems logical to briefly consider the evidence that led these scientists to such varied conclusions.

Deposits and Depositional Processes   The movement of sediment from source areas to depositional sites involves a variety of flow types, ranging from highly viscous debris or mud flows to normal water flow. The type of flow during any given event depends primarily on the lithology of the basin and its degree of weathering, and secondly on the magnitude of the precipitation causing the flow. The ephemeral nature of flow in arid regions results in spasmodic rather than continuous deposition, and the depositional site changes repeatedly. Only a limited portion of the fan surface is occupied by flow and undergoing deposition at any given time.

As flow leaves the confines of the trunk channel, deposition is initiated by changes in hydraulic geometry, not by a sudden decrease in gradient (Bull 1964). Generally, when the flow becomes unconfined on the fan surface, the width increases so dramatically that both depth and velocity decrease to a level where the flow can no longer transport the load. In segmented fans, however, areas close to the apex are commonly occupied by channels called **fan-head trenches** that are incised 7–12 m below the level of the fan surface (Bull 1964a, 1964b). These may connect to the trunk river in such a way that flow remains contained and is transmitted far downfan before it is freed to increase its width. The effect of changing hydraulic geometry is reinforced by a loss in discharge if the fan surface is permeable and water seeps into the underlying deposits. It is not unusual, for example, for the entire flow to disappear underground before it can traverse the length of the fan.

The deposits of any single flow usually form as narrow tongues, possibly up to several kilometers long, but normally only 120–700 meters wide (Bull 1968). The length of each deposit probably depends on the viscosity of the flow, the permeability of the surface, and the distance downfan that the flow is held in a distinct channel. In many cases, the flow at first follows well-established channels, but at some point along its length overtops the banks and spreads outward as diffuse flow. In the case of water flows, lateral shifting of the loci of deposition allows the braided stream system to deposit a sheet of poorly bedded sand and gravel in which individual beds can be traced laterally for only short distances. This sheetlike configuration may be interrupted by thicker deposits that represent an occasional channel entrenchment into the fan surface and subsequent backfilling. Deposits within these larger channels are generally more coarse-grained. However, even within incised, active washes a microtopography may exist that is directly related to variations in textural properties of fan deposits (Wells 1977). Channels, which are positioned in topographically lower portions of the wash, are floored with coarse sediment. Higher parts of the wash consist of berms that flank the channel and are composed of fine-grained sediment.

Debris flows or mudflows usually follow more well-defined channels because the confining limits of the channel ensure the depth of flow needed to offset the high viscosity of the fluid. During transport, however, debris flows also may overtop banks and spread out as sheets (Bull 1963). Debris flows are so dense and viscous that only the very largest particles can settle from the mass during flow. Nonetheless, they are capable of transporting extremely large boulders for considerable distances on lower gradients than normal streamflow would require. Their high viscosity, however, effectively restricts the distance of transport (in comparison with water flow), and their forward movement may simply stop even though they are still confined in a channel. Therefore, deposits from debris flow are poorly sorted with boulders embedded in a fine-grained matrix; in contrast to water-transported sediment, they are usually lobate and have well-defined margins often marked by distinct ridges.

Some fans are built almost entirely by debris flows (Beaty 1970), although their deposits may be reworked almost immediately by normal stream-flow. Beaty (1963) reports debris-flow deposits being dissected within 48 hours of their deposition by streamflow that began in the same storm but continued after the debris sediment had been deposited. Beaty (1970) also found that in the Milner Creek (White Mountains, California) fan, debris was derived from the floor of the trunk channel. Accumulation of 3–7 m of sediment is needed in the channel to provide the volume found in each major debris-flow deposit on the fan. Sediment production in the drainage basin, therefore, must be rapid enough to collect sufficient amounts on the channel floor before successive high-magnitude storms remove it. If sediment production is too slow, not enough particulate matter will accumulate within the recurrence interval of major floods to produce a debris flow, and the discharge delivered to the fan surface will be in the form of normal streamflow. Some debris flows are initiated by landslides (Johnson and Rahn 1970).

In fans composed of coarse-grained deposits, large discharges may infiltrate before crossing the entire fan (Hooke 1967). Under these conditions, coarse sediment may be deposited in lobate masses called *sieve deposits*. These resemble debris-flow deposits but lack primary fine-grained material and are highly permeable. Sieve material can be deposited when the flow is confined or unconfined, but only when the surface is permeable and the flow does not contain fine-grained sediment.

Entrenchment and Location of Deposition   It is axiomatic that lateral migration is involved in developing the convex cross-profiles and the plan-view shape of alluvial fans. It is equally important to recognize that the loci of deposition also migrate along radial lines during fan development. This longitudinal shifting is accomplished by entrenchment or backfilling of the main channel that extends from the source area onto the fan. Trenched channels have lower gradients than the fan surface, and so they are deep near the fan apex and become progressively shallower downfan until they finally emerge at the surface. In segmented fans, the trenches tend to incise until they have the same gradient as the adjacent downfan segment (fig. 7.21). Entrenchment is significant in that it provides an explanation for abnormal distribution of particle sizes on the fan, since even coarse sediment can be placed far downfan if entrenched channels confine the entire flow. In addition, entrenchment tends to enlarge fans because moderate flows that could not traverse the fan unconfined are transported farther in well-defined channels, resulting in deposition and fan construction in downslope areas.

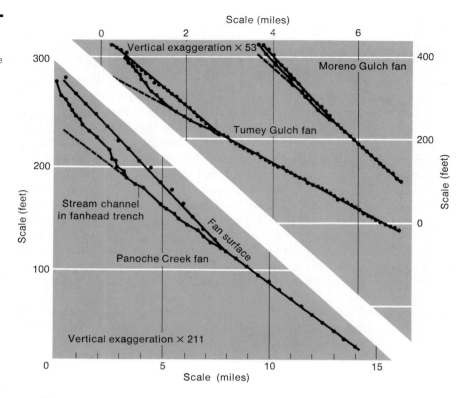

Entrenchment can be either temporary or permanent, and distinguishing
between the two seems to be critical in the analysis of fan origin. Many fan-
head trenches appear to be rather ephemeral; that is, they have evidence of
alternating episodes of trenching and filling. Temporary entrenchment sometimes occurs when debris-flow deposits plug the channel, and the flow shifts
laterally to a new position where entrenchment begins again. It also may be
that alternating trenching and filling are expected results of fan processes associated with changes in rainfall intensity (Bull 1964a, 1964b) or normal alternations of debris flows and water flows (Hooke 1967). In any case, temporary
trenching seems to be a common process on most active fans and can probably
be explained in terms of local conditions. For example, experimental studies
and field observations of wet fans have increased our understanding of temporary trenching (see discussion in Schumm 1977). After the source area
channel and the surface of a wet fan are accordant, water and sediment are
spread as a sheet over most of the area near the fan apex. Downfan deposition,
however, occurs in numerous braided channels. The depositional setting is interrupted when the fan slope near the apex reaches a threshold condition and
incision begins. This results in a fan-head trench, and flow becomes confined
within that channel rather than being widely disseminated over the apex zone.
As entrenchment migrates upstream into the source area, the trunk river is
rejuvenated, and the increased load derived from incision of the trunk river
channel is deposited in the fan-head trench. This in-filling raises the floor of

Figure 7.22.
Dissected fan in Rock Creek valley, Beartooth Mountains, Mont. Grassy areas represent original fan surface. Tree-covered zones are in portions of the fan that have been entrenched when the master river (Rock Creek) cut to lower level.

the fan-head trench until the threshold is reached again, and the process repeats itself. Thus, as Schumm (1977) points out, the entire fan may continue to grow with time, but the fan head experiences spasms of entrenchment during which sediment is reworked and moved farther down the fan.

In permanent entrenchment, channels are incised to depths that cannot be easily backfilled, often being cut to levels greater than 30 m below the fan surface. In addition, the impetus for the downcutting may be outside the fan system itself and so cannot be explained by local fan processes. On the upper fan slope, permanent entrenchment may be caused by accelerated erosion, but it also can occur naturally in a uniform environment if the trunk river continues to downcut within the mountain. In either case, the fan surface standing above the trench is no longer involved with active fan processes; soils may develop on the alluvium, and incipient drainage networks may be established. If the basin of deposition is open and base level for the fan is the floodplain of a through-flowing river, downcutting of that river may initiate a wave of fan incision that is propagated upslope from the toe of the fan. In addition, lateral migration of the master river can erode the toe of the fan and rejuvenate the streams crossing the fan surface. Eventually the entire fan is dissected when the incision reaches the apex and captures the trunk river as it emerges from the mountain. This process is especially effective on small valley-side fans of glaciated valleys (fig. 7.22) and commonly relates to climatic fluctuations and the glacial cycle (Ryder 1971a, 1971b).

Pediments

Since Gilbert first described "hills of planation" in the Henry Mountains of Utah, geomorphologists have been intrigued with their origin, and this feature, given the name *pediment,* has been discussed endlessly during the last century. Our discussion of these interesting features will necessarily be brief, but excellent reviews of the topic are available: Tator 1952, 1953; Tuan 1959; Hadley 1967; Cook and Warren 1973; Twidale 1978. Definitions of the term "pediment" are as numerous as the workers who have studied this feature, and like the landform itself, descriptions range from general to rather graphic and precise. For example, Denny (1967, p. 97) employs the term in reference "to the part of the piedmont that is more or less bare rock surface." On the other hand, R. U. Cooke (1970, p. 28) suggests that "pediments are composed of surfaces eroded across bedrock or alluvium, are usually discordant to structures, have longitudinal profiles usually concave upward or rectilinear, slope at less than 11°, and are thinly and discontinuously veneered with rock debris."

In order to understand better what it is that we are discussing, and in deference to clarity, it may be better to look briefly at those characteristics that are universally recognized as salient properties of pediments rather than to adopt a formal definition:

1. Pediments are erosional surfaces that abut against and slope away from a mountain front or escarpment.
2. They are entirely erosional in their origin and commonly form in a direction that diverges from the trend of the regional structures.
3. The surfaces are usually, but not necessarily, cut on the same rocks that make up the mountain. They may truncate both bedrock and/or alluvium, but they are best developed and preserved on bedrock, especially resistant types such as granite or related crystalline rocks.
4. Pediments may or may not have a thin covering of sediment which presumably represents load that is in transit. This characteristic has traditionally created problems because the question arises of how much alluvium can be tolerated before it must be recognized as a fan, younger in age than the pediment surface and therefore divorced from the processes of pedimentation. It is applaudable, then, that Cooke (1970) restricts the pediment to only that part of the eroded surface not continuously covered by alluvium (fig. 7.23). The erosional surface beneath the continuous debris cover is called the *suballuvial bench,* a term first used by Lawson (1915), and the cover itself is referred to as the *alluvial plain.* The pediment, then, is bounded upslope by the mountain front and downslope by the alluvial plain (fig. 7.23).
5. Pediments are usually found in arid regions, although most workers would not restrict the processes of pedimentation to that climate. Note that we said they are *found,* not formed, in arid climates. Field data suggest that some pediments in the Mojave Desert may be relict features that formed under a more humid, Tertiary

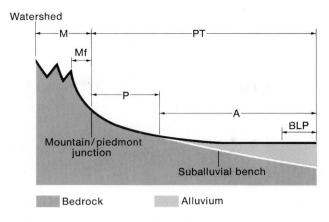

Watershed

Mountain/piedmont junction

Suballuvial bench

Bedrock   Alluvium

Figure 7.23.
Landforms in the mountain-basin geomorphic system: *M* = mountain area; *Mf* = mountain front; *P* = pediment; *PT* = piedmont plain; *A* = alluvial plain; *BLP* = base level plain.

climate, and the surfaces are at present undergoing virtually no expansion under modern desert conditions (Oberlander 1972, 1974). L. C. King (1953) suggests that pedimentation may be a basic geomorphic phenomenon, present in all hillslope developments regardless of climate.

Morphology and Topography   It is ironic that in spite of the singular attention devoted to pediments, a multitude of untested hypotheses exist concerning the processes of pedimentation but an amazingly skimpy pool of reliable data to support them. After a century of study, there is still confusion and lingering disagreement about every aspect of pedimentation. Cooke and Warren (1973, p. 188) express this succinctly in their description of the topic as "a subject dominated by almost unbridled imagination." Cooke (1970) suggests three reasons for the failure to resolve these differences of opinion: (1) we have not viewed pediments as part of an erosional-depositional system but have studied only the pediments to the exclusion of other related forms; (2) we have not collected the precise data needed to understand the system; and (3) we have been overly concerned with general, evolutionary hypotheses of pediment formation and, in many cases, have deduced processes from the genetic model rather than employing direct measurement and observation. Regardless of these past sins, we can make some definitive remarks about pediment morphology and topography.

Size and Shape   Pediments vary in size from less than 1 square kilometer to hundreds of square kilometers, probably depending on fundamental geomorphic controls. Shape is also variable, with pronounced irregularities when the rocks cut by the pediment surface have wide differences in resistance to erosion (Hadley 1967). Generally they tend to be fan-shaped in plan view, narrowing toward the mountain front and widening downslope (D. W. Johnson 1932; Rich 1935; Gilluly 1937). Across the pediment, the shape can be either convex or concave.

Figure 7.24.
View across pediment surfaces to
Bear Peak, Boulder County, Colo.,
February 29, 1972.

Surface Topography   Contrary to lay opinion, pediments are not monotonous, smooth, flat surfaces but are dissected by incised stream channels and dotted with residual bedrock knobs, called **inselbergs,** that stand above the general level of the pediment itself. Inselbergs have been investigated repeatedly with regard to their relationship to pedimentation (Twidale 1962, 1978; Kesel 1973, 1977; Twidale and Bourne 1975; and many others). In some cases, these residual hills might be the last unconsumed vestigages of a landscape that has been totally pedimented. More likely, however, most inselbergs represent areas of rock that are more resistant to weathering and erosion (Kesel 1977; Twidale 1978).

The frequency and size of both incised valleys and inselbergs seem to increase toward the mountain front, sometimes giving the topography the aspect of gently rolling hills and valleys (Gilluly 1937). Some of the channels and other depressions may be filled with alluvium up to 3 m thick (Cooke and Warren 1973), giving the false impression that the bedrock surface is smooth and perfectly planed.

The Piedmont Angle   The upper boundary of the pediment is usually marked by an abrupt change from the steep slopes of the mountain front to the low declivity of the pediment surface. In plan view the boundary is usually linear, but embayments into major valleys of the mountain front can give the trace a rounded or crenulate appearance. The angle formed by the junction of the two surfaces is the **piedmont angle** (fig. 7.24), and its development and maintenance have traditionally been cited as evidence in theoretical models of pediment origin.

In detail, the piedmont angle can take the form of a narrow zone of intense curvature rather than a distinct angle. Twidale (1967) reports mountain front slopes of 22° changing to pediment gradients of 3° over a transition zone 100 m wide. Both the magnitude of the piedmont angle and the sharpness of the angular relationship are probably related to structural or lithologic control (Denny 1967; Twidale 1967; Cooke and Reeves 1972), but other processes including weathering and several forms of corrasion have been suggested as contributing or dominant factors. The early idea of Bryan (1922) that the angular relationship represents an ajustment of the two slopes to the size of debris they are required to transport cannot be accepted without qualification since Melton (1965b) was unable to demonstrate a significant correlation between slope angle and the size of weathering products. At the present time, no one set of processes adequately explains the origin and development of all piedmont angles.

Slope   The longitudinal profile of almost all pediments is slightly concave-up, although local convexities do occur. Overall longitudinal convexities have been suggested as a theoretical possibility if the suballuvial bench is also considered (Lawson 1915), but available observation and geophysical data (Langford-Smith and Dury 1964) have not demonstrated the actual presence of such a form. Slope angles on pediments range from 0.5° to 11° but seem to average about 2.5°.

A prevailing perception concerning pediment slopes is that they are controlled entirely by the size of the material they are required to transport, and so it is widely accepted that they are "slopes of transportation" (Bryan 1922). It is true that some studies have demonstrated a strong correlation between pediment slope and the particle size in debris mantling the eroded surface (Akagi 1980). However, detailed measurements show that the relationship is not as straightforward as previously supposed (Dury 1966b; Cooke and Reeves 1972). Where particle size decreases in an orderly way downslope, the rate of decline in pediment gradients is often greater than the rate of reduction in size. Cooke and Reeves (1972) also found that only the largest particles showed a consistent decrease with distance from the mountain front. Other statistical parameters of size varied incoherently with distance and slope decrease. They attribute these anomalous relationships to differing amounts of in situ sediment being added to the total load at any given sampling locality.

It might also be logically assumed that slope should be related in a significant way to the area of the mountain drainage basin or to the length of the pediment. It has been noted, for example, that pediments have lower gradients where they are associated with large rivers or canyons. (Bryan 1922; Gilluly 1937). Although data are limited, neither of these assumptions could be substantiated in more recent morphometric studies of pediments in California and Nevada (Mammerickx 1964; Cooke 1970). In addition, Twidale (1978) showed that in sequences of pediments having the same source, pediment slopes may decrease with time. The oldest and highest pediments have steeper gradients than the younger pediments. Twidale stressed that in order for this to occur the mountain front scarp must be notably stable. He also demonstrated that slopes are related to structural and lithologic controls.

Processes   Any viable model of pediment genesis must explain the morphologic and topographic elements of the pediment association. The pediment association includes the pediment, the mountain area adjacent to it, and the area of the related alluvial plain. Because these show considerable variation, it seems unlikely that any one combination of processes will produce all pediments or that any single evolutionary model will suffice. Over the years, a number of models utilizing a few basic processes have emerged as the prime hypotheses for pediment origin. However, pediment processes are not easy to study over large areas or during short time periods, and so most of the proposed mechanics of formation are based on intuition rather than solid observational data.

There is no doubt that water flows across pediment surfaces in several different forms. The presence of drainage patterns on undissected surfaces provides unmistakable proof that true streamflow occurs on pediments. Many authors have suggested it as a dominant process of pedimentation (Gilbert 1877; Paige 1912; D. W. Johnson 1932; Rahn 1966; Warnke 1969), especially when the streams migrate laterally across the surface while simultaneously planating the rocks. Although most experts recognize the efficacy of lateral planation in developing part of the eroded surface, many believe this process is incapable of producing and maintaining the piedmont angle in areas remote from the main drainage lines. Streamflow erosion of these interfluve regions would require that rivers emerging from the mountains occasionally flow perpendicular to the pediment slope—a maneuver that defies the law of gravity, as Lustig (1969) reminds us.

Besides normal river flow, unconcentrated flows in the form of sheet and rill wash or floods also traverse pediment surfaces. The phenomenon of sheetflooding was first observed by McGee (1897), who saw the flow as the erosive mechanism in the formation of pediments. Although this concept was adopted by some later workers (Lawson 1915; Rich 1935), it was refuted by others who felt that sheetflow acting alone could not create large pediments, especially across resistant lithology, since the smooth surface itself is a necessary prerequisite for the development of unconcentrated flow. In fact, observations of storm runoffs in Arizona (Rahn 1967) indicate that large discharges on pediments occur as streamflows rather than sheetflows. It may not, in fact, be particularly important to know whether the flow at the head of the pediment is an unconfined type such as sheetwash or whether it moves in small rills or channels. What is important is that flow at the base of the mountain front appears to be a capable transporting agent and therefore plays a significant role in pedimentation by preventing the accumulation of debris there and so perpetuating the character of the piedmont angle. This fact has been documented beyond question in the development of miniature pediments where the underlying rocks are relatively nonresistant (Schumm 1962).

The processes of weathering also have been noted, with different degrees of emphasis, as important factors in pedimentation. The objection that sheetwash may be ineffective as an erosive agent is partly overcome if the initial strength of the pedimented rock is lowered by weathering. Weathering profiles of considerable thickness have been recognized on pedimented rocks (Mabbutt 1966; Twidale 1967; Oberlander 1972, 1974), leading to the suggestion that the rock surfaces beneath the weathered mantle, and even some suballuvial benches, may be produced by weathering rather than by water erosion. Mabbutt (1966) suggests that pediments developed on granite and related rocks may be formed primarily by continuing subsurface weathering that, because of slight differences in resistance, produces an uneven bedrock surface beneath the mantle. The surface on top of the mantle is kept flat by temporary alluviation. In other rock types, weathering may be more rapid at the surface, and the products are removed by sheetwash to maintain the relatively flat pediment surface. Some evidence exists to support the conclusion that subsurface weathering is most pronounced at the junction of the pediment and mountain front (Mabbutt 1966; Twidale 1967). Thus, headward extension of the pediment and the maintenance of the piedmont angle may be intimately related to the type of rocks involved and the efficiency of the weathering processes.

Formative Models   Cooke and Warren (1973) point out correctly that the critical boundary in piedmont areas lies between zones that are primarily depositional and those that are predominantly erosional. The position of this boundary and the presence or absence of pediments are determined by the amount of sediment produced in the mountain relative to the ability of processes to transport the material across the piedmont zone. If supply exceeds transportation, the boundary may abut against the mountain front, and no pediment will be found. In the opposite case, the boundary may be well down the piedmont slope, and pediments will be present. If equilibrium exists between rates of supply and removal, the boundary will be stable and not necessarily parallel to the mountain front. Once established, the boundary can change its position in response to climatic or tectonic alterations. Pediments may be modified after their formation by regrading, or they may be isolated from the processes that are adjusted to the new piedmont setting, as when the surface is entrenched or is buried beneath an alluvial cover. With these possibilities in mind, Cooke and Warren (1973) categorized situations under which pediments form, and their designations are presented here with some modification and addition.

In one situation, which we can call *headward pediment extension,* erosional processes allow the pediment to expand into the area of the mountain. The alluvial boundary may remain constant, or it may move toward the mountain in phase with the migrating mountain-pediment junction, depending on whether the sediment derived from headward erosion can be transported from the system. Cooke and Warren (1973) recognized three formative models as promoting headward growth: the lateral planation hypothesis, the parallel retreat hypothesis, and the drainage basin hypothesis. The *lateral planation* hypothesis was developed most fully by D. W. Johnson (1932), who believed that rock planes (pediments) are a natural consequence in arid regions. According to Johnson, rivers near the mountain front are essentially graded and so heavily loaded that they cannot cut vertically but tend to migrate and erode laterally.

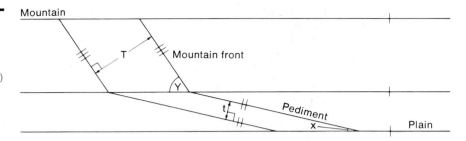

Extreme migration is presumed to trim back interstream bedrock spurs, forming and maintaining the piedmont angle. In response to changes imposed on the system, the rivers will perpetuate their equilibrium condition by regrading the pediment slope.

The *parallel retreat* mechanism was introduced by Lawson (1915) and championed by Rich (1935). It requires that after a mountain front achieves its diagnostic slope, weathering and erosion will maintain that declivity by making the surface retreat parallel to itself. The bedrock bench, produced as the front recedes, is swept clean by rill and sheetwash, processes that keep the angle between the mountain front and the original piedmont slope at a constant value. A slightly different mechanism of parallel retreat occurs in granitic regions where subsurface weathering may continue to level the rock surface at depth (Ruxton and Berry 1961; Mabbutt 1966). Mabbutt called this process *mantle-control planation.* It includes slight back-trimming at the hill base that stabilizes the piedmont angle and extends the pediment toward the mountain. It may also construct rectilinear profiles across the mountain front-pediment boundary (Ruxton and Berry 1961), which may also be preserved by back-wearing (fig. 7.25).

The *drainage basin* hypothesis rises from the apparent failure of either of the above models to explain the total morphology along most mountain fronts. There appears to be no dramatic change in gradients where master channels cross the mountain front onto the pediment surface, indicating the viability of lateral planation there and negating the necessity of parallel retreat. Contrariwise, the interfluve areas that do exhibit the pronounced piedmont angle are far from the main valley and, for reasons stated before, probably cannot be trimmed back by lateral corrasion. The drainage basin hypothesis, therefore, represents a compromise between the two extremes of the other theories. It recognizes that lateral planation will be dominant along the main drainage line, while interfluve areas evolve by weathering and wash. The recognition that planation, weathering, and wash all are involved in pediment mechanics is not new. It was first espoused in early investigations (Bryan 1922, 1935; Gilluly 1937; Sharp 1940) and more recently by Lustig (1969).

A second situation may exist where there is equilibrium between the supply of debris to the piedmont and its removal to the base level zone. In this case, piedmonts consist of both depositional features and pediments that are continuously and simultaneously being formed and destroyed. Theoretically, the amount of sediment accumulated in fans in one part of the piedmont is equal to the amount removed somewhere else by pedimentation. Stream piracies, due to the contrasting gradients between mountain and piedmont streams, repeatedly shift the zones of erosion and deposition. Gully erosion by streams heading in the piedmont area and flowing perpendicular to the mountain front forms the pediment surface. It may subsequently be buried by alluvium after piracy diverts the mountain stream onto the pedimented surface. This basic process has been employed to explain piedmont geomorphology in a variety of climatic regimes (Hunt et al. 1953; Hack 1960b; Denny 1965) and has been formalized as a widespread, pediment-producing mechanism by Denny (1967).

In the third case, erosion on the piedmont exceeds the debris supplied from the mountain, so the pediment-alluvial plain boundary is moved away from the mountain front, and new pediment zones develop farther downslope. We can call this process *basinward pediment extension.* Pediment zones already present before the boundary shift may be modified during the transition; the most common evidence cited for basinward migration of the alluvial boundary is the truncation of weathering or soil profiles developed on the pediment surface. A significant corollary to this concept is necessary. Since a basinward shift of the alluvial plain may expose older pediment surfaces by removing their sediment cover, some pediments may be truly relict features that have no sensible relationship to present conditions. Cooke and Warren (1973) refer to this as the *exhumation hypothesis* and cite Mabbutt (1966) and Tuan (1962) as its most recent proponents.

The effects of climate and time in the meaning of pediments cannot be ignored. Oberlander (1972, 1974) presented a thought-provoking argument that the granite pediments of the Mojave Desert were formed in their entirety before the region became as intensely arid as it is today. He concluded that boulders included in the mantles covering the pediment were originally isolated as corestones in very deep chemical weathering and have reached their present position by subsequent stripping of the weathered zone. The evidence for this conclusion is quite strong. The bouldery mantle can be traced into a well-developed weathering profile, including corestones in a grus matrix, which is preserved beneath basalts dated at > 8 m.y. The pediment surface, therefore, most likely represents the weathering front formed under a semiarid Tertiary climate. The mantle cover is not material in transit but is the remaining part of a weathered profile that was progressively stripped after the region became more arid in the late Pliocene and Quaternary. Oberlander's explanation differs from others in that the pediment surface is not an exhumed, rock-cut, suballuvial bench. The pediments, in fact, were not cut in rock but were formed under a soil cover by the erosion of regolith that was being developed continuously at the weathering front. Slope retreat and maintenance of the piedmont angle were probably the results of wash processes with parallel rectilinear back-wearing similar to that described by Ruxton and Berry (1961). It is even possible that stripped pediment surfaces possess considerable relief and may be "born dissected" as was proposed earlier (Gilluly 1937; Sharp 1940).

Oberlander's proposal merits special attention as geomorphologists grope for all-inclusive models. His work demonstrates that all landforms do not necessarily yield significant relationships based on analyses of process and form. The failure of pediments to reveal any morphometric consistency may be attributed to the fact that some pediments developed under different morphogenetic conditions than those of the present. They are not equilibrium forms but, in fact, may be relics from the distant past that are disequilibrium freaks in their modern surroundings. Oberlander states this message most clearly:

> . . . granitic landscapes in the Mojave Desert have been evolving in a sequential manner since the late Tertiary. This evolution has been triggered by climatic change, and reflects the instability of a landscape determined under conditions that no longer exist. Consequent changes in morphology will cease to be diagnostic of climate change only when all the weathered residuum inherited from the Tertiary morphogenetic regime is stripped from elevated portions of the landscape and transferred to adjacent basins of sedimentation. (Oberlander 1972, p. 19)

The ideas proposed by Oberlander apply only to granitic terrains in the Mojave Desert and may even be invalid in other parts of the arid, southwestern United States (Kesel 1977). They do, however, demonstrate once again the irrefutable importance of climate and geology in geomorphic systems and resurrect the spectre of time in considering equilibrium. Granites in an arid climate may require imponderable time spans before their external form reflects an adjustment between processes and geology. Thus, it may be that the absolute values of graded time (Schumm and Lichty 1965) are eminently dependent on the systemic components.

Deltas

Because sediment being transported by rivers must ultimately come to rest, deposition at or near a river mouth represents a bona fide component of the fluvial system. The most important geomorphic feature produced in that environment is called a delta. **Deltas** are also important sedimentary entities, and geologists continue to study the depositional sequence and complex facies relationships included in the deltaic mass. Our interest in deltas here is only in the geomorphic processes that develop their form. The details of delta sedimentation will be left to the sedimentologists.

The term "delta" is usually applied to a depositional plain formed by a river at its mouth, where the sediment accumulation results in an irregular progradation of a shoreline (Coleman 1968; Scott and Fisher 1969). The feature was first named 2,500 years ago by the historian Herodotus, who noted that the land created at the mouth of the Nile River resembled the Greek letter Δ (delta). Modern deltas, however, display a great variety of sizes and shapes. At the apex of a delta, the trunk river divides into a number of radiating branches, called *distributaries,* that traverse the delta surface and deliver sediment to the delta extremities. In plan view some deltas look like alluvial fans and in fact, a **fan-delta** generally means an alluvial fan prograding into

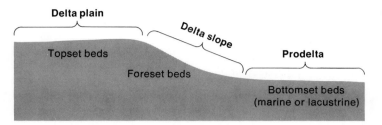

Figure 7.26.
Primary depositional environments and their associated layering in a classic delta.

a body of standing water. Sedimentologists are keenly interested in portions of the fan-delta that are continuously submerged and therefore have studied in detail the marine deposits and processes effecting the feature. Geomorphologists are more interested in the feature as being the result of fluvial processes and usually concern themselves more with the subaerial portion (for example, see Morton and Donaldson 1978). Thus, even though fans and deltas are related in form and process, they differ in several important respects: (1) Deposition on deltas is due to a reduction of river velocity as the flow enters a body of standing water; the standing water can be the ocean or lakes of any size or origin. (2) Delta expansion in a vertical sense is finite, the base-level water body being the approximate limit of upward growth. (3) The gradient on the delta surface is notably flatter than that on most fans.

Sediment, Form, and Classification

Sediment deposited in deltas is usually fine sand or silt but, depending on controlling variables, may occasionally contain gravel or clay. Deltaic sedimentation and evolution were first described by G. K. Gilbert in his study of Lake Bonneville. As envisioned by Gilbert and illustrated in figure 7.26, the feature in its classic and unmodified state consists of a **deltaic plain** standing partly above and below lake level. The plain is fronted by a **delta slope** that connects its surface to the basin floor over which the delta is advancing. The basin floor, called the **prodelta** environment, is composed of fine-grained marine or lacustrine sediments that were swept in suspension beyond the delta front. The delta plain is composed of a complex of nearly flat layers (topset beds) that truncate the strata of the delta slope (foreset beds) as the delta progrades. The dip of foreset beds varies widely. In marine deltas formed by large rivers, the beds rarely dip more than 1°, making it very difficult to recognize them in the field. In small lakes, foreset beds may approach the angle of repose, and the sequence is easily discerned.

Most major rivers of the world develop deltas, and each has its own unique properties reflecting some balance between the fluvial system, the climate, tectonic stability, and the shoreline dynamics. Deltas, therefore, come in a multitude of plan-view shapes, but in general several types serve as model forms and are used for classification purposes. One or more of the controlling factors is dominant in each of the major categories of deltas. **High-constructive deltas** develop when fluvial action is the prevalent influence on the system (Scott and Fisher 1969). As figure 7.27 shows, these deltas usually occur in one of two

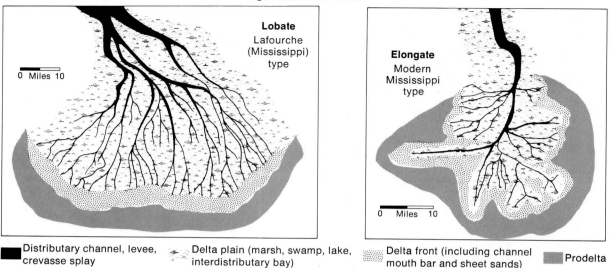

High-constructive deltas

Lobate
Lafourche
(Mississippi)
type

0 Miles 10

Elongate
Modern
Mississippi
type

0 Miles 10

■ Distributary channel, levee, crevasse splay

✤ Delta plain (marsh, swamp, lake, interdistributary bay)

▫ Delta front (including channel mouth bar and sheet sands)

▨ Prodelta

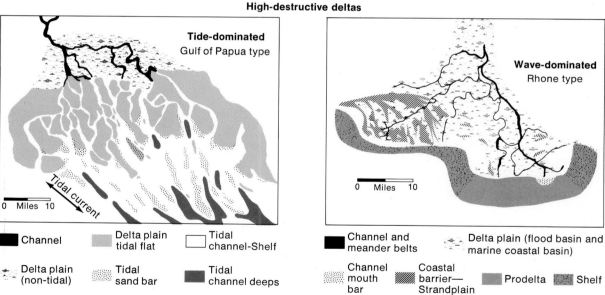

High-destructive deltas

Tide-dominated
Gulf of Papua type

Tidal current

0 Miles 10

Wave-dominated
Rhone type

0 Miles 10

■ Channel

▨ Delta plain tidal flat

□ Tidal channel-Shelf

✤ Delta plain (non-tidal)

▫ Tidal sand bar

■ Tidal channel deeps

■ Channel and meander belts

▫ Channel mouth bar

▨ Coastal barrier— Strandplain

✤ Delta plain (flood basin and marine coastal basin)

▨ Prodelta

▨ Shelf

Figure 7.27.
Classification and geomorphic characteristics of basic delta types.

forms: an *elongate* type exemplified by the modern birdfoot delta of the Mississippi River, or a *lobate* type exemplified by the now-abandoned Holocene deltas of the Mississippi River system. Both types have high sediment input relative to the marine dynamics. Elongate deltas have a higher mud content and tend to subside rapidly when they become inactive, thereby preserving their upper sand facies. Lobate deltas sink slowly upon abandonment, and much of the sand that was prograded in the upper zones is reworked by marine processes (Scott and Fisher 1969).

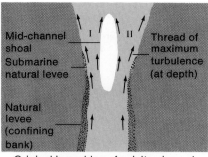
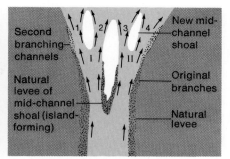

Original branching of a delta channel
(A)

Later stage of channel subdivision
(B)

Figure 7.28.
Stages in bifurcation of trunk river and creation of distributary channels of a delta.

High-destructive deltas originate where ocean or lake energy is high, and much of the fluvial sediment is reworked by waves, etc., before its final deposition. Figure 7.27 shows two types of these. In *wave-dominated* types, such as those of the Nile and Rhone rivers, sediment is accumulated as arcuate sand barriers near the mouth of the river. In *tide-dominated* types, tidal currents arrange the sediment into sand units that radiate linearly from the river mouth. Muds and silts accumulate inland from the segmented bars where extensive tidal flats or mangrove swamps evolve.

Dynamics and Delta Evolution

A sediment-laden river entering a body of standing water behaves much like a free jet of flow (Bates 1953). The jet flow is in the form of either an *axial jet* in which mixing is three-dimensional or a *plane jet* in which two-dimensional mixing prevails. Which flow type develops depends on the relative densities of the two water bodies. If the inflowing water is denser (*hyperpycnal flow*) because of its cold temperature or high sediment concentration, a plane-jet flow occurs in the form of a turbidity current moving along the basin floor. If the densities are nearly equal (*homopycnal flow*), axial-jet flow results, and complete mixing occurs very close to the river mouth. Almost all the river sediment is deposited immediately after entering the standing water. Homopycnal flow is most common in freshwater lake deltas, and its mechanics leads to the classic Gilbert-type construction (see Born 1972, for example). In the third possibility, where rivers flow into more dense ocean water (*hypopycnal flow*), mixing is rather slow and the river water spreads out laterally in a plane-jet flow.

The depositional pattern that develops at any individual river mouth depends on the intensity of spreading and turbulence and how those factors are modified by tides and waves (Wright 1977). In high-constructive deltas, distributaries develop at the mouth of the in-flowing river where longitudinal bars are deposited because bedload cannot be transported when the velocity suddenly decreases. The initial bar exerts an influence on the flow and, as figure 7.28 shows, causes the river to bifurcate into two channels immediately upstream from the bar crest (for details see Russell 1967b). The distributary channels are lined with natural levees that may begin beneath the surface

Figure 7.29.
Sequence of the development of the subdeltas that comprise the Mississippi River deltaic plain.

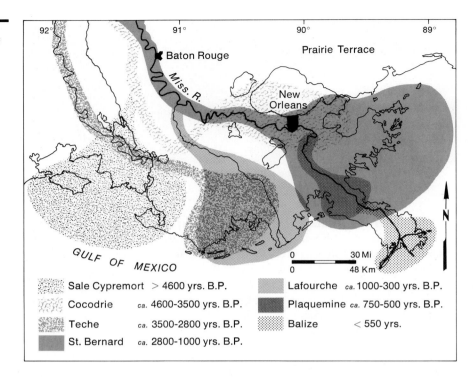

through slow accretion of suspended load as it spreads laterally under plane-jet flow (Morgan 1970). With continued deposition, combined with minor channel scouring, the levees and bars emerge (fig. 7.28), and the distributary channels extend farther into the basin. The process may be repeated frequently, giving the basin a veinlike appearance and prograding the delta front.

As the delta progrades, shorter routes to the ocean become available. These pathways often begin far inland from the delta front, usually developing when the river is diverted through a breach in the levee called a *crevasse*. The new river course shifts the locus of sedimentation and begins to form a new deltaic lobe. The Mississippi River deltaic plain, for example, actually represents the coalescence of seven major lobes that were built at different times and in various positions during the last 5000 years (fig. 7.29.)

The modern birdfoot-delta growth is only a minor portion of the entire deltaic area. We have good reason to believe that a new lobe of the Mississippi delta is being developed at the present time by the Atchafalaya River. This river, a distributary breaking off the Mississippi channel upstream from Baton Rouge, Louisiana, carries about 30 percent of the Mississippi River flow. During flood events considerable load is transported down the Atchafalaya, which has progressively filled in shallow lakes in the lower Atchafalaya basin. Since the early 1950s most of the sediment has been reaching Atchafalaya Bay (approximately 160 km west of New Orleans), where it is actively building the new deltaic lobe (Shlemon 1975; Rouse et al. 1978). Landsat satellite imagery shows clearly that the lobe is growing (fig. 7.30), with approximately

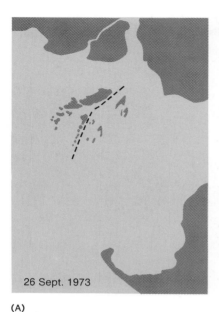

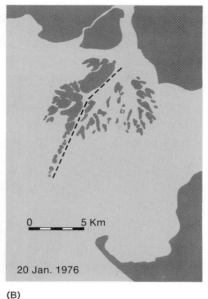

(A)　　　　　　　　　　　　　(B)

26 Sept. 1973　　　　　　　　20 Jan. 1976

0 ———— 5 Km

Figure 7.30.
Emergence of new land in Atchafalaya Bay as part of development in a modern lobe of the Mississippi delta.

6.5 km² of new land having emerged from the bay each year since the early 1970s (Rouse et al. 1978). Most scientists believe that total diversion of the Mississippi into the Atchafalaya is inevitable because that route to the ocean is about 300 km shorter than the present course. This gives the Atchafalaya a distinct advantage, and capture will occur unless humans use absolutely heroic measures to maintain the status quo. When lobes are abandoned during shifting of the river course, they are no longer fed by incoming river debris and immediately become vulnerable to erosional attack by the ocean. Thus, as a new lobe develops, older lobes are being destroyed.

The processes that produced the major lobes of the Mississipi deltaic plain are obscured by modifications since their abandonment. Within the modern delta, however, the genetic mechanics is known in considerable detail. Here crevasses in the levee system have repeatedly shifted the site of deposition. The breaks in the levee begin because of overtopping during a major flood and gradually increase in size through scouring associated with succeeding floods. Sediment diverted through a breached levee progressively builds a subdelta from deposits known as **crevasse splays.** Four such subdeltas have formed in historic times (fig. 7.31). As channels in the subdeltas bifurcate and prograde, their gradients decrease and they lose their ability to transport load. When their gradients approach that of the main trunk channel, it is no longer advantageous for the river to utilize the crevasse system. Sedimentation in the subdeltas ends, and the ocean begins to inundate the area as subsidence and compaction lower the subdelta surface. At the same time, however, a new subdelta may be developing somewhere else within the birdfoot system.

Figure 7.31.
Subdeltas of the modern birdfoot delta of the Mississippi River.

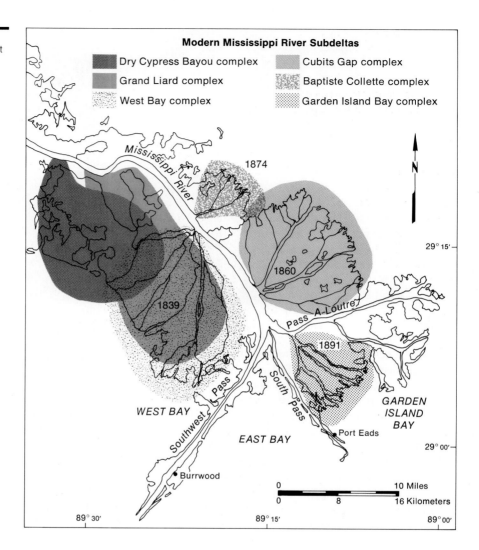

It seems clear that delta formation can be viewed on several time scales. On a short-term basis, only a limited area (subdelta) receives any sediment, but the position of the accumulation shifts repeatedly. On a longer time scale, the entire active delta (lobe) has a periodic migration. What we observe as the Mississippi River delta is in fact a monstrous area created by the coalescence of a number of major lobes over a long period of time. Active deposition occurs on only one lobe at any given time, and on only a minor portion of that lobe. Even so, the evolutionary processes on any time scale are probably similar in that they involve channel bifurcation, levee development, and crevassing.

Even though this evolutionary model is probably correct, it is valid only for deltas of the high-constructive type. The model is known in such detail only because geologists have studied the Mississippi delta for many years. Other delta types have not been so closely examined, and their mode of origin is not nearly so well known. Furthermore, we are far from understanding precisely how changes in climate, tectonics, or any of the other fundamental controls affect the mechanics of delta evolution.

Summary

The character of each major fluvial landform is due to the manner in which a particular set of geomorphic controls influences the river mechanics. The features can be predominantly erosional, predominantly depositional, or derive from the combination of both types of processes.

Floodplains originate by lateral migration of meanders and by periodic overbank flooding. The sediment composing a floodplain sequence is mainly laterally accreted point bar deposits accumulated as the river shifts its position across the valley bottom. The point bar material is usually capped by a thin layer of silt and clay deposited as vertically accreted sediment during overbank flooding of the river. The total amount of vertical accretion is probably controlled by the rate of the river's lateral migration. Because many rivers move across the valley floor rapidly, most floodplains are developed by lateral accretion, but overbank deposition may be important in the initial phase of construction.

Terraces are merely abandoned floodplains. They form when entrenchment places the river at a lower level, thereby removing the former floodplain surface from the river's hydrologic activity. The origin of a terrace usually refers to how its flat tread (the former floodplain level) was formed. The tread of an erosional terrace is produced by the lateral migration of a river and is capped by the thin point bar and overbank deposits associated with floodplains formed in that manner. In contrast, the surface of a depositional terrace represents the upper level of the sediment deposited in an episode of valley filling. The use of terraces to reconstruct geomorphic history demands some knowledge of their origins, because the salient properties of the various terrace types develop through different sequences of events.

Piedmont regions are characterized by depositional features (alluvial fans) and plains of erosion (pediments). Both features manifest some accommodation between the amount of sediment derived from a source area and the ability of the river to transport the sediment across the piedmont zone. The processes that develop fans and work on their surfaces are so complex that little agreement exists regarding their origin. Pediments also seem to defy genetic generalization. They are usually interpreted as being the result of lateral planation, weathering, and rill wash, or some combination of the various processes. As with fans, however, little agreement can be found concerning the origin of pediments; some may be relict features that are completely unrelated to modern geomorphic controls. It is clear that piedmont landforms cannot be placed into all-inclusive genetic models. Their properties vary too much to be explained by one mode of origin. Because of this, every piedmont region must be examined and interpreted according to local tectonics, climate, geology, and geomorphic history.

Deltas represent the accumulation of sediment as a transporting river enters a body of standing water. At its mouth the river bifurcates into distributaries and constructs levees. This allows debris to be transported farther into the ocean or lake basin and permits the delta to expand by prograding into the basin area. The form and size of the delta, however, depend on the balance reached between river flow and the counteracting energy of currents and waves in the ocean or lake. Detailed studies of deltas reveal a very complex growth history in which the site of active sedimentation shifts periodically through crevasses in the natural levees. Active sedimentation occurs on only a small part of the feature at any given time.

Suggested Readings

The following references provide greater detail concerning the concepts discussed in this chapter and also contain more extensive bibliographies on the various topics.

Bull, W. B. 1968. Alluvial fans. *Jour. Geol. Educ.* 16:101–06.

Cooke, R. U., and Warren, A. 1973. *Geomorphology in deserts.* London: Batsford Ltd.

Denny, C. S. 1967. Fans and pediments. *Am. Jour. Sci.* 265:81–105.

Hadley, R. F. 1967. Pediments and pediment-forming processes. *Jour. Geol. Educ.* 15:83–89.

Howard, A. D.; Fairbridge, R. W.; and Quinn, J. H. 1968. Terraces, fluvial—Introduction. In *Encyclopedia of geomorphology,* edited by R. Fairbridge, pp. 1117–23. New York: Reinhold Book Corp.

Leopold, L. B.; Wolman, M. G.; and Miller, J. P. 1964. *Fluvial processes in geomorphology.* San Francisco: W. H. Freeman.

Morgan, J. P. 1970. Deltas—A résumé. *Jour. Geol. Educ.* 18:107–17.

Oberlander, T. M. 1974. Landscape inheritance and the pediment problem in the Mojave Desert of southern California. *Am. Jour. Sci.* 274:849–75.

Schumm, S. A. 1977. *The fluvial system.* New York: Wiley-Interscience.

Twidale, C. R. 1976. *Analysis of landforms.* Sydney, Aus.: John Wiley & Sons.

Wolman, M. G., and Leopold, L. B. 1957. River floodplains; some observations on their formation. U.S. Geol. Survey Prof. Paper 282-C.

Wind Processes and Landforms

8

Introduction

Scientists have not agreed on the role of wind in geomorphology. Irrefutable proof that wind is capable of significant geomorphic work is simply not available in many instances, because definitive, quantitative data concerning eolian processes and features are woefully few. It has been described as being absolutely dominant in arid regions to being only a minor perturbation on features formed almost entirely by fluvial or slope processes. Disavowing these extreme views, it is probably safe to say that wind can be an effective geomorphic agent under certain physical conditions. Regions having sparse vegetation and unconsolidated sediment not tightly bound by rooting systems are most susceptible to wind attack. Extensive evidence of eolian processes is therefore found in those areas, such as modern deserts, where vegetation growth is stunted by lack of water or immature soil development. As is true of any process, however, the proficiency of wind to do geomorphic work depends on whether the driving force can exceed resistance of the surficial material. Thus, assigning wind processes to one particular environment alone is a gross perversion of the principles governing geomorphic processes. The emphasis placed here on deserts is simply a pedagogical tool, not a geomorphic necessity.

Detailed reviews of the physical basis of wind action and its geomorphic results are available (Bagnold 1941; Chepil and Woodruff 1963; Cooke and Warren 1973), and much of the following has been liberally excerpted from those excellent treatments.

The Resisting Environment

Contrary to popular thought, deserts are not the barren tracts of shifting sands depicted in Hollywood epics. In fact, only one-fourth to one-third of most desert surfaces is occupied by sand, which usually occurs in large, sandy plains called *ergs* (see I. G. Wilson 1973). Normally, deserts display a variety of erosional and depositional landforms imprinted on a diverse topography that ranges from flat plains to rugged mountains (table 8.1 and fig. 8.1). Deserts also exist in different temperature zones; polar deserts are common, although relatively unstudied. Specific processes differ in relative importance in polar deserts and hot deserts, and so the similarity of landforms that sometimes exists between the two regions is not infallible proof of an identical genetic history. In fact, the same external form may be a function of a myriad of basic processes combined in slightly different ways. Unless stated otherwise, the discussion following refers only to hot desert conditions.

Deserts are by definition arid. They receive less than 25 cm of annual precipitation and have enormous evaporation rates, commonly 15 to 20 times greater than the precipitation (Stone 1968). Although deserts have a meager plant cover, the diversity of vegetal types is surprising, ranging from shrubs and grasses to true woodlands. This diversity is created by minor variations in soil moisture that relate to elevation. It is important because humus content and soil binding differ with particular species and because some plants are annual and others are perennial. The type of flora influences the resisting setting; woodlands, for example, are significantly denser than other vegetation

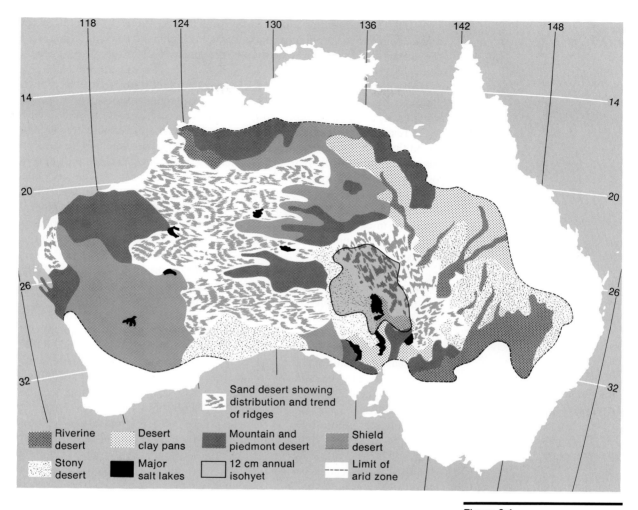

Figure 8.1.
Map of Australian arid regions showing diverse physiography of deserts.

Table 8.1   Area and percentage of total desert region of major components in the arid zone of Australia. (Compare with map in fig. 8.1.)

	Km³	Percentage of Arid Zone
Mountain and piedmont deserts	930,000	17.5
Riverine desert	210,000	4.0
Stony desert	640,000	12.0
Desert clay plains	690,000	13.0
Sand desert	1,680,000	31.0
Shield desert	1,200,000	22.5
Total	5,350,000	

From Mabbutt 1971. Used with permission of the Australian National University Press.

types and protect more surface area (Whitaker et al. 1968). In general, mountainous zones with stabilizing vegetation and thicker weathering mantles are less susceptible to wind attack than is the lower level of the desert environment.

Weathering and soil-forming processes are intimately involved in the character of a land surface and how it resists or succumbs to wind erosion. The angular, unaltered debris so prevalent on desert surfaces is presumed to be the residuum of rock disintegration, leading to a general concurrence that mechanical weathering predominates in the desert environment. This does not mean that mechanical weathering functions to the total exclusion of decomposition. Chemical processes, in fact, do function in the desert, as evidenced by the ubiquitous desert varnish and the common red coloration of desert soils, both of which probably require some degree of ion mobility. In addition, as explained in chapter 4, many processes of disintegration, such as exfoliation and grus formation, are abetted by chemical reactions that produce volume expansion.

In general, however, even though decomposition does function, maturely developed soils are rare in desert environments unless relict profiles formed during more humid intervals of the Quaternary are preserved.

Desert soils (even in their immature state) are important determinants of the facility of wind action. In general, A horizons are quite thin, mechanically weakened, and more permeable than the lower horizons (Cooke and Warren 1973); case hardening at the surface may in places counteract this tendency (Lattman 1973). Nevertheless, regardless of how slowly it proceeds, illuviation may cement B horizons to the extent that they become nonpermeable, highly indurated zones (Gile et al. 1965, 1966; Gile 1966). The long-term result of desert processes is to create a surface that is susceptible to wind erosion and a subsurface that is eminently resistant. Stripping of the surface horizon will expose the lower, resistant zones, which are preserved as stable platforms for extended periods of time. It should be noted that stripping at one place is commonly balanced by deposition of eolian debris somewhere else in the area.

Surfaces armored with a thin layer of stones that protects an underlying horizon of sand, silt, or clay are common in many environments. They are particularly evident, however, in hot deserts where they are given a variety of names such as *stone mantles, gibber, hammada, reg,* and *desert pavement.* Desert pavement, as mentioned in chapter 7, is associated with alluvial fans and other deposits of unsorted alluvium. The surface armor is usually only one or two stones thick and consists of whole clasts or disintegrated parts of the coarse fraction found in the underlying alluvium. In most cases, a thin, fine-grained layer, which is probably a product of desert weathering (fig. 8.2), exists beneath the pavement. The pebbles at the surface are most likely a lag concentrate formed as wind action progressively removes the fine-grained alluvium.

In contrast to the desert pavements, some parts of the desert are covered with abundant fine-grained debris, and there deflation is a viable erosive process. Playas, for example, are notable for their silt and clay mantles; it has often been suggested that they are subject to extensive degradation by deflation (Blackwelder 1931). There is no doubt that deflation of fine sediment occurs (fig. 8.3), but it may not be as pervasive in the desert environment as once supposed (Hooke 1968). Although most deserts do contain fine material, it is usually fabricated into a cohesive crust by a number of processes (Dury 1966a; Cooke and Warren 1973). Crustation not only gathers individual particles into aggregates larger than the wind can entrain, but it also protects

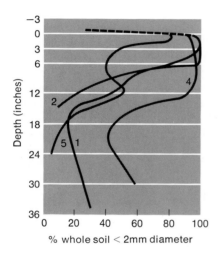
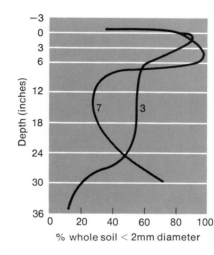

Figure 8.2.
Percentage of soil composed of particles less than 2 mm in diameter. Analyses of soils in the Lahontan Basin, Nev. (Reproduced from *Soil Science Society of America Proceedings*, Col. 22, 1958, p. 65, by permission of the Soil Science Society of America)

Figure 8.3.
Deflation of fine-grained sediment at Pyramid Lake, Nev. Bottom sediment has been exposed by recent decline in the lake level.

unconsolidated fines immediately beneath the surface. The greatest resistance occurs when sand, silt, and clay are mixed together in proportions of clay 20 to 30 percent, silt 40 to 50 percent, and sand 20 to 40 percent (Chepil and Woodruff 1963), but establishment of the most resistant size blend is not the only determinant of resistance. Variables of moisture, humus, and chemical composition may give considerable resistance to size blends that are usually susceptible to deflation. Calcium carbonate, on the other hand, may actually induce erodibility because it tends to inhibit the formation of aggregate clods (Cooke and Warren 1973, p. 241).

In summary, the geomorphic effect of wind action is regulated to a significant degree by the properties of the resisting framework. For wind processes to have any geomorphic consequence, the surface must be unvegetated and littered with noncohesive sediment smaller than gravel-size. Any region producing sediment of that type, and lacking the climatic, topographic, and geomorphic conditions needed to protect it, may be open to pronounced wind attack. Because hot deserts are most susceptible to wind action, they are taken as the model for the discussion of eolian processes. However, coastal regions, unvegetated semiarid or subpolar zones, and areas in front of active glaciers also often are modified by wind processes. In any situation, the amount of geomorphic work actually accomplished by the wind depends on the second prime variable—the character of the wind itself.

The Driving Force

Global wind systems are related to the large pressure differentials associated with worldwide circulation patterns. In contrast, local winds are generated by anomalies in the physical characteristics within a specific area. For example, in many cases winds reflect a difference in thermal properties of surface materials, and so we can expect the greatest wind activity in those environments where large temperature variations exist in the air layer immediately above ground level. Such conditions are common in deserts, along seacoasts, or in areas with pronounced diversity in elevation, such as the juncture of mountain ranges and the low plains fringing them.

Certain attributes of the wind—mainly its direction, velocity, and degree of turbulence—are responsible for most geomorphic effects. In areas with large thermal contrasts, *wind direction* is more or less predetermined by the temperature gradient of the near-surface air. However, temperature gradients often change from daytime to nighttime because of variations in rates of cooling and heating. In deserts, heating of the bare desert surfaces incites winds to move toward them from surrounding areas that have a vegetal cover. At night, however, the desert surface cools more rapidly than the other regions, and the winds reverse their direction. A similar diurnal direction change occurs along coasts because the ocean warms less slowly than the adjacent land, prompting on-shore winds during the day. The ocean, however, gives up stored heat less readily at night, and winds blow off-shore. The importance of wind direction is most evident in the development and preservation of eolian bed forms constructed from loose, transportable sand.

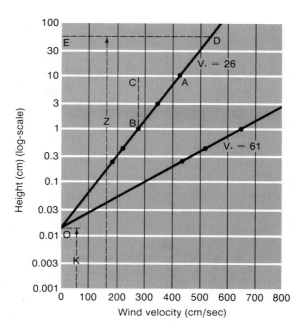

Figure 8.4.
The relationship of wind velocity,
height above surface (z) and
shear or drag velocity (V_*).

Wind velocity is important because it is the prime determinant of what material will move under wind attack and what will remain stationary. Wind velocity increases with height above the ground because it is slowed at the surface by friction. The change in velocity with height can be expressed in a number of ways exemplified by the equation (Bagnold 1941)

$$V_z = 5.75 \, V_* \log z/k,$$

where V_z is the velocity at any given height z, V_* is a parameter called **drag velocity,** and k is a constant relating to surface roughness. V_*, originally designated as a drag phenomenon, is actually a shear velocity and relates to shear stress. The parameter k is the height of a thin zone immediately above the surface in which velocity is zero. The thickness of this zone depends on the surface roughness, but on a flat, granular bed it tends to be about 1/30 of the diameter of the surface grains. The general relationship between height and velocity plots as a straight line on semilogarithmic paper (fig. 8.4).

As the wind blows harder, V_* increases and exerts a greater stress on particles exposed at the surface. This follows because the magnitude of shear (drag) across any unit area of surface is related to V_* such that

$$V_* = \sqrt{\tau/\rho}$$

or

$$\tau = \rho V_*^2,$$

where τ is drag/unit area and ρ is the density of the air. Thus, for a surface with constant particle size, a stronger wind will increase the value of V_*, and therefore the height-velocity relationship is defined by a number of straight lines, each having a different V_* value (fig. 8.4). Drag velocity in this sense represents the *rate* at which velocity increases with height and can be thought of as a velocity gradient. The value of V_* can be determined as the tangent of the straight line, as shown on figure 8.4, divided by the constant of proportionality, 5.75. For example, the tangent of line OD on figure 8.4 is $\dfrac{AC}{CB}$ or $\dfrac{150}{\log 10}$. Therefore,

$$V_* = \frac{150}{1} / 5.75 = 26 \text{ cm/sec.}$$

Interestingly, all lines of different V_* values merge at the ordinate where velocity is zero. This occurs simply because k is a function of surface conditions, such as particle size, rather than wind velocity.

Once the value of V_* is known, it is possible to calculate the wind velocity at any height above the surface. For example, if $V_* = 26$ cm/sec, the velocity at 4 cm above the surface is

$$V_z = 5.75 \ V_* \log z/k$$
$$= 5.75(26) \times \log z - \log k$$
$$= 150 \ [0.6 - (-1.82)]$$
$$= 363 \text{ cm/sec.}$$

Assisting velocity in wind attack is the factor of *turbulence,* which occurs in the form of eddy currents. Turbulence has a direct influence on the entrainment process, although it is probably capable of lifting only particles smaller than sand (Bagnold 1941; Chepil and Woodruff 1963). Sharp (1964), however, suggests that grains up to 0.125 mm may be affected by wind turbulence. Most likely, turbulence helps in the molding of desert landforms, but the types of eddy motion are so numerous and mechanically complex (Lumley and Panofski 1964), that precise measurement of turbulence is precluded, and its relationship to specific bed forms in not clear. At present little data are available to generalize about velocities and turbulence in wind processes but, in ways that are still not quantitatively clear, both factors must bear directly on the origin of wind-produced landforms.

Wind Erosion

Processes

As wind blows across a surface composed of loose sediment, a critical drag velocity exists at which particle motion begins. The ease of entrainment, however, depends not only on the size of the particles and the wind velocity but also on complicating factors such as soil moisture, packing, etc. Recognizing the inherent complexity, the velocity that initiates movement is called the **fluid threshold** (Bagnold 1941) and can be estimated by the equation

$$V_{*_t} = A \ \sqrt{\frac{\rho_s - \rho_a}{\rho_a} gD}$$

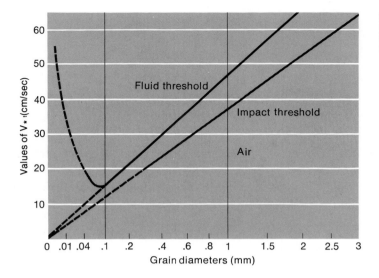

Figure 8.5.
Relation of particle size to threshold velocity.

where V_{*t} is the threshold value of drag velocity, ρ_a is the density of the air, ρ_s the density of the sediment, D is the particle diameter, g is gravity, and A is a constant of proportionality which for air is 0.1. Most desert sands have a threshold velocity of about 16 km/hr, but the precise value for any size varies with other factors such as particle shape (G. Williams 1964), sorting (Woodruff and Siddoway 1965), and surface roughness. Roughness is a function of particle size but is also controlled by vegetation. Rough surfaces reduce wind velocity; as high, dense vegetation increases roughness drastically, it lessens the erosive effect of the wind.

The influence of moisture on the critical threshold velocity was documented by Belly (1964) who suggested that a surface moistened by liquid or vapor would require a significantly higher velocity for sediment entrainment. For example, Calkin and Rutford (1974) found that when Antarctic dunes were moistened by summer melting or condensation, the threshold velocity needed to entrain their sands doubled.

The fluid threshold equations clearly indicate that entrainment velocity varies directly with particle diameter; nevertheless, as figure 8.5 shows, the relationship becomes invalid when size is less than 0.1 mm because of the greater interparticle cohesion and low roughness associated with finer particles (Bagnold 1941; Smalley 1970). Furthermore, grains greater than .84 mm are moved with great difficulty, and that size may represent a logical upper limit for unaided wind entrainment. Once motion begins, however, the surface is subjected to a continuous rain of moving particles that, on impact with stationary grains, produce entrainment at a velocity lower than the fluid threshold. This reduced threshold velocity, called the *impact threshold* (fig. 8.5), becomes progressively more significant as size increases, and particles greater than .84 mm may be entrained even when the velocity is well below the fluid

Figure 8.6.
Diagram of the saltation process. Moving grain strikes the surface and dislodges a particle to elevation *h*. The particle moves downwind and repeats the process. ℓ represents the distance of travel of the dislodged particle before striking the surface.

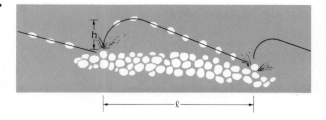

threshold. Indeed, an exceptionally strong 1977 windstorm in southern California carried particles as large as 7 mm at a height 240 cm above the surface (Sakamoto-Arnold 1981). It appears, therefore, that when grains begin to move at the fluid threshold, the entrainment process is self-generating because the wind speed is already above the impact threshold. Each moving grain is not only a product of the system but also an integral component of the entrainment mechanics through the process of saltation as depicted in figure 8.6.

Features Produced by Erosive Action

Geomorphic features manufactured by wind action are due to abrasion and deflation, the two main erosive processes. Abrasion results when sand particles carried by the wind act as grinding tools to physically wear away exposed surfaces of solid rocks or rock fragments. It is most effective where relatively weak rocks, bare of vegetation, are blasted by high-velocity winds that carry an abundant amount of hard particles (Suzuki and Takahashi 1981). In fact, Sharp (1980) has demonstrated that the abrasion rate increases significantly when more load is made available for wind transport. Thus, maximum abrasion is determined by some unique combination of velocity and particle concentration. Normally this will occur within 2 m of the ground surface.

Perhaps the most frequently cited evidence of wind abrasion is the development of faceted stones called **ventifacts** (fig. 8.7). Ventifacts are produced on pebble surfaces that are oriented perpendicular to the prevailing wind. The incessant bombardment of wind-driven sand gradually cuts a smooth face into the windward portion of the original surface and forms a sharp edge between it and the leeward side of the pebble. The slope on the faceted surface is normally between 30° and 60°, but enough variability exists in these angles to suggest that some evolutionary sequence is involved in their development. It is equally feasible, however, that various heights of pebbles above the surface relative to various levels of maximum abrasion potential in the wind can produce a number of facet inclinations and shapes (Sharp 1964). It is also possible that some faceting is accomplished by dust-sized particles (silts and clays) carried in suspension, which would make ventifact sculpturing a function of the unique aerodynamics associated with each individual situation (Whitney and Dietrich 1973; Whitney 1978).

Facets do not form on surfaces that parallel the wind direction, and therefore the ubiquitous presence of ventifacts with more than one face and edge (fig. 8.7) has created some interpretive furor. It is tempting to regard multifaceted ventifacts as evidence of a shift in the prevailing wind direction, but

Figure 8.7.
Pebbles and cobbles pitted and faceted by wind-blown sand. Ventifacts were all found in Sweetwater County, Wyo.

it is equally possible that the stones themselves are occasionally turned over or rotated (Sharp 1964, 1980). Overturning, of course, would result in a new portion of the original pebble being subjected to wind erosion, and any number of facets could form in a unidirectional wind.

Abrasion affects rock outcrops by shaping or grooving the surface. The most common feature etched by the wind, a **yardang,** is an elongate ridge, normally less than 10 m high but occasionally as high as 30 m, which is aligned parallel to the prevailing wind. Yardangs are most prominently developed in regions underlain by relatively soft rocks and tend to occur in groups in which individual ridges are separated by round-bottom, wind-eroded troughs (Haynes

1982). Blackwelder (1934) emphasized the importance of eroding the intervening "yardang trough" rather than the ridge itself. He believed that most erosion occurs at the lowest level, decreasing in magnitude and efficiency toward the ridge crest. As a result the troughs are eroded more rapidly, so that sometimes the ridge sides are slightly undercut and the ridge crests are left rather ragged.

Grooving of bedrock surfaces is also a common eolian phenomenon. The abrasive processes that result in parallel furrows seem to operate at different orders of magnitude, ranging from tiny elongate flutes on ventifact surfaces measured in centimeters (Sharp 1964), to giant hollows more than a kilometer wide and up to 10 m deep. Some question lingers as to whether the megagrooves are demonstrably erosional or merely a depressed zone between parallel ridges of deposited sand. Although the possibility of misinterpretation remains, some large erosional grooves have been reasonably well documented (D. King 1956; Shawe 1963; Stokes 1964). In some cases, grooving may have a distinct influence on regional drainage patterns (Stokes 1964; Beaty 1975b; Hallberg 1979), but care must be exercised in making such an interpretation (Rahn 1976).

The process of deflation has been suggested as the prime mechanism in the genesis of many enclosed desert basins. These basins range in size from small deflation hollows to vast expanses measured in hundreds of square kilometers. Usually the small depressions are easily related to wind erosion since they are often strung out parallel to the wind and may be associated with dunes (Flint and Bond 1968; Sharp, 1979). A deflation origin for the larger areas, called *pans,* is much more difficult to prove since they have no clear orientation. Arguments about their origin, therefore, are often negative in the sense that they merely eliminate other processes and leave deflation as a viable, but untested, alternative. Very large pans, however, must be relict to some extent since wind erosion, even if proven, functions so slowly that an enormous time span would be necessary to accomplish their formation.

Wind Transportation and Deposition

The entrainment of sediment releases particles to the processes of transportation and deposition, and the result is the array of surface forms so diagnostic in areas under wind attack. The type of transporting process is determined fundamentally by the size of the debris made available to the eolian force; silts and clays are carried in **suspension,** fine and medium sands travel by **saltation.** Coarse sand, and sometimes gravel (Sakamoto-Arnold 1981), move by rolling or sliding motions collectively referred to as **surface creep.** The most striking eolian features are made with sand because, in the absence of unusually strong storms, few winds can move larger grains, and sediment entrained by turbulence and carried in suspension often is so diffuse as to preclude its accumulation in explicit surface forms. In addition, entrainment of fine-grained sediment is not so easy as one might expect, because the surface is usually smooth and the small grains tend to retain moisture and have stronger interparticle chemical bonds. Grains larger than 0.1 mm probably cannot be transported in suspension (Bagnold 1941; Kuenen 1960; Sharp 1963), and particles

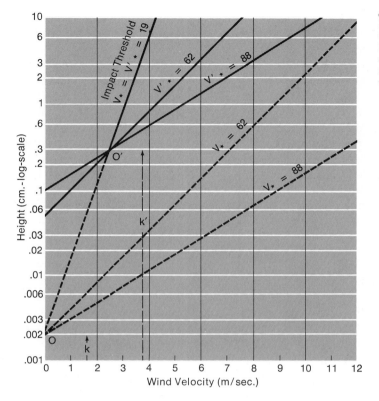

Figure 8.8.
Change in drag velocity caused by saltation. Dashed lines show drag velocity before saltation begins. Solid lines represent drag velocity after saltation starts.

both larger and smaller than sand often resist entrainment. For those reasons, our discussion will center on depositional features composed of sand, and on the mechanics of saltation.

Figure 8.8 shows that when entrainment occurs the moving sand directly influences the characteristics of the wind itself because the saltating grains exert extra drag on the system. As discussed earlier, the drag velocity (V_*) is a measure of the velocity gradient of the wind, i.e., the rate at which velocity increases with height. The stronger the wind blows, the greater the drag on the surface and the greater the divergence of the drag velocity curves away from the vertical. The slope of each line on the figure, therefore, reflects its rate of increase in velocity with height (given as drag velocity, V_*); the higher V_* values indicate a more rapid increase in velocity with ascension above the surface.

The solid lines on figure 8.8 represent the changes impressed on the system after saltation begins. Note that under saltating conditions k rises dramatically, and it no longer represents a zero velocity but assumes a constant velocity value at the height k' where all the gradient lines merge. The wind velocity at any height (z) is then defined by the equation

$$V_z = 5.75 \ V_{*}' \log z/k' + V_t,$$

where V_*' is the drag velocity under conditions of saltation, k' is the new height of the focus, and V_t is the threshold velocity (impact threshold) measured at the height k'.

The effect of saltation is obvious from the graph. For example, a wind with a velocity gradient $V_* = 62$ blowing across a completely resistant surface has a velocity of about 970 cm/sec at a height of 2 cm. Once motion begins, however, the velocity at that level, read along the gradient line $V_*' = 62$, has decreased to approximately 530 cm/sec. It is important to stress again that the value of k or k' is dependent on the roughness components of the surface, and so figure 8.8 defines curves that are valid for only one particle size. In addition, very few measurements have been made in the natural setting, and although estimates of the quantitative relationship between k values and particle diameter have been made (Cooke and Warren 1973, p. 259), enough variance exists in these for us to be cautious about their use.

Saltating grains travel in a characteristic path in which particles dislodged or bounced from a surface move at first in a predominantly vertical direction (fig. 8.6). The subsequent travel path is probably controlled by (1) the particle's initial upward velocity and (2) the velocity gradient above the surface. Bagnold (1941) suggested that most of the forward momentum is gained from the wind when a grain is near the top of its path, and from then on the actual trace of the particle movement depends on the magnitude of the wind thrust and the settling velocity of the grain. Precisely how high a grain will rise from the surface depends, of course, on the size of the particle involved and on what kind of a surface it is bouncing from. It is clear, however, that the dense curtain of saltating grains is restricted to a thin near-surface zone. Most observations (Bagnold 1941; Chepil 1945; Sharp 1964) indicate that an overwhelming percentage of the total load is carried within 2 m of the surface, although minor amounts may rise above 10 m.

Even though most sand moves by saltation, some of the load never enters the realm of airstream mechanics, but moves as surface creep. Creep results when the impact of saltating grains spasmodically shoves or rolls surface particles forward without displacing them upward. This type of motion becomes progressively more important as the grain size increases (Sharp 1964) and is probably the dominant mechanism of forward motion in very coarse sand and gravel. Nonetheless, in average wind-blown sand, ranging from 0.15 to 0.25 mm, creep seems to account for 7 to 25 percent of the total load.

Theoretically, it should be possible to estimate the total wind-traction load, because the discharge of sand must be related to the drag exerted on the wind by the saltating grains, and drag is dependent on V_* (drag velocity) and air density. Indeed, a number of equations have been derived to make these estimates. It is doubtful, however, that any equation will be very precise, for other uncontrolled factors such as particle shape (G. Williams 1964) and the ambiguities of grain motions in flight must introduce a finite uncertainty to any formula. This is perhaps best demonstrated by figure 8.9 where, for example, a drag velocity of 50 cm/sec may produce a significant difference in Q depending on which equation is employed.

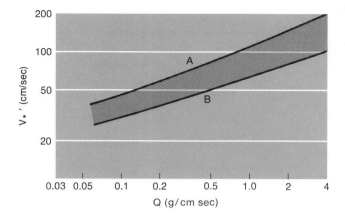

Figure 8.9.
Relation of sand movement to drag velocity. Zone between curves A and B contains a family of curves derived from numerous studies.

Deposits and Features

The most striking features associated with eolian processes occur in vast, sandy deserts called **sand seas** (or **erg** in the Sahara Desert of North Africa). These areas have an enormous supply of sand and are marked by varied assemblages of sand dunes that give a wavelike configuration to the surface. Complex dune fields, however, are not the only component of sandy deserts since large portions are often occupied by tabular bodies of sand, called **sand sheets,** which have little surface topography, or by relatively flat interdune areas of various types (for discussion see Ahlbrandt and Fryberger 1982). The largest sand sea in the Western Hemisphere is the Nebraska Sand Hills, an area of eolian features that was formed in the Holocene and covers almost 57,000 km² of northwest Nebraska (Ahlbrandt and Freyberger 1980).

On the basis of wind data and Landsat imagery, it seems certain that sand seas form where there is a significant reduction of wind energy along the direction of sand drift (Fryberger and Ahlbrandt 1979). Zones of reduced energy leading to the development of sand seas are produced in several ways: (1) Topographic barriers, such as high plateaus or mountains oriented across the direction of sand drift, tend to reduce energy by dispersing prevailing winds into variable flow directions. They also physically block the advance of drifting sand. For example, dunes piled up in the Great Sand Dunes National Monument, Colorado, were probably positioned there by the blocking action of the Sangre de Cristo Range. (2) Large bodies of water, such as lakes or oceans, also serve as interceptors of drifting sand. In these cases, eolian deposition may be prograded into the standing water. (3) Wind energy may decrease parallel to the direction of sand drift because of zonation in the regional climate. In North Africa, for example, alternating zones of climatically controlled low and high wind energy are common. Sand that drifts out of regions of high energy accumulates in the lower energy zones.

Within the sand seas, surface features commonly referred to as bed forms are spaced with pronounced regularity and, like erosional features, come in a variety of sizes ranging from tiny **ripples** to giant forms called **draa,** which apparently are restricted to the North African deserts. Intermediate in scale between ripples and draa are **dunes,** the most common depositional form in

Table 8.2 Eolian bed forms and their geometry and possible origin.

Wave-length	Height	Orientation	Possible Origin	Suggested Name
300–5500 m	20–450 m	Longitudinal or transverse	Primary aerodynamic instability	Draas
3–600 m	0.1–100 m	Longitudinal or transverse	Primary aerodynamic instability	Dunes
15–250 cm	0.2–5 cm	Longitudinal or transverse	Primary aerodynamic linstability	Aerodynamic ripples
0.5–2000 cm	0.05–100 cm	Transverse	Impact mechanism	Impact ripples
1–3000 cm	0.05–100 cm	Longitudinal	Secondary vortices	Secondary ripple sinuosity

From I. G. Wilson 1972. Used with permission of *Sedimentology,* Blackwell Scientific Publications.

Figure 8.10.
Relation of coarse-grained portion of sediment load (P_{20} = the coarse twenty-percentile) and wave-length of bedforms (A = ripples; B = dunes; C = draas). Note that there are no transitional forms between the groups.

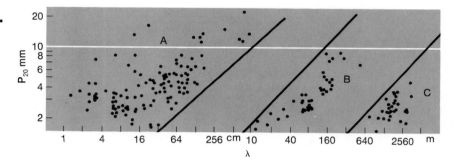

regions susceptible to wind attack. No clear dimensional boundaries separate the geometric types, which are described in table 8.2, and it is not unusual for the smaller features to be superimposed on parts of the larger ones. I. G. Wilson (1972), however, demonstrated that the dominant members of the bed form hierarchy can be differentiated by plotting their wavelength distance against a parameter of their constituent particle size (fig. 8.10). It seems clear from Wilson's work that for any given feature (ripple, dune, draa) a larger particle size is associated with a greater wavelength, and because grain size is a function of velocity, the expanded wavelength is also a reflection of a stronger formative wind. In addition, at any given particle size (and presumably wind velocity), all the types may be forming simultaneously. This strongly suggests that different process mechanics are associated with each geomorphic form. The fact that no transitional forms exist between the major varieties strengthens this supposition (Wilson 1972).

Ripples

Wind ripples range in amplitude from .01 cm to 100 cm and may be spaced up to 20 m apart. Their dimensions depend primarily on the wind velocity, the particle size (Sharp 1963), and the type of ripple. Some ripples are formed purely by the shear stress of the wind acting on the surface (aerodynamic ripples), but the overwhelming majority of ripples result from the surface bombardment of saltating grains and its associated creep.

Bagnold (1941) suggested that the process of forming impact ripples is closely allied to the angle at which saltating grains strike the surface. Since saltating grains descend in a nearly uniform manner, most variations in the angle of incidence are caused by minor topographic irregularities of the surface (fig. 8.11). The intensity of creep in any surface area is proportional to the number of impacts produced by saltating grains. Thus, in the natural surface hollow depicted in figure 8.11, the number of grain collisions is much greater on the windward face of the depression (*BC*) than the lee side (*AB*). This results in the transport of more grains up the slope *BC* than are replenished by movement down the slope *AB* into the depression. Not only is the hollow maintained, but grains also begin to accumulate at point *C* because they are delivered there faster than they can be removed on the adjacent level surface. Eventually the accumulation at *C* produces a second lee slope *CD* which reinitiates the mechanics that operated in the original hollow. Repetition of this sequence inexorably propagates the ripple form downwind. Ripple crests, however, are not always perpendicular to the direction of the wind current (Howard 1977). Because of this distinct formative process, Bagnold (1941) feels that surfaces covered by unprotected sand and having no bed forms are probably unstable.

The process just described should create an ever-rising ripple crest and a gradually deepening hollow. Actually, the height of the ripple is effectively limited because it eventually rises to a level where greater wind velocity precludes deposition. At that height, grains are transported directly over the crest and deposited in the next downwind hollow, where the wind velocity is much lower. In this way the ripples assume a consistent geometry that, for well-sorted sands between 0.19 and 0.27 mm, has a height/wavelength ratio that is normally only 1:70 and never lower than 1:30 (Bagnold 1941).

In nature, however, coarse grains are often concentrated at the ripple crest (Bagnold 1941; Sharp 1963), allowing the ripple to grow higher than expected because the large grains continue to be deposited while the smaller particles flow over the crest into the hollow. This commonly results in a height/wavelength ratio as low as 1:10, and an asymmetric ripple shape in which windward slopes are notably more gentle than lee slopes. The asymmetry, however, may be lost if a high percentage of the ripple is composed of very large grains. These "granule ripples" (Sharp 1963) have very large wavelengths and may be burdened by a high proportion of grains larger than sand size.

Dunes

Of all desert and wind phenomena, sand dunes have received the greatest scientific attention. Dunes attain a characteristic equilibrium profile that can be logically divided into three components, as in figure 8.12: the *backslope* or windward surface, the *crest,* and the *slip face* or lee slope. Measurements show that the backslope declivity, normally between 10° and 15° (McKee 1966; Sharp 1966; Inman et al. 1966), is in stark contrast to the slip face, which always stands near the angle of repose for sand, between 30° and 34°. The crest, separating zones of erosion and deposition on the dune, is usually convex-up, but on very large dunes the pronounced convexity may be lost (Hastenrath 1967).

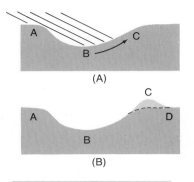

Figure 8.11.
Process of creep in the development of ripples in a natural surface hollow. (A) Minor topographic irregularity increases the incidence of saltating grains on the windward face *BC*. (B) Creep builds ripple crest at *C* and a second lee slope *CD*.

Figure 8.12.
Cross-profile of normal dune showing common geomorphic components.

Dune height increases until it stabilizes along with the equilibrium form. Most dunes range in height from less than 3 m to 100 m, but in rare cases they have been observed to be as high as 500 m (Walker 1982). The equilibrium height no doubt depends partly on the upward velocity gradient in the air, but most workers now seem to believe that height is mainly controlled by some poorly understood wave motion within the wind. This belief is reinforced by the fact that dunes normally occur in groups with distinctly regular spacing rather than as randomly placed individuals. Within any given dune field the wavelength is quite consistent, although from region to region it can show considerable variation (table 8.2).

The pronounced regularity in dune spacing hints at some prevailing atmospheric motion that is capable of maintaining dune forms and their spacing in an equilibrium state. For example, formation of an initial dune possibly interferes with the airflow in such a way that it creates eddy currents or, alternatively, fixes a regularly spaced pattern of turbulence downwind from the intruding dune (Cooper 1958). Eddy currents unquestionably exist in the lee of dunes (Hoyt 1966), but considerable question remains as to whether they are modified by other flow mechanics and, more important, whether they possess the erosive power needed to regulate the spacing or shape of dunes (Hoyt 1966; Inman et al. 1966; Sharp 1966; Glennie 1970).

Bodies of sand, which serve as the birthplace of dunes, form wherever local conditions favor deposition of sand that is moving under wind transport. This will occur when V_* (and therefore τ) is lowered, and thus deposition is commonly prompted by topographic irregularities of the surface. Once deposited, however, a sandy patch on an otherwise resistant surface becomes an integral part of the system mechanics.

Equilibrium in sand dunes represents a balance between the volume of erosion and the volume of deposition occurring on the feature (Howard et al. 1977). This balance is maintained through adjustments of the backslope and slip face angles, which in turn are controlled by particle size and velocity gradients. Theoretically, the crest zone should exist as a sharp angle formed by the equilibrium surfaces of the backslope and the slip face. Actually, the wind does not respond immediately to a change in slope, and so the locale of maximum deposition will be some finite distance downwind from the crest, creating the tendency toward a convex summit. This tendency is sometimes complicated by slumping that occurs along the slip face when deposition in the crestal zone increases the lee slope beyond the angle of repose.

It is axiomatic that maintenance of the equilibrium shape requires forward movement of the entire feature, because erosion from the rear slope must be volumetrically balanced by deposition on the lee slope. There seems to be general agreement that dunes retain their original form as they advance and

Table 8.3   Terminology for basic types of dune forms.

Form	Number of Slip Faces	Name Used in Ground Study of Form Slip Face, and Internal Structure
Circular or elliptical mound	None[a]	Dome
Crescent in plan view	1	Barchan
Row of connected crescents in plan view	1	Barchanoid ridge
Asymmetrical ridge	1	Transverse ridge
Circular rim of depression	1 or more	Blowout[b]
U shape in plan view	1 or more	Parabolic[b]
Symmetrical ridge	2	Linear (seif)
Asymmetrical ridge	2	Reversing
Central peak with 3 or more arms	3 or more	Star

Modified from McKee 1979, table 1, p. 10.
[a]Internal structures may show embryo barchan type with one slip face.
[b]Dunes controlled by vegetation.

that the shape components, especially the height, influence the rate of forward migration (Finkel 1959; Long and Sharp 1964; McKee 1966; Inman et al. 1966; Norris 1966; Hastenrath 1967). In fact, it has been repeatedly demonstrated that the horizontal displacement (c) that occurs in any time interval can be defined by the equation

$$c = Q/\gamma H,$$

where H is the height of the dune and γ is the specific weight of the sand. The absolute rates of forward migration are variable because they depend on local controlling factors, but movement measured in tens of meters per year seems to be common. This equation shows a reasonable correlation with some actual measurements (Finkel 1959), but the value of Q is subject to so many external constraints (Chepil 1959; Svasek and Terwindt 1974) that to expect precise predictions of advance rates may be too much to ask.

The cross-sectional characteristics of dunes tend to be somewhat changeable because the complexity of wind motions and the variability of surface conditions interfere with their ideal development. An even more complex matter is the attempt by students of desert landforms to categorize dunes according to their plan view, a property known as the **dune pattern.** Pattern classifications are almost as diverse and bewildering as the forms they claim to classify. Like other features, dunes have been grouped on the basis of shape, genesis, wind types involved, surface conditions, and the like, but each attempt somehow fails to account adequately for nature's incredible diversity. For our purposes, it seems best to employ a modified form of the classification devised by the U.S. Geological Survey (table 8.3; fig. 8.13).

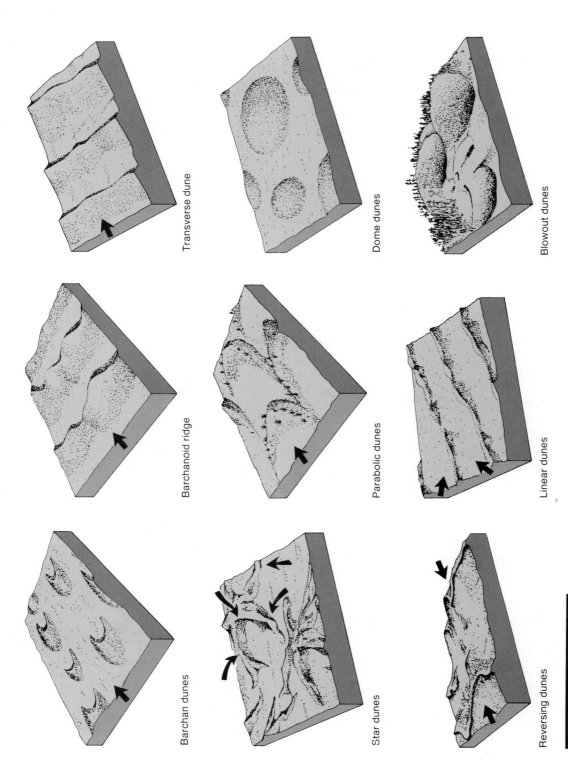

Transverse dune

Dome dunes

Blowout dunes

Barchanoid ridge

Parabolic dunes

Linear dunes

Barchan dunes

Star dunes

Reversing dunes

Figure 8.13.
Common dune types. Arrows
indicate direction of formative
winds. (From McKee 1979,
pp. 11–13)

Hack (1941) suggested that in the region he studied, three basic dune forms exist, which he called transverse, parabolic and longitudinal. Transverse dunes are usually free of vegetation; they may be the most probable dune pattern if winds are unidirectional over a limited supply of sand and the sand is free to migrate. Actually, the mechanics involved in maintaining this dune form may be controlled by the wind velocity (and its associated turbulence) needed to transport the largest particles in the sand (Lancaster 1982). The normal transverse dune is a crescent-shaped feature, called a **barchan dune** (fig. 8.14), in which tapering edges or horns of the crescent point downwind. Transverse dunes may also stand as simple ridges oriented perpendicular to the wind, or they may form a sinuous asymmetrical ridge, called a *barchanoid ridge,* which is composed of connected crescents (fig. 8.13).

Parabolic dunes differ from transverse dunes only in that they are U or V shaped, and their horns or tapered extremities point upwind. When the lateral edges of a transverse ridge become anchored by vegetation, the wind removes sand from the central zone and deposits it on the leeward slope. This, of course, allows the middle segment to advance relative to the edges and develops the characteristic parabolic shape. In addition, the dune surrounds a scoop-shaped hollow, called a *blowout,* from which the sand in the dune was derived, indicating that the pattern is partly erosional in origin.

Longitudinal forms are called **linear dunes.** They exist as narrow ridges that extend parallel to the forming wind. They are usually wider and steeper at the upwind end, gradually tapering downwind until they merge with the desert surface. In the Navajo country, linear dunes are separated from one another by sand-free flats up to 100 m wide. Both the flats and the dune flanks may be vegetated, leaving only the sand of the ridge tops bare of vegetal cover and susceptible to wind transport. Locally the dunes may form in the wind-shadow behind an obstruction, or they may spring forth as wind-rift dunes (Melton 1940; *AGI Glossary* 1972) where blowouts provide sand that is extended downwind as a linear ridge. A special variety of linear dune is called a **seif dune** (Bagnold [1941] considered it to be one of only two true dune forms, the other being the barchan). Seifs are elongate, sharp-crested ridges that often consist of a succession of oppositely oriented curved slip faces that impress a sinuous or chainlike appearance on the dune crest. In many cases seifs attain spectacular length (up to 300 km) and are really draa in terms of size. Size may be regarded as a distinguishing property of seifs, but the term has been applied to much smaller features.

In addition to the basic patterns described above, other types have been reported including *dome-shaped, reversing,* and *star* (McKee 1966, 1979), and *coppice* dunes (Melton 1940) that are fixed by clumps of vegetation. It is important to understand that all dune types may be present side-by-side in a single region with the same prevailing wind, indicating that other controls are significant in pattern development. Thus, variables of multiple wind directions, topography, size and abundance of sand, and vegetation may be so inconstant that complex patterns are the rule, and the simple patterns of our classification the exceptions. It seems reasonable, therefore, to use the simplest ideal forms—barchans and linear types—as models to demonstrate how

Figure 8.14.
(A) Barchan dune in Sherman County, Ore., September 1899. (Photo by G. K. Gilbert, U.S. Geological Survey)
(B) Asymmetric wind ripples on side of compound barchan dune. Dawson County, Mont., September 1928. (Photo by C. E. Erdmann, U.S. Geological Survey)
(C) Dune forms in the Tularosa Basin, Otero County, N.M. Wind direction is from lower right. Dunes are crenulated transverse in lower right of photo. Dune forms progressively change to individual barchans and finally become U-shaped in the downward direction.

(A)

(B)

(C)

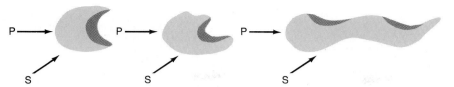

Figure 8.15.
Stages of seif development from barchan under influence of secondary wind direction: P = primary wind; S = secondary wind.

controlling factors may influence dune patterns. This choice is predicated on the generally accepted premise that one forms perpendicular to the wind and the other parallel to it. However, dunes oriented at oblique angles to prevailing winds also are found in every region where dunes are reported. Thus, even though linear and transverse dunes are most easily related to wind action, they may be less important in understanding processes than the explanations of the oblique forms.

Bagnold (1941) explained barchans as forms that develop in an area where sporadic sand patches exist within a desert pavement. It is assumed (Bagnold 1941, p. 222) that certain conditions must be met in order to mold the barchanoid form: (1) a constant, unidirectional wind; (2) a rate of sand supply that is symmetrically distributed on either side of the longitudinal axis; and (3) a slip face that is completely sheltered so that all sand crossing the crestal zone is trapped on the lee slope. Under these constraints, sand movement will be fastest near the lateral edges where the sand patch thins to meet the surface of the adjoining desert pavement. The greatest vertical growth of the sand body occurs in the middle segment, decreasing gradually to the edges. Thus, the lateral extremities will advance more rapidly than the center, and the crescentic form will develop as the equilibrium state is attained. As Cooke and Warren (1973) point out, however, Bagnold's hypothesis does not explain why barchans seem to have a regularly repeated width, nor does it account for the tremendous diversity in the length and direction of the barchan horns. Those authors suggest that divergence from the ideal form is a function of complex secondary flow patterns in the wind and of differential sand supplies. In fact, as suggested above, ideal barchans probably develop only where there is a limited supply of sand and unidirectional winds. Abundant sands lead to linear dunes rather than individual barchans.

Parallel ridges of sand, either dunes or bigger draa-sized features, are prevalent in all large sand deserts. Often the linear ridges bifurcate in the upwind direction, forming a Y or tuning-fork junction. Melton (1940) interpreted this phenomenon as the result of the wind excavating blowouts in deep sand. As such, the dunes are partly erosional, a hypothesis accepted by Folk (1971a, 1971b) for the linear dunes in northern Australia, but rejected by Cooke and Warren (1973).

The origin of linear dunes and draa has traditionally revolved around their relationship to the prevailing wind direction. Bagnold (1941) suggested that seifs develop when barchans are modified by crosswinds so that one of the horns is dramatically extended, as shown in figure 8.15. According to this idea, linear dunes are not aligned parallel to the prevailing wind, but their trend represents the resultant direction of more than one wind. The secondary wind influence may arise from storm winds, diurnal reversals, or seasonal changes

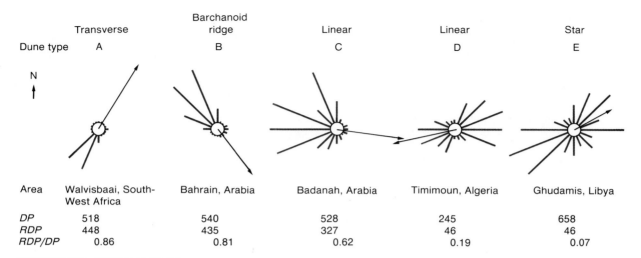

Dune type	Transverse A	Barchanoid ridge B	Linear C	Linear D	Star E
Area	Walvisbaai, South-West Africa	Bahrain, Arabia	Badanah, Arabia	Timimoun, Algeria	Ghudamis, Libya
DP	518	540	528	245	658
RDP	448	435	327	46	46
RDP/DP	0.86	0.81	0.62	0.19	0.07

Figure 8.16.
Five commonly occurring relationships of modes on sand roses and related dune patterns. (A) Narrow unimodal. (B) Wide unimodal. (C) Acute bimodal. (D) Obtuse bimodal (a special example in which the modes are almost exactly opposed). (E) Complex. D.P. (drift potential) and R.D.P. (resultant drift potential) in vector units. Arrows indicate resultant drift direction (R.D.D.). (Modified from Fryberger 1979)

in wind direction. The multiwind hypothesis has received support from workers investigating dune formation in a variety of physical environments (Cooper 1958; McKee and Tibbitts 1964; McKee 1966; Lancaster 1980).

The possibility of a resultant wind being generated by the influence of crosscurrents on the prevailing wind has never been questioned, but considerable argument remains as to how much and what type of geomorphic work it can accomplish. Substantial evidence has been reported to support the conclusion that seif chains are aligned with the resultant of several prominent wind directions (Cooper 1958; McKee and Tibbitts 1964; Brookfield 1970; Warren 1970). However, other data argue against a major genetic role for resultant winds. For example, seifs are often oriented in more than one direction within the same region, and some are distinctly oblique to the resultant wind direction (Brookfield 1970; Warren 1971).

It seems clear from the above that the relationship between wind direction and dune pattern is more complex than we might think. Nonetheless, eolian features are generally so wind dominated that landform characteristics can be utilized to make interpretations concerning wind direction and velocity. Where features are large, Landsat imagery can be employed to make such interpretations, and normal aerial photography is more suited to document wind properties from smaller features (for general discussions see McKee 1979; El-Baz and Maxwell 1982; Marrs and Kolm 1982).

A number of studies have produced rather interesting insights as to how some of the complexity might be resolved. Fryberger (1979) and Fryberger and Ahlbrandt (1979), for example, have evaluated wind data by estimating the transporting capability of the wind and plotting those estimates in sand roses (fig. 8.16). *Sand roses* are essentially circular histograms in which the lengths of arms are proportional to the amount of sand that can be transported toward the center of the circle along a given direction. For convenience, several terms were employed in these studies to indicate energy and directional properties. *Drift potential* (D.P.) is a measure of the total annual transporting capability of the wind at any locality. *Resultant drift potential* (R.D.P.) measures the net transport capability after considering all wind directions at the locality.

The significance of this technique is demonstrated in figure 8.16. Note that the type and complexity of the dune changes as the wind regime becomes more complex. Simple dune forms develop when the R.D.P./R.D. ratio is high and the winds are strongly unidirectional. As the winds become multidirectional, the R.D.P./R.D. factor decreases and more complex dune forms result. The star dune, probably the most complex, forms where significant winds come from three or more directions.

In summary, the complete understanding of depositional processes and features is veiled in the complexity of a multitude of interacting geomorphic systems. Features usually occur as ripples and dunes, but megascaled forms called draa are found in the great sandy ergs of North Africa. Each feature of the hierarchy occurs in patterns that are mainly a function of the wind properties. The major dune patterns are probably oriented parallel and perpendicular to the prevailing wind direction, but enough forms trend obliquely to suggest the possibility of a more complicated relationship involving form adjustment to multidirectional winds. In addition, vagaries of grain size and sorting, vegetation, topography, and sand supply introduce additional complications, and geomorphologists are only beginning to understand the influence of environmental changes on the genesis of dunes (Twidale 1972). Some dunes undoubtedly formed under wind and sand conditions that no longer exist, and these relict forms confuse the issue even more because their geomorphic characteristics may not be in equilibrium with the modern wind and sediment conditions.

Fine-Grained Deposits

Our examination of wind action in geomorphology logically concludes with a look at sediment that is not normally moved near to or in contact with the surface. Most silt and clay transported by the wind is carried in suspension and comes to rest as blanketlike deposits of loess (fig. 8.17). **Loess** is usually characterized as homogeneous unstratified silt, up to 100 m thick, which is highly porous and has the capacity to maintain vertical or nearly vertical slopes (Lohnes and Handy 1968). It covers all surfaces regardless of their topographic position, capping drainage divides as well as valley bottoms. Most loess is moderately well sorted, with nearly 50 percent of the deposit consisting of silt grains between 0.01 and 0.05 mm in diameter (Pesci 1968). It usually contains significant amounts of clay (5 to 30 percent) and 5 to 10 percent sand. The mineral composition is fairly consistent, with quartz being dominant and feldspars, carbonates, heavy minerals, and clay minerals present in smaller amounts. Each of the minor constituents varies in percentage according to local controls. For example, the amount of calcium carbonate in loess tends to be higher in dry regions, while clay mineral content increases with greater humidity.

The hypothesis that loess originates as wind-blown dust stems from at least a century of observations. The conditions that facilitate its wind derivation are an abundant supply of loose, fine-grained sediment, moderate to strong prevailing winds, and a surface free from a continuous vegetal cover. Deserts obviously meet these requisites (Péwé 1981), but curiously several continents with vast deserts (Africa, Australia) have almost no loess deposits, and some regions notably lacking in desert conditions have experienced major loess deposition. Chief among these is the periglacial environment (Péwé and Journaux 1983), especially where glacial meltwater has spread vast outwash debris in the path of the prevailing winds (Péwé 1955). There silt is winnowed

Figure 8.17.
Loess deposit on east side of Mississippi River valley near Chester, Ill.

from the outwash before the surface can be fixed by vegetation. In the mid-continental United States, for example, outwash is a logical source for many loess sheets, whose distribution shows a marked affinity to the outwash bodies occupying large valleys of the region. Presumably the loess can grow to its considerable thickness because the silt supply is continuously replenished during annual or even diurnal flooding by the proglacial rivers.

In either hot or cold loess sources, deflation, saltating grains, or other processes entrain the silt particles into the higher velocity zones, and with proper turbulence and lift, the grains can rise to elevations of several kilometers. When conditions are right, the amount of dust lifted into the atmosphere can be enormous. In a 1935 dust storm over the interior plains, about 5 million tons of dust was estimated to be in suspension over a 78 km² area near Wichita, Kansas, and at least 300 tons/km² of dust was deposited in one day of the same storm near Lincoln, Nebraska (Lugn 1962). The topographic effect of such loess deposition is different from that of most constructional geomorphic processes; it molds no important landforms but tends to form plains by filling in depressions and thereby smoothing out preexisting relief. Dust is carried as particulate matter suspended in the air, and its transport distance depends mainly on the constancy of the wind velocity, both vertical and horizontal, and the settling velocity of the grains. Loess constituents can be transported for hundreds of kilometers and deposited over vast areas (Van Heuklon 1977).

Although the shape of grains included in loess deposits and the markings on their surfaces suggest wind transport (Millette and Higbee 1958; Krinsley and Donahue 1968), wind-blown loess is difficult to distinguish from silts deposited in other ways. In fact, much controversy has arisen in the past because sediments bearing the characteristics of loess have been deposited by processes such as mass wasting, wash, weathering, and fluvial action. This has produced what some authors refer to as the "loess problem." Obviously some confusion results when loess is reworked by other transporting processes, and some of the clay and $CaCO_3$ comes from weathering of the original mass. In addition, "loess" often thickens towards topographic lows, indicating that downslope movement has been prevalent after the material's original deposition. Nonetheless, the relationship between widespread blanket deposits and the presumed prevailing winds is often so clear that an eolian origin for the original concentration of the material cannot be denied. In fact, arguments for a wind origin of the loess in the midcontinental United States are quite strong (Swineford and Frye 1951; Lugn 1962, 1968).

Some of the problem can be resolved if we restrict the term "loess" to those deposits that have the characteristics described and are definitely of wind-blown origin. Proof of an eolian genesis may be circumstantial in some workers' minds, but the evidence of field relationships and deposit properties is often more than convincing. In the western Great Plains, for example, the thickest loess (up to 70 m) occurs immediately downwind from the Sand Hills Region of Nebraska, which contains the largest accumulation of sand dunes in North America. In Illinois, as the map in figure 8.18 shows, the loess is thickest on the bluffs overlooking the Mississippi and Illinois valleys and thins gradually eastward into Indiana. The usual decrease in loess thickness downcurrent is

Figure 8.18.
Approximate thickness of loess on level, uneroded topography in Illinois.

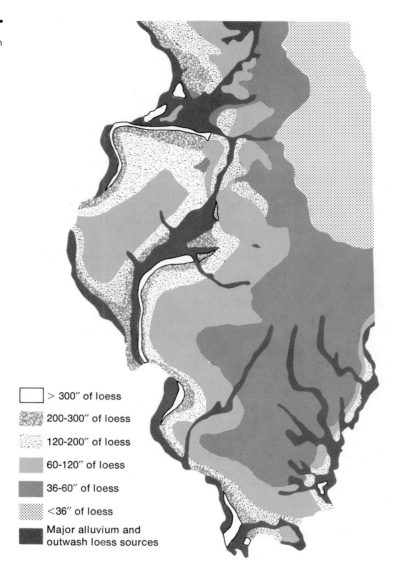

> 300″ of loess

200-300″ of loess

120-200″ of loess

60-120″ of loess

36-60″ of loess

<36″ of loess

Major alluvium and outwash loess sources

so systematic it can often be expressed mathematically. For example, Hutton (1947) showed that loess in Iowa southeast of the Missouri River thins according to the equation

$$Y = 1250.5 - 528.5 \log X,$$

where Y is thickness in inches and X distance in miles. Different equations fit other situations (G. D. Smith 1942; Frazee et al. 1970), but all demonstrate a regular, systematic decrease in thickness with increasing distance from the source. Additionally, as might be expected, the particle size also decreases

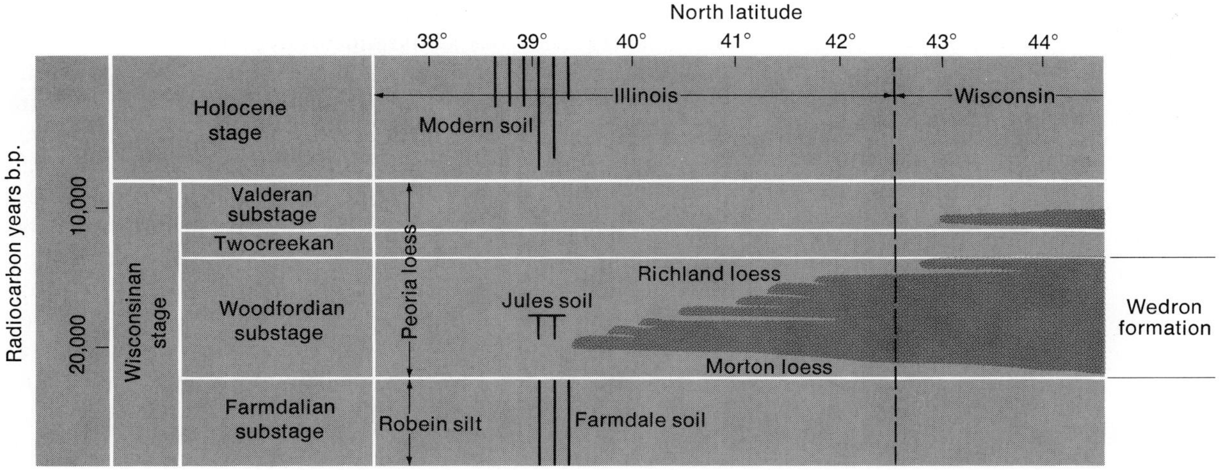

Figure 8.19.
Stratigraphic relationships between loess deposits and glacial deposits in the Late Wisconsinan of Illinois. (After Frye and Willman 1970)

downwind, again with a demonstrable regularity (Ruhe 1969). It is not safe to assume, however, that these systematic relationships are the result of one strongly prevailing wind. Indeed, Handy (1976) reminds us that significant loess deposition in the upper midwest of the United States occurred on both sides of the river valley source areas. This indicates that variable winds are responsible for loess deposition and the gradual changes in its character.

Although historical geomorphology is not the prime concern of this book, perhaps the most interesting aspect of loess is its value in deciphering Quaternary stratigraphy. Individual loess deposits are widespread geographically and blanket diverse topography, so that the deposits are ideal for stratigraphic analyses. As mentioned, loess deposition is commonly related to periods of glaciation. Where several loesses are superimposed, each deposit having been formed during a specific glacial episode, the sequence provides a useful stratigraphic framework. The upper and lower boundaries of each unit within the sequence can usually be identified because each loess has distinctive properties such as soil profiles (often buried), textural or mineralogical charactristics, or fossils. These criteria may be used for regional correlation of the deposits and therefore as bases for interpreting Quaternary history.

The use of loess in regional correlation requires caution and careful field study because loess properties, like properties of any rock substance, change laterally. In addition, in glaciated areas, single loess units may divide at the position of the former glacial margin so that loess associated with the same glacial event lies both beneath and above the till of that glaciation (fig. 8.19). Beyond the glacial margin the two loess bodies are combined into one, indicating that loess was deposited continuously during the complete cycle of glacial advance and retreat. The Peoria Loess in Illinois, for example, was deposited throughout the entire Woodfordian substage and correlates with both the Richmond Loess and the Morton Loess, which were deposited before and after the Woodfordian till (fig. 8.19).

The fossil assemblage contained in loess also provides useful information about the environmental conditions at the time of deposition. The predominant fossils in loess are land snails. They commonly are abundant when the loess is thick, but are rare in thin loess because leaching destroys the carbonate shells. Snail ecology indicates that most loess in the North American midcontinent originated when the climate was slightly cooler and moister than today. However, considering the diverse environments in which modern loess originates, it is probably safe to say that loess by itself has little climatic significance. Climatic indicators are substances such as plant fragments or mollusks included in the wind-blown debris at the time of its deposition or shortly thereafter. As shells and wood can be dated by ^{14}C, loess stratigraphy is often definable in terms of absolute time. This is especially important in the stratigraphy of the Late Wisconsinan. Older deposits may be beyond the limits of validity for the radiocarbon method.

Summary

Wind is an effective geomorphic agent in regions with sparse vegetation and an abundant supply of unconsolidated sediment. Although the most discernible evidence of wind action is found in deserts, it must be emphasized that eolian processes function in any locality having strong winds and the proper conditions of vegetation and sediment. The amount of geomorphic work actually accomplished depends also on the properties of the wind, especially its velocity and turbulence.

Particle entrainment occurs when the wind reaches a critical velocity called the fluid threshold. For any given sediment size, the value of the threshold velocity depends on a number of variables such as particle shape, sorting, soil moisture, and surface roughness. Once in motion, however, particles may strike stationary grains and cause their entrainment at wind velocities well below the fluid threshold (impact threshold). Wind-transported sediment moves in suspension, by saltation, or by surface creep. The largest portion of the load is carried within 2 m of the surface.

Geomorphic features associated with wind action are both depositional and erosional, ranging in size from microscopic to those measured in kilometers. Erosional features develop primarily from the abrasive action of wind-blown sand. The most prominent depositional features occur in a hierarchy of bed forms, including ripples, dunes, and draa, that are produced from sand-sized debris. The geometric shape and spacing of bed forms probably reflect an equilibrium condition between the wind and the characteristics of the sand. The system mechanics, however, is so complex that a precise quantitative expression of the equilibrium relationship is not yet possible. Fine-grained sediment occasionally accumulates in dune form, but most wind-blown silt and clay are deposited in sheets of loess that tend to smooth out topography rather than collect into pronounced constructional features.

The following references provide greater detail concerning the concepts discussed in this chapter.

Ahlbrandt. T. S., and Fryberger, S. G. 1982. Eolian deposits. In *Sandstone depositional environments,* edited by P. Scholle and D. Spearing, pp. 11–48. Tulsa: American Association of Petroleum Geologists.

Bagnold, R. A. 1941. *The physics of blown sand and desert dunes.* London: Methuen and Co.

Chepil, W. S., and Woodruff, N. P. 1963. The physics of wind erosion and its control. *Advances in Agron.* 15:211–302.

Cooke, R. U., and Warren, A. 1973. *Geomorphology in deserts.* London: Batford Ltd.

Marrs, R. W. and Kolm, K. E., eds. 1982. Interpretation of windflow characteristics from eolian landforms. Geol. Soc. America Spec. Paper 192.

McKee, E. D., ed. 1979. A study of global sand seas. U.S. Geol. Survey Prof. Paper 1052.

Sharp, R. P. 1964. Wind-driven sand in Coachella Valley, California. *Geol. Soc. America Bull.* 75:785–804.

Wilson, I. G. 1972, Aeolian bedforms-their development and origins. *Sedimentology* 19:173–210.

Glaciers and Glacial Mechanics

9

Introduction

Some of the most spectacular landscapes in the world are the results of the erosional and depositional action of glaciers, and every textbook of physical geology and geomorphology includes numerous photos and descriptions of these remarkable features. Nonetheless, to be true to the theme of this book, such features, regardless of their unique beauty, will be considered only because they manifest the processes that form them. It would also be tempting to stress the stratigraphic relationship between different glacial deposits and the effects exerted by glaciation on climate, life, and the physical setting. Obviously these subjects are of paramount interest and significance to geomorphologists because much of the modern landscape is closely associated with the physical events that occurred during the alternating glacial and interglacial stages of the Quaternary Era. While recognizing the importance of these topics, I realize pragmatically that an attempt to treat them in a general way is destined to fail. Therefore, our goal in this chapter will simply be to understand glaciers: how they form and how they move. Glacial erosion and deposition and the landforms resulting from these processes will be examined in the chapter following.

The study of glacial mechanics is an intimate part of the science of glaciology. Although this field is in its infancy, many of the techniques developed in recent years to study glaciers are extremely sophisticated and involve geophysics, remote sensing, and computer analyses well beyond the scope of an introductory discussion. Excellent reviews detail the basic concepts of the discipline (Embleton and King 1968, 1975a; Sharp 1960; Paterson 1969, 1981; Shumskii, 1964; Sugden and John 1976; Colbeck 1980), but those desiring greater depth and discussions of more recent advances in the field should refer to current periodicals, especially the *Journal of Glaciology,* which contains articles pertinent to all aspects of our discussion.

Glacial Origins and Types

A **glacier** is a body of moving ice that has been formed on land by compaction and recrystallization of snow. Assuming this simple definition suffices, it is obvious that two critical requirements must be met before a mass of ice can be called a glacier. First, the ice must be moving, either internally or as a sliding block, and second, the mass must be due to the accumulation and metamorphism of snow. Areas in which the winter snow is entirely lost to summer melting and other forms of dissipation cannot bring forth glaciers, even if the amount of snowfall is enormous. Where a portion of the snowfall does survive the summer melt, it is buried by the next winter's accumulation; with continued annual increments, the snow pack grows in size, changes its constituent properties, and finally, with enough mass, begins to move. In any region a specific elevation exists called the *snowline* or the *firn line,* above which some snow remains on the ground perennially and permits the formation of a glacier. In polar climates, snowlines are usually near or at sea level, and they gradually increase in elevation in climatic zones with higher annual temperatures. However, temperature is not the only determinant of snowlines since

Table 9.1 Increasing density of snow during transition to ice.

Materials	Density (gm/cc)
New snow	0.05–0.07
Firn	0.4–0.8
Glacier ice	0.85–0.9

glaciers form at lower elevations at the equator than in the desert zones near 30° north. This somewhat ironic situation arises because the annual precipitation in the two regions is vastly different.

When snow accretes over a period of years, changes in the properties of the particles mark the transition of a snow pack into true glacier ice. Newly fallen snow having a very low density (usually 0.05–0.07 gm/cc) and a delicate hexagonal crystal structure is transformed into glacier ice through a series of complex but recognizable stages (fig. 9.1). In the initial phase, points of the crystalline flakes are preferentially melted, resulting in a more spherical particle shape and tigher packing due to settling of the grains. At the same time, water produced by melting percolates to the base of the snow pack where it refreezes in the pore spaces (Woo and Heron 1981). As table 9.1 shows, this tends to decrease porosity and dramatically increase the density. The time needed for this initial change varies depending on the climate and the pressure added by continuous accumulation of snow; where temperature remains near freezing or where partial melting occurs, the transition from fluffy snow to coarse granular snow may take hours or days (Embleton and King 1968). On the other hand, extremely frigid conditions retard the process such that years may be required.

In temperate regions, snow lying on the surface for a complete year becomes granular and usually increases in density to about 0.55 gm/cc through the rounding and settling processes just mentioned. The material, now known as *firn,* is much denser than the original snow but is still permeable to percolating water and is not yet true glacier ice (table 9.1). The rate of densification beyond this point, and the mechanics of further transformations is different. The time needed to convert firn to ice is also variable, depending once again on temperature and the rate of increased load on individual grains. Where accumulation is rapid and water is plentiful, the transition from firn to ice probably takes place in less than 50 years, but in drier, colder regions the process may take several hundred years.

The mechanics involved in transforming firn to ice produces enlarged grains (in some cases up to 10 cm long) by recrystallization (fig. 9.2). When densities are low and porosities high, most of the stress exerted by the accumulating load is in the form of vertical compression. With increased density, however, the stress pattern is hydrostatic, and crystal growth occurs in any or all directions. Melted water may occupy air spaces lower in the firn pack, and as mentioned above, regions that produce free water will make the transition from firn to ice much more rapidly. Gradually, pore space is eliminated by crystal growth or by freezing of the downward permeating meltwater until the density is approximately 0.8–0.85 gm/cc. The only air remaining is trapped

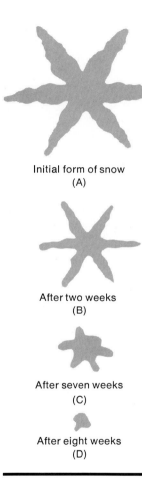

Initial form of snow
(A)

After two weeks
(B)

After seven weeks
(C)

After eight weeks
(D)

Figure 9.1.
Changes in the shape of snow crystal in the transition to firn.

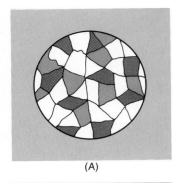

(A)

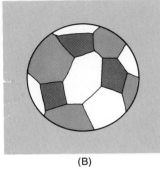

(B)

Figure 9.2.
Increase in grain size caused by
recrystallization process.
(A) Original texture of water-
soaked snow. (B) Texture after
application of stress.

as bubbles within the crystal. For all practical purposes, the creation of ice is complete at this stage, even though the bubbles continue to be slowly expelled by compaction, and the density may increase to about 0.9 gm/cc.

The rate of density increase and crystal growth is closely related to the temperature of firn. Firn in temperate regions will transform into ice faster and at a shallower depth than it will in polar regions. This fact is probably best demonstrated by the depth-density relationship in glaciers from the two regions (Sharp 1951; Behrendt 1965). In the Antarctic, a 0.85 gm/cc density is not achieved until at least 80 m of firn is accumulated. This is in drastic contrast to the 13 m depth needed to change snow to ice on the temperate Upper Seward Glacier in Alaska. Since accumulation rates are greater in glaciers of temperate regions, it follows that the greater depth to true ice in the Antarctic ice sheet represents the monumentally greater time needed to create glaciers in dry, cold climates.

Glacier ice is not preserved intact after its creation but is susceptible to further changes with continued increase in stress. For example, it seems clear from experimental work that the random orientation in polycrystalline ice cannot be maintained at higher stresses. The process known as recrystallization occurs when deformation of the polycrystalline mass exceeds several percent. This suggests that changes in the orientation of the *c* axis (known as the **fabric**) should occur in a vertical plane of a glacier. Such changes in fabric with depth have been noted in many glaciers (Gow and Williamson 1976; Russell-Head and Budd 1979; Hooke and Hudleston 1980, 1981). In most cases, the fabric changes from being a weakly oriented *c* axis in fine-grained, near-surface ice to coarser ice with a broad, single maximum fabric at some lower level. With increasing depth, the fabric usually develops multiple orientation directions. Finally, in some, but not all, glaciers the fabric returns to a single maximum orientation at the glacier base (Hooke and Hudleston 1980).

The depth at which fabric transition occurs is variable and depends on a complex interaction of density, impurities, temperature, etc. Importantly, the *c*-axis orientation may result from and be related to the overall scheme of glacier movement because the orientation of basal crystal planes is probably parallel to the direction of shear (Steinemann 1954, 1958; Rigsby 1960).

Glaciers have been classified on the basis of many salient properties, but as with any physical phenomenon, the best classification depends on the prime purpose of the groupings. The classification used here is no more astute than others; it simply serves our needs better by relating more closely to the processes of glaciation. Ahlmann (1948) suggested several bases for the classification of glaciers as *morphological, dynamic,* and *thermal.* Morphological classifications, which are based on glacier size and the environment of its growth, are most commonly employed, and the different categories, such as valley glaciers and ice sheets, are undoubtedly familiar. Although Ahlmann (1948) recognizes many morphological subdivisions, Flint (1971) suggests that glaciers can be placed in three broad categories (*cirque glaciers, valley glaciers,* and *ice sheets*) and two intermediate types (*piedmont glaciers* and *mountain ice sheets*). In this abbreviated form the classification is quite useful, especially in a descriptive sense.

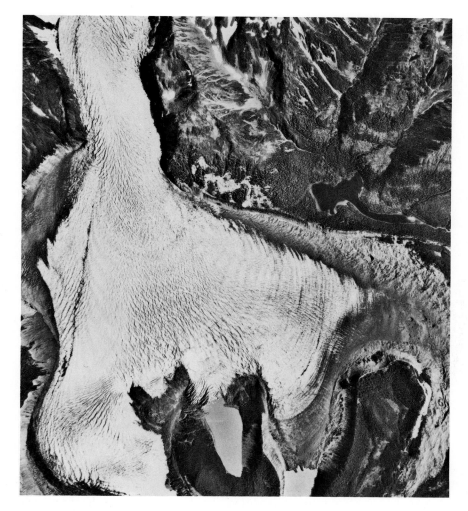

Figure 9.3.
The La Perouse Glacier in Alaska.
A very active glacier that spreads
onto lowland when it emerges
from a narrow valley. Crevasses
are distinct, and annual dark
bands (ogives) are clearly visible.

For our process orientation, however, the other two classifications are probably more useful, and we will at times utilize parts of each. The dynamic classification is based on the observed activities of glaciers and consists of three main groups: *active, passive,* and *dead* glaciers. Each type is closely related to the balance between losses and gains of ice and probably also depends somewhat on thermal properties. Active glaciers, like the one in figure 9.3, are characterized by continuous movement of ice from their accumulation zones to their edges. The movement may occur in response to normal snow accumulation, or it can be generated by avalanches or ice falls that provide the impulse for the forward motion. As you might expect, a glacier is passive when its movement is minimal. Dead ice has no discernible internal movement.

The thermal classification is based primarily on the temperature of the ice. It is well known that glacial behavior is directly influenced by ice temperature, and therefore the thermal classification should lend itself to process analysis. Two types of glaciers exist in this classification: **temperate glaciers** and **polar** or **cold glaciers.** In temperate glaciers the ice throughout the entire mass is at its pressure-melting point, although the upper 10 m may freeze in the winter. Meltwater seems to be present in abundant amounts within or beneath the ice mass and, in contact with the underlying rock, often facilitates slippage of the ice over the bed. This causes velocity and erosive action to be generally greater in temperate glaciers than in other types. Near the snout, meltwater may emerge as basal streams, or it may be temporarily dammed within the ice; both situations promote extensive fluvial removal of debris from the terminus of the glacier.

Polar glaciers were subdivided by Ahlmann into **subpolar** and **high-polar** types. In subpolar glaciers the accumulation zone is characterized by a thin layer of firn, perhaps 20 m thick, which contains some water in the summer if temperatures are warm enough to melt the surface ice. The surface of a high-polar glacier remains below freezing at all times, resulting in a completely water-free ice mass and a thick firn zone extending to at least 75 m before true ice is encountered. The absence of meltwater within polar glaciers is an extremely significant difference between polar and temperate glaciers, and its geomorphic importance cannot be overemphasized. It requires that ice at the base of polar glaciers be below its pressure-melting point and, for all practical purposes, be solidly frozen to the underlying bedrock. Since slippage cannot occur over the bed, ice movement is totally internal, and the glacier's erosive action is greatly diminished.

The suggestion that temperate ice is at its pressure-melting point is perhaps unfortunate because pressure is only one of several factors that determine the temperature within a glacier. Temperature variations can be caused by downward transfer of surficial heat or upward transfer of geothermal heat that is released at the glacier base (Sugden and John 1976; Hooke 1977). Heat is also produced inside the ice by friction when the glacier moves. Thus, it should be anticipated that a temperate glacier may have patches of frozen ice at its base (Robin 1976; Goodman et al. 1979) and that portions of polar ice sheets may have free water. In addition, the effect of climatic change may be gradually superimposed on the other types of heat variation and may not permeate through an entire glacier for centuries. For example, it seems likely that the West Antarctic Ice Sheet is still responding to a climatic change initiated at the end of the Wisconsin stage (Whillans 1978; Thomas 1979). Thus, most "temperate glaciers" have a temperature distribution that disagrees with the pressure distribution predicted by the thickness of the ice (Harrison 1975), even though the temperature generally increases with depth (Hooke et al. 1980). In some cases the temperature increase with depth may be linear (fig. 9.4); however this is not always true, and the gradient of temperature increase may vary considerably from glacier to glacier or within a single ice mass (Hooke et al. 1980). In sum, the simple model of pressure-melting may be an approximation, but it is not a precise truth.

Regardless of the difficulties discussed above, it is important to understand that the condition of basal ice with regard to melting temperature is perhaps the primary determinant of a glacier's ability to do geomorphic work.

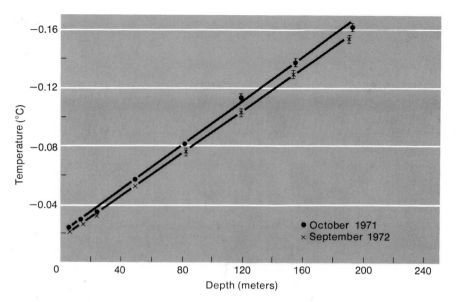

Figure 9.4.
Temperature-depth relationship on
a temperate glacier.

Glaciers with basal ice at the melting temperature tend to move faster, erode more, and carry greater loads than polar glaciers. Therefore, we can expect erosion and deposition to be more pronounced when they arise from temperate ice. In addition, many differences between modern and Pleistocene glacial activity may be due to variations in the thermal characteristics of the ice at the different times.

The Mass Balance

The mechanical behavior of glaciers and the geomorphic work they accomplish are intimately related to their **mass balance,** sometimes called the **glacial budget.** The mass balance is essentially an accounting or budgeting of the gains and losses of snow that occur on a glacier during a specific time interval. The water equivalent of ice and snow added to a glacier during the period in question is called *accumulation* and may result from a variety of processes including snowfall, rain and other water that freezes on the surface, and avalanches. Processes that remove snow or ice, collectively known as *ablation,* commonly include melting, evaporation, wind erosion, sublimation (Beaty 1975a), or the breaking off of large blocks into bodies of standing water, a process called *calving* (discussed in Holdsworth 1973). Losses by melting within or beneath the glacier are usually minor compared with surface volumes and are therefore neglected in budget studies.

The time interval used in most balance analyses is the budget year (or balance year), which is ordinarily taken as the time between two successive stages when ablation has attained its maximum yearly value. Usually these values are achieved at the end of the summer season, but the two ablation maxima may not occur on the same day, and so the budget year is not necessarily 365 days long.

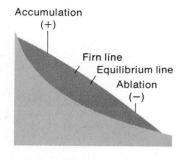

Figure 9.5.
Zones of accumulation and
ablation on a glacier as
determined by budget analyses.

On a glacier surface two values of accumulation and ablation can be determined: (1) a gross annual accumulation or ablation, representing the total volume of water equivalent added to or lost from the glacier during the budget year; and (2) a net annual accumulation or ablation, representing the difference between the gross values and indicating whether there was an actual gain or loss of mass during the year. The latter value is simply the algebraic sum of accumulation and ablation, and for the budget year at any single measurement locality is called the *net specific budget.* If net specific budgets are determined for a network of points distributed over the entire glacier surface, their values can be integrated into the mass balance. Any glacier may have a positive mass balance, meaning that more accumulation has occurred during the year than ablation, a negative mass balance, indicating an excess of ablation, or a mass balance of zero if the volumes of accumulation and of ablation have been precisely the same.

It should be obvious that even though both accumulation and ablation occur on all parts of the glacier surface, the higher elevations of the glacier will usually receive more accumulation and experience less ablation than the lower reaches of the glacier, where the opposite is true. Thus, large areas with either positive or negative net specific budgets can be identified on most glaciers, as diagrammed in figure 9.5. The two areas are separated by the **equilibrium line,** along which the annual volumes of accumulation and of ablation are equal. The equilibrium line should not be confused with the firn line or snowline because the two may not occur at the same place. On many glaciers, the firn line is clearly marked as a contact between snow above the line and dense blue ice below the line. On others, some of the dense ice downglacier from the snowline may have been formed by refreezing of meltwater and, as such, represents a net accumulation of mass (fig. 9.5). This *superimposed ice* is found in enough cases to warrant the distinction between the firn line and the equilibrium line, and it sometimes creates a rather complex transitional zone between the accumulation area and the ablation area (Müller 1962). Nevertheless, equilibrium lines and firn lines are usually close enough to approximate one another.

What does all this have to do with glacial mechanics and the resulting processes of erosion and deposition? If glaciers are viewed as open systems, both the mass balance and the absolute total amounts of accumulation and ablation are important factors in the character of glacial movement. When the net budgets are perfectly balanced (total mass balance equals zero), no expansion or shrinkage of the glacier occurs, and the glacial extremities remain stationary. This equilibrium condition, however, is seldom maintained for a long period of time, and so the glacial front and sides usually fluctuate constantly. Glaciers with a positive mass balance actively advance and characteristically maintain a relatively steep or vertical front. In contrast, an overall negative budget induces recession of the glacial front, and the snout will be gently sloping and often partially buried by debris released from the ice. The mass balance, therefore, is closely related to the position and type of morainal system constructed by the ice.

The absolute amount of snow and ice relative to the area of a glacier (gross accumulation and ablation) directly affects its internal activity. Large gross-accumulation values on small glaciers promote very rapid flow from the accumulation area to the ablation area; in most cases temperate glaciers are likely to have large accumulation and ablation values and, consequently, high flow velocities. In contrast, polar glaciers, which tend to be rather passive, usually have small values of accumulation and ablation and low internal flow velocities. It is very important to recognize that glaciers with different gross values can have the same mass balance and that, in fact, if they are in an equilibrium condition, no advance or recession of the glacier front is occurring even though the internal transfer of ice may be enormous. Figure 9.6 is a hypothetical diagram demonstrating a polar glacier with low activity and a highly active temperate glacier, both of which have a mass balance equal to zero. The snouts and lateral edges are stationary in both cases, but the temperate glacier, having a large value for accumulation and ablation, is moving at a higher rate internally. This dynamic activity causes pronounced erosion and rapid transportation of debris through the system. With the proper budgets, large moraines may result at the terminal and lateral boundaries. The polar glacier depicted in figure 9.6 will have little if any internal motion of its ice, and as a result, will probably not form any significant depositional features. Every glacier's depositional and erosional character, therefore, is fundamentally determined by the characteristics of the snow and ice added or lost from its surface.

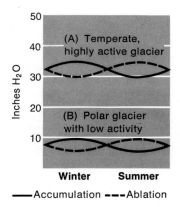

Figure 9.6.
Annual budgets on two glaciers in which the net mass budget of each equals zero. Active, temperate glacier (A) has greater total amounts of ablation and accumulation than the low-activity polar glacier (B). Even though both are in equilibrium and their fronts are stationary, (A) does more work because it must transfer more ice during the year from the accumulation zone to the snout.

The Movement of Glaciers

Internal Motion

Several hundred years ago, through direct observations, Alpine residents realized that glaciers move, and measurements of the rate of glacier flow were made as far back as the early eighteenth century. It is now generally accepted that glaciers move by two mutually independent processes: (1) internal deformation of the ice, called **creep,** and (2) **sliding** of the glacier along its base and sides.

The first real attempt to explain the physical dynamics of englacial ice motion seems to have been in the work of J. D. Forbes, who in 1843 suggested that glaciers respond to stress much like a plastic substance. By the beginning of the twentieth century, numerous models based on a plastic flow mechanism had been developed, although different ideas were also being proposed. They included flow caused by differential shearing along numerous, closely spaced planes (Phillip 1920; Chamberlain 1928); and the widely accepted notion that ice behaves as a viscous liquid that deforms in linear proportion to stress (Lagally 1934). Each of these concepts suggests that glacier flow manifests a relationship between ice deformation (strain) and the stress (force/unit area) that produces it. Stress generated at any point can be separated into two parts, hydrostatic pressure and shear stress. *Hydrostatic pressure,* which is related to the weight of the overlying ice, is exerted equally in all directions. *Shear*

Figure 9.7.
The relationship between stress and deformation. (A) Newtonian viscous fluid. Relationship is normally plotted as a function of stress and the rate of strain. (B) Perfect plastic deformation. The material experiences no deformation until stress is increased to the value of the yield stress. The material will then continue to deform as long as the stress is applied. Normally plotted as a function of stress and strain. (C) Ice creep according to Glen's flow law.

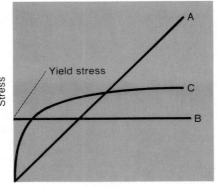

stress, however, is a function of the overlying weight and the slope of the glacier surface. Therefore, shear stress, which causes glacier motion, can be calculated as $\tau = \rho g h \sin \alpha$, where τ is shear stress, ρ is ice density, g is acceleration of gravity, h is glacier thickness and α is the slope on the glacier surface. This relationship implies that the slope at the base of the ice is the same as that on the ice surface. The relationship also defines the basic flow laws of ice. For the purposes of demonstration let us consider the two simplest possibilities: (1) that ice behaves as a viscous fluid and (2) that ice deforms as a perfectly plastic substance.

If ice behaves as a Newtonian viscous material, and we assume a constant viscosity, the application of stress should result in a linear relationship between the stress value and the strain rate (fig. 9.7). In addition, deformation will begin in the ice as soon as stress is applied and will maintain the linear proportionality regardless of the changes occurring in the stress. In contrast, as the diagram shows, a plastic substance shows no immediate response to stress but is capable of supporting a certain amount of stress without sustaining any deformation. Thus, at low stresses the strain of a plastic will be zero. As stress is increased, however, it eventually attains a value, called the *yield stress,* where the ice will experience limitless and permanent deformation. An entire glacier would behave plastically only if the shear stress along its base were equal to the yield stress (Kamb 1964).

It is now known that neither of the simple cases discussed above prevails in the mechanics of ice motion, although in many instances glacial properties predicted according to pure plasticity closely approximate the observed characteristics (Nye 1952b), and therefore, glacial movement has been referred to as pseudoplastic (Meier 1960, Johnson 1970). Laboratory studies since the late 1940s have shown that single ice crystals under stress deform as soon as stress is applied. Under any given stress the strain will increase rapidly at first (Glen 1952, 1955), but within a short time (tens of hours) it approaches an almost steady value that plots as a nearly straight line on a graph (fig. 9.7). This continuous deformation with no increased stress is the creep process that allows glacial ice to flow steadily under its own weight.

The rate of strain during the creep process is related to varying stress values by the following equation derived by Glen (1952, 1955) and now commonly referred to as the *power flow law:*

$$\dot{\epsilon} = k\tau^n,$$

where $\dot{\epsilon}$ is the strain rate $(d\epsilon/d\tau)$, τ is stress, and k and n are constants. The values of n, determined by a number of investigators, seem to vary from approximately 2 to 4 for individual crystals. In polycrystalline ice they range from 1.9 to 4.5, with the mean value being close to 3. In any case, the n values associated with the power flow law are significantly greater than 1, the value required for a linear viscous flow. The power flow law indicates that minor changes in stress can produce a major response in the strain rate. For example, if $n = 4$, doubling of τ increases the strain rate 16 times.

It is significant to note that glacier ice is always polycrystalline. The flow law, although based on single crystal deformation, still seems to predict the response of glaciers (Thomas et al. 1980), although minor modifications of the equation may be needed (Meier 1960; Colbeck and Evans 1973). The value of k in glacier flow might be reduced by interference of adjacent grains and recrystallization. In addition, as shown in the diagram in figure 9.8, cold ice deforms less readily than temperate ice, mainly because the constant k is also dependent on temperature. In Glen's experiments the value of k decreased by two orders of magnitude (0.17–0.0017) when the temperature was lowered from 0°C to −13°C. In fact, Paterson (1969) suggests that the strain produced in ice at −22°C by any stress is only 10 percent of its value when the ice is at 0°C.

Thus, ice clearly does not behave like a viscous fluid, although it may approximate a viscous response under low stress when the strain rate is still in its transient phase. At stress <1 bar in temperate ice, for example, n values as low as 1.3 have been measured (Colbeck and Evans 1973). However, when the nearly steady strain is attained or the ice is under high stress, the deformation becomes more plastic. Although this condition approximates plasticity, there is no distinct yield stress associated with the creep process, indicating that ice is not a perfectly plastic material.

Much research has been done to determine why ice under high stress takes on a nearly plastic behavior. The answer probably rests in the mechanics of strain. Ice deforms by recrystallization and by slip within the grains, a process involving a dislocation of planes of atoms inside the crystal (see Glen 1958; Weertman and Weertman 1964). The dislocation process works most efficiently through gliding on the basal plane of the ice crystal. In fact, it was suggested years ago that presumed plastic flow was closely associated with ice crystals lying in a preferred orientation (Perutz 1940). It follows that a consistent orientation of grains in polycrystalline ice mass (ice fabric) would produce strain exerted in the same direction (Russell-Head and Budd 1979; Duval 1981).

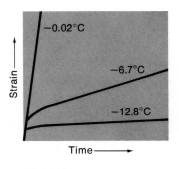

Figure 9.8.
The effect of temperature on the creep deformation curve. All curves represent deformation under a stress of 6 bars. (After Glen 1955. Used with permission of the Royal Society of London)

In sum, the power flow law derived in Glen's laboratory seems to fit the actual observed motions of glaciers. As mentioned earlier, it is too much to expect that polycrystalline ice will respond in precise agreement with the flow law, and modifications of the equation are generally required as the situation demands. In addition, where stress systems are complex because of irregular valley floors, cross-sectional shapes, or nonisothermal ice, the flow law will have to be generalized or modified (Nye 1957, 1965). Attempts to go beyond the general approach and replace the simple power law with a more sophisticated mathematical model become extremely difficult (Hutter 1982). Thus, in spite of inherent difficulties, Glen's power flow law seems to provide an excellent first approximation between theoretical and observed flow data and, with proper modification, probably best describes the internal mechanics of glaciers.

Sliding

The difficulty in applying the power flow law directly to glaciers is in part due to the fact that glaciers also move by sliding over the underlying bedrock, this motion being in addition to the creep operating within the ice mass. Sliding processes operate at the contact between the underlying bedrock and the base of the glacier, or within the lower ice layers. The processes themselves are poorly understood because direct observation of their action requires tunneling through the ice mass to the bedrock floor, an endeavor seldom tried and more rarely accomplished. Nonetheless, it is generally accepted that ice temperature and the character of the bedrock-ice interface are the prime factors that determine how sliding actually works.

Perhaps the most intuitive and acceptable process of sliding is the phenomenon of **glacial slippage** over a water layer that rests between the underlying rock surface and the base of the glacier. Where basal ice is near the pressure-melting point, a thin film of water, probably only millimeters thick, can exist at the bedrock-ice interface. This water not only lubricates the underlying surface, but if the subglacial water is under enough hydrostatic pressure, it may partly offset the weight of the overlying ice. This, of course, increases the movement produced by slipping of the glacier on the water layer (Weertman 1964).

The critical unresolved questions about slippage are whether water can exist as a continuous sheet beneath a glacier and, if not, how much of the ice-rock contact must be occupied by water to produce an influence on sliding. We know that water-filled cavities exist along the basal contact (Haefeli and Brentani 1955; J. E. Fisher 1963; Savage and Paterson 1963; Vivian and Bouquet 1973), and although they are usually isolated pockets, they might possibly merge by enlargement or coalesce as they migrate along with the moving ice. Lliboutry (1964, 1968) developed a sliding model based on the hypothesis that ice will separate from its bedrock floor downstream from a bedrock obstruction, and that the cavities thus formed will fill with subglacial water under pressure. Assuming an irregular bedrock topography, water-filled cavities may nearly submerge the obstacles, thereby reducing the area of contact between ice and rock. The reduction of overall friction thus produced will allow a drastic increase in sliding velocity.

Although the above model has been questioned, the presence of pressurized water at the base of a glacier has been demonstrated (Hodge 1976). Such water may play a fundamental role in the mechanics of sliding (Bindschadler 1983) and, in fact, may cause uplift of a glacier surface during the beginning of a melt season (Iken et al. 1983). Clearly, pressurized subglacial water can be a significant factor in slippage mechanics. It cannot be stated, however, that such water is always part of those mechanics because the generation of effective hydrostatic pressure requires that the underlying material be impermeable. Where underlying bedrock is pervasively fractured or the subsurface composed of unconsolidated sediment, it is probable that free water will drain into the underground system. Such downward release of water should lessen the possibility of creating enough hydrostatic pressure in the remaining interstitial water to facilitate slippage. It is also possible that subglacial water may drain in well-defined channels or in a sediment layer between the bedrock and ice bottom (Engelhardt et al. 1978). In either case, downslope water drainage will lower the hydrostatic pressure and reduce sliding velocity (Chadbourne et al. 1975; Engelhardt et al. 1978).

The premise that lubrication of the bedrock surface, aided by hydrostatic pressure, will initiate slippage is probably easy to accept, but the question arises as to how ice can move over an irregular surface by this mechanism alone. A glacier slipping up a slope on a film of water stretches credibility and suggests that other processes must be involved in the sliding phenomenon. Two such processes that have been recognized are **regelation** and **enhanced creep.** The process of regelation involves the melting and refreezing of ice due to fluctuating pressure conditions and usually results in a texturally distinct layer of ice, only a few centimeters thick, that rests in contact with the bedrock floor (Kamb and LaChapelle 1964).

The mechanics of regelation was first revealed by Bottomley (1872), who demonstrated that a thin wire under tension could be passed through a block of ice without splitting it apart. The pressure exerted by the wire melts the ice beneath it, and the water thus released flows in a thin layer to the upper surface of the wire where it refreezes. The temperature of the ice is lower beneath the wire than above it because the pressure depresses the melting point. The heat of fusion released by refreezing above the wire is transferred through the wire to provide the heat needed for the pressure-melting along the leading edge, and so the speed at which the wire moves through the ice block is partly dependent on the rate at which the wire conducts heat. Other objects such as cubes, spheres, and discs have been forced through ice (Kamb and LaChapelle 1964; Barnes and Tabor 1966; Townsend and Vickery 1967; Morris 1976), and wires of various size and composition have been utilized to study the regelation process (Nunn and Rowell 1967; Drake and Shreve 1973). Each experiment has reinforced the correctness of Bottomley's original interpretation of the process, although simple regelation theory seems to predict much greater velocities than those actually measured (Morris 1976). Thus, it seems that any object can be passed through ice without severing the mass and without changing the properties of the ice except along the path of transport. Along that path, the ice develops a new texture, similar in all aspsects to that observed in the thin basal layers of glaciers.

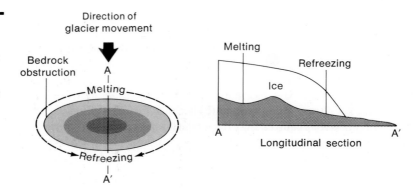

Figure 9.9.
The process of regulation as it functions beneath a glacier. Melting occurs on upstream flank of obstruction where pressure is greatest. Refreezing occurs downstream of obstruction where pressure is least, although some cavitation may form.

In glaciers the process of regelation, shown in figure 9.9, allows ice to circumvent minor irregularities of the bedrock surface over which the glacier is moving. The basal ice melts at the upstream edge of the bedrock knobs where pressure is the greatest. The meltwater thus released flows around the obstacle and refreezes in the region of least pressure, which is along the downstream edge of the barrier. The mechanism seems to be quite effective when the obstacles are small but becomes less important when protuberances become larger (Weertman 1957, 1964). This, of course, begs the question as to what magnitude of bedrock irregularity will render the regelation process inoperative and how thick the layer of regelating ice can be.

The problem is complicated by the fact that ice near the glacier base also may deform according to the flow laws explained earlier. In that case, ice at the upglacier edge of the obstacle, where pressure is greatest, will have a higher strain rate (enhanced creep) than that at the downstream edge. Larger obstacles will augment the stress and cause higher flow velocities; i.e., the velocity will be directly proportional to the size of the bedrock knob. Because of these complications, Weertman (1957, 1964) suggested that sliding over an irregular bed is produced by components of the two processes. One (regelation) is the predominent control on sliding when the barriers are small, and the other (enhanced creep) when the obstacles are large. It follows that some intermediate size, called the **controlling obstacle size,** determines which process will function. In the subglacial bedrock topography, ice moving over or around protuberances smaller than the controlling obstacle size will do so mainly by regelation; where surface bulges are larger, creep will probably dominate.

Finally, we now know that other factors have a direct influence on glacial sliding. The most important of these is the type and character of the substrate material. Boulton (1979) points out that the shear strength of this material in an unconsolidated state is

$$S = C + (P - P_w) \tan \phi,$$

where S is shear strength, P is pressure exerted by the overlying ice, P_w is pore pressure, and ϕ is the angle of internal friction. You should recognize this equation as being the same as the strength factor used when we examined

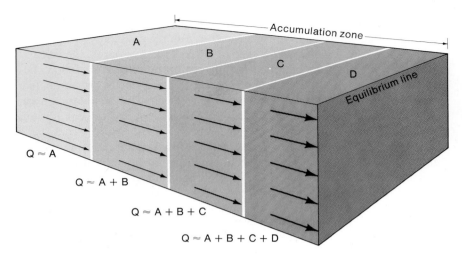

Figure 9.10.
Hypothetical diagram showing why the velocity of glacier flow reaches a maximum at the equilibrium line. If no change in cross-section area (width × thickness) is allowed, increasing discharge down-ice requires the increase in velocity shown by arrows.

slope stability in chapter 4. Investigators now realize that unconsolidated sediment beneath the ice may be highly deformable, especially when its strength is lowered by having a high porewater pressure. Deformation of the substrate may actually move the overlying ice with the mobilized material. For example, Boulton (1979) estimates that in some cases as much as 90 percent of basal ice movement is caused by deformation of unlithified substrate. The mobilization of subglacial debris is an important factor in the development of glacial features that originate beneath the ice, and we will return to this phenomenon in the next chapter.

In addition to unconsolidated substrate, the sliding velocities of temperate glaciers moving over carbonate rocks are affected by chemical reactions within the subglacial meltwater (Hallet 1976a, 1976b). This is especially true in the case of regelation because solutes in the water become concentrated on the lee side of obstacles. This lowers the freezing temperature there and inhibits heat transport away from those zones, a factor needed for the process of regelation sliding to work efficiently.

Velocity and Flow

We can now ask whether measured glacial velocities agree at all with the mechanics of ice motion discussed above. For the purpose of developing a theoretical model, let us first assume that an active glacier remains unchanged in its total and local dimensions over an entire budget year; i.e., the total accumulation equals the total ablation, and all cross sections of the glacier are equal in area and remain constant in area during the year. To maintain its area, each successive downglacier cross section must transport from its lower boundary exactly the amount of ice and snow delivered to its upstream boundary. As figure 9.10 shows, in the accumulation zone, the cross section at the highest elevation has only a small area of accumulation above it and so must discharge only a small volume of ice, equivalent to the snow accumulated in that restricted surface area. Each section farther down-ice, however, must

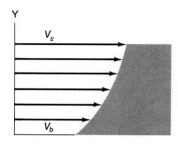

Figure 9.11.
Velocity distribution in longitudinal section of glacier. Surface velocity (V_s) is sum total of flow rates at every level within the ice. Strain rate due to shear stress is highest at base of ice and is shown as the basal velocity (V_b).

transfer a progressively larger volume of ice, since it moves not only the ice-equivalent of accumulation on its surface but also the cumulative volumes of all the higher sections. Without a change in cross-sectional area, the velocity of flow must increase to a maximum at the equilibrium line. This follows because discharge is increasing to that level and because glacier discharge, like river discharge, is equal to the area times the velocity. In the ablation area, we can similarly expect a gradual decrease in velocity from the equilibrium line to the terminus.

The model assumes that glaciers strive for and maintain some type of equilibrium and that the ice movement obeys the power flow law in some form. Mathematical treatment of even this simple glacial model is quite complicated unless we further assume, as did Nye (1952a), that the motion is two-dimensional, plastic, and laminar such that the lines of flow are parallel to the bed and surface at all places (fig. 9.11). Accepting these conditions, the shear stress at the base of a glacier, measured along the central longitudinal axis and perpendicular to the surface, is given as

$$\tau_b = \rho gh \sin \alpha,$$

where τ_b is the shear stress at the glacier base, ρ the ice density, g the acceleration of gravity, h the ice thickness, and α the slope of the surface.

By assimilating Glen's flow law into the above equation, the velocity at any depth along the central axis (xy axis) can be estimated by assuming that shear stress is proportional to depth, and that the strain rate directly relates to that stress. Theoretically, then, the internal velocity profile of a glacier in a longitudinal section should show a decreasing rate of flow from the surface to the bedrock floor. Even though the strain rate is greatest where shear stress is the highest (at maximum thickness), the velocity increases from the base to the surface because each internal layer not only moves in response to the shear stress generated at that level but is also being carried on top of, and at the speed of, the adjacent lower layer. Thus, the surface velocity (v_s) is the sum total of strain rates for all the layers within the ice mass (fig. 9.11). The equation also shows us that internal velocity is proportional to the product of surface slope and ice thickness. The product seems to be fairly consistent, meaning that where the ice is thin the surface gradient will be steep, and vice versa.

Using the same basic approach, Nye (1957, 1965) predicted the velocity distribution for an entire cross section, using different cross-sectional shapes as models, and showed that velocity should decrease from the center of the ice to the lateral boundaries both at the surface and at depth (fig. 9.12). Thus, the simplest ideal model of a glacier, assuming continuity of discharge and shape, and laminar flow, shows velocity decreasing with depth and distance from the central axis. In the downstream direction, velocity should increase toward the equilibrium line and decrease away from it.

The velocity of ice at a glacier surface is obtained by marking the position of stakes driven into the ice relative to some nearby fixed point. This can be done by normal surveying techniques or by photography repeated at some specified time interval. Actual measurements of surface velocities substantiate

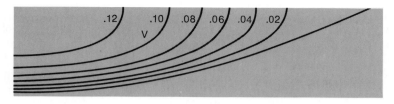

Figure 9.12.
Velocity distribution in glacier
flowing in a channel with a
parabolic shape. Diagram
represents one-half of the channel
cross section. Numbers represent
velocity expressed in any
common velocity units.

in a general way the theoretical predictions (Meier 1960; Meier et al. 1974), although few measurements have been made in accumulation zones. In most temperate valley glaciers, surface velocities range from 10 to 200 m a year, but vary locally above or below these values. Outlet glaciers and ice streams associated with ice sheets may also attain similar and, in some cases, even higher velocities (Embleton and King, 1968; Flint 1971). For example, the ice lobes around the periphery of the Pleistocene ice sheets of North America also probably advanced at rates between 10 and 200 m a year (Ruhe 1975, p. 192). As predicted, velocities tend to increase from the glacier head to a maximum near the equilibrium line (Meier and Tangborn 1965) and to decrease from there to the snout (Meier 1960; Meier et al. 1974). The transverse velocities are usually greatest along the central axis, decreasing to the lateral margins (Raymond 1971). In both patterns, however, enough variations occur to recognize that the flow model is infinitely more complex than our original assumptions allow. In fact, even though flow may show a relatively simple relationship to ice thickness and surface slope as predicted, in detail it may not do so on a local scale; other factors such as variations in ice temperature, basal sliding, and subglacial water pressure (Raymond 1971) may be involved.

Measurements of englacial velocity require that a borehole be drilled through the ice and cased to prevent its closure. The differential movement with depth can then be calculated from an *inclinometer,* which measures the angle between the axis of the borehole and the vertical. Almost every borehole that has penetrated to a glacier floor or to great depth shows a velocity profile similar to that predicted by theory (fig. 9.13). Velocity does not change significantly with depth in the upper zones of most glaciers, although Meier (1960) has demonstrated that some differential movement can occur even in the upper meters of a glacier. In the lower half of most glaciers, the velocity decreases more rapidly with depth.

Complications of the Simple Model

Real glaciers seem to possess the general traits that were predicted in idealized flow, at least in the distribution of velocity. It is too much to expect, however, that glacier characteristics will be predictable in detail. The initial assumptions are invalid on a local scale, and in fact, velocity itself is controlled by several factors that are external to the system defined by the flow laws; we know also that thickness and surface slope are free to change. We will briefly examine some of the phenomena that may cloud the relationship between predicted and observed ice motion.

Figure 9.13.
Internal velocity of two glaciers.
Horizontal displacement of
boreholes indicates the velocity.
(A) Saskatchewan Glacier (Meier
1960); (B) Malispina Glacier
(Sharp 1953).

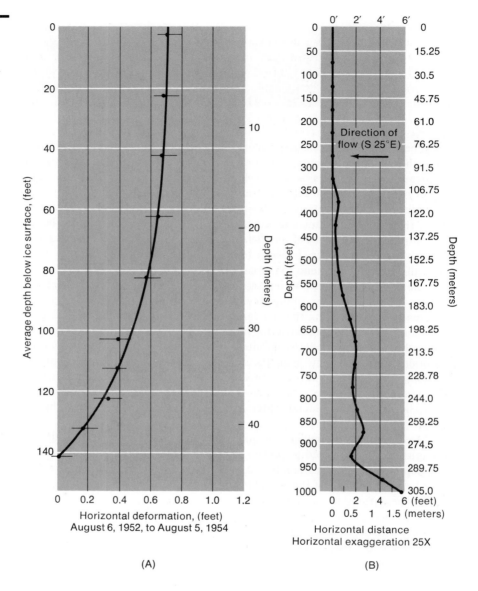

(A)

(B)

Extending and Compressive Flow   Returning to our ideal model, we see that allowing no change in cross-sectional area requires that the elevation of each point along the surface of a balanced glacier be maintained. If this is to happen, then the flow cannot consist of laminar horizons moving parallel to the surface because the ice must move slightly downward in the accumulation zone and slightly upward in the ablation zone. (Accumulation tends to elevate the surface and ablation to lower it unless these responses are counterbalanced by motion of the ice.) Thus, the pattern of flow should be as shown in figure 9.14 and not as laminar sheets.

Actually ice will tend to thicken in some places and to thin in others, processes called **compressive** and **extending flow.** Nye (1952a) first investigated these flow types and suggested that rates of accumulation and ablation, as well as changes in the slope of the underlying bedrock, will determine which type will prevail. Compressive flow results in a decrease of velocity and occurs where the underlying rock surface is concave-up or where there is a consistent loss of surficial ice. In contrast, extending flow has increased downglacier velocity and exists where ice is added at the surface or the glacier bed is convex. At a constant bedrock slope, therefore, accumulation zones should be characterized by extending flow and ablation zones by compressive flow. Within each of these zones, however, topographic irregularities on the underlying surface may produce local changes in the flow type.

This analysis generally fits the velocity distribution that we surmised in our balanced glacier and observed in real glaciers. In ablation zones, with the pervasive compressive flow, the velocity should decrease downglacier because the ice is being compressed (Nye 1952a), and in accumulation zones, with extending flow, the velocity should increase downglacier. We must ask, however, whether it fits the direction of flow predicted for a balanced glacier. Nye (1952a) showed that extending and compressive flow should each generate a pattern of potential slip planes that follow the orientation of maximum shear stresses (fig. 9.15). The family of planes are such that their resolution at any point will give the two directions of maximum shear; thus, they will be perpendicular and parallel to the bed of the glacier and form 45° angles with the surface of the ice. The potential slip planes show that in zones of extending flow (accumulation zones), the predominant downglacier slip direction will be downward at the surface; while in compressive zones, such as ablation areas, the low-angle downglacier slip will be upward.

This seems to coincide with the predicted mode of flow in our ideal glacier (fig. 9.14), but in real glaciers the pattern is much more complicated. An irregular bedrock profile will cause reversals of flow type in the ice (fig. 9.16), and where measurements have been made for an entire glacier (Meier and Tangborn 1965), the flow pattern (fig. 9.17) is not the simple one of our ideal case. Furthermore, more than one velocity maximum may occur, and the equilibrium line does not necessarily mark a zone of maximum flow velocity. In fact, on the South Cascade Glacier (Washington), the equilibrium line is usually near a zone of lower surface velocity, but the thickness is so great that the total discharge through the section is still very high.

Slip   A second complication arises because theoretical estimates of velocity do not consider the added component of basal sliding, the magnitude of which is represented by the displacement of boreholes at the bed. Although few direct measurements are available, they show that sliding may account for less than 10 percent of the surface velocity in some glaciers (Engelhardt et al. 1978) and as much as 90 percent in others (McCall 1952; Vivian 1980). Even worse, sliding velocities may be extremely variable within different portions of the same glacier (Savage and Paterson 1963). Although, as discussed earlier, the mechanics are not understood, it is clear that the influence of basal

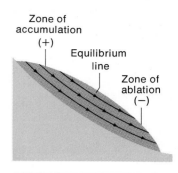

Figure 9.14.
Longitudinal flow lines in hypothetical glacier.

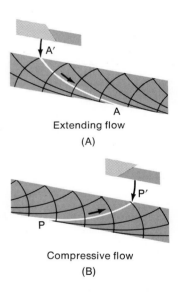

Figure 9.15.
Potential slip planes under (A) extending and (B) compressive flow. The preferred downglacier slip paths will follow *A-A'* and *P-P'*.

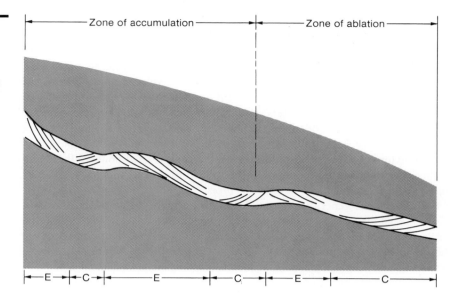

Figure 9.16.
Longitudinal section of
hypothetical glacier showing
irregular bedrock profile and
preferred slip planes within the
ice. Zones of extending flow and
compressive flow indicated by *E*
and *C*.

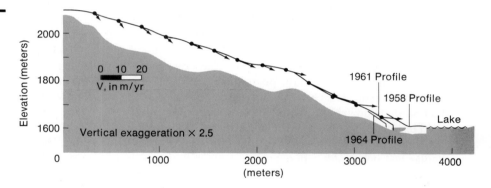

Figure 9.17.
Longitudinal section showing
average calculated bedrock
profile and surface velocity
vectors on the South Cascade
Glacier, Wash.

sliding must complicate the simple relationship between flow laws and velocity, at least in temperate glaciers. Slip also can occur along the lateral margins of glaciers, and abnormally high velocities are sometimes encountered at the glacier sides (Meier et al. 1974).

It is now apparent that our basic model of ice sliding over a bedrock surface is simplistic. As discussed earlier, a large proportion of forward glacial movement may come from deformation of underlying sediments rather than sliding or englacial creep. This may be especially true in deformable sediment with low permeability. In those cases, water builds within the sediment rather than in the subglacial zone and aids in the substrate deformation (Boulton and Jones 1979). In addition, Engelhardt and his colleagues (1978) have identified a thin (10 cm) layer of debris between the underlying bedrock surface and the base of the Blue Glacier in Washington. This layer, which they called the *active subsole drift,* is involved in decreasing the sliding velocity because its surface is rough, its movement involves friction between the included grains, and water enters and flows within the layer, thereby reducing hydrostatic pressure at the glacier base.

Variations with Time   The third factor complicating the relationship between flow theory and observed flow rates is that velocity is not constant in any given glacier. On most glaciers it varies significantly with time; to add to the confusion, the interval over which velocity variations occur is different from one glacier to another and even in specific locations on the same glacier. For example, many investigators have noted sudden, jerky motions that increase surface velocity for hours, days, or weeks (Meier 1960; Glen and Lewis 1961; Goldthwait 1973; Jacobel 1982). These spastic movements, which may involve a change in velocity of as much as several hundred percent, seem to be a surface-ice phenomenon (Goldthwait 1973) and tend to be restricted in extent. They are normally explained by local controlling factors such as weather conditions, fault slips, or a sudden release of ice that has been retarded in its flow by some obstruction.

Seasonal or yearly variations in flow also are common. Velocities in the ablation zone usually increase at or near the end of the summer. These fluctuations most logically reflect accelerated basal and side slip, which is facilitated by the abundance of free water at that time (Meier 1960; Elliston 1963; Paterson 1964; Hodge 1974; Vivian 1980; Anderson et al. 1982). Seasonal velocities may deviate by as much as 80 percent from the mean annual values. They are not restricted to temperate glaciers but also affect the movement of subpolar glaciers (Friese-Greene and Pert 1965). Yearly variations probably represent internal adjustment to accumulation rates, but they also can be influenced by basal meltwater lubrication (Meier et al. 1974).

Short-term variations in velocity are usually recognized because the span of most investigations is long enough to demonstrate their presence. However, imagine the sampling program needed to observe and document long-term fluctuations of velocity that are produced by changes in climate and the associated glacial regime. Any positive change in a budget should increase the discharge of flow, and according to theory, the mechanical response to the increased accumulation will probably involve the propagation of a **kinematic wave** moving down the glacier two to five times faster than the actual particles of ice. The wave will reach the glacier snout long before the new ice formed from the flux in accumulation could possibly be transported that far. Without going into detail, it appears that the wave motion will have little effect on ice that is extending; but in the ablation zone, where compressive flow dominates and velocity decreases down-ice, the oncoming wave accentuates the compression and the glacier becomes very unstable there. As the wave approaches any point in the ablation zone, the ice will thicken and the surface will rise dramatically. As the wave passes, it will quickly thin and subside to its former level. On some glaciers the ice surface may rise and fall more than 100 m in the sequence of events.

Velocity measurements on some glaciers do not fit the predictions based on kinematic wave theory (Lliboutry and Reynaud 1981). Nonetheless, if kinematic waves generally represent the mechanism by which glaciers respond to changes in mass balance (Nye 1960), it becomes important to recognize that the time needed for the wave transmission varies from glacier to glacier. For example, the time needed to reestablish a steady state (i.e., the **response time**) in single, temperate glaciers varies from 3 to 30 years. The significance of this fact can be seen in a hypothetical case. Suppose that during the years 1954–1958, abnormally high amounts of snow accumulated on two adjacent valley glaciers. In one with a short response time, a kinematic wave rapidly traveled the length of the glacier, and the snout advanced dramatically in 1960. The second glacier, having a longer response time, showed no visible effects of the accumulation by 1960; the kinematic wave did not reach the terminal zone until 1975, when the terminus suddenly advanced. The two glaciers were completely out of phase, although both advanced in response to the same event. In addition to problems of mechanics, measurements made on a portion of the ice experiencing the wave action are not documenting the flow of ice but rather the movement of waves.

In ice sheets, where response times are much longer, the difficulty of relating flow to a change in regime is magnified. For example, Whillans (1978) suggests that the modern thinning and accelerated flow in the central portion of the West Antarctica Ice Sheet was initiated by warming that commenced 10,000 years ago. Thinning in the interior of ice sheets requires transfer of ice to the margins, thereby thickening the marginal ice and promoting its forward movement. Thus, it is possible that the rapid advances often noted during the final phases of Pleistocene glaciations may have been in response to warming rather than cooling climates.

Kinematic wave velocities may attain catastrophic values, reaching hundreds of meters a day (Meier and Johnson 1962; A. E. Harrison 1964); they have often been cited as the mechanics governing **surging glaciers.** However, in many cases glaciers surge even though there is no discernible net accumulation of mass to the glacier or any other external stimuli for the movement (Post 1960, 1969; Meier and Post 1969). It is probably best to separate surging from kinematic wave transfer even though the two movements have many of the same characteristics (Palmer 1972).

A surging glacier, then, is one in which sudden, brief, large-scale ice displacements periodically occur. The ice moves 10 to 100 times faster than its flow rate in the quiescent periods between surges resulting in the fact that surging glaciers entrain significantly more debris than normal glaciers (Clapperton 1975). The surge periodicity seems to range from 15 to more than 100 years (Post 1969) and probably results from unique conditions that create a cyclic instability within the glacier. Some surges have no distinct periodicity and may be generated by outside events such as earthquakes or episodes of abnormally high geothermal heat.

Glacial surges are now recognized as fairly common phenomena, with 204 surging glaciers identified in western North America alone by 1969 (Post 1969). In fact, Budd (1975) has suggested that they may represent a completely separate type of glacier. However, they have no distinct size, shape, or activity, and they can contain temperate or subpolar ice. Their only unifying characteristics seem to be that surges are initiated in the ablation zone slightly down-ice from the equilibrium line, and a long and pronounced stagnation of ice occurs in the terminal zone in the interval between surges (Post 1960, 1966). During the surge the glacier surface is broken chaotically, and medial moraines and ice bands are intensely contorted.

Apparently, for some unknown reason, a reservoir of ice collects in the upper portion of the ablation zone, where ice is still active, while the terminal area is physically lowered by ablation. Eventually a critical disequilibrium condition is reached, and the response is a sudden burst of glacial movement into, and often overriding, the stagnated terminal zone. The key to understanding cyclic surging thus seems to lie in two factors: (1) how thickening of ice, with a concomitant increase in basal stress, can occur in one region of the glacier while stagnation occurs in another; and (2) what triggers the sudden release of the ice reservoir.

Several models have been proposed to explain the internal controls of periodically surging glaciers. Budd (1975) suggests that certain glaciers with the proper sliding velocity gradually develop a lubricating factor that in turn lowers the basal stress enough to induce a sudden movement. Surging glaciers, in his model, do not have sufficient mass flux to continue this rapid flow, and so they periodically oscillate from the slow flow of an ordinary glacier to sudden bursts of activity.

Another interesting model (Robin and Weertman 1973) proposes that surges end when the bed shear stress reaches a low value, a conclusion postulated first by Meier and Post (1969). During the quiet period following the surge, the glacier develops zones of vastly different basal shear stress. In the stagnating terminal area, basal shear will be relatively low. Further up-ice, however, the stress will progressively increase with time along with the thickening of the ice. Between the two regions is an area called the *trigger zone,* where the *gradient* of basal shear stress is continuously increasing and eventually reaches a critical, unstable condition. Robin and Weertman also suggest that a distinct pressure distribution, related to the basal stress is promoted in the water at the bed. The water-pressure gradient is inversely related to the shear-stress gradient, and water is dammed in the trigger zone. The next surge begins when the trapped water lowers frictional resistance so much that the sliding velocity drastically increases.

One objection to this model, clearly recognized by Robin and Weertman, is that the basal water to be dammed must be prevented from draining down-glacier through interconnected channels. This objection might be partially overcome if the glacier snout is frozen to the bedrock, a possibility suggested for subpolar surging glaciers (Jarvis and Clarke 1975). Even in temperate glaciers, patches of cold ice may exist at the glacier floor (Robin 1976).

To summarize, it has been shown theoretically that surges can be triggered by lowering basal shear stress over a period of time (Campbell and Rasmussen 1969); presumably, the lubricating effect of meltwater can accomplish the necessary decrease in stress. If that is true, as the proposed models suggest, the key to surging lies in the mechanics of basal sliding.

Ice Structures

Glaciers usually display a variety of structures that develop during growth of the glacial mass—which we can call primary structures—as well as secondary structures that develop in response to glacial movement. Primary structures are indicative of accumulation and ablation characteristics during the glacier's history, and secondary structures relate closely to the glacier's mode of flow and stress field.

Stratification

Primary structures in glaciers appear as discernible layers or bands within the ice. The layering results from processes that reflect an annual cycle of snow accumulation and ablation above the firn line. During the winter a thick pile of new snow is added to the glacier and, with time, proceeds through the phases of metamorphism that culminate in a layer of clear white ice. In the ablation season, however, the upper portion of the winter accumulation is subjected to melting and refreezing and develops a texture different from that forming lower in the snow pack. In addition, sediment and organic debris may collect in the partially ablated upper zone, making it slightly darker in color. The alternating white and "dirty" layers give the ice a stratified appearance and allow glaciologists to estimate the annual growth in the glacier thickness. The layers tend to be tilted and deformed as the glacier moves. At the surface of the ablation zone, the bands look like truncated beds of plunging folds (fig. 9.18) with the layers dipping into the glacier in an up-ice direction.

Figure 9.18.
Saskatchewan Glacier—outcrop pattern of stratification in Castleguard sector. View upglacier from cliff on south margin below Castleguard Pass. Splaying and en echelon crevasses are also visible. Province of Alberta, Canada.

Secondary Features

Foliation   A secondary type of layering, called **foliation,** is produced by shear during ice motion; it is sometimes difficult to distinguish from primary stratification because both types may display similar grain size or textures. Foliation usually appears as alternating bands of clear blue ice and white bubble-rich ice. The layers, which dip at all angles, are most prevalent near the ice margins and commonly offset or wrinkle the primary stratification (fig. 9.19). The origin of foliation is poorly understood. Some evidence suggests that it originates where shear stress is greatest, but Meier (1960) argues that it is neither formed nor preserved at great depth, where shear stress would presumably be at a maximum.

Crevasses   Crevasses are cracks in the ice surface that range in size from miniature fractures to gaps several meters wide. The fractures are important in that they provide avenues for surface meltwater to penetrate the interior of the glacier, although the openings are rarely deeper than 30 m. It is generally assumed that crevasses develop perpendicular to the direction of maximum elongation of the ice. On some glaciers, however, the crack directions do not precisely coincide with the measured surface strain rates (Meier 1960). Nonetheless, crevasse types (fig. 9.20) do reflect a local tensional stress environment. *Splay crevasses* or *radial crevasses* form near the flow centerline

Figure 9.19.
Saskatchewan Glacier—gently dipping stratification, wrinkled and intersected by nearly vertical foliation. Exposed on east wall of a crevasse, 4.5 km below firn limit in midglacier. Province of Alberta, Canada.

under compressive flow where spreading exerts a component of lateral extension (fig. 9.3). In contrast, near the ice margins the shear stress parallels the valley walls, and crevasses develop diagonally to those sides. Here the crevasses are either *chevron* or *en echelon* types.

Transverse crevasses (fig. 9.21) develop under extending flow where the ice extends in a longitudinal direction. In temperate glaciers, these cracks occur most commonly in ice falls where the glacier cascades over convex irregularities in the underlying bedrock slope. In the summer, large ice crystals may develop in the transverse cracks when water trapped there repeatedly melts and refreezes. Sediment may also accumulate in these openings. At the base of the ice fall the crevasses are closed by compression, and a band of dirt-stained ice forms. During the winter, however, ice descending the fall reconstitutes into clear, bubbly ice (King and Lewis 1961). The annual downvalley

Figure 9.20.
Types of crevasses in valley glaciers: (A) marginal or chevron—(1) old rotated crevasses, (2) newly formed crevasses; (B) transverse; (C) splaying; (D) radial splaying. Arrows show flow direction.

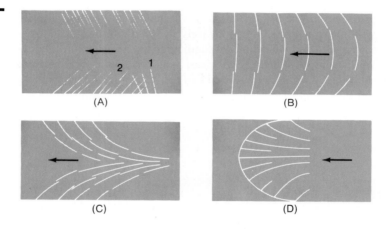

Figure 9.21.
Transverse crevasses developed at ice fall on Little Yanert Glacier, Alaska Range.

flow, therefore, tends to produce a distinct series of alternating white and dark bands, called **ogives,** that are prominent at the foot of ice falls (fig. 9.3). They are different from primary layering in that they do not dip as strata into the glacier but are purely a surface phenomenon.

Summary

In this chapter we examined the origin and movement of glaciers, as well as the factors that might explain the properties associated with the various glacier types. Glaciers can be classified on the basis of their morphology or geographic position, their activity, or their thermal characteristics. Any glacier develops when snow accumulated over a period of years is transformed into ice by compaction, recrystallization, and melting and refreezing. These processes progressively increase the density of snow and firn as the space within the original mass is removed by pressure, crystal growth, and orientation of the grains. When the ice reaches a critical thickness, it is capable of movement.

The amount and type of geomorphic work a glacier can accomplish depends on its mass balance and on the total volume of ice added and lost during a period of time. The mass balance, or budget, is the net difference between accumulation and ablation during the time period in question. A positive mass balance indicates that more ice has been added to the glacier than has been lost; and the transfer of ice from the accumulation zone to the ablation zone causes the glacier front to advance. A negative budget results in a retreat of the front because the amount of ice transferred from the accumulation zone to the ablation zone is not sufficient to replenish the volume of ice lost by ablation. The total volumes of accumulation and of ablation determine the level of glacial activity. Two glaciers may have identical mass balances, but the one having the higher total volumes of addition and loss will be more active and do more geomorphic work.

Glaciers move by internal deformation of the ice and by sliding along the bedrock floor at the base of the glacier. The internal movement occurs as a type of creep that is mechanically different from viscous flow or pure plastic flow. In most glaciers the deformation caused by flow can be estimated by an equation called the power flow law or some modification of it. The mechanics of sliding is poorly understood but probably consists of the combined effects of regelation, lubrication by water under pressure, and enhanced creep.

Velocity of flow relates well with modified versions of the power flow law, but many characteristics of glaciers complicate a direct relationship. For example, ice does not flow in parallel laminar sheets but thickens and thins along the length of the glacier. This develops the compressive and extending flow that causes ice to move downward in the accumulation zone and surfaceward in the ablation zone. In addition, flow velocity varies with time. The phenomenon of surging is difficult to fit into an all-inclusive flow model.

Primary and secondary structures in glaciers are related to the processes of accumulation and ablation and to the stress field generated during the ice movement. The major structural features are stratification, foliation, crevasses, and ogives.

Suggested Readings   The following references provide greater detail concerning the concepts discussed in this chapter.

Embleton, C., and King, C. A. M. 1968. *Glacial and periglacial geomorphology*. Edinburgh: Edward Arnold Ltd.

Engelhardt, H. F.; Harrison, W. D.; and Kamb, B. 1978. Basal sliding and conditions at the glacier bed as revealed by bore-hole photography. *Jour. Glaciol.* 20:469–508.

Glen, J. W. 1955. The creep of polycrystalline ice. *Proc. Royal Soc. London,* ser. A: 228:519–38.

Hutter, K. 1982. Glacier flow. *Am. Scientist* 70:26–34.

Meier, M. F. 1960. Mode of flow of Saskatchewan glacier, Alberta, Canada. U.S. Geol. Survey Prof. Paper 351.

Meier, M. F., and Post, A. S. 1969. What are glacier surges? *Can. J. Earth Sci.* 6:807–17.

Nye, J. F. 1965. The flow of a glacier in a channel of rectangular, elliptic, or parabolic cross-section. *Jour. Glaciol.* 5:661–90.

Sugden, D. E., and John, B. S. 1976. *Glaciers and landscape*. London: Edward Arnold Ltd.

Glacial Erosion, Deposition, and Landforms

10

Introduction

We are now ready to consider the geomorphic significance of the glacier mechanics we examined in chapter 9. It would be ideal if we could directly correlate geomorphic features with specific ice processes, but we seem to be far from such sophistication. The reasons for this involve several main factors: (1) Our conceptual models of glacier mechanics are greatly oversimplified and based on theory rather than extensive observation. (2) Much erosion and deposition take place at the base of glaciers where it is rarely feasible to document the actual process. (3) We are just beginning to understand the feedback mechanics that function between the basal ice and the geologic framework. For example, we know that glaciers have the ability to erode because we can see the results of that erosion after the ice disappears. What we don't understand is precisely how the erosive process itself is modified by remolding of the subglacial framework. Does removal of certain obstructions by erosion decrease the subsurface roughness enough to facilitate rapid sliding velocities, as suggested by Kamb (1970), and might this in turn accelerate further erosion?

Until we can resolve these deficiencies in our understanding, the process-feature relationship will be arrived at by traveling a one-way street in the wrong direction—we are interpreting the process by the character of its results. Nonetheless, glaciers have left us a variety of deposits and myriad features that demand our attention. It is not completely satisfying to reconstruct their origin without a wealth of solid observations, but this type of deductive reasoning has been a part of geology for a long time and is not necessarily incorrect. Furthermore, investigations of features and deposits can provide tangible clues about their origins that become invaluable in deciding how and where to study glaciers in the future. It seems inevitable, however, that new techniques for investigating glacier mechanics will demand continuous modification and testing of our present interpretations of features and deposits.

Erosional Processes and Features

Minor Subglacial Features

Glacial erosion is accomplished primarily by two processes, a scraping action called **abrasion** and a dislodgement or lifting action called **quarrying** or **plucking.** Since ice is not a hard mineral (1.5 on Moh's scale at 0°C), it cannot abrade most solid rock material unless it utilizes as grinding tools the fragments of rock carried in its load. Therefore, the efficiency of abrasion and the features it produces depend on the character and concentration of the debris being dragged along the base of the ice and, of course, on the properties of the bedrock being overridden. Abrasion is also influenced in a complex way by the nature of the subglacial topography and the velocity and direction of ice flow. The rate of abrasion, estimated by various methods, usually ranges from 0.06 to 5 mm a year, but it may be considerably higher under thick, high-velocity glaciers. Boulton (1974) reports that the abrasion of a marble plate inserted beneath the Glacière d' Argentiere in France proceeded at a rate up to 36 mm a year where the ice was 100 m thick and moving at 250 m a year.

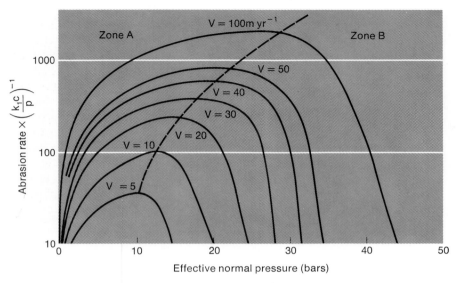

Figure 10.1.
Theoretical abrasion rates plotted against effective normal pressure for different ice velocities. The rate increases in zone A and decreases in zone B as pressure increases. At any given velocity, a pressure increase greater than that at the higher X-axis intercept causes deposition of lodgement till. k, c, and p are constants depending on relative hardness, debris concentration, and penetration hardness of the ice.

The apparent dependence of abrasion on ice thickness does not agree with theoretical analyses (McCall 1960; Hallet 1979). The reason for this is that the yield stress for ice (2 kg/cm^2 at 0°C) suggests that thickness greater than 22 m should cause ice to flow around and under rock fragments without increasing its ability to abrade the underlying bedrock surface. We know, however, that basal cavities and extensive contact between loose particles and the bed do exist beneath thick glaciers, indicating that the ice can maintain some strength. Therefore, Boulton (1974) suggested that at any given ice velocity, assuming a constant concentration of basal debris, the abrasion rate will increase with increasing normal pressure until it attains a peak value, as plotted in figure 10.1. Further increase in normal stress will cause the rate to decrease rapidly to zero, where abrasion ceases and any debris in transport will be deposited beneath the ice. The form of the curves shown in figure 10.1 have been verified by direct measurements (Boulton 1979).

Small erosional features produced by abrasion are usually in the form of linear scratches or crescentic marks that show the relationship between the size and composition of the abrading particles and the resistance of the underlying bed. Very fine particles in sufficient abundance produce a smoothly polished surface composed of microscopic scratches; as the grain size of the load increases, the scratches become larger in a transitional sequence from polish to striations to grooves or furrows.

Striations like those shown in figure 10.2 have been noted in every glaciated region of the world. They are best preserved on fine-grained rocks that have not been deeply weathered and on smooth bedding planes that dip gently away from the direction of ice movement. Striations and larger linear features also can form in unconsolidated material, such as till or loess, if it is highly compacted (Westgate 1968). Striae are only millimeters deep and are most likely eroded by sand grains or by jutting edges of larger particles carried in

Figure 10.2.
Striations on outcrop of limestone
in south central New York State.

the basal ice. Striations tend to be continuous for only relatively short distances, probably because the sliding clast itself produces a carpet of plowed debris at its forward edge. When sufficient debris accumulates, the scratching particle will ride over the material, thereby interrupting its contact with the solid bedrock until it reaches a fresh surface downstream from the debris cover (Boulton 1974). This affect, however, can be eliminated if circulating subglacial water removes the fines (Vivian 1970). It is also known that the entire scratching phenomenon will become ineffective unless particles within the ice mass continuously move downward to replace the original abrasive grains. Those grains become smoothed during the abrasive action, and failure to replenish the basal ice with new particles having sharp edges and corners will cause abrasion to end (Boulton 1979).

Increasing grain size in the load produces grooves and furrows larger than striations. Normally grooves are up to 1–2 m deep and 50–110 m long, but under the proper controlling factors they may achieve giant proportions. For example, H. T. U. Smith (1948) describes grooves in the Mackenzie River valley of Canada that are 30 m deep, 100 m wide, and several kilometers long. These are not the product of a single boulder but possibly represent gouging by a pocket of boulders solidly frozen together (Embleton and King 1975a). There may, in fact, be a limit to the size of boulder that can act in the grooving process, because large fragments, having too much surface contact with the underlying rock, will force the ice to flow over and around the boulder instead of carrying it as part of the basal load.

In addition to scratches of all types, a group of small features, generally referred to as *friction cracks* and/or *chattermarks,* are formed by chipping or grinding of the underlying rock surface (Harris 1943). Most workers believe these features result when ice flow is temporarily retarded in its forward motion and then suddenly released. This produces a jerky flow component commonly referred to as a slip-stick movement. The various cracks and marks are usually lunate in form, 10–12 cm long and 10–25 mm deep, and perpendicular to the direction of ice flow as determined by other criteria. Chattermarks are sometimes present on surfaces of minerals within glacial deposits (Folk 1975; Gravenor 1982). These are indicative of the grinding action within a glacier, even though chemical etching may produce similar features.

The features of abrasion are thought to reflect the direction of ice movement. Remember, however, that other processes such as mudflows or snowslides can form striations, and even floating ice blocks can cause them in nonresistant materials (Dionne 1974). Ice will diverge and converge over an irregular bedrock topography and basal ice may be moving in different directions than surficial ice (Engelhardt et al. 1978). Thus, minor abrasion features have somewhat limited value as indicators unless a large number of measurements are obtained and treated for statistical significance.

Quarrying differs from abrasion in that the functional success of the process depends less on the type of load being transported than on the properties of the underlying rock. In fact, fractures must exist in the bedrock if plucking is to operate at all. Intense shattering of rock in preparation for plucking probably requires some form of pressure release (Lewis 1954; Glen and Lewis 1961), crushing (Boulton 1974), or cyclic freezing and thawing. These processes, which weaken the internal cohesion of the bedrock, may occur subglacially (Sugden and John 1976; Anderson et al. 1982) or in association with periglacial conditions prior to the arrival of the glacier.

In detail, plucking has two basic requirements: first, the ice must exert a shear force on the loosened particles, and second, this force must exceed the resistance caused by friction when the particle is dragged over the residual bedrock (Boulton 1974). It is not enough simply to dislodge the particle from its original position, but the driving force must overcome any frictional resistance generated in the system. Boulton suggests that where ice flows over unconsolidated sediments, plucking is a relatively straightforward process related totally to the shear stress.

In tightly lithified materials, the mechanics are more complex. The process is directly influenced by periodic opening and closing of cavities beneath the ice, which are caused by obstructions and fluctuations of ice thickness or velocity. When cavities are open, free water may freeze to the fragmented particles or within the shattered mass. If the cavity is subsequently closed, the glacier incorporates this new ice into its basal layer; its forward motion will pluck some rock fragments away from the surface. Thus, plastic flow may also be involved in the encasement of shattered material and its subsequent removal (Boulton 1974). It is important to recognize that frictional resistance to plucking increases rapidly as the cavity is closed because it is directly proportional to the normal pressure. As Boulton suggests, plucking is probably most effective when the normal pressure is sufficient to incorporate the loose particles into the ice but not great enough to inhibit their forward movement by increasing the frictional resistance. Exactly when this condition prevails will vary from glacier to glacier, and other factors may complicate the dynamics. For example, rock fragments projecting downward from the ice may dislodge particles from the rock floor even when the interface is slightly open. These do not necessarily become encased in the ice but are shoved spasmodically along the bedrock surface until they encounter the downglacier end of the cavity.

The evidence of plucking action in glaciated landscapes is usually found in erosional features somewhat larger than those produced by abrasion, but commonly the two erosive processes are closely associated in the same landforms. Where abrasion is dominant, the landscape may be indented with smoothly curved elongate surfaces whose long axes are subparallel to the direction of ice flow. Some of these surfaces are distinctly higher at one end, and they taper laterally and longitudinally until they blend into the surrounding ground level, producing a unique teardrop or raindrop shape which Flint (1971) describes as a "whaleback form." Some whaleback forms may be related to streamlined depositional features, such as drumlins, in that their shape represents the minimum resistance to flowing ice. The composition of streamlined forms seems to be of little consequence, however, as they can be entirely bedrock, entirely sediment, or any combination of the two. Whalebacks, therefore, may be merely a transitional form in a range of streamlined features from pure bedrock to pure drift (Flint 1971).

When plucking is a significant factor, whalebacks develop a pronounced asymmetry, having a gently sloping upstream surface and a steep rock face on the down-ice side of the feature. Such a form, commonly called *roche moutonnée,* is the result of abrasion on the upstream slope and intense quarrying at the position of the steep, downstream face. Its development is due to irregular spacings of fractures within the bedrock. Where joints are widely spaced, abrasion is the dominant process, while closely spaced jointing facilities plucking and more rapid erosion (fig. 10.3). Flint (1971) objects to the term "roche moutonnée" because of its wide misuse and suggests that *stoss* and *lee topography* better describes the forms developed by the combination of abrasion and plucking.

Figure 10.3.
Relationship between joint spacing and roche moutonnée development in Yosemite Valley. (After Matthes 1930)

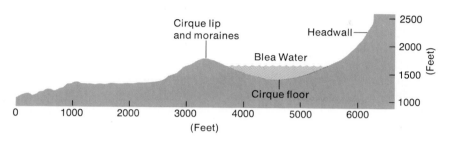

Figure 10.4.
Long profile through Blea Water corrie, a cirque.

Subglacial erosion is much more complex than our simple analysis of abrasion and plucking suggests. Many subglacial processes are involved in fracturing of the underlying bedrock (Gray 1982), and these directly affect the intensity of plucking and the character of ice-scoured topography (Gordon 1981). Flow of basal ice around obstacles often results in extreme velocities known as *ice streaming,* which produces large, elongate erosional forms (Boulton 1979; Goldthwait 1979). Water may also be involved in subglacial erosion. For example, in carbonate regions subglacial meltwater may differentially dissolve the bedrock surface (Hallet 1976a, 1976b). In addition, rapid drainage of meltwater through subglacial channels can physically erode the bedrock (Dahl 1965), as can the fluid mobilization of subglacial till (Gjessing 1967; Gray 1982).

Cirques

The striking landscapes found in the uplands of glaciated mountains have been sculptured primarily by the erosive action of ice contained in cirques. The term "cirque" was first used in the early 1800s to describe the collecting basins for valley glaciers in the Pyrenees, and locally the feature has been given a variety of names including cwm, corrie, kar, and botn. A **cirque** by any name is still a deep erosional recess with steep and shattered walls that is usually located at the head of a mountain valley. It is normally semicircular in plan view, often being described as an amphitheater, and it is floored by a distinct rock basin where the surface has been smoothed by abrasion. As figure 10.4 shows, the bowl-shaped rock basins commonly contain lakes, called *tarns,* that are dammed in by a convex-up rock lip that stands as a threshold boundary between the cirque floor and the downstream part of the valley. The cirque lip is often capped by small moraines that contribute to the damming effect. The rock basins can be of spectacular dimensions. For example, the rock basin floor of Blea Water corrie in England (shown in the figure) is 96 m (316 ft) lower than the rock lip (Lewis 1960).

Cirques range in size from shallow depressions to monstrous cavities that are kilometers wide and several thousand meters high along the rear wall. Their dimensions and their geomorphic form depend not only on the rocks into which they are cut, being larger and more perfectly developed in igneous or high-rank metamorphic rocks, but also on the rock structures (Olyphant 1981), the preglacial relief, and the time span of the formative glaciation. Most maturely developed cirques seem to possess a reasonably consistent geometry when their length to height ratios are compared, indicating that cirques probably attain some equilibrium form related to the processes of their formation.

Cirques are often preferentially oriented according to the direction of solar radiation and the prevailing winds (Graf 1976), and their elevation is probably (but not necessarily) related to the snowline at the time of their formation (Porter 1977; Trenhaile 1977). Thus, although most cirques originate in the headward reaches of stream valleys, any hollow, regardless of its origin, that stands at the proper elevation and has the ideal orientation may progressively accumulate snow and finally become a maturely developed cirque like those in the photograph in figure 10.5.

The significance of cirque processes in the development of Alpine scenery is that cirque expansion by continued erosion gradually eliminates the preglacial upland surface. As a number of cirques grow headwardly and laterally, they progressively consume much of the intervening upland region and leave as its only vestiges spectacular **horns** and **arêtes,** the features so indicative of mountain glaciation (figs. 10.5, 10.6). With prolonged headward erosion, adjacent cirques may merge, forming *col* depressions in the knife-edged arêtes.

Figure 10.5.
Cirques in the Alaska Range near Mt. McKinley, Alaska. Knife-edged ridges between cirques are arêtes.

Figure 10.6.
Schwan Glacier in the Chugach Mountains of Alaska. Note dark lobes on glacier surface where rockslide avalanches have moved onto the ice. Medial moraines are displayed as long linear dark bands. Horns and arêtes are shown in mountain uplands.

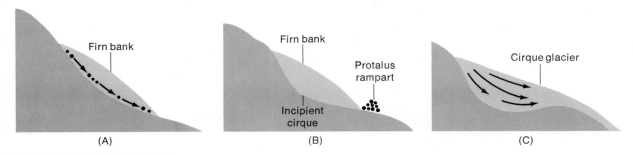

Figure 10.7.
Stages of cirque development.
(A) Nivation beneath firn bank.
(B) Nivation cirque. (C) Cirque
with fully developed cirque
glacier.

The erosive work in cirques lends itself to the suggestion that Alpine topography can be classified according to stages of its development, and such hypotheses are not uncommon (see Embleton and King 1975a). Our purpose here, however, is to understand the processes that create the Alpine topography, not the sequence of its development. Therefore, we must examine the origin of cirques and the erosive mechanics that functions within their boundaries. Of utmost importance are the processes that scour the basin in the cirque floor and those that cause the recession of the cirque walls.

Cirques result from two separate groups of processes: (1) mechanical weathering and mass wasting, and (2) erosion by cirque glaciers. The development of a cirque, diagrammed in stages in figure 10.7, begins in a patch of firn that fills a small depression and stands near the regional snowline. In the ablation season, meltwater released during the day percolates into fractures of the bedrock beneath the firn bank and refreezes there at night. The repeated pressure associated with freezing and thawing presumably wedges out particles of rock that are then moved slowly downslope by creep and by water flowing at the base of the firn. The combined processes are commonly referred to as nivation (Matthes 1900). Actually, *nivation* refers to a set of geomorphic processes, including chemical weathering (Thorn 1976), each of which may function more effectively under different controlling factors (Thorn and Hall 1980). Nonetheless, the shape of the original depression is gradually deepened and widened, and eventually it approaches a semicircular form that can logically be called a nivation cirque. The nivation hypothesis also contends that with continued accumulation, the firn changes into true glacier ice, and erosion by cirque glaciers rather than nivation becomes the dominant process in further development of the cirque. Exactly when this transition occurs is not clear, but it is virtually impossible for rock basins cut into cirque floors to be formed by physical nivation processes alone since they cannot carry particles upslope to the cirque lip.

Cirque Glaciers   Observations made in tunnels excavated into cirque glaciers indicate that such glaciers move by a process known as rotational sliding, in which ice slides over the arcuate bedrock floor, rotating at the same time around a horizontal axis. The ice exposed in the tunnels displays recognizable yearly accumulation layers that are separated by marked ablation surfaces, giving the entire glacier a banded stratigraphy (Grove 1960). Some of the ablation

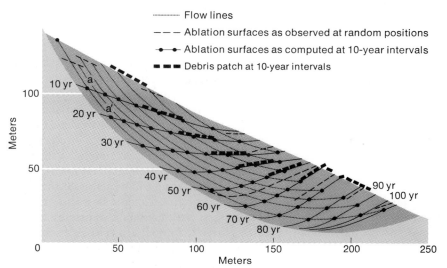

Flow lines
− − − Ablation surfaces as observed at random positions
−•−• Ablation surfaces as computed at 10-year intervals
▪ ▪ ▪ Debris patch at 10-year intervals

Figure 10.8.
Long section through a cirque glacier in Norway, showing ablation surfaces and debris patches at 10-year intervals. Note rotation of ablation surfaces in the down-ice direction.

zones are laden with debris that fell onto the ice surface near the cirque headwall. As figure 10.8 shows, these layers originally dip downglacier at the angle of the ice surface, and with time each layer is incorporated into the ice mass as more snow accumulates in the headwall area. As the ice moves, however, the layers are reoriented so that (as in the figure) near the equilibrium line they are almost horizontal; while close to the terminus they dip steeply up-glacier. The deformation of the layers, combined with changes in the position of stakes in the tunnels and on the surface (McCall 1952, 1960), makes it clear that flow lines are moving downward near the headwall, parallel to the surface at the firn line, and upward near the terminus.

Rotational sliding is an appealing mechanism to explain the scouring of the bowl-shaped depression in the cirque floor, for this process should be capable of carrying the products of abrasion or frost wedging upslope and over the cirque lip. We should emphasize, however, that not all cirque glaciers exhibit rotational sliding as the dominant flow mechanism. The rate of cirque erosion and growth seems to vary with the geologic and climatic setting. In polar or subpolar regions, erosion rates have been estimated as 8–76 mm/1000 years (Anderson 1978), and plucking seems to be the dominant erosional process. In temperate settings where abrasion is more important, the rate has been estimated as 95–165 mm/1000 years but could have been much greater during full glacial conditions (Reheis 1975). Actually, the rate of cirque growth produced by erosion is difficult to ascertain because the time span over which erosion acted is not precisely known, and the influence of preglacial topography is not usually considered (Olyphant 1981b). Regardless, because the length-to-depth ratio of cirques in widespread locations is normally 2:1, headwall retreat probably occurs at a faster rate than deepening of the cirque floor (Gordon 1977; Olyphant 1981a).

Headwall Erosion   The creation of Alpine topography requires removal of large amounts of bedrock from the head and side walls rather than the millimeter-by-millimeter abrasion occurring on the cirque floor. The surfaces of the cirque walls are assumed to be prepared for erosion by severe frost action associated with climates in high elevations. In addition to the shattering caused by freezing and thawing, joints often develop parallel to the wall faces when pressure is released by the removal of outer layers of rock (Glen and Lewis 1961). These *dilatation joints* aid in the fracturing process by providing avenues for percolating water and by isolating rock material into discrete units.

The actual processes of frost shattering are not so simple as they first appear and, like many processes, seem to be accepted more on faith than on solid evidence. The phenomenon of frost shattering will be discussed in greater detail in chapter 11. At this juncture, however, you should be aware that laboratory experiments (Battle 1960) suggest that maximum shattering by frost action occurs only when the temperature falls rapidly to between $-5°C$ and $-10°C$. A slow decrease in temperature will have little if any effect. In addition, freezing of water in a crack must proceed from the top of the opening downward to the bottom. An ice plug must form first at the top of the fracture in order to produce the closed system that will allow pressure from ice growth to exceed the tensile strength of most rocks.

Since W. D. Johnson descended into a deep **bergschrund** in 1904, many workers have investigated the role played by bergschrunds in the mechanics of headwall retreat. A **bergschrund** is a crevasselike opening near the headwall that separates actively moving ice of the glacier from nonactive ice frozen to the headwall. Johnson (1904) suggested that surface meltwater gained access to the base of the headwall by percolating down the bergschund and thus produced extensive frost shattering when water that permeated into rock fractures was alternately frozen and melted.

Although the bergschrund hypothesis is attractive, it suffers because actual measurements of temperature fluctuations in bergschrunds (Battle 1960) are not severe enough to produce shattering. In addition, some cirque glaciers do not have bergschrunds, and in others the gap does not always penetrate deeply enough to intersect the rock of the headwall.

In light of the above, we must entertain the possibility that we have misinterpreted the driving mechanics. S. E. White (1976a) suggests that hydration may exert almost as much pressure as that produced by ice expansion at $-22°C$. As minerals adsorb water, expansion and contraction of this nonfreeable water are produced by temperature fluctuations in the range of freezing and thawing. The shattering seen at the base of the headwall might then take place beneath a glacier without the stringent conditions necessary to generate forces by ice growth. The hydration-shattering proposal is thought provoking and deserves careful investigation. The process not only can explain shattering where temperature fluctuations are minimal, but it also dispels the problems associated with the bergschrund hypothesis. In spite of these, experimental evidence suggests that hydration by itself may not be as effective as frost action in promoting disintegration (Fahey 1983), and Washburn (1980) reminds us that most large products of disintegration are found in periglacial environments, the point being that hydration shattering should operate in any climatic zone as long as water is available. Clearly, more work needs to be done before the relationship between frost action and hydration shattering is completely understood.

Glacial Troughs

Glacial erosion is not limited to the cirque environment, for ice passing over the cirque lip can also remold the preglacial valley topography into a characteristic glaciated form. The ability of ice to remove rock protuberances tends to produce valleys with steep, nearly vertical, sides and relatively wide, flat bottoms (fig. 10.9). The transformation of a V-shaped river valley into a U-shaped glacial valley has been explained mechanically by A. Johnson (1970). Assuming ice to be pseudoplastic and using Nye's (1965) analysis of ice behavior in a triangular form, Johnson suggests that dead regions (areas of no

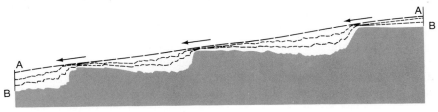

Figure 10.9.
(A) U-shaped cross-profile of glaciated valley and a hanging valley—view up Yosemite Valley from vicinity of Artist Point. El Capitan at left, the Cathedral Rocks and the Bridalveil Falls at right. Yosemite National Park, Mariposa County, Cal.
(B) Longitudinal section of glacial valley in Yosemite National Park illustrating staircase profile. AA is preglacial valley floor; BB is present valley floor. Broken lines represent intermediate stages of development. (After Matthes 1930)

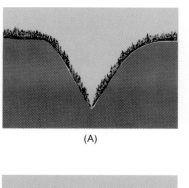

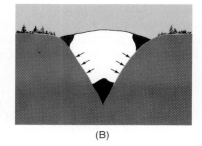

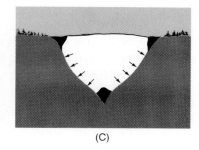

(A) (B) (C)

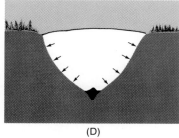

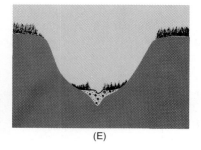

(D) (E)

Figure 10.10.
Possible sequence of events
leading from V-shaped mountain
canyon to U-shaped glacial valley.
(A) V-shaped mountain canyon.
(B) V-shaped canyon visited by
glacier. (C) Glacier erodes sides of
canyon. (D) "Dead" regions
disappear and entire side of
canyon is rasped by rock-studded
glacier. (E) Glacier disappears,
leaving U-shaped valley. (From
Johnson 1970, *Physical Processes
in Geology.* With permission of
Freeman, Cooper and Co.)

flow or shear stress) should exist in a glacier that invades a mountain stream
valley. The dead regions should be present in the ice at the top of the valley
sides and along the valley bottom (fig. 10.10). Erosion would be negligible in
these zones but appreciable along other parts of the valley sides where shear
stress and velocity would be high. As differential erosion causes the sides to
bow outward, the dead regions are progressively removed because the shear
stress and velocity distribution changes with the approach to a more parabolic
cross-profile (fig. 10.10). Eventually, erosion affects all parts of the valley sides,
resulting in the characteristic U-shaped glacial valley (fig. 10.10).

A parabolic cross-profile aids glacial movement because it probably exerts
the minimum resistance to glacier flow (Flint 1971). The precise width-depth
dimensions, however, probably depend on the intensity of the glacial dynamics
(Graf 1970) and the properties of the geologic framework. The formation of
the cross-sectional shape occurs by both lateral and vertical erosion of the
preexisting valley. Whether the parabolic form is derived predominantly by
widening or predominantly by deepening depends on the properties of the rocks
and the ice, as well as the extent of preglacial weathering and the amount of
load available within the ice to be utilized as cutting tools. In any case, the
erosion leads to the truncation of rock spurs jutting into the valley and the
formation of **hanging valleys.** These occur particularly where trunk valleys
carry more ice and are more extensively eroded than their tributaries.

Glaciated troughs are also characterized by uniquely irregular longitudinal profiles that essentially represent a series of interconnected basins and
steps; the diagram in figure 10.9 shows this typical profile in Yosemite National Park. The basins may contain lakes, a series of which are sometimes

called *paternoster lakes*. Immediately downvalley from the basins holding the lakes are rock bars or steps that commonly show the effects of intense abrasional smoothing on their relatively flat surfaces. At the distal end of the step, a sharp break in slope occurs where plucking has produced a steep scarp that connects the rock bar to the next lower basin or step.

The staircase profile has intrigued geologists for years and has been the subject of considerable speculation. Hypotheses for its origin include (1) variation of rock structures, especially spacing of joints, causing differential erosion, (2) preparation of weak zones in the rock by preglacial weathering, (3) irregularities in the preexisting valley topography, and (4) increased erosive power at the confluence of tributary valleys with the main valley (see Bakker 1965). It is possible that rock bars producing a steplike profile require no special conditions for their formation other than the normal slip-plane orientation noted in glacier flow (Nye and Martin 1967). Theoretically, rock bars can form where there is no obviously harder rock, and conversely, zones of resistant rocks do not necessarily evolve into rock bars under glacial erosion.

The abrasion prevalent on the treadlike steps and the quarrying at the position of the steep scarps led Lewis (1947) to note that each step resembles a large roche moutonnée and so probably has a similar origin. He concluded that the mechanics probably worked best beneath thin glaciers where meltwater could easily penetrate to the glacier floor and shatter the rock by refreezing. Although freeze-thaw mechanics might be a reasonable accomplice in the formation of staircase profiles, we cannot ignore the fact that development of the associated basins probably required some differential ice movement, capable of flowing uphill and carrying load in the process. Such a requirement does not encourage the belief that the ice was thin; in fact, it is common knowledge that many valley glaciers have thickened to the point of overtopping their divides and spreading into adjacent valleys. In addition, as discussed earlier, the freeze-thaw component in rock shattering is somewhat suspect because we do not completely understand the thermal conditions at the base of glaciers. What we are left with, then, is the problem that while plucking is necessary to form the scarps, the rock shattering required for this process to function is probably not a result of frost action. Furthermore, some scarps form where no regional joint system is present to facilitate the shattering process.

Boulton (1974) suggests an alternative hypothesis of shattering that deserves our consideration because it not only resolves the above problem, but also relates directly to glacier mechanics. When a glacier moves across a horizontal bedrock surface, the effective normal pressure will be $pgh - w_p$ where p,g,h and w_p are, respectively, density, gravity, ice thickness and water pressure at the bedrock interface. The dashed line in figure 10.11 represents the normal pressure on the horizontal bed and is estimated by $\rho_i\, gh$ (where $\rho_i = \rho$). ΔP in the figure represents the increase or decrease of normal pressure produced by a bedrock obstruction; the total normal pressure at any point becoming $\rho_i gh + \Delta P$. Therefore, if the bed is irregular the normal pressure will fluctuate so that it will be higher than average on the upglacier side of a bedrock obstruction and lower than average on the lee side. Boulton shows that

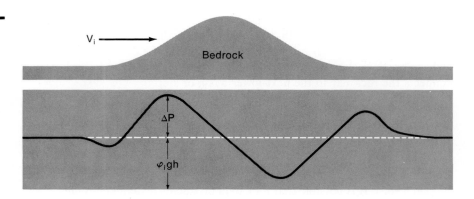

Figure 10.11.
Schematic view of normal pressure distribution at glacier bed as ice flows over a bedrock obstruction. ΔP represents the increase or decrease of normal pressure produced by a bedrock obstruction. The normal pressure where no obstruction exists is represented by the dashed horizontal line and has a value equal to $\rho_i gh$.

the shear stresses included in the bedrock will be greatest where the normal pressure is lowest, i.e., down-ice from the crest of the obstruction. The exact position and absolute magnitude of the maximum shear stress depend on whether cavitation occurs downglacier from the bedrock knob. Furthermore, Boulton points out that the shear strength of the rocks will be least in the downglacier position where the shear stress is greatest, creating the ideal situation for rock failure at that locale. Assuming this analysis is correct, the inevitable conclusion is that with the proper bedrock configuration, even hard unfractured rocks may be crushed beneath the ice, and the rocks already jointed can be intensely shattered. Preglacial irregularities in the long profile, structural weaknesses, and lithologic variations all will be accentuated by this process because shattering will prepare those zones for the plucking action that produces the scarps in the staircase profiles.

A special type of glacial trough exists mainly in high-latitude coastal regions that are underlain by resistant rocks, so that the general land surface stands at considerable elevation above the nearby ocean. These troughs are called *fiords,* and they differ from other types only because they are partially submerged by the ocean. The inundated bottoms of fiords have the same variety of topographic elements, both erosional and depositional, that exist in a normal continental glaciated valley (Holtedahl 1967). Their history may include components of both glacial and fluvial processes, and they may be partly controlled by tectonic and lithologic factors. For these reasons, it is unwise to make sweeping generalizations about their origin.

Perhaps the most salient property of fiords is that part of their development took place when the ice was physically beneath the ocean (Crary 1966). Flint (1971) reminds us that a glacier 1000 m thick with a density of 0.9 will remain in contact with its bed and be fully capable of erosion at water depths up to 900 m. Even at greater depths, when the snout begins to float, high topographic irregularities of the valley might still be eroded (Crary 1966). The water depths over fiords, several hundred meters in many and greater than 1000 m in some, are well beyond that which can be attributed to a postglacial rise in sea level (see Flint 1971), adding credence to the suggestion that much fiord erosion was accomplished in a submarine environment.

Before glaciers were recognized as viable geomorphic agents, deposits containing boulders that obviously came from a distant source were called *drift*. This term arose because elimination of known processes led to the belief that the anomalous boulders reached their site of deposition by riding on top of floating ice. After glaciers were recognized as the transporting vehicles, the term "drift," or **glacial drift,** was retained and expanded to include all deposits associated with glaciation. It is estimated that drift covers 8 percent of the Earth's surface above sea level and almost 25 percent of the North American continent. The thickness of this cover varies greatly. In the United States, for example, only a thin layer of drift ($<$ 20m) covers upland areas in most of New England, although drift may be several hundred meters thick in buried valleys. The drift in the central United States is generally from 10 to 60 m thick, but once again these are average values. In some places the drift is merely a thin mantle on top of bedrock, and in other regions, such as parts of Michigan, it exceeds 200 m in thickness. The exact volume of drift deposited depends on the time span of glacier activity, but with high velocities and loads, as much as 30 m of drift can be accumulated in less than 10 years (Flint 1971).

Through the years geologists have been intrigued with the amazing variety of glacial drift and the complex interrelationships that exist between the different types. This complexity arises because (1) drift may be deposited from mediums that contain vastly different amounts of water; (2) deposition occurs beneath, within, or on top of ice, at the glacier margins, in bodies of standing water, or in fluvial settings far from the glacier, the debris being transported there by streams rising in the ice mass itself; (3) the depositional sites and environments and the drift composition all change with time because glaciers themselves are not constant in their properties or fixed in their position; and (4) the glacier may be active or stagnant. Because of these complicating realities, any discussion of glacial drift and depositional features is difficult to organize in a way that is entirely satisfactory. In this approach, we will first briefly examine the varieties of glacial drift and then examine features according to the depositional environment in which they originate. An attempt will be made to relate the morphology of features and their sedimentary properties to the dynamics of the system. Throughout this discussion it is important to recognize the distinction between the sedimentological character of drift and the morphology of features that result from its deposition. Morphological terms such as moraines and kames are not to be interpreted as implying a particular drift type. Many features with similar morphology are composed of a number of drift varieties, especially where the environment of deposition is subject to repeated change.

Drift Types

Over the years glacial deposits have normally been divided into two categories based on their sedimentary characteristics. Particularly important are the presence or absence of layers and the degree of sorting in the deposit. In our discussion we will separate drift into stratified and nonstratified types.

Nonstratified Drift   Sediment originating directly from glacial ice characteristically has no discernible stratification. The material is usually called *till* and typically is a nonstratified mass of unsorted debris that contains angular particles composed of a wide variety of rock types. In addition, the term "till" usually connotes material that has been transported and deposited by the ice itself, a process often indicated by striations or microscopic fractures on the grains (Krinsley and Donahue 1968). Many examples justify this description of till; absence of layering and poor sorting (especially an almost universal bimodal size distribution) seem to be the most reliably consistent properties. The bimodality observed in most tills probably is due to the differences in grains produced by abrasion and those derived by plucking. In an excellent review, Goldthwait (1971, p. 4) points out that till is probably more variable than any other sediment that is described by a single name.

Any of the identifying criteria in our definition may be missing at a particular till locality as a result of varying transporting and depositing mechanics and of heterogeneity in the rocks over which the ice has passed. For example, many clasts in till have a subangular pentagonal or triangular shape, but these forms may be significantly altered by rounding during their transportation. This is especially true in till that was transported subglacially. In that environment, parent debris is rounded by attrition of edges and corners (Boulton 1978), although the degree of rounding is partly dependent on rock type (Holmes 1960; Vagners 1966), the original clast shape (Drake 1968, 1974), and the distance of transport. In addition, glaciers that override older stream deposits may incorporate in their load boulders that have already been rounded, resulting in a till that is notably less angular than one would expect. Figure 10.12 shows till that contains rounded boulders.

Figure 10.12.
Late Wisconsinan (Pinedale) till in Rock Creek valley, Beartooth Mountains, Mont. Note rounded boulders in till.

Overall particle size tends to be reduced by attrition in subglacially transported till (Mills 1977; Boulton 1978). Each particular source rock tends to produce a characteristic texture depending on how easily its clasts can be crushed and how far they have been transported (Mills 1977; Dreimanis and Vagners 1971). Thus, the overall angularity and texture of till depend on the lithologic heterogeneity of the source rocks as well as the distance of transport from the location of their outcrop.

The unpredictability of till increases still more because the material is carried in different parts of the glacier where the ice is characterized by diverse mechanics. In ice sheets, debris is carried as englacial load in layers 1–100 m above the base of the ice, as indicated by the fact that the surface is free of any load except near the margin (Goldthwait 1971). In valley glaciers, on the other hand, considerable debris may be shed onto the surface from the valley sides and be transported as supraglacial load. Each transport subenvironment produces till with different characteristics. Supraglacial till has a texture dominated by coarse, angular clasts because the particles are not crushed during transport, and fines tend to be washed away as the surface ablates. Subglacial till is more compact and contains a higher percentage of fine-grained sediment. Englacial load can be deposited directly on the bedrock floor as subglacial matter, and a considerable thickness of the glacier will remain above it at the moment of deposition. It can also rise within the ice along shear planes or along the normal upward flow lines in the terminal zone. This material finally emerges as supraglacial debris when ablation of the surface ice releases the contained particles. In this case the till will be emplaced on the bedrock surface as the ice beneath it dissipates, and no ice will remain above the till at the time of its final deposition. Till laid down in the subglacial environment under pressure of the overlying ice is usually referred to as basal till or **lodgement till.** In contrast, **ablation till** occurs where the debris is concentrated at or near the surface and is gradually let down onto the bedrock as the ice disappears.

Ablation till forms as one of two types, **flow-till** (Hartshorn 1958; Boulton 1968) and **melt-out till** (Boulton 1970b). When englacial debris is released at the surface, the till tends to be highly mobile and likely to move downslope as a quasi-mudflow, by creep, or as a semiplastic slide. However, when the thickness of the surface cover reaches about 3 cm, the ablation process is severely retarded, and if the surface layer exceeds 1 or 2 m, ablation ceases. Further increase in the thickness of the supraglacial cover can occur only if more debris moves to the site from higher levels of the ice surface, or if the ice melts beneath the layer already concentrated at the surface. Boulton (1971, 1972b) asserts that supraglacial till may have two components: an upper zone accumulated from debris transported from a higher source (flow-till), and a lower zone accumulating in situ and never actually exposed at the surface (melt-out till). Melt-out tills rarely exceed 3 m in thickness, but the total supraglacial layer can be much greater if flow-till continues to accumulate on the surface. Little is known about the rate of accumulation of ablation tills, but Mickelson (1973) has calculated that the accumulation rate of basal melt-out till of the Burroughs Glacier in Alaska ranges from 0.5 to 2.8 cm a year.

Stratified Drift The second major category of glacial drift is distinguished by the fact that the sediment was transported by moving water before its final deposition, thereby acquiring a degree of stratification not normally seen in tills. Such drift is often referred to as **fluvioglacial** because running water is involved in its origin, even though the water may not always be confined in discrete channels. Fluvioglacial deposits are also distinguished from till in that they are usually sorted and the clasts contained in the mass are more rounded. However, the demarcation between some types of fluvioglacial deposits and thoroughly washed ablation tills is a matter of degree rather than substance. Exactly where the line between the two is drawn becomes somewhat arbitrary. Highly saturated flow-till, for example, might move in a nearly fluvial manner.

The layering and sorting in a fluvioglacial deposit depends on precisely where it is formed with respect to the ice that provides the transporting meltwater. Sorting is also partly a function of the energy possessed by the meltwater, the distance of transport, and the continuity of the sorting process. Because the discharge of meltwater is notoriously inconstant, varying drastically with time of day, local climate, and the characteristics of the ice, significant differences in the sedimentology of fluvioglacial deposits can be noted over short distances. These are especially evident where deposits are formed in contact with the ice and the free circulation of meltwater is restricted (see Shaw 1972 for sedimentary characteristics). If debris is transported away from the glacier terminus, the sedimentary characteristics tend to vary more regularly.

We should stress again that the prime requisite of stratified drift is transport by water, much of which is released from melting ice. However, this puts no constraints on the environment in which the sediment is deposited. Fluvioglacial debris can come to rest in stream channels, floodplains, lake or ocean floors (Rust 1977), deltaic plains, in contact with ice, or in any other place where sediment-laden running water loses its transporting energy. In some cases sedimentologic properties in the deposit, such as the degree of roundness and mean grain size, can help identify the depositional environment (King and Buckley 1968), but normally there are too many variables in the system to rely on these critiera alone (R. J. Price 1973). Detailed field study is almost always necessary to reach a firm conclusion about environments of deposition.

Deposits that originate in contact with the ice often contain interbedded bodies of ablation till, and their particles tend to be less well rounded and sorted because of the limited distance of transport. The characteristics of such a deposit often vary from the bottom of the sequence to the top because the environment of deposition in contact with the ice repeatedly changes with time, especially if the ice is stagnating. In any deposit, then, a particular layer may be superseded vertically by one with different sedimentary properties that reflect a new depositional environment. The stratigraphy is complicated by the fact that much of the drift is physically supported by ice during its deposition, and when the ice dissipates, the support is removed and the sediment collapses. Such a process, as figure 10.13 shows, leads to flexures in some of the layers, minor faults, and beds dipping at angles well beyond the angle of repose for such material. Overall, the ice-contact setting at places produces an interconnected maze of stratified and nonstratified drift in which every conceivable process and environment is possible and probably has been present at some time during the depositional history. Moreover, the manner in which the ice ablates influences the resulting deposit. Drift produced while the ice is still active may be quite different, especially in distribution, from that derived from a large mass of stagnant ice that is simply downwasting as it melts and is not influenced by internal glacial movement.

Sediment deposited beyond the terminal margin of the ice is formed in the proglacial environment and is often referred to as outwash. **Outwash** is usually well sorted and normally consists of rounded sands and gravel representing bedload carried and deposited in stream channels. Silt and clay are usually carried as suspended load and are commonly removed from the system unless, as in the lower Mississippi River valley, the transport distance is so great that some of the outwash is silty in texture. It is important to understand that streams transporting outwash do not usually head at the glacier terminus but begin on top of or within the ice, well upglacier from the margin. Proglacial features and deposits often can be traced into and through the maze of ice-contact deposits, increasing the complexity of the depositional sequence developed near the ice margin.

The Depositional Framework

Before considering what geomorphic features might be produced during deposition, we must first establish a realistic framework within which we can give some order to the subtle variations of depositional features associated with glaciation. This is done in table 10.1. From our brief look at the nature

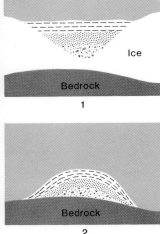

Figure 10.13.
(A) Gravel pit cut in kame, south central New York, shows stratification in kame deposits. (B) Structures and deformation of strata in ice-contact stratified drift develop as ice melts and debris collapses and is lowered onto bedrock floor.

Table 10.1 The depositional framework and associated features.

Setting	Features	Type of Drift
Ice contact		
Marginal	End moraines	Till and fluvioglacial
	Kames and kame terraces	Fluvioglacial
	Kettle holes	Fluvioglacial
	Eskers	Fluvioglacial
Interior	Medial and interlobate moraines	Till
	Ground moraines	Lodgement till
	Fluted surfaces	Lodgement till
	Drumlins	Lodgement till
Proglacial environment	Sandar	Fluvioglacial (outwash)
	Kettled sandar[a]	Fluvioglacial (outwash)

[a]May merge with marginal environment.

of drift, it is obvious that both stratified and nonstratified deposits can be formed in contact with the ice. They differ only in whether the ice alone was the primary transporting and depositing agent or whether a stage of water transport intervened between the release of particles from the ice and their final deposition. As a result, the **ice-contact environment** must be considered as one of the major settings of our depositional framework. The ice-contact environment can be subdivided geographically into **marginal** and **interior zones** depending on where the drift was originally deposited (table 10.1).

Features formed in the ice-contact environment can be composed of either stratified or nonstratified drift (till or fluvioglacial sediment) or a combination of both types. The region beyond the terminal edge of the glacier is classified as the **proglacial environment;** in contrast to the ice-contact setting, features formed there are composed entirely of fluvioglacial sediments (outwash).

The distinction of marginal and interior zones in the ice-contact environment is problematical for several reasons. First, glacial margins migrate forward and backward with time according to the mass budget. An active glacier with a negative mass balance should have a terminal margin that is progressively receding toward its source; marginal features will be formed in regions that were interior when the ice was at its greatest extent, and deposits of the two zones may be complexly intertwined. Second, the process operating near the contact between the marginal and interior zones are dependent upon the characteristics of the subglacial environment. For example, the outer 2 to 3 km of ice sheets are commonly frozen to the underlying surface even though farther upglacier the basal ice may be temperate. Where marginal ice is frozen to the bed and high porewater pressure exists, thrusting along preexisting planes of weakness in the substrate can inject large blocks of subglacial material into the ice (Moran et al. 1980). These thrusted blocks tend to concentrate where ice margins rest on aquifers and on the upslope edges of upland areas. Proceeding upglacier toward the interior zone, the ice is free to slide because the ice-bed contact is unfrozen. Thrust blocks in the transitional area between the

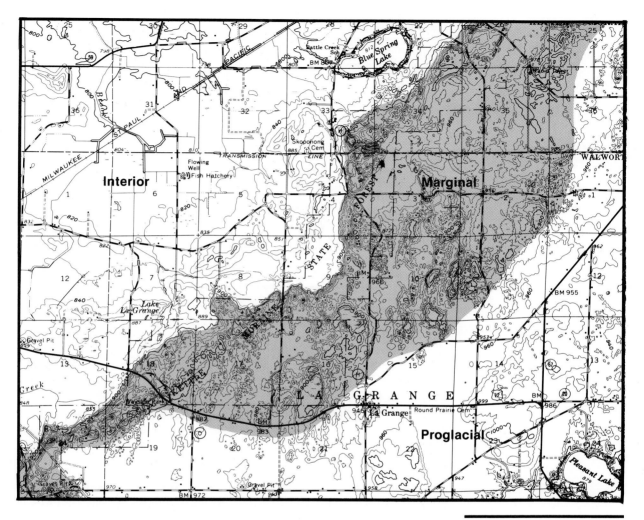

Figure 10.14.
Map view of the depositional
framework after ice has
disappeared. Part of the
Whitewater, Wis. quadrangle
(U.S.G.S. 15′). Marginal zone
contains terminal moraine and ice-
contact stratified features, and
interior zone is characterized by
ground moraine. Proglacial zone is
a large outwash plain or sandur.

marginal and interior zone are smaller and smoothed over by the sliding ice. In the true interior zone, no thrusting will occur, and the subglacial terrain is characterized by streamlined forms (Moran et al. 1980). Clearly, the position of the frozen ice/thawed zone contact may migrate with time and changes in the subglacial environment. Thus, in any given glaciated area, the inner portion of the marginal zone may be transitional into the interior zone rather than marked by a well-defined contact between the two.

Regardless of the problems associated with precise boundary locations, the depositional framework fits our purposes because the suite of features found in each zone is a direct reflection of the genetic processes involved. The utility of the classification is shown in the map in figure 10.14. In this region the terminal moraine, marking the ice-contact marginal zone, is characterized by a maze of small hills and depressions (kames and kettles) that distinguish deposits formed near the boundary between active and stagnating ice. North of

Table 10.2 Moraine types.

End Moraines	Moraines produced at front or sides of an actively flowing glacier.[a]
Terminal moraines	Mark the farthest advance of an important glacial episode.
Lateral moraines	Deposited at or near the side margin of a mountain glacier.
Recessional moraines	Formed at glacier front during temporary halt or readvance of ice in a period of general recession.
Ground Moraine	Gently rolling surface formed of debris released from beneath the ice.
Interior and Minor Varieties	
Washboard moraines	Small, parallel ridges oriented transverse to direction of ice movement. Also called moraine ridges or cross-valley moraines.
Interlobate moraines	Formed at junction of two ice lobes.
Medial moraines	Elongate ridge developed at junction of two coalescing valley glaciers.
Rogen Moraines	Large sequence of ridges transverse to ice flow. Formed in the interior zone.

[a]Lateral moraines may be excluded by some geomorphologists because end moraines are commonly considered only as topographic features developed at the front of a glacier.

the moraine the relatively low, flat area is underlain by material that was deposited beneath the active glacier (ground moraine); it represents the ice-contact interior zone. South of the moraine, the proglacial zone is marked as a plain composed of debris (outwash) transported and deposited by meltwater streams heading within or on top of the ice.

Marginal Ice-Contact Features

Moraines   The term "moraine" originated several hundred years ago as a local name for ridges of debris found at the edges of glaciers in the French Alps. Since then many definitions have appeared, but for our purposes we can think of a **moraine** as a depositional feature whose form is independent of the subjacent topography and that is constructed by the accumulation of drift, most of which is ice-deposited (Flint 1971). A precise morphological definition is not possible since, as table 10.2 shows, moraines take many different forms and have a variety of dimensions. The term "moraine" is not synonymous with "till," as many have suggested, but in reality refers to a suite of topographic forms on which the only restriction is that they must be composed of drift (and even that rule has been violated in some cases).

The most spectacular moraines develop at or near the edges of active glaciers and so are designated as **end moraines.** The end moraine constructed at the downstream edge of the ice at the farthest point of advance is called a **terminal moraine.** In valley glacier systems, it merges imperceptibly into **lateral moraines** on both sides of the valley because ablating ice deposits debris along the glacier's lateral extremities as well as at its terminus (fig. 10.15).

(A)

(B)

(C)

Figure 10.15.
(A) Terminal moraine of Pinedale glaciation, East Rosebud valley near Roscoe, Mont. Low ridge above road in left center of photo is lateral moraine merging with terminal zone. (B) Recently formed lateral moraine in Rocky Mountains of Canada.
(C) Terminal moraine on southeast end of Devil's Lake, Wis. Moraine dams old valley of the Wisconsin River.

Figure 10.16.
Diagram showing retreating
margin of Barnes ice cap. Material
moves up inclined shear planes
and accumulates as till cover on
core of ice.

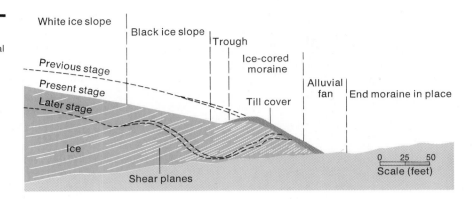

Ice sheets form terminal moraines that tend to be long (often hundreds of kilometers), linear, topographic highs marking the forward boundary of the ice, but lateral moraines are absent because there is no side to the ice sheet. Where ice sheet movement was distinctly lobate, however, end moraines called *interlobate moraines* (Flint 1971) may develop along the junction of two lobes.

Ideally, end moraines assume a rather narrow, ridgelike shape, but actually the form and size depend directly on the amount of glacial load, the mass budget, and the volume of meltwater circulating in the system. Temperate valley glaciers tend to build higher and more massive terminal moraines (some reaching 300 m in height) because these glaciers have higher flow velocities, loads, and total budgets. Ice sheets seem to generate moraines that are less dramatic in size, seldom exceeding 50 m in height. Where moraine ridges are transverse to the flow direction, it is difficult to distinguish between those that develop at the glacier front (end moraines) and others that are formed subglacially. This difficulty arises because ridges in both geographical positions result from similar processes. R. J. Price (1973) distinguishes these processes as dumping, squeezing, and pushing.

The mechanics of the *dumping* process were first described in detail by Goldthwait (1951), who observed that debris in the Barnes ice cap is contained in a series of shear planes near the terminus of the ice (fig. 10.16). The sheared zone extends only approximately 150 m from the snout, and the attitude of the planes suggest they cannot penetrate the ice to depths greater than 75 m. This indicates that, if the debris is brought surfaceward from the basal ice layers, the process occurs only in the outer 500 m of the glacier.

Figure 10.16 shows that as the ice fringe thins by ablation, dirt contained in the shear zones is released, creating a film of sediment on the ice surface that Goldthwait refers to as "black ice." Some of the debris moves downslope, as described earlier, where it blankets the lower ice up to a thickness of a meter, thereby retarding the ablation process. Black ice with only a thin dirt cover ablates more rapidly than the white ice upglacier from the sheared zone, and a trough is produced parallel to the ice front that separates the active black ice zone from ice-cored debris at the front edge of the glacier. The final hummocky surface, so characteristic of moraines, results from debris sliding off the ice-cored ridge, debris collapsing as the underlying ice melts, and from fluvioglacial action.

The shear-plane hypothesis is attractive when the outer fringe of the ice is stagnant. It is possible, however, that active ice can also move basal debris surfaceward along normal flow lines that consistently intersect the ice surface near the terminus (Boulton 1967). Regardless of the precise mechanics by which load rises to a glacier surface, the evolution of topography in a morainal belt proceeds in a complex manner depending on (1) the character of the ice, (2) the spacing of the debris bands as they emerge at the surface, and (3) the thickness of the supraglacial cover. These factors produce differential ablation rates and widespread zones of drift that are held up by cores of solid ice.

In contrast to moraines formed by the dumping process, moraine ridges are also developed by *squeezing* drift originally deposited beneath the ice. Ridges generated by squeezing are usually smaller than dumped moraines, normally standing less than 10 m high, although they can be higher where the subglacial till is highly saturated. Theoretically, the process involves the response of water-soaked lodgement till to pressure exerted by the weight of the overlying glacier. The till will move from under the ice and emerge along the ice front, or it will be squeezed into zones of low pressure within the ice, represented by crevasse openings (R. J. Price 1970; Mickelson and Berkson 1974). In either case the ridges seem to be typified by (1) steeper distal slopes than proximal slopes, (2) a plan-view outline consisting of linked arcuate segments that parallel the ice front, and (3) a till fabric with pebbles oriented perpendicular (or nearly so) to the ridge crests, though occasionally oriented parallel to the trend of the ridge (Mickelson and Berkson 1974).

R. J. Price (1970) examined a series of small morainal ridges in Iceland that formed during the last 70 years as the ice retreated from the main, and larger, terminal moraine. The ridges apparently rise quickly (in one area 15 ridges formed in 16 years), prompting Price to suggest that they develop annually when summer meltwater descends to the glacier base and saturates the subglacial till (a proposal made earlier by Andrews and Smithson 1966). Squeezed ridges also seem to originate beneath bodies of standing water (Mickelson and Berkson 1974); since saturation of the debris would be rather complete in such an environment, this adds credence to the proposed mechanism. Hydrologic processes, however, may be involved in the origin of sublacustrine or submarine ridge development (Barnett and Holdsworth 1974).

A third mechanism capable of forming a moraine ridge is the collision of advancing ice with older deposits, which deform them into a ridgelike feature. Such end moraines, called *push moraines,* are only partly composed of sediment carried by the ice and may in fact include blocks of older rocks of a nonglacial origin (Kaye 1964b; Andrews 1980). Push moraines usually have a steep distal slope, and they often display a sharp break in slope at the contact between the moraine and the proglacial deposits (Embleton and King 1975a). They are more distinctive, however, in their internal structure. Here individual layers, sheared away from the ground lying in front of the ice, are intensely deformed into large (sometimes overturned) folds and faults of all kinds (Mills and Wells 1974). Thrusting, with plates as thick as 30 m, is not uncommon and may occur as imbricate slices or even as an underthrust phenomenon (Kaye 1964a, 1964b). The exact style of internal deformation depends on the rigidity of the bedrock being shoved by the ice front or on how tightly unconsolidated material is held together by freezing around the ice perimeter.

Most, if not all, of the processes that function at the glacier front can also create moraine ridges subglacially (Sugden and John 1976). For example, large transverse ridges, called *Rogen moraines* or *ribbed moraines,* are 10–30 m high, > 1 km long, and tend to develop as a series of separate hills spaced 100–300 m apart. They probably form by shearing or thrusting behind the glacier front, usually in broad depressions of older till sheets or in local bedrock valleys. Significantly, the ridges are often in direct assocation with flutes or drumlins, and therefore, it is reasonable to assume that they all have a common origin. This suggests that the ridges reflect an interaction of basal debris, porewater pressure, and ice temperature and that the Rogen system develops where transverse variations in stress exist at the glacier bed (Sugden and John 1976).

The mechanics associated with Rogen moraines are remarkably similar to the thrusting phenomena proposed by Moran and his colleagues (1980). As discussed earlier, the thrusted blocks seem to occur along the inner part of the marginal zone and often assume a morainic topography. This topography may be subdued, however, where temperate ice slides over the blocks.

The pushing and squeezing phenomena may also relate to smaller varieties of subglacial moraines. Within the marginal zone the locus of deposition shifts periodically, giving rise to ridges, mounds, and depressions of varying size. The ridges in the sequence, usually small and composed of till, commonly parallel the orientation of the ice front. They have been given many names, such as *cross-valley moraines* (Andrews 1963; Andrews and Smithson 1966) and *washboard moraines* or *moraine ridges* (R. J. Price 1970), and several theories have been suggested for their origin (Elson 1968). The ridges are usually segmented or interwoven with many deposits of stratified drift, giving the entire topography a chaotic expression rather than a regular undulation of ridge crests and intervening sags.

Moraines are not always characterized by distinct ridges that have developed transverse to the direction of ice flow. Instead, moraines may exist as a belt of interspersed mounds and depressions that are merged into a chaotic topography completely devoid of linear ridges. These moraines, commonly called *disintegration moraines* (Gravenor and Kupsch 1959), have local relief up to 70 m and develop from supraglacial drift in the lower part of the ablation zone. When the ice in glacier margins stagnates, it often breaks into isolated blocks of wasting ice covered by ablation till (fig. 10.17). Thus, although the depositional environment is ice-contact marginal, the morainic topography gradually develops as the glacier surface downwastes over dissipating ice cores. This is different from ridges formed from active ice that oscillates back and forth or from those created by pushing, squeezing, or thrusting. In stagnant marginal zones, flow till, melt-out till, and fluvioglacial deposits can all coexist on the wasting surfaces and, because the ice cores melt at different rates, may

Figure 10.17.
Ice-cored moraine, Yanert Glacier, Alaska.

become shifted and mixed within the chaotic surficial expression. For example, slow melting allows till to flow into and fill hollows and trenches interspersed between the ice-cored mounds. When the ice finally disappears, the topography is inverted so that the former depressions, now filled with considerable thickness of drift, stand as circular or irregular hills (fig. 10.18).

Stratified Marginal Features   As pointed out earlier, many glaciers are characterized by marginal zones of thin, stagnating ice. Within these zones a suite of genetically related ice-contact features is developed, composed predominantly of stratified drift and morphologically distinct from the moraines just described. These features form by deposition of drift (1) where water flows through openings in and beneath the ice, or in ice-surface channels, (2) in spaces between the ice and the bedrock of the valley sides, and (3) where disseminated sediment is passively concentrated by melting of the encasing ice.

Many of the channelways for the flowing water originate when stagnating ice breaks into individual segments along planes of structural weakness in the ice, and so the features are genetically related to many of the ice-disintegration forms described by Gravenor and Kupsch (1959), Stalker (1960), Clayton (1967), Parizek (1969), and many others. Our discussion, however, is restricted to those features that consist primarily of fluvioglacial drift, even though ablation till may be a minor ingredient. Their morphology is entirely constructional, although they can be affected by slumping as the ice walls that supported the drift during its deposition are melted away.

Figure 10.18.
Hummocky topography in terminal moraine. East Rosebud valley at front of Beartooth Mountains near Roscoe, Mont.

Kames and Kettles   *Kames* are moundlike hills of layered sand and gravel that vary in size from minor swells to conical protuberances standing up to 50 m high and extending 400 m along their base. Kame material can accumulate at the ice-substratum interface, and also in cavities located within stagnating ice or on its surface. Englacial or supraglacial debris can be lowered onto the ground surface as the ice dissipates (Cook 1946; Holmes 1947). The moundlike shape results only if the accumulate was originally isolated and nonlinear (fig. 10.13), because sediment deposited in linear openings is ultimately converted to ridges rather than mounds. In addition, some kames form when debris collects as fans or small deltas built against the ice or outward from the ice, with the apex resting at the stagnant margin. In either case, melting of the supporting ice allows the drift to collapse into a kame, a process evidenced by slumped strata within the deposit.

Kames are only one of many forms with essentially the same origin. They are transitional into eskers or minor ridge and circular forms that Parizek (1969) calls ice-contact rings and ridges, or into other features whose names utilize the term "kame" as a descriptive adjective, i.e., kame delta, kame moraine, etc. Perhaps the most common of the latter type is a *kame terrace.* Kame terraces originate from drift deposited in narrow lakes or stream channels between the valley side and the lateral edge of the stagnating ice. When the supportive ice disappears, the inner edge of the deposit collapses into the terrace scarp. Kame terraces differ from normal river terraces in that they are restricted in their longitudinal extent and are usually narrow; the tread may slope gently into the valley, and the surface may be dimpled by kettle holes (McKenzie 1969).

Kettle holes are circular depressions that are formed in a variety of ways (Fuller 1914), most commonly by the burial of isolated blocks of ice by stratified drift. The gradual ablation of the ice leads to a gentle downward flexing of the layers as they settle over the dissipating mass. Some kettles are almost 50 m deep and up to 13 km in diameter (Flint 1971), but these giants are exceptions to the normal kettle size of less than 8 m deep and 2 km wide. Kettles usually form in association with kames and other related features, producing an irregular surface described as "kame-and-kettle topography," but similar surfaces can develop in the absence of either or both of these features.

Eskers   The term "esker," evidently stemming from the Gaelic word for "crooked" or "winding," has been applied to a wide variety of ridged ice-contact features (fig. 10.19). *Eskers* range in shape from the single, narrow, sinuous ridge that is the classic form to a complexly intertwined maze of branching and joining ridges (Huddart and Lister 1981). Eskers seem to form most commonly in stagnating margins of large ice sheets where the underlying surface is broad and relatively flat. The ridges are not necessarily continuous but may consist of crudely connected linear segments. In broad valleys the eskers usually parallel the slope of the valley floor, but this is not always the case. Some ridges, for example, ascend the valley sides, transect the divide, and descend the flanks of the adjacent valley.

Eskers show dramatic inconsistencies in dimension. They range from 2 m to more than 200 m in height, from several meters to as much as 3 km in width, and from tens of meters up to 500 km in length (if gaps are considered in the total distance). In cross-profile, they usually have rather sharp crests and steeply inclined sides (up to 30°), but broad eskers can maintain rather flat upper surfaces or may be pitted by kettle development. R. J. Price (1973) suggests that the height and width of eskers may be directly related to their overall length, longer ridges being proportionately higher and wider than shorter ones.

It seems certain that eskers result from sediment accumulation in a variety of openings, such as (1) ice channel fillings (crevasse fillings, fillings between stagnant blocks, etc.), (2) tunnels beneath or within the ice, (3) supraglacial channels or even, in rare cases, (4) narrow longitudinal embayments of the ice front (Cheel 1982). The subglacial origin is made possible by meltwater that descends from the ice surface to the base through fractures and holes in the ice. Along the subglacial surface a network of interconnected tunnels passes water and sediment toward the terminus, and normal fluvial deposition fills or partially fills the openings. Filling probably takes place during the last stage of deglaciation when the ice is stagnant and thin, for basal tunnels could not remain open if the glacier were still moving or if the ice were thick enough to cause pressure flow (Flint 1971). In addition to subglacial channels, field studies and photographic analyses of esker formation by the Casement Glacier in Alaska (Price 1966; Petrie and Price 1966) and the Breidamerkurjökull Glacier in Iceland (Price 1969) provide rather conclusive evidence that englacial and supraglacial deposits can be transformed into eskers.

Figure 10.19.
Esker near Whitewater, Wis.

In those areas, some of the eskers are definitely ice-cored and were measurably lowered in elevation during the period 1948–1963 by wastage of the buried ice. Presumably the final melting of the ice will rest the esker on the original subglacial floor.

Interior Ice-Contact Features

Behind the marginal zone in the interior portion of a glacier, the predominant features are deposited at the base of the ice. There pressure from the overlying ice either spreads till rather evenly across the ground surface or molds water-soaked till already in that position into distinct morphologic shapes. Supraglacial drift is somewhat rare in the interior zone, but where two valley glaciers join, the debris dragged along their lateral edges may coalesce into *medial moraines*. Although these moraines appear as striking linear belts on the surface of the ice (fig. 10.6), they are superficial in that the deposits are shallow. Not all of these linear features represent former ice margins. Some may have an interior origin (Small et al. 1979) and sediment sources that are both surficial and englacial (Eyles and Rogerson 1978). Medial moraines are rarely preserved on the ground surface because they are let down in the middle portions of valleys where meltwater streams are likely to destroy them. The dominant forms are those deposited from beneath the ice such as ground moraines, drumlins, and fluted surfaces.

Ground Moraine   In contrast to end moraines, **ground moraine** is distinguished by its apparent lack of topographic expression. It is accepted as a moraine despite its low relief and a complete absence of transverse ridges (Flint 1971) because its surface expression is independent of the topography it covers. Ground moraine occupies much of the surface in North America and Europe that was covered by major Pleistocene ice sheets. The moraines usually exist as smoothly undulating plains, like that in figure 10.20 seldom exceeding 10 m in total relief; they range in size from small areas interspersed among younger marginal features to regions covering thousands of square kilometers behind the terminal moraine.

Although in most cases the primary building material of ground moraine is lodgement till, it is a mistake to consider ground moraine and lodgement till as synonymous. The deposits from which ground moraine is constructed also include ablation till and interbeds of fluvioglacial deposits that originate in the same glacial advance (R. P. Kirkby 1969; Boulton 1972a, 1972b) or in more than one episode of glaciation.

Fluted Surfaces and Drumlins   The monotony of ground moraine topography is sometimes broken by ridges or elongate hills that manifest the mechanics functioning at the base of glaciers, especially in temperate ice sheets. The most common features developed in the subglacial environment, **fluted surfaces** and **drumlins,** appear to be fashioned by moving ice; Flint (1971) refers to them as streamlined molded forms. Others (Sugden and John 1976) consider them as special types of moraine ridges that form parallel to the direction of ice flow. Regardless of classification, it appears certain that the two forms are

Figure 10.20.
Flat till and loess plain in southern Illinois. Gently undulating topography on ground moraine has been smoothed by influx of younger loess.

related by their common subglacial origin, their orientation parallel to the direction of ice flow, and the fact that mobilization of subglacial debris is involved in their origins.

Fluted surfaces consist of narrow, regularly spaced, parallel ridges. The ridges are normally less than 5 m high and several hundred meters long, although individual ridges may be considerably larger. Small ridges are usually composed of till, and their origin may be related to pressure-squeezing of saturated debris into longitudinal cavities at the base of the ice, a process similar to that forming some of the minor moraine ridges discussed earlier. There is no question that till is mobilized and arranged in ridges where cavities open on the lee side of large boulders (Hoppe and Schytt 1953). In fact, Boulton (1976) defines flutes in a genetic sense as being formed when deformable subglacial material is intruded into ice tunnels on the lee side of boulders or other rigid obstructions. This occurs because unloading in the lee of obstructions sets up a pressure gradient in the till that causes it to flow into the cavity.

Some larger ridges are composed of material other than till (Lemke 1958; Gravenor and Meneley 1958), and the surfaces between adjacent ridges are often noticeably grooved. These characteristics have led many workers to believe that fluted surfaces are both erosional and depositional in origin. Gravenor and Meneley (1958), for example, suggest that alternating high and low pressure zones in the basal ice produce the unique fluted surface. In their model, grooving occurs in the material beneath the high-pressure zones, and debris eroded from there is moved not only downglacier but also upward into regions of low pressure. Boulton (1976) rejects the notion that fluted surfaces require periodic pressure distributions in the ice and suggests that the features represent postdepositional deformation of preexisting materials. As such, they are neither erosional nor depositional.

Drumlins also are elongated parallel to the direction of ice flow, their long axes deviating only slightly from the average trend of the glacier movement (fig. 10.21). These distinctive forms have received attention from a large number of geomorphologists during the last century, but as yet their origin has not been fully explained (see Muller 1974 for an excellent review). Drumlins have been described as having a plan-view shape that is similar to a lemniscate loop (Chorley 1959) or an ellipsoid (Reed et al. 1962) and in long profile a form that Flint (1971) likens to an inverted bowl of a spoon (fig. 10.21). The exact shape, however, is probably variable enough that no particular model will fit all drumlins. In any case, it can be stated that some assurance that drumlins are higher and wider near their rear edges and that they narrow and thin downstream until they merge imperceptibly with the surrounding surface. Drumlins average in size from 1 to 2 km in length and from 400 to 600 m in width, and stand anywhere from 5 to 50 m high; individuals can be smaller or larger. Their length-width ratio seems to be reasonably consistent, ranging from 2 to 3.5 (Reed et al. 1962; Vernon 1966; Trenhaile 1971), even though length and width may be controlled by different factors (Mills 1980).

Drumlins display a variable internal composition; many are fabricated entirely from clay-rich till, but others have obvious cores of solid rock or preexisting drift that may or may not be stratified. The spatial character and distribution of where drumlins develop are critical in any hypothesis concerning their origin. First, drumlins rarely exist as individuals but instead cluster together in fields that are commonly wider than most morainal belts (Gravenor 1953). The density of drumlins within a field seems to be inversely related to their sizes; that is, very dense clusters are composed of relatively small drumlins (Doornkamp and King 1971). Second, many studies show a strong probability that drumlins within any field are spaced in a nonrandom manner; that is, spacing between neighboring individuals is somewhat regular (Reed et al. 1962; Vernon 1966; Smalley and Unwin 1968; Trenhaile 1971). Third, most glaciated areas have no drumlin development. Fourth, drumlin fields seem to be located in zones close to but behind the terminal moraines that mark the limit of a particular glaciation.

Figure 10.21.
(A) A portion of the drumlin field
located near Weedsport, N.Y.
From the northeast quarter of the
Weedsport, N.Y., quadrangle
(U.S.G.S. 15'). Contour interval 20
feet. (B) Drumlin shown in
longitudinal profile. Looking east,
four miles northwest of Kalispell,
Flathead County, Mont.

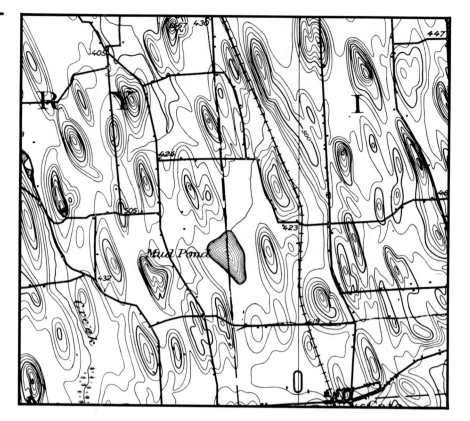

(A)

(B)

The models used to explain drumlin genesis fall into two main groups: (1) drumlins are erosional features developed when moving ice streamlines preexisting drift or rock, or (2) drumlins are depositional features formed when a moving glacier deposits till and molds the material as the ice continues its forward motion. There is reasonable evidence to support both theories. For example, when the internal composition of drumlins is stratified drift or bedrock that predates the ice advance in which the drumlins were formed, it is rather difficult to ignore erosional processes in the drumlin origin (Gravenor 1953; Kupsch 1955; Lemke 1958; Boulton 1979; Whittecar and Mickelson 1979). On the other hand, many drumlins have no cores of older material but consist entirely of lodgement till of the same age as the surrounding drift. They often display an internal concentric banding that suggests some processes of gradual accretion (Hill 1971). Drumlins such as these can hardly be doubted as being depositional.

In light of all this, we should probably accept a multiple origin for drumlins, as Embleton and King (1975a) suggest, and not pay undue attention to hypotheses that utilize one process to the exclusion of all others. Yet it is bothersome that drumlins have such unique distributional properties. It seems, somehow, that any explanation of drumlin development, whether by erosion or deposition, should have its roots in the characteristics of the glacier and the subglacial debris. One attempt to integrate all drumlins into a single genetic model (Smalley 1966; Smalley and Unwin 1968) is based on the fact that many tills have *dilatant* characteristics. In simplest terms, dilatant materials increase in volume (expand) under stress. As stress increases on till at rest, the debris will resist deformation until the stress produces the loose packing attained in the dilatant condition. When dilatancy prevails, deformation proceeds more easily and will continue even if stress is decreasing, until finally the stress becomes too low to maintain the dilatant property. Conceivably, then, dilatancy occurs only within certain well-defined upper and lower stress limits.

Smalley and Unwin (1968) suggest that the proper stress conditions for dilatancy are related to the thickness of the ice mass that exerts pressure on the underlying till. Where the glacier is thick, the till is highly mobile because the weight of the ice exceeds the load limit needed to initiate dilatance, and all the interior basal drift will be carried forward. Near the ice margin, however, the glacier thins to a point where the stress exerted on the basal load is less than the minimum required to maintain dilatance, and the material will revert to a more tightly packed texture, thereby becoming stable and resistant to further deformation. In essence, this implies that the ratio between stress exerted by ice thickness and strength of the subglacial material increases in the upstream direction (total deformation upstream and none at the margin). Drumlins, therefore, develop in the zone between the thin marginal ice and the thick interior ice. There the loading stress is less than that required to initiate dilatancy but greater than the pressure at which the till returns to a stable form (fig. 10.22). Some material collapses into stable cores in this zone because of its particle size distribution, but the till that remains dilatant is accreted and streamlined over the cores.

Boulton (1979) suggests that dilatant subglacial debris is not a prerequisite for drumlin formation. In his model, the shear *strength* of this debris increases upglacier because ice pressure is increasing over pore pressure (shear strength is $S = C + (P - P_w)\tan\phi$; see chapter 9). Strength values could also change by altering other factors (see Smalley 1981 for a review). At some point a critical value of strength is reached where the drag force exerted by ice upstream from that point is not enough to deform the subglacial till. Downstream, the ratio of ice pressure to pore pressure decreases until the strength reaches another (lower) critical value, and all subglacial till downstream from there will readily deform. Drumlins form between the critical strength levels where some undeformed debris serve as the nuclei needed for drumlin growth.

Other than dilatance, the essential difference between the preceding two models rests in whether the ratio stress/strength increases or decreases upglacier and how subglacial till behaves above and below the critical values of stress and strength. They both, however, provide an intermediate zone in which drumlin formation should occur. Therefore, each explains the common distribution pattern associated with the feature.

Proglacial Features

The large volume of water released from glaciers carries with it a tremendous quantity of sediment that is deposited in a number of environments beyond the margin of the ice. This debris usually accumulates in stream channels and associated floodplains that, because of their continuous lateral shifting, spread the sediment into a large plain called a **sandur** (from Icelandic; plural *sandar* or *sandurs*). Downstream from the sandar, the meltwater streams may empty into bodies of standing water and construct deltas, beaches, and other geomorphic features from the fluvioglacial sediment. These forms do not differ appreciably from their counterparts developed in normal fluvial, lacustrine, or marine environments and so will not be discussed here. Sandar are somewhat analogous to alluvial fans, but they are unique in that the hydrology of streams that form them is controlled by intense seasonal variations in the melting of ice. In addition, because each glacial advance develops its own related sandur, the distribution and age of the alluvial surfaces are extremely helpful in unraveling Pleistocene history. Thus, the origin of sandar and the geomorphic criteria for recognizing their form in ancient settings deserve our attention.

Figure 10.22.
Cross section at edge of ice sheet showing most likely stress conditions for the formation of drumlins according to the dilatant hypothesis. (A) Stress level above that required to initiate dilatance; all material moving. (B) Stress decreasing but drift is mobile. (C) Stress below level needed to maintain dilatance.

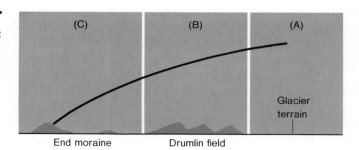

The study of sandar began in the nineteenth century in both Europe and North America. They are now recognized as consisting of two primary types (Krigstrom 1962). A **valley sandur,** which originates within well-defined valleys, is created by one main river and its anabranches; the entire system rarely occupies the total valley bottom at any given time. In the United States, valley sandar have been called **valley trains** and are usually associated with individual mountain glaciers (fig. 10.23). The second type of sandur, called a **plain sandur,** differs in that it develops with no lateral constraints but represents the coalescence of many braided rivers that spread debris across wide areas in the form of a massive plain. In North America, these are often referred to as **outwash plains** and are usually associated with large ice sheets.

Most sandar are composed of gravel, although the deposits may include lenses of sand. A general decrease in particle size is sometimes, although not always, apparent in the downstream direction (Fahnestock 1963; Church 1972). The deflation of very fine material on active sandar that are unvegetated helps to produce a coarse-grained surface and simultaneously produces loess (see chapter 8).

In general, the long profiles of sandar are similar to river profiles, being concave-up in form and expressible as a simple exponential function (Church 1972). The concavity, however, may not be perfect because of the presence of linear segments similar to those characterizing alluvial fans. Cross-profiles tend to be convex-up, but the exact form may be irregular or may slope continuously in one direction. In addition, the cross-profile shape seems to depend on where it is measured in relation to the ice margin.

Figure 10.23.
Valley sandur near Mt. McKinley, Alaska.

Krigstrom (1962) has recognized on sandar three distinct zones relative to the ice front, called the proximal, intermediate, and distal zones, each of which has different surficial characteristics. The *proximal zone,* closest to the ice, is usually transversed by only a few main rivers that flow in well-defined entrenched channels. These rivers and their deposits may pass continuously onto the ice mass itself, where the drift sometimes buries extensive areas of stagnating ice. As the ice subsequently melts, kettle holes form and the proximal sandur takes on a rough, pitted configuration. These kettled sandar (or pitted outwash plains) are difficult to place in our framework classification because they form in the marginal ice-contact environment, but they are continuous with the surface of the proglacial sandur and develop with the same original mechanics (described in Price 1973). In addition, on many sandar the proximal surface stands well above the elevation of rivers emerging from the ice. Several hypotheses have been suggested to explain why rivers in the proximal zone are so entrenched in their own deposits: (1) the rivers may be regrading in response to hydrologic and load characteristics (Fahnestock 1969); (2) the sediment is supraglacially deposited and left elevated as the ice front recedes and rivers emerge at a continuously lowered level; or (3) modest incision may be normal in the proximal zone, with the sandur surface simply representing the high flow level. It is conceivable that each of these interpretations is correct.

In the *intermediate zone,* the channels become wide and shallow and distinctly braided, and the entire depositional network shifts its position rapidly from side to side. This active lateral migration leaves a maze of abandoned channels with a relief of one or two meters impressed on the surface topography of the plain. Commonly the main channel is aggraded to a higher level than the smaller channels, facilitating rapid changes in the position of the river. Downstream the system changes gradually into the *distal zone,* where channels become so shallow that the rivers may merge into a single sheet of water during high flow. The flow here commonly feeds deltaic growth when the river enters a body of standing water. However, the sandur may extend itself downstream by growing over the rear portion of the delta, while the deltaic front simultaneously progrades (Church 1972).

Sandar originate from the combined effects of a large sediment supply and the high floods associated with melting ice. Most of the abundant load is derived from older drift, morainal deposits, and the continual delivery of new debris to the ablation zone and its release from the ablating ice. The greatest fluvioglacial work occurs near the ice margin where floods are produced by summer melting or as *jökulhlaups* (sudden release of lake water dammed within the ice; an Icelandic word pronounced "yokel-lawp"). These floods are characterized by rapid and drastic increases in discharge (Church 1972; Waitt 1980).

The bulk of aggradation on a sandur takes place during high-flow events as channel fills, sandur levee deposits, and overbank sedimentation. Overbank deposits are more prevalent in the intermediate zone where channels are shallower and interchannel reaches are covered more frequently by floods. Although high flow does initiate pronounced channel scouring, the amount of aggradation during the peak and waning stages of a flood simply obliterates the scour channels. Thus, aggradation may be rather rapid. For example, Fahnestock (1963) measured a net elevation gain of 0.36 m in a two-year period on the sandur produced by the Emmons Glacier (Washington).

In general, then, sandar can be considered as transport surfaces that aggrade during high flows but are probably eroded and changed in form when discharge and load are at normal volumes. The seasonal variations in load and discharge may also be accompanied by changes in the river pattern (Fahnestock 1963). Therefore, the ultimate size and properties of a sandur are probably related to a quasi-equilibrium condition established by the balance between meltwater volumes and the quantity and size of the sediment made available for transportation. The surface will always be a montage of flood sediments, but the exact topography will change incessantly.

Summary

In this chapter we examined the landforms developed by the process of glacial erosion and deposition. Erosional features range in size from minor embellishments of exposed bedrock to major forms that dominate the landscape. Minor features such as striations, grooves, roches moutonnées, and friction cracks are a function of the subglacial mechanics that control abrasion and plucking. Major erosional landforms develop in two environments. In mountain uplands, the expansion of cirques results in the creation of features such as arêtes and horns that give glaciated mountains their characteristically rugged appearance. Cirques are created by nivation and rotational sliding of cirque glaciers. Cirques increase in size by erosional retreat of their walls facilitated by repeated freezing and thawing or possibly by hydration shattering. Glaciated valleys are created by large-scale abrasion and plucking that cause the dominant staircase longitudinal profile and the U-shaped cross-profile.

Deposits associated with glaciation, called drift, consist of either stratified or nonstratified material. Stratification requires that some sediment be transported by meltwater after the debris is released from the ice. Nonstratified drift is deposited directly by the ice. Certain types of depositional landforms tend to accumulate in particular geographic positions with respect to the ice front. In the marginal zone, moraines and stagnant ice features are most common. Mobilization and/or immobilization of subglacial drift at stress/strength thresholds determines where these depositional zones will occur. Frozen bed conditions and high pore pressure near the margins of temperate glaciers may produce thrusting of subglacial material. In the interior zone behind the ice margin, ground moraine, fluted surfaces, and drumlins are most conspicuous. Downstream from the glacial margin, in the proglacial zone, all drift is fluvioglacial and usually accumulates in the form of large plains called sandar. Processes operating in each of these depositional environments have been discussed with regard to how they might generate the landforms developed in the specific regions.

Suggested Readings   The following references provide greater detail concerning the concepts discussed in this chapter.

Boulton, G. S. 1974. Processes and patterns of glacial erosion. In *Glacial geomorphology,* edited by D. R. Coates, pp. 48–87. *Proc. 5th Ann. Symposium,* S.U.N.Y., Binghamton.

———. 1979. Processes of glacier erosion on different substrata. *Jour. Glaciol.* 23:15–36.

Church, M. 1972. Baffin Island sandurs: A study of arctic fluvial processes. *Canada Geol. Survey Bull.* 216.

Embleton, C., and King, C. A. M. 1975. *Glacial geomorphology.* New York: Halsted Press.

Flint, R. F. 1971. *Glacial and Quaternary geology.* New York: John Wiley & Sons.

Goldthwait, R. P., ed. 1971. *Till: A symposium.* Columbus: Ohio State Univ. Press.

Lewis, W. V., ed. 1960. Norwegian cirque glaciers. *Royal Geogr. Soc. Res.,* Ser. 4.

Moran, S.; Clayton, L.; Hooke, R.; Fenton, M.; and Andriashek, L. 1980. Glacier-bed landforms of the prairie region of North America. *Jour. Glaciol.* 25:457–76.

Muller, E. H. 1974. Origins of drumlins. In *Glacial geomorphology,* edited by D. R. Coates, pp. 187–204. *Proc. 5th Ann. Symposium,* S.U.N.Y., Binghamton.

Price, R. J. 1973. *Glacial and fluvioglacial landforms.* New York: Hafner.

Sugden, D., and John, B. 1976. *Glaciers and landscape.* London: Edward Arnold Ltd.

Periglacial Processes and Landforms

11

Introduction

A group of processes and features called periglacial characterize regions having extremely cold climates. The term "periglacial" was first used (Lozinski 1912) to describe the processes operating and the features developed in zones adjacent to ancient or modern ice sheets. Since then, the original connotation of "near-glacial" has been expanded, and most workers now accept the term as encompassing all nonglacial phenomena that function in cold climates, even if glaciers are not present (Dylik 1964; Butzer 1964; Washburn 1973). Washburn 1980 adopts the term "geocryology" to indicate the study of frozen ground processes and phenomena. However, he suggests continued use of the term "periglacial" as a descriptive adjective. The **periglacial** environment is difficult to define in terms of precise temperature and precipitation values, although several attempts have been made to do so. For example, Peltier (1950) suggested that average annual temperatures from −15°C to −1°C (5°F–30°F) and an average annual rainfall between 127 mm and 1,397 mm would constitute a periglacial morphogenetic region. L. Wilson (1968, 1969) suggests somewhat different physical limits of −12°C to 2°C (10°F–35°F) temperature, and 50 mm to 1,250 mm for the precipitation values.

Another question as to what constitutes a periglacial regime arises because some scientists believe that permanently frozen ground, or **permafrost,** is an essential ingredient (if not a prerequisite) for periglacial conditions (Tricart 1967; Péwé 1969). Certainly some periglacial features occur in close association with frozen ground, and a complete understanding of periglacial systems therefore requires some fundamental perception of permafrost. However, periglacial processes also function where no permafrost is known to exist. Most students of periglacial phenomena, then, would not view permafrost as a requirement of a periglacial system. The two points that all investigators seem to agree on are that (1) nearby glaciers are not necessary for the processes to function, and (2) the fundamental controlling factors of **geocryology** are intense frost action and a ground surface that is free of snow cover for part of the year. Many regions meet these requirements, ranging from polar to subpolar lowlands to high-elevation mountains that may rise in any regional climate including temperate or even tropical zones.

It is not new for humans to occupy the high latitude regions of the world; Eskimos and other nomadic groups have wandered these areas for millenia. These inhabitants, however, lived in rather simple social organizations, and their life styles brought them into little conflict with their natural surroundings. In more recent times, the emplacement of defense installations in periglacial regions, the discovery of oil in the Arctic Circle, and the increased exploration for mineral wealth brought with them the realization that developing a highly technical society in these regions is an engineering and scientific nightmare. For inhabitants of temperate zones, shattered highways and frozen or broken water pipes are inconveniences of a rigorous winter and the subsequent spring thaw. Multiply these problems by a thousand; add disruption of services, construction difficulties, and complications of all the trappings associated with a modern civilization; and you will begin to understand the meaning of life in a periglacial environment.

Governmental agencies of several countries, including the United States and the Soviet Union conduct research related to the problems of cold climates, and symposia and reports in scientific journals increasingly treat such topics. Nonetheless, detailed understanding of the processes involved is lacking, and faced with an inevitable human migration into cold regions, it is imperative that geomorphologists devote additional efforts to this discipline. We have the basics to do so because the processes responsible for periglacial effects are not greatly different from those we already know. In fact, frost action and mass movements form the core of periglacial processes. The main difference between periglacial and temperate phenomena seems to lie in the magnitude of the processes and the manner in which they are combined in the systemic operations. Thus, as L. W. Price (1972) stresses, if we are ever completely to understand the periglacial environment and provide viable information for future planning, we must cast aside our provincial, mid-latitude approaches to the study of this system.

Our treatment of geocryology will be necessarily brief, examining only the major processes and features. For greater detail, several excellent texts and reviews are available: Embleton and King 1968, 1975b; Price 1972; Washburn 1973, 1980; Tricart 1969; Péwé 1969; National Research Council of Canada 1978; National Academy of Sciences 1983.

Definition and Thermal Characteristics

Permafrost

Although permafrost is not a requirement for the functioning of periglacial processes, the most troublesome engineering problems occur in regions underlain by zones of perennially frozen ground, and many periglacial features are related to abundant ground ice. Permafrost was originally defined in terms of temperature only (Muller 1947) as being soil or rock that remained below 0°C continuously for more than two years. This definition has been accepted in many subsequent discussions of the topic (Washburn 1980). By this definition water is not a necessary component of permafrost, and in fact, "dry" permafrost has been recognized. In most cases, however, ice is so important in the mechanics of periglacial processes that some workers (Stearns 1966) believe moisture must be present before any material can be considered as true permafrost. Other investigators have placed the temperature criterion for permafrost below 0°C (Ferrians 1965; Brown 1970). In a geomorphic sense the presence of ice, regardless of the temperature value, seems to be critical. The ice may exist as a cement between soil particles or as larger masses of pure ice. Where distinct ice masses are prevalent, they usually occur as horizontal lenses, but they may also fill cracks in the parent material and stand as vertical veins or wedges (fig. 11.1).

Figure 11.2 shows that permafrost is usually a subsurface phenomenon existing as a zone of permanently frozen ground that can extend downward to incredible depths. The upper surface of the permafrost, the **permafrost table,** is overlain by a thin layer (15 cm to 5 m thick) of material that freezes and thaws on a seasonal basis. This uppermost layer, the **active layer,** is thickest

Figure 11.1.
Ice wedge (ground ice) in permafrost exposed by placer mining near Livengood, about 50 miles northwest of Fairbanks, Alaska. Tolovana district, Yukon region, Sept. 1949.

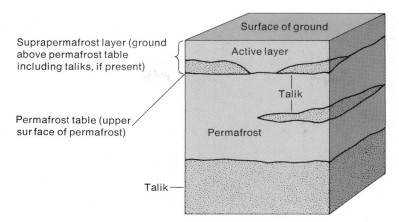

Suprapermafrost layer (ground above permafrost table including taliks, if present)

Permafrost table (upper surface of permafrost)

Surface of ground

Active layer

Talik

Permafrost

Talik

Figure 11.2.
Profile in a region underlain by permafrost. Taliks are zones of unfrozen ground within or beneath the permafrost or between the permafrost table and the base of the active layer. (After Ferrians et al. 1969)

in the subarctic region (Price 1972), becoming shallower toward both the pole and the mid-latitudes. In any area, however, significant variations in active layer thickness can occur in short distances depending on environmental conditions (Owens and Harper 1977), and it is usually thicker in sands and gravels than in more fine-grained soil materials. The mechanics in the active layer is similar in most respects to the normal seasonal freezing and thawing in temperate climate zones. In permafrost regions, however, water released by thawing cannot percolate into the solidly frozen substrate, and this tends to accentuate the effects of frost action and mass movements.

Even within the permafrost itself, the temperature fluctuates with the seasons. Temperature variations, however, become less radical with depth until some level is reached, generally at 20 to 30 m deep, where the temperature never changes (fig. 11.3). This level is known as the **zero annual amplitude.**

The thermal operations in permafrost are affected by so many factors that they are understood only in general terms. For example, thermal conductivity varies with material composition and may cause freezing or thawing to progress unevenly from the surface downward. Thus, as freezing proceeds after a thaw season, unfrozen lenses may develop between the permafrost table and the solidly frozen ground surface. This phenomenon causes unusual pressures in the soil water and depresses its freezing point. In addition, heat of fusion is generated when the zone around the ice-free lenses freezes. The result is that pockets of free water may exist in the ground for considerable time after the surface is frozen, possibly for several months.

Further complications in the distribution of frozen ground occur because the permafrost table tends to mirror the surface topography, rising beneath hills and lowering under valleys. In addition, the permafrost table is influenced in various ways by the position of surface water (fig. 11.4), depending on whether the rivers or lakes freeze over in the winter (Ferrians et al. 1969).

Distribution, Thickness, and Origin

Where annual temperatures average 0° or below 0°C, ground freezing during the winter will penetrate deeper than the depth of summer thawing. Each passing year will produce another increment of perpetually frozen ground, and

Figure 11.3.
Ground temperature change with
depth in permafrost regions.
(After J. R. Williams 1970)

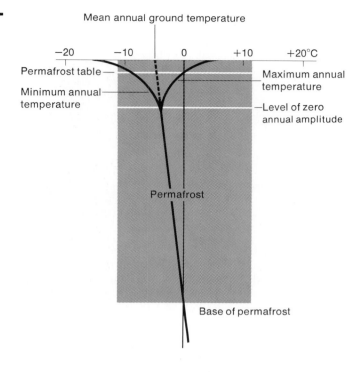

Mean annual ground temperature

−20   −10   0   +10   +20°C

Permafrost table —

Minimum annual —
temperature

Maximum annual
temperature

—Level of zero
annual amplitude

Permafrost

Base of permafrost

Figure 11.4.
Schematic cross section showing
the effect of surface water on the
distribution of permafrost. (PF is
permafrost; T are taliks.)

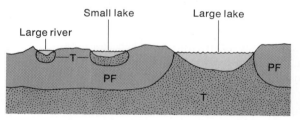

Small lake   Large lake

Large river

T

PF

PF

T

the permafrost zone will gradually thicken. The process, however, cannot go on indefinitely because the Earth's thermal regime will exert controls on the depth to which freezing can penetrate. Heat affecting permafrost issues from the sun and from the interior of the Earth as *geothermal heat flow*. Therefore, cold filtering downward from the surface is counteracted by heat escaping upward from inside the Earth. The interaction results in the **geothermal gradient,** i.e., the rate of temperature increase with depth below the surface. Thus, under a constant climate, at some depth the temperature will be kept above 0°C by the geothermal heat flow (fig. 11.5), and the total thickness of permafrost will relate to the position of this temperature level.

The geothermal gradient, and therefore thickness of permafrost, can be quite variable under areas having identical surficial climates (fig. 11.5). This occurs because the geothermal gradient is affected by the thermal conductivity of the parent material. Where mean surficial temperatures are the same, permafrost will extend to greater depths in materials of higher conductivity.

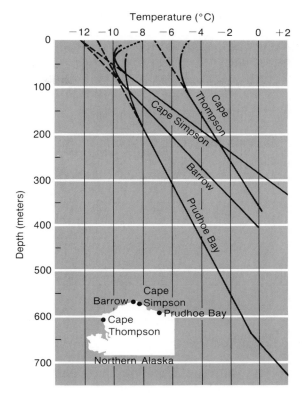

Figure 11.5.
Generalized profiles of measured temperature on the Alaskan arctic coast (solid lines). Dashed lines represent extrapolations.

The variations in geothermal gradients along northern Alaska, shown in figure 11.5, are explained by this phenomenon.

Permafrost today may extend to formidable depths (table 11.1). It averages between 245 and 356 m in North America and tends to be slightly thicker in Eurasia. The thickest known permafrost is in Siberia, where a depth of approximately 1500 m has been reported. The great thicknesses of permafrost probably reflect both past and present climate conditions.

Several lines of evidence indicate that some permafrost must have originated during the Pleistocene (see Washburn 1980, pp. 60–61), especially in areas that were not glaciated. In areas covered by Pleistocene glaciers, permafrost probably formed after the ice dissipated. In fact, in polar regions permafrost is forming today where glacial retreat has exposed unfrozen ground (Washburn 1980). This indicates that some present-day climates are cold enough to accumulate modern permafrost and certainly maintain ancient permafrost. Permafrost does, however, respond to climatic change (Mackay 1975b). For example, in the upper parts of the temperature profiles from northern Alaska (fig. 11.5), the distinct curvatures shown are caused by climatic warming after the permafrost was formed. The equilibrium geothermal gradient is shown by the straight parts of the curves, and their upward projection indicates the prevailing surficial temperatures at the time of permafrost formation (Lachenbruch and Marshall 1969; Gold and Lachenbruch 1973).

Table 11.1 Selected Northern Hemisphere permafrost thicknesses.

Location	Mean Annual Air Temperature	Thickness of Permafrost
Alaska		
Prudhoe Bay (70°N, 148°W)	−7 to 0°C (20 to 32°F)	609 m (2000 ft)
Barrow (71°N, 157°W)	−12 to 7°C (10 to 20°F)	405 m[a] (1330 ft) 16 km (10 miles) inland
Umiat (69°N, 152°W)	−12 to −7°C (10 to 20°F)	322 m (1055 ft) 235 m (770 ft) under Colville River
Cape Thompson (68°N, 166°W)	−12 to −7°C (10 to 20°F)	306 m[a] (1000 ft)
Bethel (60°N, 161°W)	−7 to 0°C (20 to 32°F)	184 m (603 ft) 13 m (42 ft) under Kuskokwim River
Ft. Yukon (66°N, 145°W)	−7 to 0°C (20 to 32°F)	119 m (390 ft) 5.5 m (18 ft) under Yukon River
Fairbanks (64°N, 147°W)	−7 to 0°C (20 to 32°F)	81 m (265 ft)
Kotzebue (67°N, 162°W)	−7 to 0°C (20 to 32°F)	73 m (238 ft)
Nome (64°N, 165°W)	−12 to −7°C (10 to 20°F)	37 m (120 ft)
McKinley Natl. Park-East Side (64°N, 149°W)	extremely variable	30 m (100 ft)
Canada		
Melville Island, N.W.T. (75°N, 111°W)	———	548 m (1800 ft) near coast, probably thicker interiorward
Resolute, N.W.T. (75°N, 95°W)	−16.2°C (2.8°F)	396 m (1300 ft)
Port Radium, N.W.T. (66°N, 118°W)	−7.1°C (19.2°F)	106 m (350 ft)
Ft. Simpson, N.W.T. (61°N, 121°W)	−3.9°C (25.0°F)	91 m (300 ft)
Yellowknife, N.W.T. (62°N, 114°W)	−5.4°C (22.2°F)	61–91 m (200–300 ft)
Schefferville, P.Q. (54°N, 67°W)	−4.5°C (23.9°F)	76 m (250 ft)
Dawson, Y.T. (64°N, 139°W)	−4.6°C (23.6°F)	61 m (200 ft)
Norman Wells, N.W.T. (65°N, 127°W)	−6.2°C (20.8°F)	46–61 m (150–200 ft)
Churchill, Man. (58°N, 94°W)	−7.1°C (19.2°F)	30–61 m (100–200 ft)
U.S.S.R.		
Upper Reaches of Markha River (66°N, 111°E)	———	1500 m (4920 ft)
Udokan (57°N, 120°E)	−12°C (10.4°F)	900 m[a] (2950 ft)
Bakhynay (66°N, 124°E)	−12°C (10.4°F)	650 m (2130 ft)
Isksi (71°N, 129°E)	−14°C (6.8°F)	630 m (2070 ft)
Mirnyy (63°N, 114°E)	−9°C (15.8°F)	550 m (1805 ft)
Ust'-Port (69°N, 84°E)	−11°C (12.2°F)	425 m (1395 ft)
Salekhard (67°N, 67°E)	−7°C (19.4°F)	350 m (1150 ft)
Noril'sk (69°N, 88°E)	−8°C (17.6°F)	325 m (1070 ft)
Yakutsk (62°N, 129°E)	———	195–250 m (650–820 ft)
Vorkuta (67°N, 64°E)	———	130 m (430 ft)

Used with permission of The Association of American Geographers; Commission on College Geography. Resource Paper #14, 1972, L. W. Price.
[a]A calculated depth not actually measured.

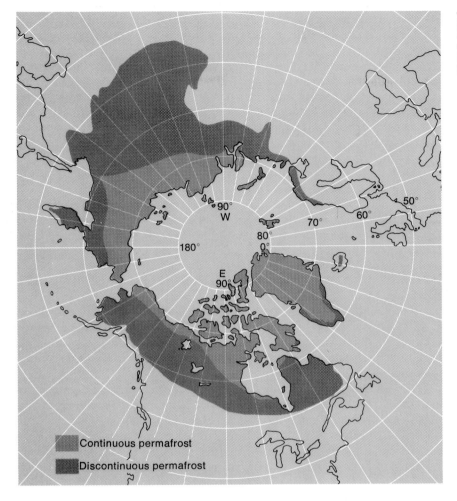

Figure 11.6.
Extent of continuous and discontinuous permafrost in the Northern Hemisphere. (After Ferrians et al. 1969)

The destruction and/or protection of permafrost varies with factors other than climate, such as the type of surface cover. The presence of ice or dense vegetation has a decided influence on permafrost thickness. In Antarctica, for example, some areas beneath the ice sheet may have no permafrost (Ueta and Garfield 1968), while zones that are glacier free have permafrost up to 150 m thick (J. R. Williams 1970).

Permafrost underlies 26 percent of the world's land surface (Black 1954) and so is not, as we may tend to think, an unusual phenomenon. In fact, it is known to exist beneath the ocean in many nearshore polar areas, although it probably formed on land and was submerged during a subsequent rise in sea level. As the map in figure 11.6 shows, most of the known permafrost exists in the polar regions of the Northern Hemisphere, extending to a southernmost limit at a latitude of approximately 55°N in both North America and Eurasia. In China, however, latitudinally controlled permafrost is known to exist as far south as 46°N. Permafrost in these regions is usually divided into continuous

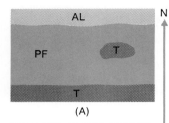

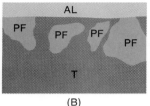

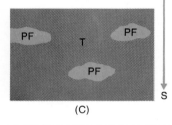

Figure 11.7.
Types of permafrost zones:
(A) continuous, (B) discontinuous,
(C) sporadic. AL = active layer;
PF = permafrost; T = talik.

and discontinuous types (Ray 1951). **Continuous permafrost** usually consists of thick layers of perennially frozen ground that are spread rather evenly under a wide areal surface. The continuous nature of the permafrost alters only where it thins under deep, wide lakes or rivers (fig. 11.4). In contrast, **discontinuous permafrost** is shallower and contains unfrozen zones within the frozen ground or wide gaps that remain unfrozen (fig. 11.7). These unfrozen areas, called **taliks,** increase in size and number southward until true permafrost exists only as isolated patches. Taliks appear as islands exposed at the surface, lenses or layers within the permafrost, or unfrozen ground beneath the permafrost (figs. 11.2, 11.6). The southern limit of continuous permafrost coincides in a general way with the $-6°C$ annual isotherm and the discontinuous boundary with the $-1°C$ isotherm (Brown 1970).

It should not be assumed from the above that permafrost exists only in high latitudes. Isolated zones of frozen ground, called **sporadic permafrost** (see fig. 11.7), can be found far south of latitude 55°N and are probably relicts of a once colder climate. Pockets of permafrost persist also where elevations are sufficiently high. In the United States, for example, permafrost is found near the summit of Mt. Washington in New Hampshire (Goldthwait 1969) and in the Rocky Mountains above the tree line (Ives and Fahey 1971). Permafrost has even been reported near the summit of Mauna Kea, Hawaii (Woodcock et al. 1970; Woodcock 1974), at an elevation of 4170 m. In some areas (for example, the high-altitude plateaus of China), permafrost is so widespread that it is appropriately considered as being continuous or discontinuous rather than sporadic.

The general but imprecise relationship between average air temperature and permafrost boundaries suggests that modern climate determines the distribution of frozen ground (Harris 1983) even though most of the included ice formed at an earlier time. This conclusion fits the interpretation that the recent retreat of the discontinuous permafrost boundary in Manitoba (Canada) is related to a climatic shift that began only about 120 years ago (Thie 1974). As Washburn (1980) suggests, this probably indicates that discontinuous permafrost may be in such a delicate equilibrium with the present climate that only minor changes in climate or surface condition will produce drastic effects.

All other things being equal, the climate most amenable to preservation of permafrost is one with cold, long winters followed by cool, short summers and low precipitation in all seasons (Muller 1947). Other variables, such as vegetation type and density, composition of surface materials, topography, and surface water, exert an influence on the spatial character of permafrost. For example, most permafrost lies beneath the northern boreal forests or, north of the tree line, under a tundra vegetation dominated by low sedges, grasses, and mosses. The boundary between continuous and discontinuous permafrost often parallels the southern limit of the tundra. Because the thermal properties of vegetation control how efficiently temperature can penetrate the ground (Corte 1969; Price 1972), it is not clear whether the climate or the vegetation produced by the climate is the determining factor of permafrost distribution. Probably all factors exert some control, and therefore we cannot expect a precise correlation between air temperature and permafrost boundaries.

Among students of periglacial phenomena, the consensus is that the regime is dominated by the processes of frost action and mass movement. Although these were discussed in an earlier chapter, an examination of their prowess in the periglacial setting can give a different perception and a greater appreciation of their geomorphic significance. Frost action encompasses a group of processes—wedging, heaving, thrusting, cracking—that all serve to prepare bedrock or soil for erosion. Mass movments transport the loosened debris. The two groups of processes function in periglacial environments as they do in temperate zones. As suggested earlier, however, major differences arise because frost action is considerably more severe in periglacial zones, and because mass movements may be intensified during thawing because the material is saturated with excessive water that cannot drain downward through the system. The combination of these factors results in geomorphic features that are unique to periglacial regions.

Frost Action

The driving force in all processes included in the realm of frost action is the growth of ice within a soil or rock. Intuitively we might expect the explanation of freezing in a porous substance to be relatively simple, but in fact, the thermodynamics of the process are very complex (see Miller 1966; Anderson 1968, 1970). In addition, other factors within the system influence the final geomorphic effect of frost action because they respond to freezing in different ways. For example, segregation of ice into discrete lenses generates a stress field that differs from that produced by freezing of disseminated porewater, and the resulting processes also are dissimilar. Whether ice lenses actually form depends mainly on (1) the rate of freezing, (2) the ability of water to be drawn to a central freezing plane or point, and (3) the size of the pore spaces.

Generally, lenses of ice are smaller and less common as depth increases because the greater pressure lowers the freezing point. However, there are exceptions to even this generalization when the soil properties are conducive to ice-lens development. The water in fine-grained sediment (small pore spaces) freezes at lower temperatures and also has a greater propensity to suck water to a central freezing plane. Because of these factors, along with the tendency of fines to hold more water, segregated ice masses are most common in fine-grained sediment, and coarse sediment usually contains interstitial ice. On the other hand, extremely fine-grained sediment may be impermeable. Therefore, ice lenses form preferentially where suction potential and permeability are most advantageously combined, and this usually occurs in deposits of uncemented silt (discussed in Washburn 1980).

In detail, the movement of water to a freezing plane is more complicated than the foregoing discussion might suggest and depends on many factors other than grain size (Konrad and Morgenstern 1983; Rieke et al. 1983). As a result, Konrad and Morgenstern have identified a parameter called the segregation potential to indicate whether a soil will readily form ice lenses (see Konrad and Morgenstern 1983 for discussion and references). The **segregation potential** is the ratio of the rate of water migration to the temperature gradient near the frost front. It assimilates all of the controlling factors and provides an index that can be applied in predicting the susceptibility of soils to frost-related problems.

The expansion associated with freezing exerts pressures that can produce results as variable as shattering of solid rock or physical lifting of the ground surface. Precisely what event will transpire during freezing and subsequent thawing is determined by a multitude of complex interdependent variables that operate within the system.

Frost Wedging   **Frost wedging** is the prying apart of solid material by ice (Washburn 1973). It is synonymous with splitting, riving, and shattering. The precise mechanisms of **frost shattering** are extremely complex (for a discussion, see McGreevey 1981) and well beyond our introductory treatment. Suffice it to say that shattering will occur when the stress generated by changing water into ice exceeds the tensile strength of rocks. Exactly what conditions (rate of freezing and absolute temperatures) will result in shattering are quite variable because the process is also controlled by the internal properties of the rocks (Douglas et al. 1983). Most evidence for the wedging process has been derived from laboratory experiments or is based on theoretical models. It is doubtful that these are directly applicable to the field situation; in fact, Thorn (1979) found little correlation between the experimental work and measurements of shattering in the natural setting. Furthermore, he suggests the possibility that porosity, saturation in the rocks, and freezing intensity may be combined to generate a number of disruptive mechanisms. This tends to reinforce the idea that hydration shattering (White 1976a; see chapter 10) may be an important partner in frost wedging mechanics.

In porous substances that are saturated, frost wedging is facilitated by rapid freezing of water in the near-surface pores. This tends to close the system, allowing the buildup of some extra pressure that is needed to cause shattering. Slow freezing seems to be an effective wedging process only if water is free to migrate to a freezing plane where excess pressure is generated by growth of large ice crystals. As discussed in chapter 10, wedging in cracks takes place only if freezing proceeds from the surface downward (Battle 1960).

Other factors influence frost wedging besides the rate of freezing and the closure of the system. It is known, for example, that a high water content increases the strain on rocks (Mellor 1970) and that most rocks shatter more readily if they are immersed in the fluid (Potts 1970). This probably explains why wedging is often more pronounced at the base of sheer cliffs where groundwater is available than at the top.

Given an abundant supply of water, characteristics of parent material will control the extent of wedging. For example, larger pore size and porosity in a rock will increase its susceptibility to wedging, and sedimentary rocks with fissility of micaceous minerals generally allow easier migration of water and promote greater wedging. In addition, the effect of frost wedging seems to be linked to the number of freeze-thaw cycles. Some caution is required, however, in correlating freeze-thaw frequency and wedging because air temperature is not always a good predictor of rock surface temperature or the temperature within cracks in those rocks (Douglas et al. 1983). Furthermore, a knowledge of rock properties is important because water may freeze at a temperature other than 0°C depending upon those characteristics. Thus, prolonged freezing with extremely low temperature may have more geomorphic significance than frequent freeze-thaw cycles (Rapp 1960; Washburn 1973).

The products of frost wedging are notably angular and range in size from blocks as large as buildings to fine-grained debris. It has been traditionally accepted that the terminal size of frost shattering is silt (Hopkins and Sigafoos 1951; Taber 1953). Although this is probably true in most situations, some workers now believe that clay-sized particles may be formed under certain conditions (McDowall 1960) or that any terminal size is only rarely attained (Potts 1970). In any case, coarse angular debris is common in periglacial regions, accounting for the prevalence of talus rubble at the base of alpine slopes. Wedging also breaks particles loose from bedrock covered by a soil, and these progressively make their way to the surface, creating additional problems for farmers working the land.

Frost Heaving and Thrusting   **Frost heaving** refers to vertical displacement of matter in response to freezing, while **frost thrusting** connotes horizontal movement (Eakin 1916). In the natural setting the two processes are virtually indistinguishable, and their designation as separate processes is probably more imagined than real. Heaving is directly responsible for several phenomena that are commonplace and accentuated in periglacial environments. First, heaving causes the ground surface to move vertically as ice formation expands the ground material. The extent of this displacement depends on the physical variables within the system (Rieke et al. 1983), and the surfaces of adjacent local areas may be lifted at disparate velocities and amounts. Such differential heaving causes building foundations and other types of construction to crack. Second, heaving has the ability to sort heterogeneous debris by forcing larger particles to migrate surfaceward relative to their finer counterparts. The particles may move upward as much as 5 cm a year (Price 1972). It is well to recognize that any object inserted into the ground or dispersed within the near-surface mass is subject to the mechanics of heaving—not only stones of various size that emerge in farmlands or on slopes but also fenceposts, telephone poles, pilings, or anything else that people might conceivably shove into the ground.

The heaving process also functions in indurated solids and is capable of displacing joint-bounded blocks of bedrock, even though these are held rather tightly by the surrounding mass. These upheaved blocks may project up to 1.5 m above the ground and, where soil is absent, give the surface a jagged appearance.

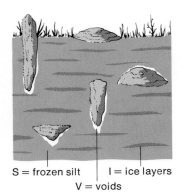

S = frozen silt   I = ice layers
V = voids

Figure 11.8.
Displacement of stones during the freezing process. (After Taber 1943. Courtesy of The Geological Society of America)

In several classic studies of heaving mechanics, Taber (1929, 1930) showed that heaving in a closed system is limited to that produced by the 9 percent volume increase caused when water freezes. In open systems, heaving is much greater because additional force is gained from crystal growth along the freezing plane. The amount of extra heaving depends on how much water can be drawn to the point of freezing from the surrounding material, a function of the segregation potential. Taber also showed that heaving stress due to crystal growth is exerted normal to the freezing isotherm and not in the direction of least resistance. Larger particles of varying compositions will have different thermal conductivities and thus may locally influence the orientation of the cooling surface. This certainly introduces lateral movements in the expansion that are not perpendicular to the freezing isotherm. Some workers also disagree with Taber's conclusion that differences in resistance to expansion do not influence the direction of the heave (Beskow 1947).

Any theory attempting to explain why large particles are moved surfaceward in a nonsorted soil mass must consider not only the forces that lift the clast but also the reasons it does not return to its original position during contraction. Details of the two plausible theories concerning heaving mechanics, called *frost pull* and *frost push,* are discussed by Washburn (1973). Briefly, the frost-pull hypothesis suggests that stones are lifted vertically along with the fines when the ground expands during freezing. The fine sediment, being more cohesive, is brought downward quickly upon thawing and collapses around the larger clasts while the bases of the stones are still frozen. Cavities formed when a large particle is heaved (fig. 11.8) may also be partly closed by thrusting into this zone of lowered resistance.

The frost-push hypothesis is based on the fact that individual stones are better conductors of heat than porous soil. As a result, stones will cool more quickly, and the first ice to form along the freezing plane will be adjacent to and at the base of the stones, thereby pushing them upward. The phenomenon of ice forming preferentially near stones embedded in a silty soil has been observed and is not conjecture; nor is it conjecture that thawing occurs first around the stones. Presumably, however, material thawed adjacent to the top of the stone will collapse against the upper part of the clast while the base remains frozen and thus prevent its return to the original position. The shape of the stone may become important in this procedure; for instance, wedge-shaped particles with the narrow edge projecting downward will have greater difficulty returning to their original level.

Evidence exists to support both the frost-pull and frost-push theories, and probably both processes function simultaneously in a heaving environment. Washburn feels, however, that rapid heaving or movement that breaks a vegetal cover probably requires that frost push be dominant. Slow ejection of clasts can probably be accomplished by frost pull with little necessity for rapid ice buildup beneath the stones.

Other factors affect the magnitude of frost heave and its continuation in the upper part of the soil. Experiments conducted in Greenland by Washburn (1967) showed that moisture content, vegetation, and depth are all critical variables in the heaving process. The greatest heave occurred in zones that had abundant moisture, and deeper objects in an active layer were displaced farther than shallow ones. However, if objects are inserted into the permafrost itself, heaving becomes negligible. In addition, ice lensing and heave are probably dependent on soil properties (thermal conductivity and grain size) and external factors other than surface temperature, such as overburden pressure (Gilpin 1980). Heaving is generally less when the vegetation cover is dense, presumably because of the insulation provided by the vegetal mat. At or near the surface another form of ice, called *needle ice* (or piprake), aids in the lifting process. Needle ice consists of groups of slender ice crystals that are usually 1–3 cm long, although they have been observed to attain lengths up to 40 cm (Troll 1958). The ice clusters are capable of lifting stones up to cobble size. Needle ice, which forms best in moist, loamy soils with no vegetation, also is very important as an aid in mass movement.

Frost Cracking   **Frost cracking** is the development of fractures at very low temperatures. The process seems to function best in permafrost regions, although it has been reported from other environments. Frost cracking is usually considered as a frost action phenomenon, but Price (1972) points out that in fact it is different because it results from thermal contraction rather than expansion associated with freezing.

At very low temperatures frozen ground often evolves into a polygonal network of contraction fractures, but cracking may depend more on the rate of cooling than on the absolute temperature value at the moment of fracture (Lachenbruch 1966; Black 1963, 1969). Initial cracking at the surface may extend to a depth of 3 m and can become progressively wider and deeper. The results of frost cracking will be discussed more in a later section when we examine polygonal ground features.

Mass Movements

Any variety of mass movement (discussed in chapter 4) can occur in a periglacial environment, but there is little doubt that two types, frost creep and solifluction, dominate the cold regime. The movement of debris occurs simultaneously with the frost action processes, and it is questionable how effective the erosion would be if it operated alone. Mass movements, however, are responsible for a variety of landforms that are unique to periglacial regions.

Frost Creep   **Frost creep,** like any form of near-surface creep, is the downslope movement of particles in response to expansion and contraction and under the influence of gravity. It is unique only because freezing and thawing generate the cycle of expanding and contracting. As explained earlier, the freezing front in the soil usually parallels the ground surface. Individual soil particles are heaved perpendicular to the surface, but during thawing they are affected only by gravity and should contract in a vertical direction (shown earlier in fig. 4.20). Actually, a component of upslope migration has been observed in the contracting phase (Washburn 1967; Benedict 1970). This motion, called

Figure 11.9.
Route of downslope movement followed by a particle under combined frost creep and solifluction.

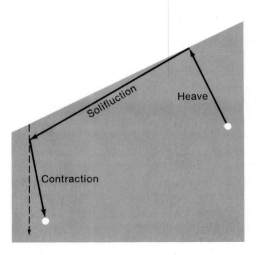

retrograde movement, probably stems from the attraction of fine-grained particles for one another and the cohesive strength developed between them. On the other hand, some downslope flowage may occur during the thaw, and so the overall route followed by any soil particle involved in frost creep probably resembles the path shown in figure 11.9.

Detailed studies in northeast Greenland (Washburn 1967) showed frost creep to be the dominant process where soils are silty and slopes stand between 10° and 14°. It was noted, however, that in any given year other processes could dominate the system. Therefore, it may not be safe to generalize about the conditions that promote frost creep because soil texture, moisture content, vegetation, and freeze-thaw frequency all have some bearing on the efficacy of the process.

Solifluction (Gelifluction)   The term **solifluction** was first proposed to describe the action of slow flowage in saturated soils. The original meaning carried with it no climatic restrictions, but over the years the term has generally been used to connote a process that functions in periglacial regions. We will follow this practice here, although you should realize that it is technically incorrect because solifluction can occur anywhere. The term **gelifluction** (Baulig 1957) refers to soil flowage associated with frozen ground and as such is a specific type of solifluction.

Solifluction is most dramatic in permafrost zones because water released in the active layer during thawing cannot penetrate below the permafrost table. Soils in the summer are often saturated, and the concomitant loss of friction and cohesion causes them to behave like viscous fluids. Such materials can flow on slopes as gentle as 1°, but maximum displacement seems to occur where the gradient is between 5° and 20°. Soils on slopes steeper than 20° tend to drain easily, and water escapes as surface runoff.

It is clear that abundant moisture is the overriding factor in the solifluction process (Washburn 1967; Chambers 1970; Price 1972). Many students of periglacial geomorphology feel that significant flow can occur only when the moisture content equals or exceeds the liquid limit, but this interpretation

Table 11.2 Representative values of combined frost creep and gelifluction.[a]

Rate (cm/yr)	Slope	Source
2.0	15°	Rapp 1960
0.9–3.7	10°–14°	Washburn 1967
1.0–3.0	3°–4°	Jahn 1960
5.0–12.0	7°–15°	Jahn 1960
10.0	20°	P. J. Williams 1966[b]
2.5	6°–13°	Benedict 1970
2.6	2.5°–4°	French 1974

[a]List is intended as random sampling and is not complete. Great variability exists depending on differences in ground moisture, vegetation, depth, slope face direction, grain size and soil texture, etc.
[b]Pebbles on bare slope surface.

cannot be accepted unequivocally because solifluction has been observed at lower moisture contents (Fitze 1971). Obviously, other factors such as grain size, slope angle, and vegetation play a modifying role on a local basis (Harris, 1973). For example, gravel and coarse sands are so permeable and easily drained that they virtually never flow. Clays, on the other hand, are extremely cohesive. Thus, silty soils, especially those with a bimodal size distribution, are most susceptible to solifluction. Nonetheless, the overwhelming importance of water content was clearly demonstrated in Greenland (Washburn 1967), where the highest rates of movement commonly occur on the most densely vegetated areas of the slopes and at gradients that are lower than the dry portions of the slopes. This indicates that moisture can overcome both the binding effect of vegetation and the apparent deficiency of a transporting gradient.

The rates of combined frost creep and gelifluction in northeast Greenland and other regions are shown in table 11.2. It is difficult to ascertain how much of the total movement is accomplished by frost creep and how much can be attributed solely to flow.

Landforms Associated with Permafrost

Periglacial Landforms

Some periglacial features are so closely allied to the distribution of permafrost that a genetic relationship can hardly be denied, and observation of these features preserved in temperate regions affords the best evidence known for reconstructing the areal extent of ancient permafrost boundaries. The basic processes involved in the development of these features are no different from those that form other periglacial landforms; that is, they rely on frost action and/or mass movement. The permafrost layer, however, places a distinctive imprint on the system and allows us to consider the geomorphic features as a separate group.

Ice Wedges and Ice-Wedge Polygons   When thermal tension exceeds the strength of the surface materials, frost cracking begins and fractures penetrate into the active layer and the upper portion of the permafrost. At first the cracks are only millimeters wide, but they may extend vertically downward to depths of several meters. The actual moment of cracking seems to occur sometime between January and March when temperatures in the ground reach their annual minima (Mackay 1974) and the rate of temperature drop may be extreme. During spring and early summer the crack may be partially or totally filled with snow that has filtered down from the surface and with water released by the onset of thawing in the active layer. As the diagrams in figure 11.9 show, this mixture freezes below the permafrost table and occupies the initial crack as an ice veinlet. Because the initial fracture now represents a zone of weakness within the permafrost, subsequent cold winters produce cracking along the trace of the original opening, and new ice is added as before. Over the years, this accumulation of relatively pure and vertically oriented ice takes the form of a downward-tapering wedge, called an **ice wedge,** shown in the diagrams (fig. 11.10) and pictured in figure 11.1. In dry permafrost, the material from the contraction fissures may be sediment, especially wind-transported sand or loess associated with extreme aridity (for example, see Carter 1983). The resulting feature is commonly called a *sand wedge* (Péwé 1959).

Ice wedges range from 1 cm to 3 m wide and from 1 to 10 m deep, with the depth to top-width ratio normally ranging from 3:1 to 6:1. They are best

Figure 11.10.
Evolution of an ice wedge by thermal contraction. (After Lachenbruch 1966)

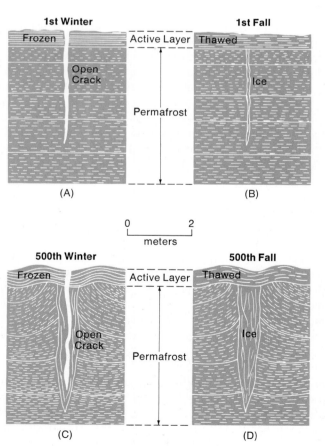

developed in fine-grained soils with a high ice content (Black 1976). The host sediments are usually upturned within 3 m of the wedge, in response to the horizontal compression generated as the wedge grows. Ice in wedges contains oriented air bubbles and a complicated fabric of ice crystals (Black 1974).

The growth of an ice wedge does not necessarily proceed on an annual basis. Mackay (1974, 1975a) observed that only 40 percent of the ice wedges on Garry Island (Northwest Territories, Canada) cracked in any given year, the frequency of cracking apparently controlled by the depth of snow cover. Even if cracking does occur, the openings often narrow before ice-veinlet growth begins. The size of a veinlet thus is dependent on the timing between closure of the winter cracks and the appearance of meltwater (Mackay 1975). In most cases the thickness of an annual increment is considerably less than the size of the thermal crack.

Ice wedges are commonly connected in a polygonal pattern that is similar in most respects to the more familiar mud cracks. They differ mainly in that **ice-wedge polygons** are much larger, having diameters that range from several meters to more than 100 meters (fig. 11.11), and they contract and expand according to the laws that govern temperature effects on ice (Black 1974).

Figure 11.11.
Polygonal markings on ground surface caused by ice-wedge polygons in the vicinity of Meade River, about 35 miles southeast of Barrow, Alaska. Barrow district, Northern Alaska region, July 1949.

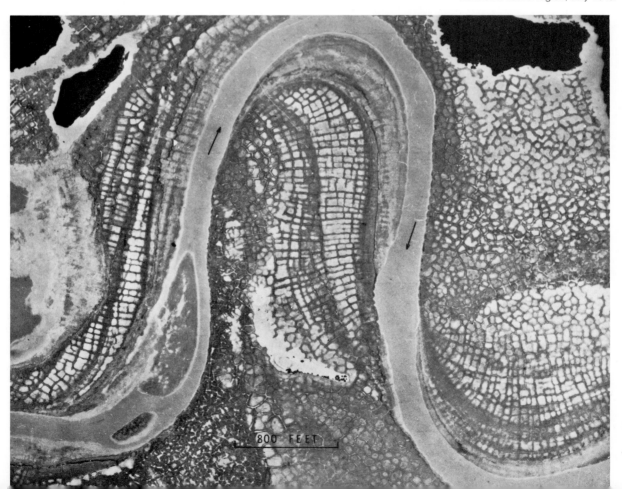

The perimeters of some polygons are accentuated as minor ridges, so that the boundary is elevated relative to the center of the feature *(low-center ice-wedge polygons)*. Others *(high-center polygons)* are depressed along their edges by melting and erosion of the ice wedges, leaving the central core higher than the rim (Péwé 1966; Black 1974, 1976).

Actively growing ice wedges are restricted to areas of continuous permafrost, and their southern boundary seems to be the $-6°C$ to $-8°C$ isotherms (Péwé 1973). Farther south in the discontinuous permafrost zone, they become inactive and disappear. As ice wedges melt, the surrounding sediment may collapse into the space opened by the disappearance of the ice, thereby creating an ice-wedge cast, a feature that is best preserved in gravels (Black 1976). Although ice-wedge casts are probably the best proof of an earlier permafrost condition, other wedgelike features unrelated to permafrost look very similar. For example, features called **soil wedges** are commonly reported in the Soviet literature. They apparently originate by frost cracking in the active layer and in-filling with local sediment. Therefore, it is easy to misinterpret these features as ice-wedge casts, which incorrectly suggests that a permafrost environment existed in the past and has subsequently been destroyed. In fact, it is possible for an old ice-wedge cast to become part of a modern soil wedge if frost cracking operates in the present active layer. To avoid confusion, Washburn (1980) suggests that any wedge or wedge-cast type should be designated as being either fossil or contemporaneous.

Pingos   The term **pingo** (Eskimo for "mound" or "hill") was first suggested in 1938 by Porslid for large, ice-cored, domelike features that exist only in permafrost regions. They are most perfectly developed in areas of continuous permafrost, although they do form in the discontinuous zone (Holmes et al. 1968). Active pingos rise from a few meters up to heights of 70 meters and have basal diameters up to 600 meters (Washburn 1980). They are oval or circular in plan view, as in the photograph in figure 11.12. The ice core in a pingo is typically massive and is often exposed by tension fractures that develop at the summit of the mound as it rises. Sometimes the exposed ice melts, and small freshwater lakes occupy the craters bounded by the tensional cracks.

Many varieties of pingos are known (Pissart 1970; Flemal 1976) but most can be placed in two major genetic categories. *Closed-system* pingos (Mackenzie type) develop in level, poorly drained, shallow lake basins. They are most distinctly preserved in the Mackenzie delta (Canada) where a clutch of 1,450 pingos gives evidence that pingos tend to form in fields rather than as single individuals. In general, the draining of a lake allows the permafrost table to rise to the level of the former lake floor. As the table rises, water is trapped in the saturated soil. The cryostatic pressure (pressure from ice formation) displaces the water upward until it also freezes and becomes the core of the pingo (see Flemal 1976). The size and shape of the pingo often reflect that of the original residual pond; the pingo may grow in distinct stages (Mackay 1973).

Figure 11.12.
The Discovery pingo. A dense
stand of large birch trees on the
pingo contrasts sharply with the
open stand of small black spruce
in the surrounding muskeg.
Goodpaster district, Yukon region,
Alaska, Aug. 1960.

Open-system pingos (East Greenland type) develop most readily on slopes where free water under artesian pressure is injected into the site of the pingo. As the water approaches the surface, it freezes. The continuing introduction of water from below under hydrostatic pressure builds up the ice mass and domes the surface material into the pingo shape (see Müller 1963). The initial growth of some pingos may involve ice lenses above the water table, but their continued growth requires a proper combination of hydrology and soil texture (Ryckborst 1975). Open-system pingos are much more common in discontinuous permafrost where taliks and freely circulating water are not unusual.

The rate of pingo growth is variable, ranging from a few centimeters to more than a meter a year (Washburn 1980; Mackay 1973). This indicates that many large pingos are quite old. In fact, ^{14}C dates have shown two closed-system pingos to be 4,000 and 7,000–10,000 years old (Müller 1962). On the other hand, some pingos are growing today (Mackay 1973), suggesting that pingo formation did not occur in a unique or specific time in postglacial history but is probably a continuing process.

Fossil pingos (Wayne 1967; Mullenders and Gullentops 1969; Flemal et al. 1976) provide good evidence of an extinct permafrost environment in regions that are now temperate. Proving that features are fossil pingos, however, requires detailed field and laboratory analyses. The DeKalb mounds (northern Illinois), for example, consist of approximately 500 circular or elliptical mounds shaped from late Pleistocene sediments. Good evidence for their pingo origin is the large cluster of features and the fact that their centers consist of lacustrine silts and clays. Topographically, however, most of the mounds are less than 5 m higher than the surrounding ground level, and many are up to 300 m in diameter. Their recognition as fossil pingos could not have been suggested without careful field examination.

Thermokarst   Thermokarst features are a variety of topographic depressions that result from thawing of ground ice. They are more abundant where ice wedges are present and where the thermal equilibrium is somehow disturbed. The destruction of ground ice is due either to lateral degradation (backwearing) where lateral erosion of surface water exposes the ice, or to vertical degradation (down-wearing) when the surface thermal properties are altered (discussed in Czudek and Demek 1970). These features are similar in many respects to those found in normal karst regions, but the fundamental process is melting rather than solution of rock material.

Many of the processes that culminate in thermokarst features are initiated by broad climatic changes, but minor events such as fires, clearing of forest vegetation for farming, shifting stream channels, etc., probably also can upset the thermal balance enough to be the impetus. Mackay (1970), for example, reports a remarkable case in which trampling of vegetation by a dog led to subsidence of 18–23 cm in a small area within two years. Observations such as this demonstrate the remarkably fragile equilibrium of the periglacial system.

Patterned Ground

Periglacial regions are often characterized by a peculiar arrangement of surface materials into distinct geometric shapes. The features, collectively known as **patterned ground,** include such diverse shapes as polygons, circles, and stripes. They are common in permafrost regions, but perennially frozen ground is not a prerequisite for their development. They can be found anywhere within the periglacial regime or even in other morphogenetic regimes. Since patterned ground was first described in the late 1800s, almost every worker in periglacial areas has noted its striking appearance, and theories concerning its origin are almost as numerous as the features themselves. Perhaps the most extensive review of patterned ground and its origin was provided by A. L. Washburn in 1956, and that excellent work still stands as the cornerstone of our knowledge about these forms.

Classification   According to Washburn (1956), patterned ground features can be placed in a descriptive classification on the basis of two primary criteria: (1) the geometric shape (circle, polygon, etc.), and (2) whether the material composing the feature has been sorted (table 11.3). Sorting separates the larger-size fractions from the fine particles and usually generates a feature rimmed by stones and centered with fines (for a review see Goldthwait 1976). The most common forms seem to be circles, polygons, and stripes, with steps and nets as minor transitional types. For a more detailed discussion see Washburn (1980).

Table 11.3 Simplified classification of patterned ground features.

Types	Subtypes	Processes
Circles	Sorted Nonsorted	Features subdivided on basis of necessity of cracking (frost cracking, permafrost cracking, dilation, joint control). Where cracking not essential, heave and mass displacement important.
Polygons	Sorted Nonsorted	Includes ice and sand wedges. Cracking of all types. Heave, mass displacement, and thaw processes important in noncracked types.
Nets	Sorted Nonsorted	Includes earth hummocks. Cracking of all types except joint controlled types. Heave and mass displacement in noncracked types. Thaw also important in sorted varieties.
Steps	Sorted Nonsorted	Cracking unimportant. Heave, mass wasting, and displacement in nonsorted. Frost sorting and thaw also important in sorted varieties. Terracette form.
Stripes	Sorted Nonsorted	All types of cracking important. Heaving, mass wasting, and displacement in nonsorted. Frost sorting and thaw also important in sorted types.

After Washburn 1970, in Acta Geographica Lodziensia. Used with permission of Institute of Geography, Poland.

Polygons   Polygons are best developed on flat, nearly horizontal, surfaces. The sorted variety is bounded by straight segments composed of stones that surround a central core of finer material. They range in diameter from a few centimeters to over 10 m and always occur in groups rather than individually. The stones marking the boundary are often oriented parallel to the direction of the rim. The stones increase in size with the dimension of the polygon and decrease in size with depth.

Nonsorted polygons differ in several ways from the sorted type: (1) the polygonal borders are devoid of stones, the geometric form being marked by furrows, (2) they can be considerably larger than the sorted variety, often reaching 100 m in diameter, and (3) they have been found on slopes as steep as 31°. Nonsorted polygons are usually associated with the ice-wedge polygons discussed earlier. As such they are primarily related to permafrost. However, they can also form by dessication and even in solid rock (Walters 1978).

Circles   Sorted circles are stone-rimmed circular forms that range from a few centimeters to several meters in diameter (fig. 11.13). As with polygons, the stone size correlates with the dimension of the circle and decreases with depth. Circles, unlike polygons, can exist singly or in groups. Nonsorted circles lack the coarse, bouldery rim and are bounded instead by vegetation surrounding a central core of bare soil. In some cases the barren core develops because scouring and winnowing action by the wind preferentially erodes and initiates the circular form (Fahey 1975).

Stripes   Stripes are linear alignments of stones, vegetation, or soil on slopes. The sorted stripes are elongated strips of stones separated by intervening zones of fine sediment or vegetation. In general, the pattern changes as the slope increases, from polygons through sorted nets or steps into stripes. Stripes usually range in width from several centimeters to several meters. Their length varies but is sometimes greater than 100 m. As in other sorted features, larger stones are associated with bigger stripes, and the largest particles are found at the surface of the accumulation. Nonsorted stripes consist of vegetation or soil with intervening zones of bare ground. They are usually not as long as the sorted variety and are sometimes discontinuous.

Origin   Patterned ground probably originates in a variety of ways; in fact, Washburn (1956) discusses 19 major hypotheses and concludes that patterned ground is polygenetic and that some forms result from a special combination of processes. Despite the conflicting opinions, the properties of the major features do imply that patterned ground exists in a range of transitional types. The characteristics of the different types reflect variations in a few controlling factors rather than completely different origins. Although details are still missing, it does seem certain that several processes are basic to the formation

Figure 11.13.
Sorted stone circles caused by frost action, in Alaska Range, south central Alaska. June 1968.

of patterned ground: (1) cracking or dessication, (2) heaving and its associated sorting, and (3) gravity as expressed by the slope inclination. Cracking is instrumental in developing polygonal geometry and is primary to both sorted and nonsorted types. Heaving separates the coarse and fine sediment fractions. In the usual case, repeated freezing and thawing move the stones upward and outward. This creates a fine-grained nucleus as the smaller grains with greater cohesion contract farther inward and downward during the thaw phase. The process continues until adjacent cells coalesce, and the coarse stones are concentrated as polygonal or circular rims along the lines of interference. This model is accepted by most periglacial students although some evidence exists to suggest that texturally unsorted debris does not always respond to frost action in the same way (Ballard 1973).

The rate of lateral movement of large clasts is not well documented. Recent detailed work by Vitek (1983) suggests that particles larger than 1.3 cm in length are transported to the polygonal margins at a rate ranging from 0.45 to 0.89 cm/yr. Larger particles move at a slower rate. The clasts are moved by frost action and the growth of needle ice and/or lacustrine ice.

In recent analyses (Ray et al. 1983a, 1983b), the regularity in sorted patterned ground has been theoretically explained as resulting from convection of water in the active layer. Convection cells are presumed to form when ice melts in an uneven manner during the summer thaw. This creates an undulating frozen ground surface beneath the active zone and temperature and density gradients in the water of the active layer. The model does not suggest that stones are physically moved by convecting water; instead, it contends that circulating water influences the shape of the underlying ice front. This, in turn, determines where sorting mechanisms will work most effectively until the surface pattern is a mirror image of the undulating ice front. In limited field studies, the model has correctly predicted the width to depth-of-sorting ratio for sorted polygons.

The origin of stripes is clearly related to slope angle. Polygons and circles are much more prevalent on flat surfaces, and stripes become dominant as slope angle increases. This does not mean that stripes never form on gentle slopes; they do (Evans 1976). However, such occurrences are rare, and as gradients increase, stripes are the most common form of patterned ground. Usually transitional forms of nets or steps develop at angles between 2° and 7°; the precise angle at which stripes develop depends on other factors such as soil texture and moisture and the formation of needle ice (Mackay and Matthews 1974; Washburn 1980).

Landforms Associated with Mass Movement

In contrast to patterned ground features, which are shaped primarily by frost action, certain periglacial forms result when the dominant mechanism is transportation. This does not mean that frost action is absent. On the contrary, frost action is a necessary prelude to the mass movement, and probably the two processes function simultaneously to fabricate the ultimate landform. Nonetheless, the final geometry of these features reflects the motion of rock and soil particles rather than a static, in situ disintegration of the parent material (for reviews see Benedict 1976 and S. E. White 1976b).

Gelifluction Features   The process of gelifluction operates on gradients as low as 1° to 2°. Where the process is active a number of deposits and surficial forms result *(gelifluction sheets, gelifluction benches, gelifluction lobes* and *gelifluction streams).* Although these features can be separated on the basis of individual morphology (see Washburn 1980), they share common genetic mechanics. Variations in form result from differences in texture, gradient, and soil moisture. Material in gelifluction deposits is usually poorly sorted, but a crude stratification may be present in some deposits. Angular clasts are normally oriented with long axes arranged parallel to the direction of movement.

In regions covered by arctic or alpine tundra the most common feature seems to be the **gelifluction** (solifluction) **lobe.** The large, tonguelike masses of surface debris may occur as a single lobe, usually 30 to 50 m wide, or as one of many individuals that are joined laterally into a much broader field. In long profile an individual lobe is marked by a pronounced scarp at its leading edge that ranges from 1 to 6 m high. Commonly, however, a succession of lobes develops on a slope in which each upstream lobe overlaps the rear of the next downslope lobe, giving the entire sequence a staircase profile. Internally the lobes consist of angular unsorted debris, with particle sizes varying from silt to coarse boulders. The long axes of the boulders are usually oriented parallel to the direction of movement. Various types of gelifluction lobes and related forms (turf-banked lobes and terraces, stone-banked lobes and terraces) are discussed in detail by Benedict (1970, 1976).

Gelifluction lobes, as you might expect, move primarily by gelifluction, but other processes are undoubtedly involved. Flow mechanics is indicated by the fact that the greatest movement occurs near the surface, and essentially no movement is perceived below a depth of 25 cm (Price 1972). As the material moves downslope, it encounters zones of high resistance to flowage. These retard the forward motion and cause the soliflucting layer immediately upslope from the resistant zone to bulge slightly, producing the scarplike front. As Price (1972) points out, however, the front is often stabilized, and only a thin surface layer slides to and cascades down the stationary front. The advance of the lobe in this manner is much slower than one might expect, probably averaging 1–3 cm a year, but the rate varies according to where the measurement is made. Maximum velocities are found along the longitudinal axis and decrease progressively toward the lateral margins (Washburn 1973), but rates are significantly affected by moisture content and gradient (Benedict 1970) in addition to vegetation cover and depth of freezing (Gamper 1983).

Blockfields   Blockfields (often known as *felsenmeer*) are usually thought of as broad, relatively level areas covered by moderate to large angular blocks of rock (Sharpe 1938). Similar accumulations of blocky debris also are found on slopes and are variously referred to as *block slopes, block streams, rubble sheets,* and *rubble streams* (see Washburn 1973). All these terms have been used interchangeably, and some semantic confusion exists. Our concern here is with block accumulations that occur on slopes, either concentrated in valleys or disseminated across a wide area. Although we may employ the term "blockfield" in reference to these accumulations (as have many others), some

workers would consider this technically incorrect. True blockfields as originally described are not the product of mass movement but result in situ from
frost wedging and heaving of the underlying bedrock. In this discussion we
will actually be treating block streams or block slopes, but the term "blockfield" has been so widely applied to these features that we will employ it to
mean any accumulation of blocky debris, including those found on slopes and
subjected to mass movement.

Blockfields on slopes tend to be elongate with widths averaging between
60 m and 120 m and lengths from 350 m to 1.3 km. Maximum and minimum
values vary drastically depending on the block size, the slope angle, and the
distribution of the bedrock source. The gradient on the accumulation ranges
widely, from 1° to 20°, although it most commonly is between 3° and 12°.
Some workers feel that an appropriate upper limit for these features is 10°
to 15° (Caine 1968; White 1976). Blocky slope accumulations normally display an internal fabric indicative of movement; i.e., long axes of individual
blocks are oriented downslope. The orientation may become transverse to the
slope near the distal end of the deposit (Caine 1968), or it may diverge locally
from the norm. Accumulations seem to be 4–20 m thick, but very few observations of the total thickness have been made. One deposit that was totally
exposed displayed an internal layering characterized by an open-textured surface zone and a general increase in matrix with depth (Caine 1968). This type
of fabric distribution, however, cannot be accepted as a general trait until more
observations become available. Individual blocks contained in boulder masses
are striking in their enormity, most being 1–3 m in intermediate diameter and
some as large as 6 m (fig. 11.14).

A number of mass-movement processes—ranging from avalanches or
landslides to catastrophic floods—have been proposed for the development of
block slopes or streams. Precisely which mechanism or combination of processes leads to the features is mostly conjecture, and the distinct probability
remains that block accumulations have more than one mode of origin. The
most commonly cited transporting mechanism is gelifluction. Caine (1968),
for example, concluded that in Tasmania the block slope material moved when
the entire deposit contained abundant matrix. The open texture in the surface
zone (up to 3 m deep) could have been created after movement ceased when
the matrix was removed by ground or surface water (Caine 1968; Potter and
Moss 1968). The gelifluction model will work efficiently only if the lower portions of the accumulation contain a high content of fine-grained matrix. Such
a layer was observed by Caine (1968) and suggested by Denny (1956) and
Sevon (1969). However, it has not been proven as a universal property of these
features, and until more conclusive data are found, the solifluction mechanism
can be only a viable hypothesis. Nonetheless, it would explain the terrace and
lobe development noted on some block streams (Denny 1956; Potter and Moss
1968). Further, the orientation of blocks is similar to that characteristically
produced by gelifluction transport.

Some investigators have suggested that slope blockfields are emplaced by movement as **rock glaciers,** i.e., an accumulation containing interstitial ice rather than a soil matrix (Patton 1910; Kesseli 1941; Blagborough and Farkas 1968). Such a mechanism has the advantage of easily accounting for the open texture in the blocky debris, since no matrix sediment must be removed. The ice simply melts away as the rock glacier becomes inactive. A rock glacier transport, however, fails to explain why slope blockfields lack such normal characteristics of rock glacier deposits as steep fronts and pronounced transverse ridges and furrows. In addition, a greater shear stress is required in moving a rock glacier than in gelifluction of a fine soil, and so, to attain the proper mobility, rock glacier accumulation should be considerably thicker than blockfield deposits (Wahrhaftig and Cox 1959; Potter and Moss 1968).

In comparison with inactive deposits, active blockfields conspicuously lack lichen cover, and the clasts show no signs of weathering. Because of the wide distribution of active block features in modern periglacial regions, inactive deposits in temperate zones or at altitudes well below functioning periglaciation have been widely interpreted as evidence of earlier periglacial conditions. For example, many blockfields have been described in the middle Appalachian Mountains (H. T. U. Smith 1953; Denny 1956; Potter and Moss 1968) where present climates are patently not periglacial.

Figure 11.14.
River of Rocks blockfield (block stream), Berks County, Pa. Slope is 5°. Blocks in foreground are as much as 5 m long.

Rock Glaciers   Rock glaciers are features having a bouldery surface composed of angular rock fragments that are commonly up to 3 m in diameter. This texture is somewhat misleading because many rock glaciers contain finer-grained debris at depth. The features usually originate in cirques or in high, steep-walled recesses.

Rock glaciers are usually classified as **tongue-shaped** or **lobate** on the basis of their plan-view shape (see White 1976b for discussion). Tongue-shaped forms often head in cirques and extend downvalley, maintaining an external shape that is similar to a true glacier. In contast, lobate rock glaciers are short and broad and develop below talus accumulations along valley sides having steep cliffs.

Tongue-shaped rock glaciers are several hundred meters to more than a kilometer long and vary in width, narrowing toward their source where the valley sides are more constricting (fig. 11.15). The deposits contain ice several meters below the surface. According to Potter (1972), the ice exists either as an interstitial cement (ice-cemented type) or as an extensive mass that contains widely dispersed rock fragments (ice-cored type). In either case the features may represent a transitional phenomenon between a nonglacial process and a glacier with a thick mantle of surface debris or, alternatively, an ice-cored moraine (for discussion see Barsch 1971; Ostrem 1971). Topographi-

Figure 11.15.
Rock glacier on Sourdough Mountain near McCarthy, Wrangell Mountains, Alaska.

cally, any rock glacier may have furrows, crevasses, and lobate or transverse ridges similar to those found on true glaciers. Their front is normally near the angle of repose and therefore quite steep.

Modern, active rock glaciers are known in polar or subpolar regions and in the high mountains of the mid-latitudes. They vary in thickness from 15 to 50 m and move continuously at rates usually ranging from centimeters to a meter per year, depending on local conditions. In some cases they are known to be advancing in areas where glaciers are presently retreating, perhaps because the surface rubble insulates the underlying ice and slows its response to climatic change (Osborn 1975). Inactive or fossil rock glaciers are widely distributed in regions that now have temperate climates. Whether they have paleoclimatic significance depends on precisely how such features originate. Although a variety of origins have been suggested, only those processes that produce a continuous forward motion are acceptable. The main controversy surrounding the genesis of rock glaciers revolves around the question of whether or not a true glacier is a required precursor (Outcalt and Benedict 1965). In other words, is the ice contained within the rock glacier necessarily glacier ice, or could it have accreted in a periglacial setting by gradual freezing of porewater? The fact that some fossil rock glaciers in the southwestern United States developed in regions having no evidence of glaciation (Blagborough and Farkas 1968; Barsch and Updike 1971) seems to support a completely periglacial origin for at least some of the features. It seems reasonable, however, that most tongue-shaped rock glaciers are somehow related to true glaciers. Lobate types may be ice-cemented and move by creep generated by the weight of the bouldery mass on the interstitial ice (White 1976b; Washburn 1980; Wayne 1981). Nonetheless, enough disagreement exists about the origin and transport mechanics of these deposits to refrain from sweeping conclusions at this time (see Whalley 1983 for discussion).

Environmental and Engineering Considerations

As mentioned earlier, human expansion into the arctic and subarctic regions is inevitable. This impending population growth has generated a growing concern as to whether our perception of geomorphic systems operating in these areas is sophisticated enough to develop the regions and still preserve their fragile environmental balance. The concerns are most evident in the engineering community that deals with construction problems in arctic regions on a day-to-day basis (for many references see National Academy of Sciences 1983). The technical problems are real; our ability to cope with them depends on how well we understand the processes functioning within the system and how those processes are ultimately related to the environmental variables.

Most engineering problems are associated with permafrost. According to Brown (1970), any one of four basic engineering approaches may be employed when dealing with the permafrost condition: (1) disregard the permafrost; (2) use an *active* approach in which the permafrost in the near surface is eliminated by keeping the soil continuously thawed or by removing the natural soil and replacing it with permafrost-resistant material; (3) use a *passive* approach in which permafrost is preserved by keeping the soil frozen at all times and thereby eliminating the annual thaw; and (4) design structures to withstand frost action.

Which of the above methods is best depends on the local situation. In discontinuous permafrost, it may be possible to ignore the condition if the parent material is bedrock or a sandy or gravelly soil. Active approaches, such as simply removing the vegetation and its insulating effect, are also feasible on a local scale. In zones of continuous permafrost, the potential problems cannot be disregarded. There the most practical method is to keep the area continuously frozen, primarily by ventilation or insulation.

Some heaving and settling will occur in the first several years after construction regardless of the method employed, because the permafrost requires some time to adjust into a new equilibrium condition. Thus, any project should prepare for the initial response in the system with a design that can withstand the stresses generated. Often the periglacial features described earlier give tangible clues about the near-surface ground conditions at any locality (table 11.4), and their recognition is an important first step in a site analysis (Thomas and Ferrell 1983). Air photos and remote sensing data interpreted by a geomorphologist can provide a valuable contribution in the planning stage.

Building Foundations

The bearing strength (ability to support load without plastic deformation) of a surface varies significantly with the type of material, both when it is frozen and when it thaws (Swinzow 1969). The loss of strength during thaw is most dramatic in clay and silt soils that are not permeable and so retain a large portion of the water released by melting. Bedrock or coarse-grained deposits are reasonably stable during the thaw phase because they drain easily. Because water cannot permeate below the permafrost table, the active layer in fine-grained soils becomes a quagmire that cannot support weight and therefore allows differential settling of overlying buildings, bridge abutments, etc. In addition, the lack of lateral drainage and the high segregation potential will accentuate differential heaving during the next freeze. Building foundations, therefore, present enormous problems. In the case of small dwellings it may be adequate simply to live with the frost activity by pursuing an annual maintenance program. Large buildings, however, become unsafe if they are allowed to endure substantial heaving and settling, and preventive measures must be taken. Lobacz and Quinn (1963) were able to show that residual thaw zones persist beneath buildings that have no ventilation at their base, and that when buildings are elevated, seasonal frost will extend from the surface to the permafrost table.

Because of these conditions, most building constructions today in areas of continuous permafrost follow the passive approach and attempt to maintain the permafrost intact. The floor of the building is built on top of pilings driven into the permafrost layer, so that it normally stands 1–2 m above the ground

Table 11.4 Landforms that indicate ground conditions in arctic and near-arctic regions.

Feature and Description	Associated Ground Conditions
Polygonal ground (ice wedges)—Usually indicates the presence of a network of ice wedges—vertical wedge-shaped ice masses that form by the accumulation of snow, hoarfrost, and meltwater in ground cracks that form owing to contraction during the winter. Wedge networks are also common in wet tundra where no surface expression occurs. (Subject to extreme differential settlement when surface disturbed.)	Typically indicates relatively fine-grained unconsolidated segments with permafrost table near the ground surface; also known from coarser sediments and gravels where wedge ice is less extensive.
Stone nets, garlands, and stripes—Frost heaving in granular soils produces netlike concentrations of the coarser rocks present. If the area is gently sloped, the net is distorted into garlands by downslope movement. If the slope is steep, the coarse rocks lie in stripes that point down hill.	Indicate strong frost action in moderate well-drained granule sediments that vary from silty fine gravel to boulders. Surficial material commonly susceptible to flowage.
Solifluction sheets and lobes—Sheets or lobe-shaped masses of unconsolidated sediment that range from less than a foot to hundreds of feet in width and may cover entire valley walls; found on slopes that vary from steep to less than 3°.	Indicate an unstable mantle or poorly drained, often saturated sediment that is moving downslope largely by seasonal frost heaving. On steeper slopes they often indicate bedrock near the surface, and on gentle slopes, a shallow permafrost table.
Thaw lakes and thaw pits—Surface depressions form when local melting of permafrost decreases the volume of ice-rich sediments. Water accumulates in the depressions and may accelerate melting of the permafrost. Often form impassible bogs.	Usually indicate poorly drained, fine-grained unconsolidated sediments (fine sand to clay) with permafrost table near the surface.
Beaded drainage—Short, often straight minor streams that join pools or small lakes. Streams follow the tops of melted ice wedges, and pools develop where melting of permafrost has been more extensive.	Indicates a permafrost area with silt-rich sediments or peat overlying buried ice wedges.
Pingos—Small ice-cored circular or elliptical hills that occur in tundra and forested parts of the continuous and discontinuous permafrost areas. They often lie at the juncture of south and southeast-facing slopes and valley floors and in former lakebeds.	Indicate silty sediments derived from the slope or valley, also groundwater with some hydraulic head that is confined between the seasonal frost and permafrost table or is flowing in a thawed zone within the permafrost.Those in former lakebeds indicate saturated fine-grained sediments.

From Ferrians et al. 1969.

surface, and air can circulate freely in the open space between the floor and the ground. This keeps the ground frozen in the winter because the heat radiating from the building is removed by cold air moving through the open space. Thawing does occur in the summer, but it is minimized because the surface is shaded by the building. Nonetheless, the piles supporting the building are susceptible to some heaving in the active layer; Johnston (1963) suggests they should be driven to a depth at least twice the thickness of the active layer in order to keep the heaving effect at a minimum. Other methods of air cooling or liquid cooling (Cronin 1983) have been utilized to maintain the frozen condition.

Whenever possible, buildings should be located on a sand and gravel substrate, for reasons explained above. In coastal regions some communities, built prior to any understanding of periglacial processes, have been built on silty, deltaic plains. Further development may be impossible in such a setting; in fact, some small towns may have to be relocated if the problems associated with the permafrost cannot be controlled. Relocation was actually accomplished for one community that originally existed on the deltaic plain of the Mackenzie River. The Canadian government physically moved the community 56 km to a new townsite that was chosen because it had abundant sand and gravel for building sites and roads. Although construction of the new town was costly, its success demonstrated that a reasoned approach and proper regulations can allow development to proceed without harmful degradation of the system (for details see Pritchard 1962; Price 1972; Cooke and Doornkamp 1974).

Roads and Airfields

The outbreak of World War II made the U.S. and Canadian governments aware that defending Alaska and the Northwest Territories from attack would be a monumental task. In an attempt to rectify this vulnerability, many necessary but hastily conceived projects were completed to facilitate the transportation of men and supplies to the area. Railroads, pipelines, and airfields were constructed, and the Alcan Highway, stretching almost 3000 km, was completed in only eight months, a truly remarkable engineering feat. The highway, however, proved to be a case study in permafrost mechanics, as the construction and maintenance problems were often baffling and disruptive (Richardson 1942, 1943).

We now know that problems associated with roads or airfields vary greatly because they are affected by microenvironmental conditions. In general, the most common ailments are (1) differential heaving and settling that creates a washboard surface (fig. 11.16), (2) sinking of portions of the surface, (3) destruction of bridges by spring floods, (4) burial of the pavement by slumping or other mass movements, and (5) icing. Icing is by far the worst problem. It occurs when water runs onto road surfaces during the winter, quickly freezes over large areas, and often accretes to a considerable thickness. Groundwater under hydrostatic pressure may continue to flow through most of the winter and, with improper planning, may also run onto the road surface if road cuts open springs and seeps. Even something as innocuous as a river freezing may result in icing of a nearby highway surface. This happens when the river freezes from the surface downward and creates a pressure head on the bottom water. The water, thus mobilized, moves to the channel sides where it rises to the surface and overflows the channel banks onto the highway.

The best road protection requires wise planning and careful pavement design. These can reduce frost action in the active layer and prevent excess water from reaching the road surface. The highway route should avoid seeps and springs by diverting around hills. If cuts must be placed in hillslopes, the pavement can be protected from icing by integrating a system of culverts and drainage ditches that will divert free water away from the road.

Pavement design usually begins with a layer of gravel 0.6–1.5 m thick spread directly on the tundra to insulate the underlying surface and so minimize thawing. A layer of insulating material such as peat (McHattie and Esch 1983) or polystyrene (Johnston 1983) may also be inserted within or

Figure 11.16.
Gravel road near the Umiat Airstrip showing severe differential subsidence caused by thawing of ice-wedge polygons in permafrost. Anaktuvuk district, Northern Alaska region, Alaska, Aug. 1958.

beneath the gravel cover. Some melting must be expected, nonetheless, and the design should allow for reduced strength in the summer and for heaving in the winter (Linell and Johnston 1973; Hennion and Lobacz 1973). Even painting the pavement surface white may help to reduce the thaw of the substrate.

Utilities—Water and Sewage

Some of the most burdensome problems facing communities in permafrost regions involve utilities and services that are taken for granted in the mid-latitudes. Water supply, for example, is complicated by deep winter freezing, and large lakes may be the only sufficient, continuous, and dependable water source. Taliks within the permafrost may provide some water, but their distribution is not regular and they may not contain enough water to support even a small settlement. Although taliks beneath the permafrost can provide large amounts of water, it is very expensive. Drilling through the permafrost is costly, and pipes must be cased to prevent freezing. In addition, some deep subpermafrost water is brackish or highly mineralized (J. R. Williams 1970).

Locating an adequate, continuous water supply solves only half the problem, because the water must still be delivered to the place where it is to be used. People in tiny settlements have traditionally obtained water by carrying it in tanks from a nearby river or lake. In the winter, when the source freezes over, blocks of ice are melted. Larger communities cannot rely on such primitive methods and most have piped-in water. Either the pipes must be insulated or the water must be continuously circulated to prevent freezing. Many areas now utilize an enclosed heated and insulated conduit, called a **utilidor,** for water delivery. The utilidor complex, placed above the ground, is designed to hold water pipes, electric cables, and sewage disposal pipes (see Zirjacks and Hwang 1983), but its installation is very expensive.

Sewage disposal is still another perplexing problem. Some small villages utilize individual buckets as collectors of all kinds of waste, human and otherwise. These are carried to a specified location on a lake or river where they are dumped in expectation of the spring thaw. Essentially, nature cleans the system at winter's end. For sanitation alone, larger communities must be more sophisticated and most now employ the utilidor system. In practice, however, utilidors solve only the immediate necessity of removing wastes from individual doorsteps. The final disposal of wastes is a most difficult problem. Few communities have efficient treatment plants, and most wastes are ultimately dumped as raw sewage into the nearest waterway. The problem is not nearly as acute in discontinuous permafrost areas, where septic tanks and cesspools are common. In continuous permafrost zones, however, alternative techniques must be developed. Sewage lagoons (large dug-out hollows) do allow waste to decompose anaerobically (Brown 1970) and are becoming a common disposal method. One novel idea, tried at a military base, was to use fuel oil instead of water in the sewage system. When the wastes are collected, they are sufficiently mixed with the oil to be injected directly into a furnace where incineration eliminates the problem and simultaneously generates heat and electricity for the base (Alter 1966).

Pipelines

It is perhaps fitting to end our discussion of periglacial geomorphology with a brief description of the one project that focused so much attention on this environment in recent years. The discovery of oil on the north shore of Alaska near Prudhoe Bay was exciting news to an energy-addicted nation. The conflict between energy and environment that ensued created a furor that ran from reasoned arguments to pure emotionalism. Nonetheless, it did force us to analyze in a systematic way the problems associated with construction of a pipeline in a permafrost region.

The Trans-Alaska pipeline was not the first pipeline constructed by the United States in a periglacial environment. The Canol pipeline system was built during World War II across 2575 km of discontinuous permafrost. Its purpose was to pump crude oil along the main pipe to a refinery located along the Alcan Highway near Whitehorse, B.C. From there gasoline was pumped to airfields and other military establishments. The Canol project, like the Trans-Alaska pipeline, was steeped in controversy (Richardson 1944), and its use was discontinued in May 1945 after only 13 months of operation, presumably because of the high cost of maintenance and its decreased importance in the war effort.

The Trans-Alaska pipeline is different in most respects from the older Canol project, and little applicable knowledge was derived from the earlier work. The Trans-Alaska line is much farther north, and much of its 1270 km route from Prudhoe Bay to Valdez crosses zones of continuous permafrost. The 1.2 m diameter pipe is larger than the Canol pipe and, fully loaded with oil, weighs over 900 kg per meter (Harwood 1969). The crude oil enters the pipe at a temperature of about 58°C, but friction along the flow path increases this temperature considerably. Cooling devices are placed along the route to keep the temperature at a constant 63°C, but the oil is still hot and so presents the most difficult problem in preserving the permafrost environment intact.

Given the above, the planning had to consider the possibility of major changes in the physical environment as well as a host of engineering problems (Lachenbruch 1970; Kachadoorian and Ferrians 1973). A pipe resting on the surface and carrying hot fluids would certainly cause thawing of the underlying permafrost. Without preventive measures, the thaw would probably expand outward with time as shown in figure 11.17, perhaps at a decreasing rate, but possibly never reaching an equilibrium state (Lachenbruch 1970). The soil could also become liquefied and unable to support the heavy pipe, and any gelifluction induced by the liquefaction might carry the pipe along with the soil, producing ruptures and oil spills. As the pipe crosses materials of diverse texture and composition, differential heaving and settling are inevitable, and stresses generated in this manner could bend and rupture the pipe. This problem is especially pertinent where the pipeline crosses slopes underlain by ice wedges that melt more quickly and undermine the support rapidly. In addition, the actual construction could have inadvertently altered the delicate environmental balance. For example, destruction of the surface vegetation or diversion of river courses would have disrupted the thermal regime and changed the thickness of the active layer.

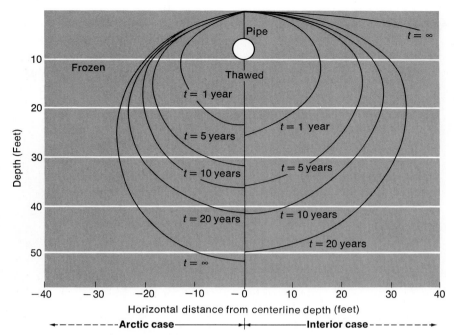

Figure 11.17.
Theoretical growth of thawed cylinder around heated pipe placed in silty soil. Pipe is 1.2 m in diameter and kept at a temperature of 80°C. The curves at the left apply to conditions near the Arctic coast, those on the right to the southern limit of permafrost. (After Lachenbruch 1970)

Faced with these potential hazards, the designers of the pipeline anticipated the problems based upon the best available knowledge of how the system would respond. In general three approaches are employed in different places: (1) the pipe is buried, (2) the pipe is suspended above the ground, and (3) the pipe is placed along the edge of a road that parallels the route of the pipeline. Burying the pipe is perfectly safe in coarse-grained soils that are well drained and contain little ground ice. In other zones a refrigerant is run along the pipe to keep the ground frozen. Suspending the pipe above the ground is necessary where the soil is particularly susceptible to thawing and also where the route crosses the path of migratory animals such as caribou. The above-ground suspension requires a dense network of pilings to provide the needed strength, and these are inserted to great depth. In addition, some of the supporting piles are refrigerated and others are distributed in a zigzag fashion to absorb the effect of differential movement by heaving, gelifluction, or earthquakes (fig. 11.18).

Only time can tell us whether the design of the pipeline and the careful planning of the overall project will provide a continuous supply of crude oil with a minimum of hazards to the environment. Initial operation of the pipeline has been encouraging, although some minor leaks have occurred (Stanley and Cronin 1983).

Figure 11.18.
Trans-Alaska pipeline crossing the
Tolovana River valley near
Livengood, Alaska.

Summary

The processes and landforms found in periglacial environments have been briefly examined. In general, periglacial conditions are those existing in any cold, nonglacial setting regardless of latitude, but most regions of intense periglaciation are polar or subpolar. In many cases periglacial zones possess a unique property of permanently frozen ground, called permafrost. Permafrost adds a complicating factor in the activity of periglacial processes and leads to the formation of diagnostic features such as ice-wedge polygons, thermokarst, and pingos that develop only where permafrost is present.

The driving processes in periglaciation are frost action and mass movements. Frost action includes wedging, heaving, and cracking. Mass movements usually involve frost creep and a form of soil flowage called gelifluction. Both processes act in the development of most periglacial features, although certain forms seem to be more common where one or the other process is dominant.

The rigorous climate and the presence of permafrost lead to unusual difficulties in urban development that require engineering techniques not employed in other regions. The expected population expansion in arctic zones demands that we develop a sophisticated understanding of periglacial processes in order to maintain the fragile environmental balance.

The following references provide greater detail concerning the concepts discussed in this chapter.

Suggested Readings

Benedict, J. B. 1976. Frost creep and gelifluction features: A review. *Quat. Res.* 6:55–77.

Black, R. F. 1976. Periglacial features indicative of permafrost: Ice and soil wedges. *Quat. Res.* 6:3–26.

Embleton, C., and King, C. A. M. 1975. *Periglacial geomorphology.* New York: Halsted Press.

Ferrians, O. J.; Kachadoorian, R.; and Greene, G. W. 1969. Permafrost and related engineering problems in Alaska. U.S. Geol. Survey Prof. Paper 678.

Flemal, R. C. 1976. Pingos and pingo scars: Their characteristics, distribution, and utility in reconstructing former permafrost environments. *Quat. Res.* 6:37–53.

Goldthwait, R. P. 1976. Frost sorted patterned ground: A review. *Quat Res.* 6:27–35.

National Academy of Sciences. 1983. *Proc. Permafrost 4th Internat. Conf.* Washington, D.C.: National Academy Press.

National Research Council of Canada. 1978. *Proc. Permafrost 3rd Internat. Conf.* Ottawa: National Research Council of Canada.

Péwé, T. L. 1969. *The periglacial environment.* Montreal: McGill-Queen's Univ. Press.

Price, L. W. 1972. The periglacial environment, permafrost, and man. Assoc. Am. Geog., Comm. on College Geog. Resource Paper 14.

Tricart, J. 1969. *Geomorphology of cold environments.* Translated by Edward Watson. New York: St. Martin's Press.

Washburn, A. L. 1980. *Geocryology.* New York: John Wiley & Sons.

Karst—Processes and Landforms

12

Introduction

In regions of carbonate rocks and evaporites, weathering and erosion produce unique landforms called **karst** or, when widespread, **karst topography.** In contrast to most processes studied before, **karstification** (the processes that develop karst topography) is not easily observed because much of the geomorphic work is accomplished well below the ground surface. In fact, some modern textbooks of geomorphology completely ignore karst or treat it in a few descriptive paragraphs because it is primarily driven by the solution process and can justifiably be considered under the topic of chemical weathering. However, the magnitude of solution in the development of karst is so great, and the unique topography resulting from the process so widespread, that it seems to deserve special treatment.

Several excellent books deal specifically with the topic of karst (Jennings 1971; Herak and Stringfield 1972; Sweeting 1973, 1981; Bögli 1980; Milanović 1981), and information contained in those sources was used as a framework for this chapter. In general, however, syntheses of literature dealing with karst, and published in English, are rather uncommon.

Definitions and Characteristics

Karst is defined by Jennings (1971) as "terrain with distinctive characteristics of relief and drainage arising primarily from a higher degree of rock solubility in natural water than is found elsewhere." The definition stresses two main points: (1) distinctive landforms and other surface characteristics developed on highly soluble rocks, and (2) a unique type of drainage pattern resulting from the karst processes. As all rocks are soluble to some extent, karst must develop only on those rocks that are particularly susceptible to solution. In such situations, the solution process can create and enlarge cavities within the rocks. This leads to the progressive integration of voids beneath the surface and allows large amounts of water to be funneled into an underground drainage system while simultaneously disrupting the pattern of surface flow. Because hydrology exists in physical "symbiosis" with solution, it becomes a very important aspect of karst phenomena. Solution integrates spaces, allowing pronounced underground circulation of water that, in turn, promotes further solution. A true karst area, therefore, possesses a predominantly underground drainage with a poorly developed surface network of streams. As our interest is primarily in processes, the basic concepts may be applicable in many areas that have surface drainage and have not developed a strong enough topography to be considered as karst. Importantly, however, the area may still be affected by karst-forming processes.

The term "karst" is a German adaptation of the Slavic word *kras* or *krs* and the Italian word *carso,* which literally mean "a bleak waterless place" (Monroe 1970) and also connote a bare rock surface. The early and classical description of karst was derived from a high plateau area near the Adriatic Sea between northwest Italy and Yugoslavia. This region is characterized by irregular topography containing many closed depressions and interrupted stream valleys; thus, early observers of the region considered karst as a geomorphic freak with chaotic and disordered topographic expression. Most geologists and geographers thought of karst as a curiosity rather than a topic

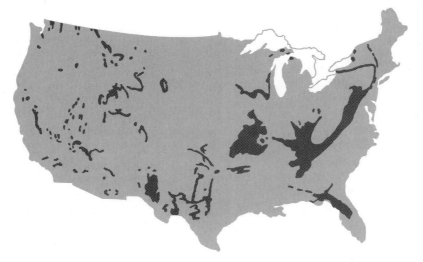

Figure 12.1.
Major karst areas of the United
States. (Boundaries are
generalized, and some regions
shown in solid black contain some
nonkarst areas.)

for serious scientific investigations. It was not until 1893, when Jovan Cvijić
(name rhymes with "screech") published his book *Das Karstphanomen,* that
karst geomorphology was given true scientific status. Cvijić's work clearly defines karst landforms and demonstrates the predominance of solution in their
development, even though the solution effect had been alluded to in earlier
studies (Sawkins 1869; Cox 1874).

Karstlands (or karst landforms) are found in almost every region of the
world, including arctic and arid zones, but they are most likely to occur in
temperate or tropical climates. In the United States, karst has developed
wherever conditions are favorable, but the major concentrations of features
exist in several general areas (fig. 12.1): (1) the Valley and Ridge province of
the Appalachian Mountains (Pennsylvania, Maryland), (2) central Florida,
(3) the plateau region of east-central Missouri, (4) a belt extending from south-
central Indiana into west-central Kentucky, (5) the Edwards Plateau (Texas).

The field of karst geomorphology has its roots in central and eastern Europe, and contributions by English-speaking geomorphologists were rather
meager during the formative years of the discipline. In recent years, however,
these geomorphologists have added greatly to our knowledge of the discipline,
especially by their recognition that analyses of karst processes and karst landscapes require the integration of hydrologic, chemical, and mathematical concepts as well as traditional geomorphology (Palmer 1984). In addition, the
modern emphasis in karst geomorphology is largely on process and the application of conceptual models on practical problems involving engineering,
water supply, and economic geology. Many articles espousing this thrust are
now appearing in scientific journals such as the *Journal of Hydrology, Water
Resources Research,* and the *National Speleological Society Bulletin.*

The historical development of karst science explains the proliferation of
terminology that has arisen for karst features and processes. Although the
Yugoslavian terrain inspired the study of karst, not all the names for individual karst features come from there, and almost every country has developed

Table 12.1 Abbreviated glossary of common karst terminology.

Term	Definition
Aggressive water	Water having the ability to dissolve rocks. Especially water containing dissolved CO_2.
Blind valley	A valley that ends suddenly where its stream disappears underground.
Cave	A natural underground room or series of rooms and passages large enough to be entered by a person.
Chamber	The largest order of cavity in a cave or cave system.
Closed depression	Any closed topographic basin having no external drainage, regardless of origin or size.
Cockpit	1. Any closed depression having steep sides. 2. A star-shaped depression having a conical or slightly concave floor.
Cockpit karst	Tropical karst topography containing many closed depressions surrounded by conical hills. Similar to cone karst.
Cone karst	Tropical karst with star-shaped depressions at base of many steep-sided, cone-shaped hills.
Doline	A basin or funnel-shaped hollow in limestone ranging in diameter from a few meters to a kilometer and in depth from a few to several hundred meters. May be distinguished as "solution" or "collapse" if precise origin is known. In U.S., most dolines referred to as sinks or sinkholes.
Exsurgence	Point at which underground stream reaches the surface if the stream has no known surface headwaters.
Karren	Channels or furrows caused by solution on massive bare limestone surfaces. Synonym *lapiés*.
Karst plain	A plain on which closed depressions, subterranean drainage, and other karst features may be developed. Also called karst plateau.
Karst topography	Topography dominated by features of solutional origin.
Karst valley	1. Elongate solution valley. 2. Valley produced by collapse of a cavern roof.
Karst window	Depression revealing a part of a subterranean river flowing across its floor, or an unroofed part of a cave.
Karstic	Adjective form of karst.
Karstification	Action by water, mainly chemical but also mechanical, that produces features of a karst topography.
Mogote	A steep-sided hill of limestone generally surrounded by nearly flat alluviated plains. Generally used for karst residual hills in the tropics. Synonym *pepino*.
Polje	A very large closed depression in areas of karst topography, having flat floors and steep perimeter walls.
Resurgence	Point at which underground stream reaches the surface. Reemergence of a river that has earlier sunk upstream.
Room	A part of a cave system that is wider than a normal passage. Similar to chamber.
Speleothem	A secondary mineral deposit formed in caves.
Swallet, swallow hole	A place where water disappears underground in a limestone region. A swallow hole generally implies water loss in a closed depression or blind valley. A swallet may refer to water loss in a streambed even though there is no depression. Also *ponor, sink, sinkhole, stream sink*.
Terra rossa	Reddish-brown soil mantling limestone bedrock; may be residual in some places.
Tower karst	Karst topography characterized by isolated limestone hills separated by areas of alluvium. Towers generally steep-sided and forest-covered hills, often with flat tops.
Uvala	Large closed depression formed by the coalescence of several dolines; compound doline.

After Monroe 1970.

a particular terminology of karst geomorphology. Table 12.1 presents an abbreviated glossary of karst terminology adopted by the U.S. Geological Survey and followed in this discussion. The complete listing of terms can be found in Monroe (1970).

Karst Rocks—The Resisting Framework

The Processes and Their Controls

It is tempting to say that karst develops primarily on limestones and leave it there without further elaboration. Although technically correct, such a statement is grossly misleading because some limestones are not potential harbingers of karst topography. Only certain limestones have the unique combination of properties that allow them to succumb to karstification and foster karst topography. We will briefly examine what rock properties are most conducive to karstification and why.

Lithology   As a general rock group, limestones show great variability, but the accepted definition is that a limestone is a rock containing at least 50 percent carbonate minerals, most of which occur in the form of calcite ($CaCO_3$). Although very young limestones may contain some aragonite, the two most common carbonate minerals in limestones are a low-magnesium *calcite,* containing 1–4 percent magnesium, and *dolomite* (Sweeting 1973). If more than 50 percent of the carbonate minerals are calcite, the rock is called limestone; if more than 50 percent of the carbonate minerals are dolomite, the rock is called dolomite (Leighton and Pendexter 1962). The purer the limestone is with respect to $CaCO_3$, the greater will be its tendency to form karst. Corbel (1957), for example, suggests that 60 percent $CaCO_3$ is needed before any karst will form, and about 90 percent is required to expect a fully developed karst region. It should be noted, however, that even pure limestones may not produce a karst terrain because the processes also depend on other factors.

Rocks other than limestone can produce karst if ancillary conditions are proper and the material is sufficiently soluble. Dolomites are commonly karstified, but unless they are very pure their porosity and permeability tend to be somewhat low. Evaporites such as gypsum and halite are also prone to karstification. In general, however, the occurrence of karst in dolomites and evaporites is minor compared to the widespread distribution of limestone karst regions.

Porosity and Permeability   In addition to lithology, the creation of karst also depends on how much water any rock can hold and how easily the water moves through the rock system. Porosity is a measure of the water-storing capacity of a given rock, and it is usually expressed as the percentage of void spaces in the rock:

$$P = \frac{V_v}{V} \times 100,$$

where P is porosity in percent, V_v is the volume of voids, and V is the total volume of the material. Presumably, open textures and higher porosities should facilitate the solution process and the development of karst because the rock will hold more water. Some evidence exists to support this contention (Sweeting 1973). Porosity, however, consists of two types. **Primary porosity** relates to intergranular void spaces created during formation of the rock. This type of porosity, however, is commonly decreased with time by precipitation of cement, recrystallization, and change in mineralogy. In older limestones, for example, calcite tends to be replaced by dolomite, a process that usually decreases the primary porosity (Powers 1962). In addition, metamorphism may decrease carbonate porosity by inducing an increase in calcite grain size and reducing void size by pressure. Thus, primary porosity may be an ephemeral characteristic of limestones and perhaps is not as important in karstification as secondary porosity and permeability.

Secondary porosity comes from openings in rocks that occur along bedding-plane partings or as fractures, such as joints and fault zones. It is generally agreed that the nature and pattern of these openings may be the single most important factor in karstification. Not only do they allow the rocks to hold more water, but they also promote circulation within the system by increasing permeability. Because permeability (the capacity to transmit water) depends on the continuity of voids, even rocks with high primary porosity may develop little karst if no secondary avenues of flow are present (Tricart 1968).

Porosity, then, has real importance only if the system is also permeable. Zones of weakness increase porosity but, more important, they have a decided influence on the permeability. This explains why permeability rates vary by as much as five orders of magnitude depending on the size and interconnection of fractures and partings.

Joints are important avenues of water transport and enhanced permeability. Usually joints occur in patterns with one dominant direction and a secondary set of joints intersecting the main set at angles between 70° and 90°. The spacing of the joint sets can be very significant in the genesis of karst. If intersecting planes are too close, the rock may be highly permeable but too weak to allow the full development of karst.

Faults also transmit water effectively, but their precise role in karstification varies according to local conditions (Stringfield and LeGrand 1969). For example, fault zones sometimes have low permeability where voids are occupied by secondary mineralization associated with ore deposits. Such a deterrent to karstification, however, may be partly offset if the ore includes sulfide minerals, because oxidation produces sulfuric acid, which makes the permeating fluids more aggressive in the solution process (Pohl and White 1965; Morehouse 1968).

In summary, it seems likely that the full development of karst depends primarily on whether water capable of solution passes through a rock sequence along discrete flow paths with enough discharge to create significant solution openings. Given that requirement, the process is aided by having a thick sequence of pure, crystalline limestones that is not interrupted by major insoluble beds. In addition, some relief should be available to permit free circulation of the water in the system.

The Driving Mechanics and Controls

Climate, Vegetation, and Biogenic CO_2   Karstification requires abundant water that is free to circulate through the karst rocks. The water not only serves as the solvent in the development of karst but also encourages the growth of vegetation and the soil microbial activity that add extra CO_2 to the system. Regions of low rainfall coupled with high temperature and evaporation rates will, therefore, be less susceptible to karst development. This is not to say that karst features never form in arid or semiarid regions. They do—Carlsbad Caverns in New Mexico and karst of the Nullarbor Plain in Australia attest to that fact. Normally, however, the topography produced is quite subdued, and features such as collapses and deranged surface drainage are not as striking. It is also possible that the karst in these areas formed at an earlier time when the climate was more humid than at present or, in the case of Carlsbad, may be the result of deep-seated acidic fluids rising into the rocks (Davis 1980; Egemeir 1981; Hill 1981).

In extremely cold arctic or subarctic regions, the full development of karst is hindered by the presence of permafrost and by the low vegetal productivity that retards microbiologic activity (D. I. Smith 1969). Although water is present, it is often frozen and no free circulation exists. The important aspect of biogenic CO_2 (which we will discuss later) is usually missing. Karst is therefore a rather anomalous phenomenon in these areas.

At the other extreme, a tropical humid climate, with the ideal combination of temperature and precipitation to drive the solution process, should be most conducive to karstification provided enough relief exists to promote downward and circulatory movement of groundwater. Chemical reactions proceed more rapidly at higher temperatures, and lush vegetal cover combined with intense microbial activity impresses the tropical soil water with high partial pressures of carbon dioxide (P_{CO_2}). The subsurface water in such regions is very aggressive in solution, and a great abundance and variety of karst features are found in tropical regions (Jennings and Bik 1962; Monroe 1976).

The large variation in the controlling factors of karst in different climatic regimes has perpetuated the climatic geomorphology thrust initiated in the 1930s (Lehmann 1936). For example, Corbel (1959) analyzes rates of karst erosion in relation to climatic parameters, and others (P. W. Williams 1963; Douglas 1964) have modified the details of Corbel's original approach while still maintaining its dependence on climatic variables. Nonetheless, it seems certain that climatic factors are important mainly because they provide free water and create the soil and vegetal characteristics that produce high biogenic CO_2 values. Thus, even though enhanced solution is generally related to climate, it is the production of biogenic CO_2 that represents the important control (Smith and Atkinson 1976; Trainer and Heath 1976; Drake 1980; Brook et al. 1983). In fact, minor vegetation changes within the same climatic zone (even arctic or subarctic) will generate areas of greater solution (Woo and Marsh 1977; Brook and Ford 1982). Consequently, recent work has placed less emphasis on climatic variables and more on solution mechanics.

The Solution Process   The solution process itself is in reality the critical function in the entire analysis of karst. Regardless of how conducive climate, lithology, fractures, and other variables are to karstification, karst topography would never develop if the solution process were somehow rendered inoperative. Its function or malfunction is fundamental to the topic, and we must attempt to understand its mechanics, at least in its simplest terms.

Laboratory studies tell us that the mineral calcite, like all common minerals, is soluble in pure water. At saturation it is soluble to the extent of about 12–15 ppm depending on the temperature of the water. This solubility is rather startling when compared to natural river waters where concentrations of Ca^{+2} and bicarbonate (HCO_3^-) are much greater (Livingstone 1963), indicating a substantial increase in the solubility. Since we are considering the same substance (calcite), it is obvious that the solvent in the natural system is not pure water. Rainwater, in fact, is not pure because it incorporates a variety of chemical constituents as it passes through the atmosphere. The most important of these is carbon dioxide (CO_2), which is soluble in pure water; some of the dissolved CO_2 reacts rapidly with the water to form a weak acid (H_2CO_3), called carbonic acid

$$CO_{2(dissolved)} + H_2O \rightleftharpoons H_2CO_3. \tag{1}$$

This acid, however, is always dissociated into its ionic state, and the above reaction can be expressed more realistically as

$$CO_{2(dissolved)} + H_2O \rightleftharpoons H^+ + HCO_3^-. \tag{2}$$

The amount of CO_2 actually dissolved in water depends on the partial pressure of carbon dioxide (P_{CO_2}) in the air standing at the air-water interface and on the temperature of the water. The air in contact with the water can be in the atmosphere, or in spaces within the soil, or in subterranean cavities such as caves. In any case, the amount of CO_2 dissolved in the water increases as the P_{CO_2} of the air increases and as the temperature of the water decreases. Colder water will dissolve more CO_2 than warm water at any given P_{CO_2} value. In the atmosphere P_{CO_2} is rather small, having values averaging about 0.03 percent of volume (3 $\times$ 10^{-4} bar); similar values are found in most caves (Holland et al. 1964). Anomalously high P_{CO_2} values are found in air contained within soil or the vegetal litter covering it. Values of 1–2 percent of volume are common, and some poorly ventilated tropical soils may contain as much as 20–25 percent (Jennings 1971). The abnormal CO_2 values in soil air, and the resulting large amounts of dissolved CO_2 in the soil water, stem from microbial action involved in the decomposition of vegetal matter. This represents the biogenic CO_2 briefly discussed in the previous section. It is regarded by most karst experts as being the prime ingredient in the solution process.

Calcite itself is dissociated into an ionic state such that

$$CaCO_{3(calcite)} \rightleftharpoons Ca^{+2} + CO_3^{-2}. \tag{3}$$

However, the CO_3 ion produced quickly reacts with the H^+ formed when CO_2 is dissolved in water (reaction 2), and thus the dissociation of calcite also produces a bicarbonate ion because

$$CO_3^{-2} + H^+ \rightleftharpoons HCO_3^-. \tag{4}$$

It is clear from these reactions that the solution of limestone revolves around the $CaCO_3$–CO_2–H_2O chemical system. This system is extremely complicated, and its mechanics much more sophisticated than this introductory treatment can show. We are not dealing with a single reaction that produces solution of calcite, but a process that involves a series of reversible and mutually interdependent reactions, all proceeding at different rates, and each regulated by different equilibrium constraints. We will therefore explain the process only in the most general terms.

If we combine reaction *(2)* and *(3)*, a general form of the process can be expressed by the following:

$$CaCO_3 + H_2O + CO_{2(dissolved)} \rightleftharpoons Ca^{+2} + 2HCO_3^-. \tag{5}$$

The bicarbonate ions (HCO_3^-) are derived from two sources shown in equations *(2)* and *(4)*. In equation *(4)*, the reaction of carbonate ions (from the dissociation of $CaCO_3$) and the H^+ (from the dissolving of CO_2 in water) produces a disequilibrium between the P_{CO_2} in the air and the P_{CO_2} in the water. This causes more CO_2 to diffuse from the air into the water and allows further solution of the calcite because reaction *(5)* is driven to the right side of the reversible equilibrium.

To a large extent, the amount of CO_2 dissolved in water regulates the solubility of limestone. Where the amount of CO_2 dissolved in water is high, the fluid will agressively attack the calcite. Soil water with its high P_{CO_2} is therefore the most effective solution agent (Drake and Wigley 1975).

Another important mechanism in the solution process is **mixing corrosion** (Bögli 1964). Essentially, when two bodies of water at equilibrium with different CO_2 contents are mixed, the resulting fluid needs less CO_2 to establish equilibrium than the sum of the CO_2 contained in the two original fluids. Some CO_2 is released and becomes available to promote further solution. The process is very significant where seawater mixes with fresh water in carbonate regions. In that situation, mixing produces brackish water that is subsaturated with respect to calcite and enhances the solution process (Runnells 1969; Plummer 1975; Hanshaw and Back 1980). Even there, however, the phenomenon is very complex, and the degree of saturation with respect to calcite may depend on other factors (Plummer et al. 1976).

In normal karst situations, the effect of mixing really depends on the type of water masses being mixed (Thrailkill 1968). For example, water slowly percolating through the zone of aeration, a fluid called **vadose seepage,** may not show a significant change when it mixes with phreatic water at the water table. However, two types of vadose seepage are known (Thrailkill and Robl 1981), and each may respond differently upon mixing. In addition, water rapidly introduced to the subsurface, called **vadose flow,** may trigger considerable mixing corrosion when it meets the liquid standing at the water table (see Thrailkill 1968). Mixing of vadose flow and the proper vadose seepage may cause pronounced undersaturation of the underground water and promote the solution process (Thrailkill 1972).

Solution rates are usually estimated by direct measurement of the dissolved load in streams leaving a karst area, although other techniques can be employed (for an excellent discussion, see Jennings 1983). The rate is equal to the concentration of dissolved load multiplied by the stream discharge such that

$$S = CQ,$$

where S is the solution rate in units of mass/time, C is the solute concentration (mass/volume), and Q is stream discharge (volume/time). Many measurements can be integrated over time and converted into rates of surface lowering or **karst denudation.** Values of karst denudation, adjusted for subsurface contribution, seem to range from about 10 mm/1000 yr to over 100 mm/1000 yr, depending primarily on the amount of runoff (Atkinson and Smith 1976; Jennings 1983).

The controls on solution rates involve a complex maze of interrelated factors. Sweeting (1973) suggests that nearly all the possible limestone solution is accomplished within the first minute of contact, but for all the reactions to establish equilibrium throughout the system it may take anywhere from 24 to 60 hours. In addition, Jennings (1983) shows that higher discharge produces greater solute load. In light of this, it is generally assumed that water moving through the system rapidly and with some turbulence will ultimately dissolve more limestone, whereas stagnant water may become supersaturated with respect to calcite as equilibrium conditions are attained throughout the various reactions in the system (Kaye 1957; Weyl 1958). Other analyses, however, suggest that the dependence of solution rates on velocity and turbulence only holds when the fluid is very acidic (see Palmer 1984). Where pH is greater than 4, the solution rate is primarily dependent on the reaction rate at the bedrock surface and may be almost independent of velocity.

In addition to flow velocity and pH, as water moves through underground fractures the solution rate decreases with flow distance. This occurs because the water gets closer to saturation with respect to calcite the farther it travels. In the past, arguments have been made that the decrease in solution rate is linear with percentage of saturation and therefore travel distance. Palmer (1984), however, points out that under such controls the transport distance for complete saturation would be so short that long, linear caves would be difficult, if not impossible, to develop. Based on earlier studies of calcite solution by water near saturation (Berner and Morse 1974; Plummer et al. 1978), Palmer (1984) presents cogent arguments that solution rates do not decrease linearly with percentage of saturation except at lower values. At some point above 65 percent saturation, the solution rate drops rapidly (fig. 12.2). This allows solution to continue for a greater time and over a greater distance, and it helps to explain the origin of elongated cave systems.

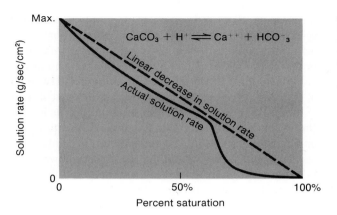

Figure 12.2.
Solution rate versus degree of saturation. Instead of decreasing linearly, the solution rate drops sharply to a low level at 65–90 percent saturation.

Surface Flow

Karst regions are unique because their drainage networks characteristically are disrupted, and few rivers can traverse such areas in a continuous and unsegmented manner. The reason for this strange fluvial behavior is the facility with which surface flow can be diverted into the underground system. Depending on soil types, vegetation, and joint spacing, overland flow on hillslopes may be drastically reduced by infiltration, and in extreme cases no water will enter the nearby channels.

Rivers also lose water when some of the flow descends into **swallow holes** or **swallets,** shown on the map in figure 12.3. These are nothing more than open cavities on the channel floor that are capable of pirating a portion of a river's discharge or even the entire river (especially during low flow) into the underground system. Thus, a large part of the total flow of rivers in karst regions may follow a subsurface route that may or may not parallel the path of the river valley on the surface. Rivers that are able to cross a karst terrain as continuous surface entities have distinct hydrologic characteristics. For example, flood records for 114 basins of different sizes show that the mean annual floods (recurrence interval of 2.33 years) in carbonate basins of Pennsylvania were considerably lower than those in basins underlain by different rock types (White and Reich 1970). Commonly, however, the peak flow is spread over a longer period as the subsurface water is slowly released to the rivers. It appears likely that the precise hydrologic character of surface rivers in karstic areas depends greatly on the state of development of the underground drainage, especially the degree of interconnection between subsurface passageways (Ede 1975).

Karst Aquifers and Groundwater

Our brief look at normal hydrogeology (chapter 5) allows us now to examine the differences between karst aquifers and the groundwater conditions in other porous and permeable materials.

Karst Hydrology and Drainage Characteristics

Historically, karst geomorphologists have viewed the hydrology in karstic systems in discordant ways. Most conflicting opinions concerned the type of water movement, the depth to which groundwater would penetrate, and most important, whether or not a water table as we normally conceive it exists in karstic terrains. In the first two decades of the twentieth century, two schools of thought developed concerning the hydrologic system.

In one model, groundwater in karst regions is considered to be similar to the normal groundwater found in unconfined aquifers composed of other rocks or unconsolidated debris. The water presumably exists in a vertical stratification. Water in the zone of aeration moves downward to a pronounced water table by seepage or by flow in various types of fissures and cavities. Below the water table the phreatic zone is completely saturated, although the mechanics of how water moves through this zone are very controversial. The water table itself is subject to significant fluctuations, rising and falling on a seasonal basis.

Figure 12.3.
The distribution of swallets, sinks, and springs in a karst region of southeast New York. (From Baker 1976)

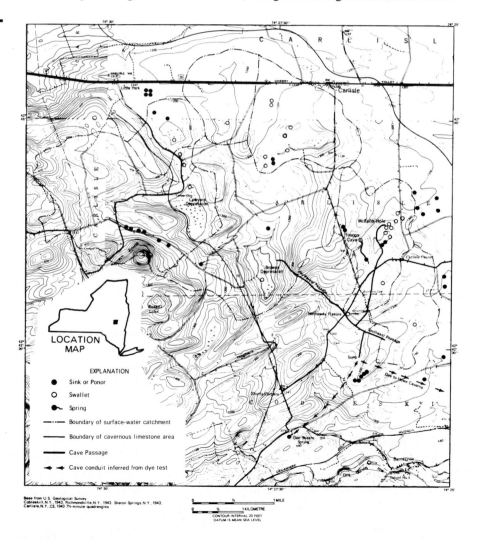

The second major conceptual model of karstic groundwater completely denies the existence of a water table and with it the distinct vertical zonation of the water. Workers accepting this view believe that the distribution and movement of groundwater are controlled entirely by the spatial characteristics of the network of interconnected passageways within the rocks. Several points support this conclusion: (1) adjacent wells drilled into limestone often have hydrostatic levels that are significantly different in elevation; (2) tunnels reveal dry fissures immediately next to cracks that are filled with water; (3) tracing of water with dyes shows that paths of movement often cross one another (obviously flowing at different levels) and even pass under a surface stream from one side of a valley to the other, a physical impossibility in normal unconfined aquifers; and (4) poljes (see table 12.1) at the same level do not behave in a similar manner, some flooding in winter while others remain dry. Thus, the underground system is thought to be a collection of conduits functioning like three-dimensional rivers. The passages may be totally interconnected, or in some cases they may function like a single confined aquifer having a recharge area, a discernible pathway, and a separate point where the water emerges once again to the surface. Most geomorphologists believe that these two basic models can simultaneously exist in the same karst area.

A relatively recent development has been to recognize the importance of a hydrologic zone existing in the weathered, upper portion of the bedrock but beneath the soil cover (see Williams 1983). This zone, the **subcutaneous zone,** stands above the phreatic zone, but it stores water and is periodically saturated, especially after storms. The subcutaneous zone develops by enhanced solution immediately beneath the soil; this enlarges cavities and fractures and creates high permeability and porosity. The solution enlargement decreases with depth, however, and at some level, openings are too small to transmit water at the same rate as in the overlying more permeable rock (fig. 12.4).

Figure 12.4.
Subcutaneous zone in limestone along Route 55 near Perryville, Mo.

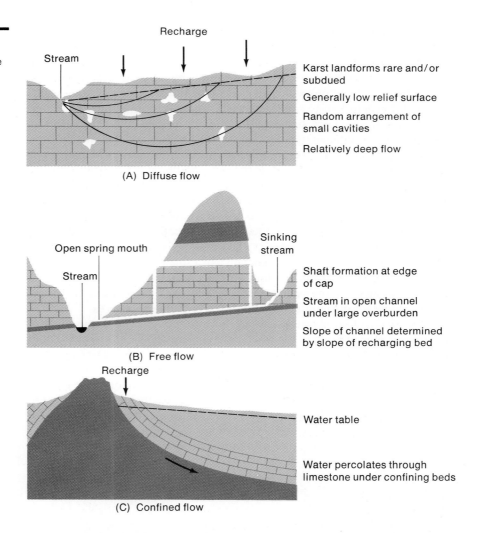

Figure 12.5.
Classification of flow types in karst aquifers. (From W. B. White 1969)

Recharge

Stream

Karst landforms rare and/or subdued

Generally low relief surface

Random arrangement of small cavities

Relatively deep flow

(A) Diffuse flow

Open spring mouth

Sinking stream

Stream

Shaft formation at edge of cap

Stream in open channel under large overburden

Slope of channel determined by slope of recharging bed

(B) Free flow

Recharge

Water table

Water percolates through limestone under confining beds

(C) Confined flow

As a result, water is stored in the form of a perched water table. The perched water table slopes toward points of rapid vertical percolation (major joints, faults, shafts, etc.), and a lateral component of flow develops in the subcutaneous zone and converges at those points. Subcutaneous flow has a significant influence on the hydrologic characteristics of the system and on the pattern of landforms developed in karst regions.

W. B. White (1969) has classified carbonate aquifers according to their hydrogeologic properties as **diffuse flow, free flow** or **confined flow,** each type having subtypes; the main classifications are depicted in figure 12.5. In diffuse flow, cavities are limited in size and numbers, caves are rare, a well-defined water table is present, and flow obeys or nearly obeys Darcy's law. In free-flow aquifers, the water moves through integrated conduits under the influence of gravity, often attaining turbulent flow. The flow is capable of transporting sediment as discharge is enlarged by surface runoff sinking into fractures (swallets). Discharge of the groundwater is usually through large springs that accumulate, at a single outlet, the water flowing through vast areas of underground drainage. Confined-flow aquifers are characterized by water that moves in response to pressure. They may be true confined aquifers as described earlier and may contain water under considerable hydrostatic head; thus, artesian flow conditions are possible in these aquifers.

It is now becoming clear that the different types of aquifers recognized by White may also be distinguishable on the basis of the chemical behavior of their water. Shuster and White (1971), analyzing 14 springs in the central Appalachians, suggest that diffuse-flow aquifers maintain a constant hardness and have spring water nearly saturated with respect to $CaCO_3$. Aquifers that utilize conduits are undersaturated and show considerable variability in hardness (Thrailkill 1972; Drake and Harmon 1973). More recently, Kroethe and Libra (1983) have identified two different flow types in the southern Indiana karst region on the basis of sulfur chemistry in spring waters.

The Relation Between Surface and Groundwater

Assuming that flow through conduits is the unique characteristic of karst hydrology, we should briefly examine how water is exchanged between the surface and the underground system and vice versa, and how the properties of the passageways control the movement. In karst areas, water is diverted into the groundwater system through vertical openings and swallets. In addition, interconnected spaces in the zone of aeration permit slow percolation of vadose seepage, and this water is added to the total accumulation of groundwater. In closed depressions of a karst landscape, water input to the underground system is through a highly permeable point at the base of the depression. The point input usually represents an intersection of vertical joints or cylindrical solution openings called **shafts.** Precipitated water reaches the input point by various transmission routes acting on the sloping surfaces of the enclosed depression. These routes are similar to those discussed in chapter 5, with the added process of subcutaneous flow. Gunn (1981, 1983) suggests that various possible combinations of flow types will have a decided effect on the extent of solution and the morphologic evolution of the karst.

Springs   After entry into the underground system, water moves as one or more of the types introduced above. The manner in which groundwater leaves the system is also dependent on the aquifer and flow type. In diffuse aquifers, water will emerge according to the rules that govern normal groundwater flow. Some question lingers about the depth to which phreatic water will penetrate, and we will discuss this problem when we review the origin of caves. In conduit systems, however, water usually reaches the ground surface in the form of springs, and the characteristics of these interesting geomorphic features commonly reveal the distribution and properties of the aquifer itself.

Springs in karst areas are usually larger and more permanent than those in other regions because of the high infiltration associated with karst rocks. Those springs or seeps stemming from flow in diffuse aquifers are called **exsurgences** (fed by seepage), in contrast to those fed by groundwater moving

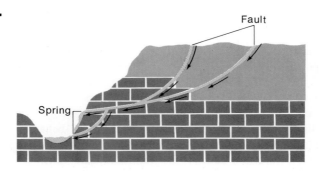

Figure 12.6.
Fault control on the position of springs. Rainwater enters shales exposed at the surface along faults. Water follows fault zones into and through limestones until it emerges as springs. (Arrows show movement paths of water.)

through distinct conduits, which are called **resurgences.** This distinction is often confused since normal diffused flow may encounter cavities during its movement and ultimately emerge mixed with water derived from sinking streams (Thrailkill 1972). In general, however, resurgences are much more inconsistent in their discharge and chemistry. In some humid areas, springs respond quickly to changes in surface flow (White and Schmidt 1966; Baker 1973a). Baker (1973a), for example, found high, turbid discharges in New York springs during the spring season when snowmelt and rainfall produced high flow in the surface streams. During fall and winter, however, the discharge from springs was less by two orders of magnitude, and the water was clear. In other cases, however, the response of spring discharge to precipitation is significantly affected by storage and flow in the subcutaneous zone (Williams 1983).

The physical controls on springs that rise from conduit systems are themselves quite variable. Some springs represent the emergence of an aquifer stream that flows through a cave system under the influence of gravity. Other springs reach the surface under pressure by rising up cavities or fractures. In most cases, however, springs are ultimately controlled by large structural features or by stratigraphic relationships. Faults sometimes serve as the locus of large artesian springs, especially if the structural relationships produce a large potential difference between the recharge area and the fault (Burdon and Safadi 1963). In other situations, faults may deflect underground flow until it emerges into a valley that stands well below the level of the recharge zone (fig. 12.6). An excellent example of this is in the area of the Kaibab Plateau in Arizona (Huntoon 1974). The plateau north of the Grand Canyon supports no surface streams, and all precipitation not consumed by evaporation infiltrates into porous surface rocks. Huntoon (1974) observes that water collected in those rocks is transmitted to faults that act as drains, funneling the water downward to carbonate rocks that stand 3000 feet below the plateau surface. The water is then discharged through springs at the base of the canyon.

Regional stratigraphy and the dip of beds also may control the position of springs. Baker (1973a) showed that in New York State, spring placement often results from water moving down dip in well-integrated conduits floored by rocks with low permeability (fig. 12.7).

Morphometry of Karst Drainage   A trend in karst geomorphology has been the attempt to describe karst drainage in quantitative terms (P. W. Williams 1966, 1971, 1972a, 1972b; LaValle 1967, 1968; Baker 1973a). Results of these

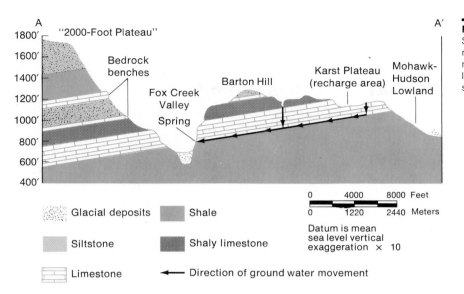

Figure 12.7.
Stratigraphic control on the movement of groundwater. Water moves down dip at contact of limestones and impermeable shale. (From Baker 1976)

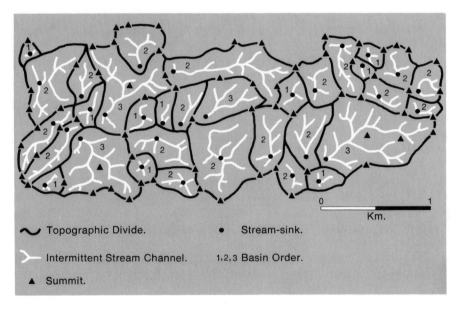

Figure 12.8.
Components of karst morphometry and hydrology in New Guinea as ordered by P. W. Williams.

studies show karst drainage to be well organized rather than having the chaotic nature assumed by early investigators. Williams devised a method, shown in figure 12.8, of ordering stream segments following the Strahler method such that every swallet would accept the drainage of a particular order. After each sink is ordered, measurements of morphometric parameters are made and plotted against the order hierarchy to demonstrate a statistical relationship. Baker (1973a) extended the Williams methodology by including in the analysis underground links between the swallets and the springs that were identified by mapping and dye tracing. Stream orders so designated also showed

Figure 12.9.
Morphometric relationships of the karst hydrology, including subsurface segments, in an area of eastern New York state.

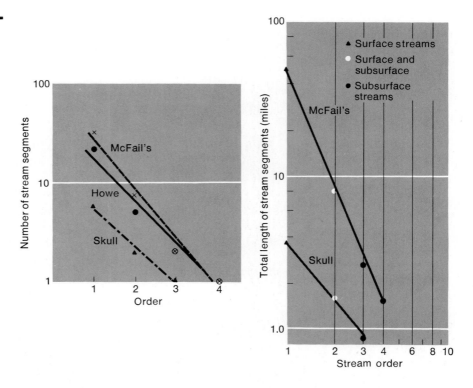

a high correlation to the number of streams and total length of streams (fig. 12.9). In addition, streams flowing in conduit systems seem to maintain longitudinal profiles that are similar to those of surface stream channels in the same area (White and White 1983).

Surficial Landforms

The solution process, working on rocks with diverse properties, results in a number of surficial landforms that define a true karst. The features range in size from tiny modifications of exposed limestone outcrops to large depressions and hills that dominate the topography. However, they all manifest the corrosional process. Unless otherwise noted, the features discussed originate in the humid-temperate climatic regime; a special section will consider humid-tropical karst because of the widespread occurrence and anomalous topographic forms generated under tropical conditions.

Closed Depressions

If someone were to ask what kind of landform best typifies a karst terrain, the answer would have to be closed depressions. Although these depressions range in size from tiny holes to those covering wide areas, they all have in common the property of supporting no external surficial drainage. In addition, it seems unlikely that widespread surficial depressions can develop unless they are connected to an underground conduit system in which water is free to flow to a spring outlet at a lower level (Palmer 1984).

Figure 12.10.
Ground surface pitted by dolines.
Monroe County, Ill.

Dolines   By far the most common karst landform is a closed hollow of small
or moderate size called a **doline** (in the United States often called a **sink** or
sinkhole). Dolines are usually wider than they are deep, having diameters
ranging from 10 to 100 m and depths between 2 m and 100 m. In plan they
are circular or elliptical, but their cross-profile shape can vary considerably
from the normal funnel-like form to shapes resembling a disc, bowl, or cylinder. Occasionally isolated dolines occur, but more commonly they are abundant enough to provide a karst terrain, such as the one in figure 12.10, with
a strongly pitted appearance. For example, Malott (1939) estimates that as
many as 300,000 dolines exist in the karst region of southern Indiana. As figure
12.11 illustrates, dolines can justifiably be considered as the fundamental element of karst because when present in large numbers they substitute for the
valleys that dominate the normal fluvial environment.

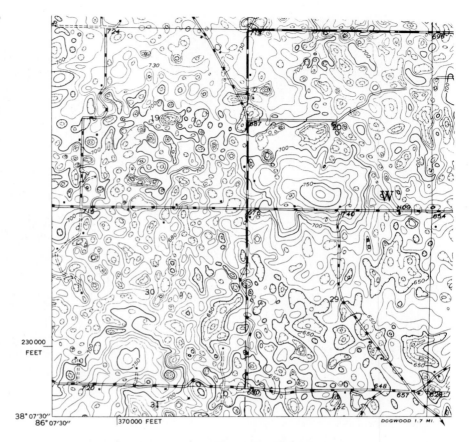

Figure 12.11.
Area of southern Indiana showing well-developed karst topography. Note the predominance of dolines and the absence of surface drainage. (Corydon East Quadrangle, Indiana. U.S Geol. Survey 7 1/2′ quadrangle. Contour interval = 10′)

The term "doline" has suffered through enough different connotations to prompt some authors to call for its elimination, but its use is so widespread that its removal is virtually impossible. In fact, Cvijić (1893) used the term in his classic book on karst, and classifications of doline types are firmly established (Cramer 1941). A variety of doline types have been described, including *solutional dolines, collapse dolines, alluvial dolines,* and *solution subsidences* (Cramer 1941). We will focus on the solutional and collapse dolines, which are the predominant forms. These are treated as separate features even though most dolines are a combination of the solutional and collapse types.

Solutional Dolines   Waters infiltrating into joints and fissures enlarge the cracks by solution and create a closed surface depression called a solutional doline. Many reports have documented the surface-downward origin, and it seems clear that this form represents the paramount doline type. As joints enlarge, especially at the intersection of joint trends, the surface is depressed, and internal drainage fosters continued development of the feature.

Like all geomorphic features, solutional dolines develop best where controlling factors are combined in a particular way. The factors most conducive to the formation of solutional dolines are these:

Figure 12.12.
Doline near Dongola, Union
County, Ill.

1. *Slope.* Since ponding or retardation of flow accelerates infiltration, the frequency of solutional dolines is inversely proportional to the surface gradient. Steep slopes promote rapid flow across the surface, and so valley floors or gently undulating plains are the best places for solvent action to initiate the process. Dolines formed on steeper slopes tend to be asymmetric; however, that phenomenon can also be produced in other ways.

2. *Lithology and structure.* Porous limestones are less susceptible to solutional doline formation than dense limestones that are well jointed. The joints allow selective solution rather than a uniform corrosion over the entire surface. Solutional dolines can form in dolomites, but the features are usually deeper with steep, rocky sidewalls. Structures tend to align and elongate the dolines parallel to the major trends (LaValle 1967; Matschinski 1968; Kemmerly 1976), but the degree of control depends on many variables.

3. *Vegetation and soil cover.* Soil and vegetal cover usually increase solution activity because of the CO_2 factor. Other factors being equal, solutional dolines will develop more rapidly under an organic-rich cover than where surfaces are bare. Trees seem to be especially important in this process (fig. 12.12).

Figure 12.13.
Sequential development of
collapse doline.

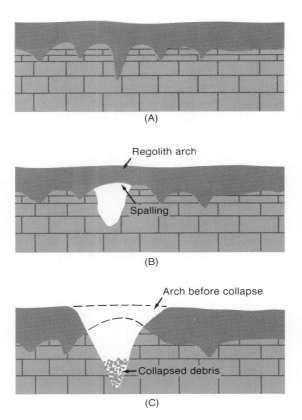

(A)

Regolith arch

Spalling

(B)

Arch before collapse

Collapsed debris

(C)

Collapse Dolines   Collapse dolines differ from solutional dolines in that the depressions are initiated by solution that occurs beneath the surface. Expansion of caverns, caused by corrosion and by the roof material falling under gravity, decreases the support of the overlying rock material. In a study of Tennessee karst, Kemmerly (1980a) suggested that most collapse dolines occur in the partially weathered residuum overlying the solid bedrock. In this process, vertical fractures beneath the residuum are gradually widened by solution. This creates a bridge of unconsolidated debris, a **regolith arch,** that is supported by pinnacles of the underlying bedrock (fig. 12.13). Further widening of the arch supports by solution and simultaneous spalling of debris from beneath the arch of sediment makes it impossible to support the overlying mass, and collapse occurs. Collapse dolines tend to have greater depth/width ratios than solutional dolines. Their sidewalls are characteristically steep and rocky, and the bottom is filled with fragments of the collapsed debris.

The collapse process is facilitated in humid environments where underground drainage is well established, and especially in the downstream portion of that drainage, where dolines seem to be deeper and more numerous (LaValle 1967). Some evidence suggests that collapsing can be initiated by rapid lowering or repeated fluctuation of the water table (Kemmerly 1980a). Although geological processes such as river entrenchment can cause this, the association of drawdown with collapse dolines demonstrates that human action can be a geomorphic agent, commonly with catastrophic results. A classic example of human intervention into the natural balance was reported by Foose (1967). Near Johannesburg, South Africa, a mining company lowered the water table by an extensive pumping program in order to have access to deeper ore that existed in the phreatic zone. Several years after the project's completion, large collapse dolines, some 125 m in diameter and 50 m deep, began to form suddenly and with disastrous results. In December 1962, an ore refining plant dropped 30 m into a doline as the surface suddenly collapsed, and 29 men were killed in one terrifying moment.

Doline Morphometry   The belief that karst forms are controlled by climate was the prevailing concept in karst geomorphology after World War I. This reliance on climate to explain karst processes and form maintained its dominance until investigators became concerned that too much variability of form existed in the same local region where climatic controls were identical (Jennings and Bik 1962; Verstappen 1964). Since the 1960s, investigators have attempted to explain karst in terms of geologic and hydrologic variations, and studies of spatial morphology of the karst features have appeared. For example, Hack concluded as early as 1960 that doline density varies significantly with rock type in the Shenandoah Valley of Virginia (Hack 1960a).

Although analyses of linear karst features have been made (White and White 1979), most morphometric work has concentrated on the spatial characteristics of dolines, relating those to some geologic or hydrologic variable (LaValle 1967, 1968; Matschinski 1968; Williams 1971, 1972a, 1972b; Kemmerly 1976; Palmquist 1979; White and White 1979; Mills and Starnes 1983). These studies have been made in many different parts of the world and under a variety of geologic and climatic conditions. For example, Williams (1966b, 1971, 1972a, 1972b) developed a sophisticated spatial analysis of tropical karst topography. Using aerial photographs or topographic maps, the karst terrain was separated into divides, summits, channels, and stream sinks. Each closed depression was given a number representing the highest stream order of the drainage that disappears into the sink (fig. 12.8). The topographic divides surrounding the depressions form a polygonal network, indicating that the terrain is completely partitioned into separate and adjoining basins. This topography, which Williams calls **polygonal karst,** is dominated by the hills but dynamically controlled by the position of the sinks. Williams analyzed the pattern of sinks by measuring the average distance from each sink to its closest neighbor and comparing that value to the expected mean distance, which is determined from a density analysis of the sink population. The index ratio $\overline{La}/\overline{Le}$ (where $\overline{La}$ is the mean actual distance and $\overline{Le}$ is the mean expected distance) tells whether the sinks have a random or uniform distribution. In New Guinea the stream-sink dispersion is highly uniform, which Williams (1971, 1972a,b) interprets as the best accommodation that can be made as depressions compete for space when the topography evolves under the processes of doline formation.

The example just given reveals the primary assumption of all morphometric studies—that derived relationships will provide insight into the basic factors controlling the morphometry and, through that insight, an understanding of process. In fact, Palmquist (1979) identified the independent variables that initiate and enlarge dolines as (1) groundwater recharge, (2) secondary permeability, (3) regolith thickness and shear strength, and (4) hydraulic gradient. From these he proposed a process-response model in which the initiation and enlargement of primary dolines would subsequently lead to the generation of secondary dolines. The implication is that mixed doline populations can exist within the same karst area.

In a recent study, Kemmerly (1982) demonstrated that such a mixture of primary and secondary dolines is present in the Western Highland Rim area of Kentucky and Tennessee. In that region, one population consists of large, joint-controlled dolines having second-order (or higher) internal drainage and wide spacing. A second population has smaller, first-order swallet depressions that exhibit no joint control. The large dolines are apparently primary in nature and reflect high groundwater recharge, permeability, and hydraulic gradients. Presumably, these may even develop in a manner similar to the subcutaneous flow model proposed by Williams (1983) and depicted in figure 12.14. The smaller dolines, however, were probably developed where recharge, permeability and hydraulic gradient were considerably less.

Assuming the preceding models are correct, we are led to several pertinent observations about karst depressions. First, the initiation of secondary dolines in the manner suggested by Palmquist (1979) requires that those depressions must be linked to an unbroken conduit system. It also follows that no depression can form until sediment covering the swallet is flushed into and through that system (Palmer 1984). Second, the distinction of primary and secondary dolines indicates that time is an important factor in the development of a full karst, a fact that has been emphasized by many researchers (Cooke 1973; Palmer and Palmer 1975; Wells 1976; Kemmerly and Towe 1978; Kemmerly 1980).

The emphasis on morphometry does not mean that climate is irrelevant in karst processes. Obviously, it is an important factor, mainly because it produces the water needed for karst processes to work. Given that, it is local quirks of the environmental setting that determine karst characteristics. What morphometry should demonstrate is that given sufficient time, karst landforms, like most surface features, may approach and possibly attain a dynamic equilibrium, the properties of which are determined by the relationship between process and local controlling factors.

Figure 12.14.
Formation of solutional doline aided by subcutaneous flow that is directed to zones having enlarged fractures and high permeability.

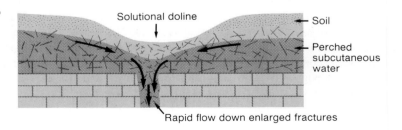

Uvalas and Poljes   Closed depressions of larger size than dolines are called uvalas or poljes. **Uvalas** form as dolines enlarge and coalesce into hollows with undulating floors, the irregularity being produced by the differences in size of the integrated dolines. Jennings (1967) reports a single uvala that was constructed from 14 separate dolines of diverse size and shape. Uvalas have no specific size requirements as they range from 5 to 1000 m in diameter and from 1 to 200 m in depth. Their plan shape can be highly irregular as a result of their unique origin.

Poljes are relatively large closed depressions with flat bottoms and steep sides. They are irregular in plan and usually are elongated along the strike of bedding or some zone of structural weakness. Thus, they can be structurally or lithologically controlled, and some expand by pronounced lateral corrosion when they are temporarily filled with water. Gams (1969) places a rather vague minimum size requirement on poljes, suggesting that they must be at least several square kilometers in area.

Poljes often abut against impermeable and nonsoluble rocks, and rivers flowing through those rocks may extend partly across the polje surface before they sink. In times of high flow the sink may not be able to absorb the discharge, and shallow lakes occupy the polje basin. In dry seasons evaporation may destroy the lakes. As discussed earlier, polje lakes may also form and disappear with changes in the underground hydrology.

Karst Valleys

The second major topographic group in karst topography is karst valleys, which show a variety of charcteristics. In general they can be divided into several types with clearly discernible properties.

Allogenic Valleys   **Allogenic valleys** head in impermeable rocks adjacent to the karstic area. As the surface flow originating in the nonkarstic rocks enters the karst region, it forms spectacular gorges with steep, canyonlike walls. For instance, the Tarn gorge in the Grands Causses area of France is 2 km wide and 300 m deep. Such magnificent valleys obviously require considerable discharge to develop, with a combination of solution and fluvial abrasion as the driving mechanics.

Blind and Dry Valleys   Most surface rivers traversing a karst surface, no matter where they originate, eventually sink into the underground system. They may disappear into holes within the channel (swallets or swallow holes), or the entire drainage may be inside a doline and so all surface water (confined in a channel or not) sinks into the base of the doline. As figure 12.15 shows, rivers flowing across a karst surface in well-defined valleys tend to lower the valley floor upstream from the sink faster than the reach downstream from the sink because the flow is drastically diminished at the point of infiltration. Eventually a limestone scarp may develop, separating the two channel reaches. This scarp can vary in size from a few meters in small streams to tens of meters in larger rivers (Malott 1939). Usually the cliff increases in height with age. Rivers terminating at the cliff face are said to be occupying **blind valleys.**

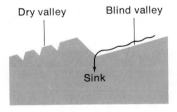

Figure 12.15.
Longitudinal profile of a blind valley and dry valley.

Dry valleys have all the properties of normal fluvial valleys except, as the name implies, they have no well-defined watercourses or carry only ephemeral flow in response to massive floods upstream. They probably represent the most common form of karst valleys and, as mentioned earlier, are the most favorable locale for doline formation because rainwater tends to pond on the valley surface and sink into the rock fractures. Like all karst valleys, their sides tend to be steep, but factors of lithology and age cause some variability in their cross-profile shape. The origin of dry valleys can be complex and varied, but in the simplest and most common case they represent the downstream reach of a blind valley that absorbs all the surface flow at a particular sink (fig. 12.15).

Pocket Valleys   **Pocket valleys** are essentially the opposite of blind valleys in that they begin where groundwater resurges rather than where it sinks. They are normally associated with large springs that resurge on top of an impermeable substrate at the foot of a thick exposure of karst limestone.

Pocket valleys, sometimes called *steepheads,* are usually U-shaped in cross-profile, having steep sidewalls. They characteristically have a steeply inclined headwall that may be recessed by spring sapping that undermines the overlying limestone. Their dimensions vary depending on the spring discharge and the nature of the karst rocks, but some are 8 km long, 1000 m wide and 300–400 m deep (Sweeting 1973).

Tropical Karst

The normal karst features just described can be compared with the spectacular karst topography developed in humid-tropical climates. Once again, we find that studies of tropical karst carried on in diverse regions of the world have resulted in a confusing array of terms describing the same features. Nonetheless, it is safe to say that every landform recognized in karst terrains of temperate regions is present in tropical karst. What sets tropical karst apart seems to be the fact that the general landscape is dominated by residual hills rather than the closed depressions so characteristic of the temperate karsts. Most of the residual forms occur as steep-sided, cone-shaped hills that, combined with surrounding depressions, constitute a particular type of karst topography known as **cone karst** (Ger. *kegelkarst*). The subforms within the general category of cone karst are numerous and transitional, but two topographic components called **cockpits** and **towers** seem to be the uniquely diagnostic elements of tropical karst; in fact the terms "cockpit karst" and "tower karst" are deeply entrenched in the literature.

Cockpits are similar to temperate-climate dolines except that they are usually irregular or star-shaped depressions that surround the residual hills (fig. 12.16). Cockpit karst was first described and named in Jamaica and was originally ascribed to solution along joints and faults (Lehmann 1936; Sweeting 1958). Later work by Aub (as quoted by Sweeting 1973) and P. W. Williams (1971) suggests that the star shape of cockpits arises from gulley erosion of centripetal streams flowing into the depressions. It seems certain, however, that cockpits, like most normal dolines, develop by solution from the surface

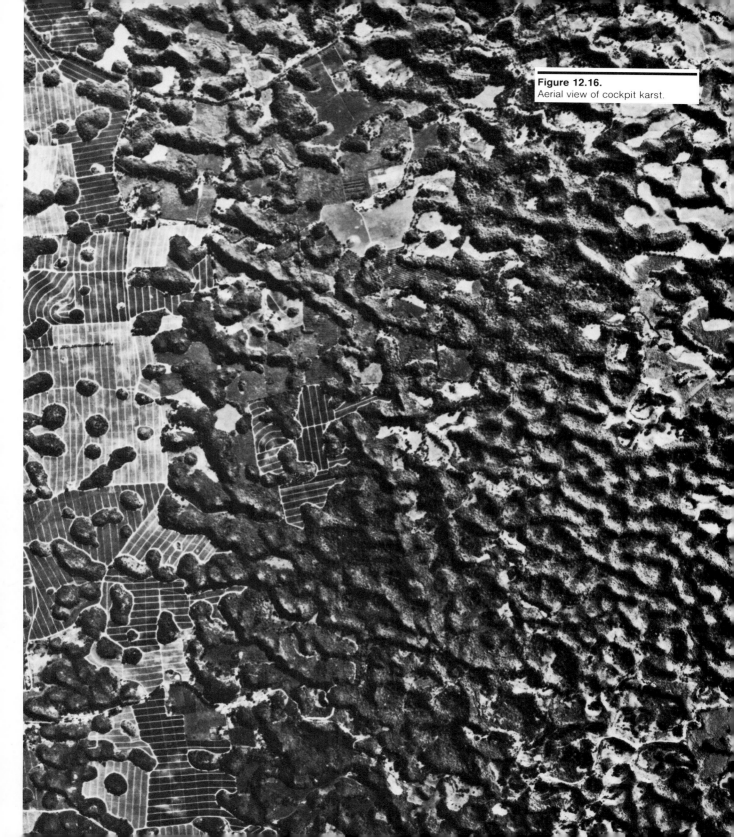

Figure 12.16.
Aerial view of cockpit karst.

downward. Indeed, Monroe (1976) found that circular solutional dolines are the most common closed depression in Puerto Rico, although collapse dolines and uvalas are also present. Where cockpit karst ("cone karst" of Monroe 1976) is prevalent, the residual hills are often joined together to form linear ridges that in turn are cut by gullies into a sawtooth configuration. The alignment of the ridges and the cockpits are apparently so random that joint control seems to be ruled out, although the gullies may be related to zones of structural weakness.

Towers are steep-sided hills and can be vertical or even overhanging. The steep inclination can change where surface erosion piles talus at the slope base (McDonald 1975). According to Jennings (1971), tower karst differs from cockpit karst in the steepness of the residual hills and the presence of swampy, alluvial plains (often similar to poljes) surrounding the towers rather than the depressed cockpits. Towers can be of dramatic size, sometimes rising several hundred meters above the surrounding plain (Wilford and Wall 1965). Towers have also been called *pepinos, haystacks,* and commonly, **mogotes,** a term used in Cuba, Vietnam (Silar 1965), and Puerto Rico (Monroe 1976). In fact, Monroe (1976) suggests that the term "mogote" is probably more appropriate for the feature and should be universally accepted.

Mogotes in Puerto Rico rise from blanket sand deposits as hills between 30 m and 50 m high (figs. 12.17, 12.18). Some have solution rock shelters etched into the mogote sides, but caves passing entirely through the hill, a

Figure 12.17.
Large mogote (tower) in Puerto Rico.

common phenomenon in towers of other regions, are rare. Many of the Puerto Rican mogotes are asymmetric in their cross-profiles (fig. 12.19), which Monroe (1976) attributes to case-hardening of the limestones on the windward slope by repeated solution and reprecipitation of $CaCO_3$. The downwind slopes, being less resistant, are oversteepened and even overhanging due to slumping of the less-indurated rocks. Day (1978), however, observes that asymmetry occurs in only 35 percent of the mogotes and is probably related primarily to erosion at the tower base, with induration being a secondary factor. The basal erosion is most likely solutional, but mechanical erosion may be possible (McDonald 1979).

Some controversy exists as to what process causes mogotes or towers to rise above the alluvial plain. Theories suggested have been collapse of caves, river erosion, and differential solution of the karst rocks; probably the features are polygenetic (Panoś and Štelcl 1968). In Puerto Rico, studies show clearly that the residual hills are composed of the same limestone that underlies the blanket sands. The one significant factor, however, is that the material holding up the mogotes has been indurated by solution and reprecipitation of $CaCO_3$ during alternating episodes of wetting and drying (Monroe 1969, 1976; Miotke 1973). The original limestone surface beneath the alluvial plain was probably irregular, and so the sand deposits preferentially collected in minor depressions during their initial accumulation. As soon as plants became fixed on the sands, the limestones were corroded more rapidly, and this, combined with the case-hardening of exposed limestone knobs, produced the mogote topography.

Figure 12.18.
Mogotes rising from pineapple fields in Puerto Rico.

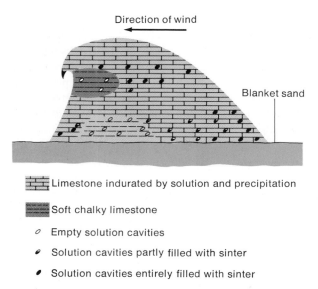

Figure 12.19.
Diagram showing characteristics of an asymmetric mogote in Puerto Rico. (From Monroe 1976)

Direction of wind

Blanket sand

Limestone indurated by solution and precipitation

Soft chalky limestone

o Empty solution cavities

e Solution cavities partly filled with sinter

• Solution cavities entirely filled with sinter

Stalactite

Limestone Caves

Any survey of karst processes and landforms must include a brief discussion of limestone caves. Caves are natural underground cavities that include entrances, passages, and rooms that can be traversed by a human explorer (W. B. White 1976). Technically, caves, being underground features, are not part of karst topography. They may, however, create surface topography by facilitating collapse, and they both influence and reflect the mode of karst hydrology that exists in any particular region. So caves can justly be considered as part of the karst system.

Cave Physiography
As defined, caves have entrances, passages, rooms and blockages called terminations. The assemblage of these components in different combinations produces caves with a variety of shapes and patterns.

Entrances and Terminations   Cave entrances can be found in places such as doline bottoms, hillsides, spring mouths, roadcuts, and quarries. Perhaps the most spectacular entrances to cave systems are openings into vertical voids called *shafts* or *chimneys*. Shafts (Pohl 1955) are cylindrical in shape and evidently formed by solution when water moves rapidly down their walls (Brucker et al. 1972). They are most common in flat-lying rock sequences such as those in the Appalachian Plateau, where shafts sometimes are more than 100 m deep and 15 m wide. Shafts characteristically drain at the bottom through a narrow opening that is connected to a larger cave system. "Chimney" is a term used by White (1976) to encompass all vertical or nearly vertical openings that do not have the cylindrical shape of a shaft. Commonly chimneys follow steeply dipping bedding planes.

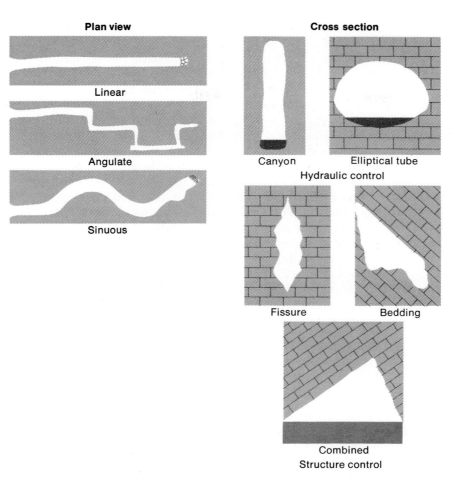

Plan view

Linear

Angulate

Sinuous

Cross section

Canyon

Elliptical tube

Hydraulic control

Fissure

Bedding

Combined

Structure control

Figure 12.20.
Plan view and cross-section shapes of cave passages. (Adapted from W. B. White 1976)

Caves terminate when the passages narrow to the point that a person can no longer follow the opening. These terminations may be due to collapse of overlying rocks, narrowing of the voids into permeable but impassable units, or clay and silt deposits that fill to the ceiling of the cave. Recognize, however, that all caves are part of an integrated network of flow paths even if terminations exist. Water following caves routes still continues through terminations to a spring outlet.

Passages, Rooms, and Patterns   Passages are the main physiographic component of caves. They have various shapes, sizes, and patterns, shown in figure 12.20, that are developed at the same or different times by a single source of water or by a complexly integrated drainage network. In plan view individual passages are either *linear, angulate,* or *sinuous* depending on the structural and hydrologic controls.

In cross section, the shape of a passage represents an accommodation between hydraulics and rock properties such as lithology, bedding, and jointing. Where flow velocities are high and limestones are thick, the passages are usually controlled by hydraulics and exist mainly as narrow, vertical slits called *canyons* or more circular voids up to 30 m in diameter known as *elliptical tubes*. Canyons form in the vadose zone, and tubes usually develop in the phreatic zone. Passages that are controlled by rock variations are much more irregular in cross section (fig. 12.20). Realistically, all variations exist in between the different types, and strict categorization of passage shapes is probably meaningless. *Rooms* are usually nothing more than zones of enlarged passages caused by intersection or enhanced solution.

Almost all passages show a variety of small sculptured markings on their walls, ceilings, and floors. The features are too numerous to describe here (see Bretz 1942 for details). Some are formed by solution and others by abrasion. In addition, many rooms and passages have an abundance of speleothems (chemically precipitated dripstone deposits) that inspire the popular fascination with caves. These features (stalactites, stalagmites, etc.) are composed of a $CaCO_3$ substance called *travertine*. They form when downward-permeating water, saturated with respect to calcite, reaches the cave passage. At this point CO_2 is diffused from the water to the cave atmosphere because the P_{CO_2} in the water is considerably greater (Holland et al. 1964). Some water may also be lost by evaporation. In either case (loss of water or loss of CO_2), the $CaCO_3$ must precipitate as shown in the general reversible reaction *(5)* shown earlier in the chapter.

Passages are usually integrated into a three-dimensional conduit system referred to as cave patterns. In general, the two major types are the **branchwork pattern** and the **maze pattern.** Branchwork patterns, which are by far the most common, are formed by tubular or canyon passages that join as tributaries in the downflow direction much like a normal surficial stream network. The pattern develops because, of all the possible flow routes, only a few are sufficiently open to transport high discharges on low hydraulic gradients. These evolve into the major passageways. Smaller openings are less efficient in hydraulic transmission, and the slow, diffuse nature of the flow causes saturation in distances that are too short for significant passage enlargement (Palmer 1975). An excellent example of the sequence involved in the development of a branchwork pattern is given in Palmer (1981b).

In contrast to the branchwork pattern, the maze system develops as a result of simultaneous, rather than sequential, enlargement of openings. A maze pattern consists of many intersecting passages that form closed loops. The pattern has been divided on the basis of geometry into three subtypes (fig. 12.21): *network maze, anastomotic maze,* and *spongework maze* (Palmer 1975). The specific maze type developed seems to depend on local groundwater recharge type and structural setting. Maze patterns develop only when the branching tendency is somehow suppressed. Palmer (1975) suggests that this occurs mainly in two situations: (1) where aggressive recharge takes place uniformly in all fractures and (2) where floodwater recharge is so variable in character that no stable passage configuration is allowed to develop.

Network
maze

Anastomotic
maze

Spongework
maze

Figure 12.21.
Subtypes of the maze pattern of
cave development.

The Origin of Limestone Caves

The origin of limestone caves has been steeped in controversy since the features were first recognized, and disagreement still rages today. The differences of opinion probably stem from several factors. (1) Historically, theories about cave formation have tried to fit all caves into genetic models with little regard for geologic differences. For example, there is no compelling reason that caves in highly deformed rocks of the Alpine region should develop in precisely the same way as those in the plateau areas of southern Indiana. (2) Most cave models are based on hydrological schemes, yet until recently there has been a remarkable lack of hydrologic data in the arguments. In fact, most theories were founded on physiographic evidence. (3) Very few caves have been examined in great enough detail to substantiate a local origin, let alone an all-encompassing genetic model.

All classical theories of cave development are concerned primarily with the position of the solvent water relative to a water table. Three main ideas have evolved over the years, suggesting that caves form (1) above the water table by the corrosive action of vadose water, (2) beneath the water table by deep circulation of phreatic water, or (3) at the water table or in the shallow phreatic zone, often associated with fluctuations of the water table itself.

The concept that caves develop above the water table by vadose water appeared early in this century and has been complicated by many researchers adopting slightly different variations of the theme (Martel 1921; Malott 1921, 1938; Piper 1932; Gardner 1935). In general, most workers recognized the solution effect of vadose flow, and many suggested that rivers flowing with some velocity under hydrostatic head are able to enlarge caves by abrasion. Some saw importance in the water table level since free-flowing water is essentially on the top of the water table, but others (Malott 1921; Addington 1927) denied its importance.

In 1930 W. M. Davis published his classic "two-cycle theory" of cave development, which depends on deeply circulating phreatic water. Davis argued that most caves are now in a phase of active deposition rather than solution, as evinced by the abundance of dripstone. Thus, cave erosion must have occurred at some earlier time when the water table was higher than the present. According to Davis, solution of the caves took place along curving flow lines in the phreatic zone (fig. 12.22a) that descended deep below the surface and emerged under the major river. Subsequently, the region was rejuvenated by

Figure 12.22.
Models of groundwater flow in the development of caves.

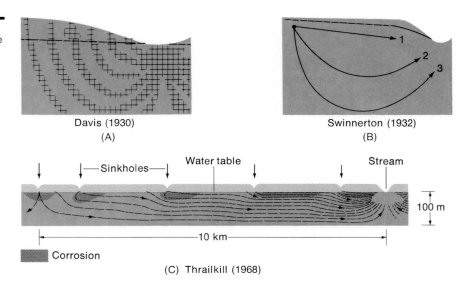

Davis (1930)
(A)

Swinnerton (1932)
(B)

Water table   Stream

↓ |—Sinkholes—| ↓ ↓ ↓

100 m

|—————————————10 km—————————————|

Corrosion

(C) Thrailkill (1968)

uplift, rivers entrenched into the surface, and the water table was lowered to join with the new level of the rivers. This, of course, exposed the already formed caves and allowed dripstone deposition to begin.

The third major model, and its many variations, place the main zone of cave enlargement in the reach where the water table fluctuates with the seasons. Swinnerton (1932) accepted earlier ideas that stressed the importance of the water table. He believed that if water moves along all possible paths beneath the water table, the zone just below the table would be the most undersaturated and would have the greatest discharge (fig. 12.22b). Thus, cave formation would proceed most efficiently in the shallow phreatic zone and should be related to the surface stream level. Other studies seem to support this model (Sweeting 1950; Davies 1960; White 1960; Wolfe 1964) but the most convincing evidence was presented by Thrailkill (1968).

Thrailkill provided a sound theoretical base for the shallow phreatic model by pointing out that where no water crosses the water table, the flow will be nearly horizontal and water just beneath the water table will follow very shallow flow routes, especially if the recharge occurs from a number of discrete point sources (fig. 12.22c). The water at the water table becomes undersaturated when it mixes with vadose water (mixing corrosion), when it becomes cooled (dissolves more CO_2), or when water is introduced by backflooding of nearby rivers. The combination of lateral flow and undersaturation dissolves more limestone during floods, and therefore, the position of cave formation is most likely to be the average level of the flood water table, which he considers to be the upper limit of the shallow phreatic zone.

There seems to be a general sympathy for the shallow-phreatic mode of cave origin (Moore 1960; LeGrand 1983). However, any attempt to restrict cave origin to a specific groundwater zone may be too limiting. Caves can probably form anywhere sufficient groundwater flow and suitable lithologic characteristics exist (White 1969, 1977). Given sufficient groundwater, whether a cave develops in a shallow or deep situation may depend on local controlling factors (for example, Ford and Ewers 1978). The complete integration of caves may also be time dependent; i.e., the location of major solution may shift with time. In addition, it now seems certain, as discussed earlier, that the transport distance of groundwater flow and the percentage of saturation also control where caves may and may not develop. Palmer (1981a) suggests that a great disparity exists in solution rates of karst rocks. This eventually leads to the creation of only a few major passages, and the result is a dendritic (branch-work) pattern. However, beyond a certain discharge there is a limit to the solution rate. Where many different fractures are exposed to flow at or greater than this limiting discharge value, a maze pattern will develop. Therefore, the type of cave pattern and the position of its development may be a function of the mode of groundwater discharge and the local structural condition.

Summary

Karst develops by extensive solution of limestones and other soluble rock types. Limestones that are most susceptible to karstification are crystalline, high in calcite content, and intensely fractured. The limestones should also have no impermeable units in a stratigraphic sequence that stands well above base level. The karst processes function best in humid-temperate and humid-tropical climates where abundant water is available and biogenic CO_2 is freely added to the water from a thick vegetal cover.

Karst areas are characterized by a distinct hydrology in which surface flow is partially or totally disrupted when water is diverted into the underground system. Groundwater movement in karst aquifers differs from normal situations in that the flow is often controlled by the orientation of fractures and interconnected solution cavities.

The most common karst landforms are enclosed depressions called dolines (sinks or sinkholes in the U.S.). Other distinct forms are larger depressions and a variety of surface valleys that are unique to karst regions. Tropical karst differs from temperate karst in that hills rather than depressions dominate the topography. In most regions, karst landforms and hydrologic parameters seem to possess regular morphometric relationships that probably reveal some type of dynamic equilibrium between process and form. Limestone caves are subsurface manifestations of karstification. The mechanics of cave formation are not completely understood. Most interpretive disagreement has centered on where the solution process is most efficient in relation to the water table.

The cave pattern developed depends primarily on the mode of groundwater recharge and the local structures in the rocks.

Suggested Readings

The following references provide greater detail concerning the concepts discussed in this chapter.

Jennings, J. N. 1971. *Karst*. Cambridge, Mass.: M.I.T. Press.

Kemmerly, P. R. 1982. Spatial analysis of a karst depression population: Clues to genesis. *Geol. Soc. America Bull*. 93:1078–86.

Monroe, W. H. 1976. The karst landforms of Puerto Rico. U.S. Geol. Survey Prof. Paper 899.

Palmer, A. N. 1984. Recent trends in karst geomorphology. *Jour. Geol. Educ*. 32:247–53.

Sweeting, M. M. 1973. *Karst landforms*. New York: Columbia Univ. Press.

Thrailkill, J. 1968. Chemical and hydrologic factors in the excavation of limestone caves. *Geol. Soc. America Bull*. 79:19–45.

Williams, P. W. 1983. The role of the subcutaneous zone in karst hydrology. *J. Hydrol*. 61:45–67.

Coastal Zones—Processes and Landforms

13

Introduction

In the preceding chapters we have examined a variety of processes that mold the exposed surface of the Earth into diagnostic landforms. However, 70 percent of our planet is covered by ocean water, and the sea possesses immense energy. The application of this energy on the adjacent land can produce rapid and enormous changes in the nearshore physical environment. Therefore, the erosional and depositional processes that function at the interface between the ocean and the land represent another major realm of geomorphology.

Coastal zones respond to forces like any other geomorphic system. Viewed on a short-term basis, parts of the coast might be considered as quasi-equilibrium forms. In that sense, repeated movement of sediment and water constructs a beach profile that reflects some balance between the average daily or seasonal wave forces and the resistance of the landmass to wave action. Considered over a longer time span (graded time), however, beaches or entire coastlines may be imperceptibly changing toward a larger equilibrium form. On the even larger geologic time scale, marine transgressions and regressions may represent dramatic alterations of the position and character of the coast. Our attention here will focus mainly on the coastal zone during steady and graded time spans.

The study of beaches and coasts is not strictly academic. Most of the largest U.S. cities are near the ocean, and three-fourths of the people in the country live in coastal states. The concentration of people in coastal regions and the pressures involved in using this land for recreation and development have placed a genuine strain on the system. Whether we can utilize the environment completely and still prevent damaging changes in its character depends on how well we understand the functional processes. Engineers, geographers, oceanographers, geologists, and other scientists have made significant contributions to our present understanding of the coastal environment. As usual, however, interdisciplinary communication has not been great because each discipline is engrossed in specific problems. Nonetheless, if demands on the coastal system continue to increase, cooperative efforts of all coastal workers will be needed to avert environmental disaster.

For those interested in coasts or beaches, there exist a monumental wealth of data and many exceptionally good syntheses. The following references treat the topic with a number of different approaches and at many levels of sophistication: D. W. Johnson 1919; Guilcher 1958; Steers 1962; Shepard 1963; Wiegel 1964; Bascom 1964; Ippen 1966; Zenkovich 1967; Manley and Manley 1968; Bird 1969; Muir Wood 1969; Shepard and Wanless 1971; King 1972; Davies 1973; CERC 1973; Komar 1976, 1983b; Davis 1978. Much of the material presented here has been drawn from these excellent sources.

A coast is a relatively large physiographic zone that extends for hundreds of kilometers along a shoreline and often several kilometers inland from the shore. Like all large physiographic entities, coasts have not escaped the irrepressible urge of scientists to classify, and many attempts have been made to categorize their features (see King 1972). Coasts are particularly susceptible to classification because their regional properties are clearly documented on available maps and aerial photographs. In some cases, no extensive field study is needed to systematize coastal properties into a viable classification scheme.

Some of the commonly used classifications of coasts combine, in different ways, a few basic ingredients that serve as the fundamental criteria. Most important of these are (1) form of the land-sea contact, i.e., the configuration of the shoreline; (2) stability or relative movement of sea level; and (3) influence of marine processes. Some of the classifications are purely descriptive and may have little application in dynamic geomorphology. Others are genetic and so are allied closely with the processes involved in developing the diagnostic coastal properties. For example, Inman and Nordstrom (1971) classify the morphology of coasts on the basis of modern plate tectonic theory. On the other hand, as Russell (1967a) suggests, perhaps we have not yet collected enough precise data concerning coastal properties to warrant any classification at this time. In addition, all coasts have a past as well as a present, and so time also becomes an important consideration in coastal classification (Bloom 1965). That is, coasts may reflect evolution and contain relict parts that are not in equilibrium with modern, or even Holocene, processes. For instance, marine terraces standing well above the ocean are not related to modern wave attack but do indicate a coast that has undergone movement relative to sea level.

Because of all these considerations, no specific classification will be advocated here, and no attempt will be made to analyze entire coastal regions. Instead, we will examine only parts of the entire coast that are being actively affected by modern processes. In that sense we will talk about "erosional" or "depositional" coasts, but these terms are applied only on a local basis and have no regional connotation.

Coastal Processes

The processes that initiate change in the coastal zone are extremely difficult to study because they are driven by interrelated forces of high energy, each of which may produce a different response in the same coastal environment. Thus, different processes are not easily studied without permanent installations designed to provide precise measurements over various time intervals. An example of such a research center is operated and maintained at Duck, North Carolina, by the U.S. Army Corps of Engineers. This installation, known as the Coastal Engineering Research Center-Field Research Facility (CERC-FRF) collects data on various oceanographic and meteorological factors, in addition to conducting regular and event-initiated bathymetric surveys of the nearshore environment (table 13.1). Measurements are made from a 561 m steel pier oriented perpendicular to the shoreline (fig. 13.1) and a mobile sampling machine called the CRAB (Coastal Research Amphibious Buggy). The CRAB (fig. 13.2) can be driven as far as a kilometer offshore into water depths up to 9 m.

Table 13.1 Data type and frequency of measurements at the Coastal Engineering Research Center-Field Research Facility, Duck, North Carolina.

Data Type and Collection Technique	Frequency of Measurement
Meteorological	
Temperature	Continuous
Rainfall	Continuous
Barometric pressure	Continuous
Wind velocity and direction	Continuous
Wave Data	
Baylor staff gages on pier	20 min. every 6 hr
Wave rider buoy 2500 m off pier	20 min. every 6 hr
Radar derived wave height, period, angle	On call; at least 1/da
Currents	
Longshore surface current direction determined by dye packets at pier end, nearshore and beach	Daily
Oceanographic	
Surface water temperature	Daily
Water density	
Water visibility	
Water Levels	
Tide levels at pier end gage	Every 6 minutes
Nearshore Bathymetry	
Pier lead lines	Daily
CRAB surveys	
Four lines parallel to pier; 500–600 m N and S of pier; up to 1 km offshore	Biweekly and after storms
Complete surveys along 14 lines within ± 600 m of pier	Monthly

Figure 13.1.
Sampling pier at the Coastal Engineering Research Center, Duck, N.C.

(A)

(B)

Figure 13.2.
(A) Coastal Research Amphibious
Buggy (CRAB) used at CERC for
offshore sampling. (B) CRAB
taking bottom samples near the
CERC pier at Duck, N.C.

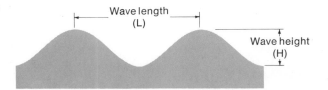

Figure 13.3.
Fundamental parameters of
waves.

The description of CERC-FRF is not meant to suggest that coastal research in the absence of such heroic efforts is inconsequential. On the contrary,
many excellent studies have been accomplished without access to such research facilities. Realistically, however, continuous data collection at permanent stations may be needed to integrate information about diverse phenomena
into reliable syntheses.

Waves

The paramount driving forces in shoreline processes are waves, which expend
the energy they obtained in the ocean against the margins of the land. A thorough analysis of wave theory is steeped in complex physics and mathematics.
Here we can introduce only a simplified version of the basic concepts needed
to understand the geomorphic role of waves, recognizing that such a discussion
is a broad generalization of a very complicated phenomenon.

Wave Generation   Most waves that do geomorphic work are generated by
strong winds blowing across large portions of the open ocean. Precisely how
energy is transferred from the wind into ocean waves is not completely understood (discussed in Komar 1976). Nonetheless, empirical studies indicate that
wave properties reflecting this energy exchange (fig. 13.3) depend primarily
on the wind velocity, the wind duration, and the **fetch** (the distance over which
the wind blows). The fetch is particularly important in determining the height
of waves and their **period,** a parameter that is merely the time interval between
two successive wave crests passing a fixed point. Extremely high waves with
long periods can be generated only when all three controlling factors are at a
maximum.

In the area where they are generated, waves are irregular, and individual
crests are discernible for only short distances before they disappear in a maze
of interfering waves. This chaotic state occurs mainly because small and large
waves generated during the same storm are out of phase. Waves passing through
the system tend to interfere with one another, adding to the height of some
waves and subtracting from others. When waves are enhanced, the heights
developed can be truly awesome. Storm waves are commonly more than 20 m
high under severe winds. The greatest documented wave height, 34 m (112 ft),
was measured in a February 1933 storm generated in the open water in the
deep part of the South Pacific.

As waves move away from their source area, they begin to separate from
one another according to their various periods. This process, called **wave dispersion,** causes regularly spaced successions of waves with rounded crests to

Wave direction

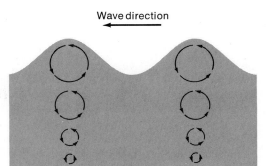

Figure 13.4.
The orbital motion of water particles in a wave of oscillation.

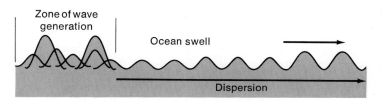

Figure 13.5.
Dispersion of waves from an area of wave generation. Waves having different periods separate from one another to create ocean swell.

migrate from the source zone. Emerging waves typically have a low ratio of wave height to wave length (a parameter known as the **wave steepness**) and appear as the long, low waves commonly referred to as **swell.** In swell, water particles assume the circular orbital paths that characterize deep-water **waves of oscillation** (fig. 13.4). Although individual water particles in oscillatory waves have little forward motion, the waves themselves advance with a typical trochoid form. That is, the wave form is moving forward, not the ocean water.

The dispersive process, shown diagrammatically in figure 13.5, works because wave length and velocity are both a function of the period such that

$$L = \frac{gT^2}{2\pi}$$

and

$$V = \frac{gT}{2\pi},$$

where g and π are the well-known constants, L is the wave length in meters, V is velocity in m/sec, and T is the period in seconds. Because of these relationships, long-period waves with greater length and velocity will separate from short-period waves. For example, a wave with a period of 6 seconds will have a length of 56 m (184 ft) and a velocity of 9 m/sec (30.7 ft/sec). Waves generated in the same storm that have a period of 14 seconds have a length of 303 m (1000 ft) and a velocity of about 22 m/sec (71.5 ft/sec). It is apparent that waves disperse from the generation zone simply because longer waves outrun the shorter ones during any given interval of time.

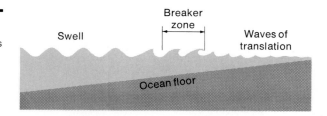

Figure 13.6.
The transformation of oscillation waves into waves of translation as swell approaches the nearshore environment.

Waves with identical periods travel away from their source as distinct groups. It is important to note that storm-generated waves are not the same as waves produced by a point-source impulse such as a pebble tossed into a pond or a tsunami created by rock displacement beneath the ocean floor. In such cases, waves radiate in all directions from the point source. In contrast, ocean swell follows a pathway that encompasses a substantial area of the ocean surface but also has finite lateral boundaries; the direction of the well-defined corridor is determined by the direction of the generating wind. Although the wave path does spread somewhat, it is quite possible for swell to strike a few score kilometers along a coast while leaving nearby reaches virtually unaffected by the storm.

Studies show that swell can travel great distances across the open ocean without losing much of the original energy (Snodgrass et al. 1966). Most energy loss, especially in the shorter period waves, occurs near the generation zone. Once the swell condition is attained, the long-period waves experience little additional dissipation of energy and may traverse an entire ocean basin.

Waves and Shoaling   As swell approaches a landmass, the waves begin to "feel" the effect of the ocean bottom. When the water depth is approximately half the wave length, the oscillatory waves begin a transformation into steplike forms called **waves of translation** (fig. 13.6). Translation waves develop when the orbital path of water particles is intercepted by the ocean floor. The orbits begin to flatten noticeably and their axes of rotation rise to higher levels. Eventually the orbital path is destroyed, producing the different wave type. Waves of translation are different from waves of oscillation in that the water particles have a distinct forward motion without the corresponding backward movement that characterizes oscillation. Once in the form of translation, the velocity and length are determined as

$$V = \sqrt{gh}$$

and

$$L = T \sqrt{gh},$$

where h is the depth of the water.

The change in wave types is accompanied by the phenomenon of breaking waves. Deep-water oscillatory waves remain stable only if the wave steepness (H/L) is lower than 1/7 (.14). As waves approach the shore, water depth is progressively decreased (a phenomenon called **shoaling**), and waves begin to "feel bottom." Their heights increase as the rounded crests become more

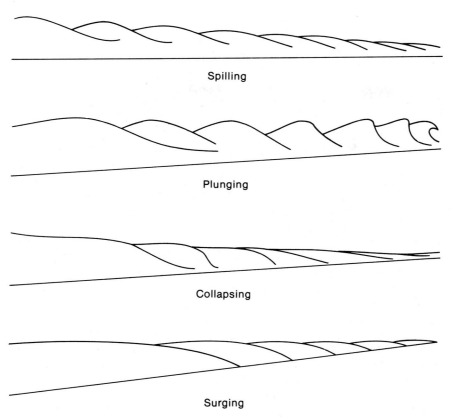

Spilling

Plunging

Collapsing

Surging

Figure 13.7.
Types of breaking waves.

peaked, and their velocities and lengths decrease as the wave crests bunch up; only the wave periods remain constant (fig. 13.6). The combined effect of an increasing height (H) and decreasing length (L) causes oversteepening of the wave; the critical H/L value is attained and the wave breaks. The distance over which breaking occurs depends primarily on the bottom slope, i.e., how rapidly the water depth decreases toward the shoreline.

Three common types of breakers, called *spilling, plunging,* and *surging,* have been observed and are depicted in figure 13.7 (Wiegel 1964). In spilling breakers, the top of the wave crest becomes unstable and flows down the wave front as an irregular foam. In the plunging type, the wave crest curls over the front face and falls with a splashing action into the base of the wave. In surging, the wave crest remains essentially unbroken, but the base of the wave front advances up the beach. A fourth type of breaker, called *collapsing,* has been identified by Galvin (1968); its characteristics are intermediate between those of the plunging and the surging types (fig. 13.7). Actually, breaking waves probably occur in a complete spectrum of types that depend on the bottom slope, H/L, and the period. In addition, breaker style at any given locality varies from periods of storms to periods of relative quiet. This indicates that the type of breaker may be subject to change irrespective of bottom topography.

The significance of breaker style rests in the fact that most types tend to push sediment toward the shore and therefore are relatively benign with regard to beach erosion. Plunging breakers, however, are dominantly erosive and will produce large (often destructive) changes in the beach environment.

Breakers mark the oceanward limit of a zone, called the **surf,** in which the original energy given to the waves receives its final transformation. High velocities and substantial impacts occur under breakers, and thus the creation of translation waves provides the water with kinetic energy that is capable of doing geomorphic work. Once formed, the waves of translation (and the water they contain) move forward to their inevitable collision with the landmass. The surf ends when the wave form is lost as it impinges on the beach face. From there, water carried by its own momentum continues to slide up the beach as **swash** until, at its highest encroachment, the force gathered in the open ocean is finally and totally dissipated. The *swash zone* is alternately covered as the water rushes up the beach face (swash) and exposed as the water moves back down the beach (backwash) under the influence of gravity.

The position of the surf zone may change frequently because breaking waves not only reflect the bottom topography but also use their energy to realign it. That is, swell approaching a shoreline will break according to the existing shoaling configuration, but at the same time, the breakers may shift the bottom sediment so as to produce a new sea-floor topography for the next train of storm-generated waves. In addition, the surf zone changes in response to vertical displacement of the ocean level due to tides, storms, and so forth.

In some cases two sets of swell generated in different areas arrive at a shore simultaneously. This produces a systematic variation in the heights of waves striking the shore, a phenomenon known as **surf beat** (Munk 1949; Tucker 1950). In surf beat, successive waves gradually increase in height until they reach a maximum, then systematically decrease in height to a minimum value. Thus a pronounced periodicity develops whereby one large wave will appear with predictable regularity. In many cases the dynamics are such that every sixth to eighth wave will be at the maximum height, but the precise beat depends on the wave periods of the two swell systems and their resulting harmonics. The variation of breaker heights produced by surf beat affects beach processes because it changes the prevailing water level and notably alters the velocity of nearshore currents.

In addition to these changes in wave mechanics, ocean swell entering shallow water also changes direction. Because deflection of the wave crests is a function of the water depth, the waves adjust to the contours of the bottom topography and so bend according to the configuration of the shoreline. The process, known as **refraction,** works because some sections of a wave crest approaching a coast obliquely are moving in shallow water and at lower velocities than other portions of the same wave, which are traveling in deeper water (Silvester 1966). As shown in figure 13.8, the waves converge on a landmass that juts into the ocean. When convergence occurs, the energy per unit length of wave is increased, causing higher waves and increased energy to impinge on the headland region. Waves can also diverge over embayments or submarine canyons, resulting in lower waves and a concomitant spreading of the energy.

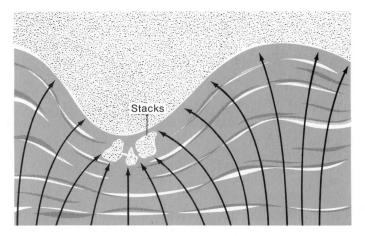

Stacks

Figure 13.8.
Refraction of waves along an
irregular coastline. Wave energy
converges on landmasses that
project oceanward and diverges
in recessed areas.

Tsunamis and Seiches   Like the waves produced in the open ocean by winds, other waves generated in different ways may have geomorphic significance. **Tsunamis** are waves formed by sudden impulses beneath the ocean that cause trains of waves to radiate in all directions from the point source. Tsunamis are usually initiated by earthquakes of more than 6.5 magnitude on the Richter scale with foci located less than 50 km beneath the ocean floor (Van Dorn 1966). Submarine landslides, volcanic eruptions, and slumping have also been cited as causes. In any case, sudden movement displaces the overlying water column, causing it to oscillate up and down as the water tries to reestablish mean sea level.

The waves leaving the source zone of a tsunami have distinct characteristics. They are extremely long (as much as 240 km), their period may be as much as 1000 sec, and in open water they may be only a meter high. Tsunamis obey the same laws that control shallow-water waves, and therefore their velocity is proportional to the water depth. In 10,000 feet (3000 m) of water, the wave velocity will be

$$V = \sqrt{gh}$$
$$= \sqrt{32 \cdot 10,000}$$
$$= 566 \text{ ft/sec or } 386 \text{ mph (approximately 618 km/hr).}$$

In the open ocean, such waves pass quickly and with little notice. As waves approach the shore, however, they seem to trigger a harmonic oscillation that does not follow the bottom configuration. In a typical event, a moderate rise or recession of sea level is followed by three to five major wave fronts that are tens of meters high and capable of great destruction. For example, N. H. Heck (as quoted by Bascom 1964) describes a remarkable case in which a U.S. warship anchored in a Peruvian port in 1868 was picked up by a tsunami wave, carried over the top of the small port city (Iquique), and finally dropped 400 meters inland. After the major waves pass, the nearshore system gradually returns to normal (Van Dorn 1965, 1966). Geomorphic effects of tsunamis are dramatic but probably short-lived; tsunamis occur rarely, and normal waves rework the coast according to more prevalent controls.

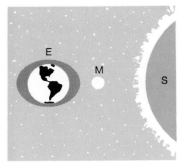

Spring tide

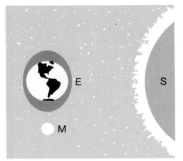

Neap tide

Figure 13.9.
Tidal distribution at various stages of the lunar cycle (S = sun; E = Earth; M = moon). During spring tide the moon and sun are aligned and cause larger tides. During neap tide the moon and sun are not aligned, and tides are lower than average. (Not to scale)

A **seiche** is another wave type that is not directly related to a prevailing open ocean wind. Seiches are free oscillations of water in enclosed or semienclosed basins. Although originally observed in lakes, the seiche phenomenon occurs also in harbors, where it is often called surging, and along open coasts with a broad, shallow continental shelf. A seiche is recognized as a repeated rise and fall of the water level; the oscillatory motion begins when some force displaces the water from its equilibrium position. The driving impulse may be heavy rainfall, flood discharge from nearby rivers, long-period waves such as tsunamis or surf beat, or rapid pressure fluctuations associated with storms (B. W. Wilson 1966). In any case, when the initiating force passes, the oscillations gradually decrease until the equilibrium level is once again attained.

Tides and Currents

Although waves are the dominant force influencing the coastal environment, they are not the only significant water motion. Tides and currents each constitute a type of movement that can modify coastal properties. Although our treatment of these forces will be necessarily brief, a complete understanding of nearshore geomorphic mechanics requires detailed consideration of these factors.

Tides   As any dedicated beachcomber knows, tides usually occur as a twice-daily rise and fall of sea level. Away from coastal areas, this movement is of little concern to most laypeople, but the tidal effect is of consequence to the coastal geomorphologist for several reasons. First, because of the continuous change in water level, the position of wave attack migrates through a notable vertical range and a corresponding lateral shift, increasing the size of the beach and complicating our understanding of beach processes. Second, tides initiate currents that flow into and out of constricted reaches of the shoreline such as bays or lagoons. Many times this ebb and flow can keep drifting beach debris from closing the entrance to the embayment, and in fact, some tidal currents are capable of eroding coastal rocks. This is especially true in narrow bays such as the Bay of Fundy in eastern Canada, where the inland constriction produces a maximum range of 15.6 m, the highest value on Earth. In some constricted estuaries tides move inland with a pronounced wave front called a *tidal bore* (see Lynch 1982). Often the bores actually break, providing observers with a spectacular show. For example, the tidal bore on the Amazon River looks like an 8 m high waterfall advancing upstream for 480 km at a rate of 12 knots.

The tides, of course, are driven by the gravitational effect exerted on the Earth by the sun and the moon (for a nonmathematical discussion see Bascom 1964; for an understandable mathematical discussion, Komar 1976). Here we need only recognize that in most coastal regions the lunar influence results in two high tides and two low tides daily. The gravitational attraction of the sun complements or detracts from that of the moon. As figure 13.9 illustrates, every two weeks the moon and sun are aligned, causing a higher tide than normal, called the **spring tide.** Midway between spring tides, the moon and sun reach positions that are 90° apart. The solar pull detracts from the lunar effect, and the tide is lower than normal, the so-called **neap tide.** Spring tides are about 20 percent greater than the average tidal range and neap tides about 20 percent less.

Not all places on Earth experience the tidal motion that pure astronomical and gravitational theory predicts. Some areas, for example, have only one high and low tide, which occur at the "wrong" times. The tide at any locality also varies in magnitude because of other factors, such as perturbation of the lunar orbit, tilting of the Earth's axis, and ocean bottom topography. Davis (1964) has classified tides according to their tidal range as *microtidal* (0–6 ft), *mesotidal* (6–12 ft), and *macrotidal* (> 12 ft).

Normal tides are not usually destructive. However, some unique tidal events can be devastating, especially if they coincide with storms that produce strong onshore winds. For example, because the orbital path of the moon is elliptical, the moon periodically reaches a position where it is closer to the Earth than at any other time. This point is known as *perigee.* Occasionally the location of perigee is perfectly aligned with the celestial orientation during spring tide, producing what is known as a **perigean spring tide.** These unique tides not only raise the normal tide levels but, more important, increase the rate at which the tide rises (Wood 1978). In March 1962 the chance combination of a perigean spring tide and a large offshore storm generated enormous flooding and erosion along the Atlantic coast from the Carolinas to Cape Cod (fig. 13.10). The event resulted in a loss of 40 lives and $500 million in property damage; in some areas destruction was almost total.

Nearshore Currents   Another type of water motion complicates the nearshore system and our understanding of the mechanics in the surf and swash zones. Other than wave action itself, two wave-induced currents control water movement in the beach zone: (1) a cell circulation consisting of *rip currents* and their associated longshore currents; and (2) longshore currents that usually are generated by waves striking at an angle to the prevailing direction of the shoreline but may also be caused by tides or storms.

Rip Currents and Cell Circulation   Rip currents are narrow zones of strong flow that move seaward through the surf (fig. 13.11). They return to the offshore zone water that was moved toward the beach by waves of translation. The rips are caused by small currents that move on the beach face parallel to the shoreline; these originate about halfway between two adjacent rips. The velocity of the feeder current reaches a maximum at the entrance to the rip zone. Velocity in the rip current itself can be significant, with known speeds reaching approximately 2 m/sec (Sonu 1972).

The development of the cell formed by a longshore current and a rip is related to variations in the rise of mean water level above the level normally attained under still water, a phenomenon called **wave set-up.** Circulation begins because the height of incoming waves varies in a longshore direction, and

Figure 13.10.
Overwash flooding of the New
Jersey coast during the March
1962 offshore storm combined
with the perigean spring tide.

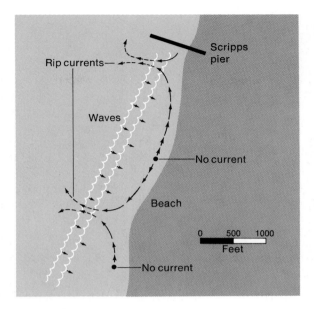

Figure 13.11.
Nearshore circulation pattern at Scripps Beach, La Jolla, Cal., showing rip currents and longshore currents that feed the rips.

higher waves create greater wave set-up. Longshore currents therefore flow from zones of the highest breakers and return oceanward at the position of the lowest breakers. Waves of similar height also initiate cells, but the process probably depends on the type of breakers involved (Sonu 1972).

Shepard and his coworkers (1941) realized long ago that the position of rip currents is controlled by the bottom topography in the surf zone. We can expect, therefore, that areas of wave convergence, where waves characteristically are higher, are the starting points of the longshore component in the cell. Rips occur away from these zones where breaker heights are smaller. Although this analysis has been shown to be correct, rip currents also exist on long, straight beaches with smooth bottoms. Therefore, bottom topography is not the only cause of cell circulation. For example, it is now known that ocean swell by itself produces secondary waves in the surf zone (Bowen and Inman 1969; Huntley and Bowen 1973). These waves, called **edge waves,** have crests normal to the shoreline and wave lengths parallel to the shore; they oscillate with an up-and-down motion. Some authors have suggested that edge waves may be the triggering devices of rip currents (see Komar 1976), but this suggestion is by no means universally accepted (Hino 1975). Regardless of their precise origin, rip currents tend to perpetuate themselves because they can modify the bottom topography so that it enhances the circulation pattern.

Longshore Currents from Oblique Waves   Longshore currents are also generated by waves that strike the beach obliquely. These currents are extremely important in beach mechanics because they tend to move sediment parallel to the shoreline for considerable distances and thereby present coastal engineers with innumerable problems. A number of attempts have been made to derive equations that relate wave properties to the velocity of the longshore current (for reviews, see Galvin 1967; Komar 1976). In general, most equations fail in a predictive sense because they either (1) do not distinguish the longshore current associated with cell circulation from that produced by obliquely striking waves, or (2) employ invalid criteria as the fundamental theoretical base.

One approach, based on momentum analysis, has produced good results and probably represents the best hope for a usable predictive model. Momentum, unlike energy, is preserved as waves break and is separated into two components, one directed toward the shoreline and one directed parallel to it. Thus, the flux in momentum directed along the shoreline should be proportionately related to the velocity of the longshore current. This concept has been developed thoroughly by many authors but most completely by Bowen (1969), Longuet-Higgins (1970), Komar and Inman (1970), and Komar (1975). Komar (1976) suggests that the best estimate of longshore currents at the mid-surf position follows the equation

$$\overline{V_l} = 2.7u_m \sin \alpha \cos \alpha,$$

where u_m is the maximum orbital velocity at the breaker zone, $\overline{V_l}$ is mean velocity in cm/sec, and α is the breaker angle, i.e., the angle of incidence with a line parallel to the shoreline (details of the derivation have been omitted).

It should be pointed out that tides and winds blowing in the longshore direction may complicate the system. At low tide, for example, water can be trapped in troughs that parallel the shoreline. Continued spilling of waves into the troughs drives a circulation pattern in which water moves alongshore confined within the troughs until it reaches a rip channel cut through an offshore bar. There the water turns seaward as part of the rip current. In these cases the longshore current may have a velocity that obeys a mass continuity law rather than a momentum law (see Inman and Bagnold 1963; Bruun 1963; Galvin and Eagleson 1965).

Coastal Storms

In addition to the driving factors discussed above, the dynamics of coasts are directly affected by storms that impinge on the shoreline. Coastal storms occur as two types, tropical and extratropical. Both types are cyclones, meaning that they involve a circular wind pattern, which in the Northern Hemisphere moves in a counterclockwise direction. Extratropical storms derive their energy mechanically in a process associated with the interaction of air motion between zones of high and low pressure. In contrast, tropical storms are fueled by latent heat from the evaporation of water. They evolve as deep, atmospheric low-pressure systems that originate and gain intensity over warm-water marine areas. Tropical cyclones are given a variety of names depending on their geographical location and/or their wind velocities. In the Atlantic and Gulf coasts of the United States, the term "hurricane" (fig. 13.12) is employed when wind velocity exceeds 74 mph (119 km/hr).

Figure 13.12.
Hurricane Diana, September 1984.

Table 13.2 Characteristics of tropical and extratropical storms.

	Tropical Storms	Extratropical Storms
Wind Speed	Great, > 75 mph	Less, usually < 50 mph
Duration[a]	Short, few hours	Longer, many hours to days
Size	Small, 50–80 km	Large, 100's km
Shape	Circular	Often elongate
Surge	Large, > 15 ft	Small, < 5 ft
Barometric pressure	Low central; greater pressure differential	Higher central; lesser pressure differential
Fetch	Small, 10's km	Large, 100's km
Occurrence	June-Sept.	Oct.-Apr.

[a]Duration refers to time of effect on coast.

In the United States, hurricanes are predominant south of Cape Hatteras, North Carolina, whereas extratropical storms are much more common north of Cape Hatteras. This frequency distribution is important because the characteristics of the storms are quite different (see table 13.2). A major difference between the storm types is the magnitude of storm surges that they generate. A **storm surge** is an elevation of normal water level in response to a passing storm. Storm surges are produced when strong onshore winds push and hold water against the coast, thereby "setting up" the mean water level. This occurs because any low pressure promotes a compensating upward bulge of the ocean level (a 1-inch drop in pressure causes about a 13-inch rise in water level). In hurricane-induced surges, the extremely low pressure in the storm center compared to extratropical storms significantly enhances the surge magnitude.

Extreme surges combined with high surface waves allow water to penetrate inland beyond the beach (a process known as **overwash**) and cover areas normally immune from wave attack (see fig. 13.10). Historically, hurricane-related surges are much more damaging than those associated with extratropical storms. This does not imply that major flooding and destruction are impossible in extratropical storm surges. On the contrary, the March 1962 storm discussed earlier is excellent proof that extratropical storms can be significant events. Usually, however, they require involvement of some ancillary factor to reach their full destructive potential.

Beaches

The difficulties of associating entire coasts with formative processes tend to direct our attention to smaller components of the coast where processes are more easily understood. The feature most amenable to process analysis is the **beach,** which is most simply defined as the relatively narrow portion of a coast that is directly affected by wave action. It usually terminates inland at a sea cliff, a dune field, or at the boundaries of permanent vegetation. Oceanward, under the constraints of our definition, part of the beach is continuously submerged because it lies beneath low-tide sea level. It is still part of the beach,

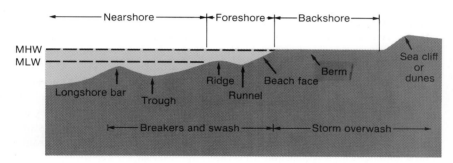

Figure 13.13.
Typical beach profile.

however, because the bottom is subjected to wave action. In that sense the term "beach" is synonymous with the **littoral zone,** an expression used commonly in geological work.

The location of major beach forms as significant components of a coastline depends, to a large degree, on the availability of sand. The primary source of such debris is the enormous detrital load delivered to the coastal domain by major rivers. For example, the Mississippi River provides more than 3 billion tons of solid material to the Gulf Coast region annually. Sea cliff erosion adds a significant amount of sediment, but normally it is less than 10 percent of the total debris available for accumulation in beaches (Komar 1976). Small additions are made to the detrital load by the wind or by slow onshore transport of sand eroded from unconsolidated shelf regions.

The Beach Profile

Beaches represent dynamic systems where loose granular debris is moving steadily under the attack of waves and currents. If water motion could be held constant, this debris would be molded into a characteristic profile that would reflect an equilibrium between the driving forces (waves and currents) and the properties of the beach sediment. We can thus visualize an equilibrium beach profile for any set of water and sediment conditions. Waves and currents, however, do not remain constant but change their properties on a daily or seasonal basis, requiring some response in the process of sediment transportation and ultimately in the beach profile. Although we can think of the equilibrium profile as the ideal case, under the dynamics of the natural setting its properties are constantly changing along with the driving forces.

The beach profile, shown in figure 13.13, consists of a number of component parts, each of which develops its own diagnostic characteristics. In its entirety the profile represents a topographic form that induces waves to dissipate energy by breaking. The exact location of breakers is important because it determines where the greatest amount of energy is expended and which part of the beach will be subjected to the greatest change. The **berm** is a nearly horizontal surface on the backshore portion of the beach. Some beaches have more than one berm; others have none, especially if the sediment is coarser than sand. Landward the highest berm terminates at the base of the sea cliff, and oceanward it joins the **beach face** (fig. 13.13). Because the berm is formed by deposition of sediment during backwash, its elevation is determined by how high the swash runs up the beach and by the grain size of the sediment. As swash moves up the beach, it loses velocity because of friction and the loss of water that permeates into the beach debris. Continued upward growth of the berm surface would require ever higher waves and swash run-up. In this sense berms are analogous to vertically accreted floodplains, and their rate of upward growth must decrease with time, for only infrequent storm waves can add sediment to the surface. As Bascom (1964) points out, however, storm waves also erode the front edge of the berm, thereby reducing its horizontal length while simultaneously building the remaining surface to a higher level.

The beach face is the sloping section of the beach profile immediately seaward of the berm. The slope of the beach face is controlled by many factors and so may range from nearly horizontal to gradients that approach the angle of repose for unconsolidated sediment. Sediment moves up the beach face with the swash and down the beach face during the less powerful backwash. The beach face therefore represents a surface striving to attain some balance between onshore and offshore sediment transport. Its slope adjusts to provide the balance.

Beach face slopes are directly proportional to the size of the particles being moved in the swash zone. Large particle sizes tend to maintain steep slopes and vice versa (Bascom 1951; Wiegel 1964; McLean and Kirk 1969; DuBois 1972; Wright et al. 1979). We would expect such a relationship because slope provides the backwash velocity needed to transport sediment of any given size, but the phenomenon is much more complicated, and beach slopes probably relate to many factors involved in shoreline dynamics.

In the zone seaward from the beach face, a submerged **longshore bar** is commonly, but not always, present. An associated trough develops between the bar and the beach face. Longshore bars may be absent from the profile, especially if the beach is steep. On the other hand, where the beach gradient is low, several bars may be present. The creation of multiple bar systems may be accomplished in a variety of ways: (1) each bar may reflect the breaker position for waves of different sizes; (2) the bars may relate to shifting breaker positions associated with high and low tide levels; or (3) oscillation waves that break far offshore on low-gradient beaches may re-form over the trough and break again closer to shore, each episode of breaking creating an associated bar.

These hypotheses concerning bar formation are all based on the assumption that breaking waves are intimately involved in the formative mechanics. A number of early laboratory and field studies support that contention (Evans 1940; King and Williams 1949; Shepard 1950), and the relationship is now generally accepted as being real. Usually breakers establish the size, position, and depth of the bars and troughs. Larger breakers normally produce features in deeper water. These generalizations, however, must be conditioned by evidence showing that bar positions may migrate shoreward or seaward according to variations in wave height, steepness, and breaker type.

Some bars are continuous parallel ridges that extend unbroken for tens of kilometers. Bars do not usually display such regularity, however, because shifting wave directions realign parts of the bar system and break ridges into segments with a chaotic distribution. Violent storms also produce drastic changes in bar directions and patterns.

The maximum depth of bar formation depends on the depth to which wave action is able to agitate the bottom sediment. Under average wave conditions, this depth is approximately 10 m below low tide, but large storm waves are known to move bottom sediment in water as deep as 25 m. The deepest bars, molded in the fury of violent storms, maintain their position and shape for long intervals during which normal waves do not influence the bottom. In considering bars as products of wave action and as controlling factors in the breaking process, the shallow shoreward bars and troughs are therefore decidedly more important in nearshore wave dynamics.

Beach Morphodynamics

The ideal beach profile is subject to pronounced changes resulting from deposition of materials that are added to the profile or by eliminating components during erosion. In addition, the position of the major parts of the profile may be changed according to the dynamics of the modifying factors.

It has generally been assumed that the most pronounced changes in the ideal profile are brought about by the dynamics associated with storm waves. Where storms are common, the large and vigorous waves tend to destroy or drastically limit the extent of the berm. The eroded material is simultaneously shifted to an offshore position where it collects in a series of longshore bars. In contrast, where storm waves are unusual, the movement of beach sand is shoreward. This tends to destroy the offshore bars and transport the sediment to the upper part of the beach where it rebuilds a broad, flat berm. Because the frequency of storms is a seasonal phenomenon, we can expect the beach profile to change dramatically from one part of the year to another. Many authors, therefore, refer to a *summer profile,* characterized by the absence of bars and a wide berm, and a *winter profile,* with no berm and a series of longshore bars; an example is shown in figure 13.14. It is probably incorrect to think of the summer and winter profiles as occurring specifically during those seasons. What we are really talking about is a profile formed by storm waves versus a profile generated by swell. Many field studies document the offshore movement of sands during storm wave conditions and the landward transport under swell (Shepard 1950; Bascom 1964; Strahler 1966; Gorsline 1966), and the fact of the process cannot be questioned.

We now know, however, that the summer and winter profiles (fig. 13.14) probably apply best in certain environmental conditions such as those along the California coast. In other areas (eastern U.S. beaches and many Australian beaches), the morphodynamics are much more complex. For example, numerous studies of Australian beaches (Short 1979; Wright et al. 1979; Short

Figure 13.14.
Generalized summer and winter profiles along Scripps Pier, La Jolla, Cal. (Adapted from Shepard 1950)

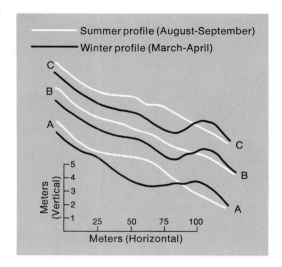

and Wright 1981; Wright et al. 1982a; Wright et al. 1982b; Wright and Short 1983) have shown how preexisting beach topography will affect transporting mechanisms in the surf zone and how fluid motions other than storm waves, such as rip currents and low-frequency standing oscillations, are extremely important in determining the beach character.

The detailed studies of Australian beaches has led to the perception that beaches and surf zones can be classified into six **morphodynamic states,** each of which has a distinctly different association of morphology, water motions, and sediment characteristics (Short 1979; Wright et al. 1979; Wright and Short 1983).

The end-member states in this classification are called reflective and dissipative. **Reflective** beaches are characterized by steep, linear beach faces and well-developed beach cusps (discussed later) and berms (fig. 13.15). They tend to associate with surging breakers, high runup, and minimum setup of sea level. **Dissipative** beaches have low-angle, concave-up beach faces attached to a wide, flat surf zone. Topographically, they have one or more subtle offshore bars (fig. 13.15). Beach cusps are absent, and berms, if present, are poorly defined.

In simplest terms, the reflective and dissipative systems differ dynamically because the accumulated wave energy is expended at different places. In the reflective system, most of the incident wave energy is expended at the beach face. In dissipative systems, most wave energy is expended offshore, where energy is lost in turbulence as waves break over the bars. This pronounced morphodynamic distinction can be demonstrated on the basis of the **surf-scaling parameter** (Guza and Inman 1975), which is expressed as

$$\epsilon = a_b \, \omega^2/g \, \tan^2\beta,$$

where a_b is breaker amplitude, ω is the incident wave radiant frequency ($2\pi/T$; T = period), β is beach/surf zone gradient, and g is acceleration of gravity.

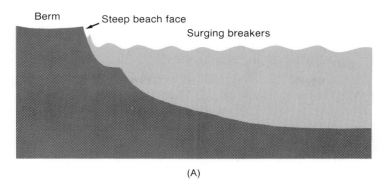

Berm   Steep beach face
Surging breakers

(A)

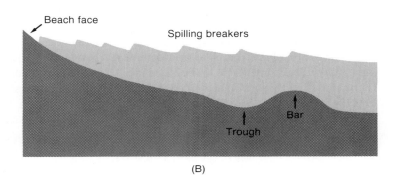

Beach face
Spilling breakers

Bar
Trough

(B)

Figure 13.15.
Generalized profile characteristics and wave properties for reflective and dissipative morphodynamic states. (A) Reflective. (B) Dissipative.

Strongly reflective beaches occur when $\epsilon < 2.5$. In these instances, surging waves prevail. Waves reach the beach face without breaking and surge up the beach or collapse over a step at its base. Turbulence is therefore confined to the zone of runup on the beach face. Wave energy that is reflected may be trapped and initiate strong, standing-wave motion, commonly in the form of edge waves. This motion promotes the formation of beach cusps. In extreme dissipative systems, high ϵ values (> 30) prevail. Spilling breakers occur 75–300 m seaward of the beach face, and the wave bores become smaller and lose much of their energy before they reach the inshore zone.

In Australia, four intermediate states have been identified between the two distinct end-member states. The intermediate states are much more difficult to characterize because each contains both dissipative and reflective elements. In some cases, the morphodynamic state may change from the offshore to nearshore position or with the high and low tides. This complicates our understanding enormously because different features may develop during the tidal cycle or with distance from the shore. It is also apparent that the tidal range may be significant. In general terms, however, dissipativeness increases with increasing values of ϵ, increasing swell-wave heights, and decreasing bed slope. Greatest dissipativeness also prevails in areas with the largest amount of inshore sediment and during and immediately following a severe storm. As longshore bars migrate shoreward, transitional states develop, leading to a steepened beach face and reflective conditions.

Because morphodynamic states are transient from one to another, it is probably not correct to think of an ideal or equilibrium beach profile. However, when considered over a long term, any beach and surf zone will assume a most frequently occurring condition, which Wright and Short (1983) refer to as the **modal state.** The modal state is dependent on the forcing **wave climate** (the most frequently occurring wave conditions) and the input of the local environmental setting. Thus, although there may be no all-encompassing equilibrium profile, any given beach will attain its own particular modal state, about which variations will occur as the water motions and bottom configurations continuously change.

In addition to the morphodynamics just discussed, large beaches are usually spatially and temporally altered by currents moving in the longshore direction. Although these currents can be generated in a variety of ways, the majority of longshore flow derives from the momentum carried by waves striking the beach in an oblique direction. This phenomenon takes place because waves are seldom completely refracted and, therefore, usually make their final approach at some angle to the shoreline. The velocity of a longshore current varies directly with wave height and the angle between the shoreline and the approaching waves. Velocity also increases with distance from shore, reaching maximum values near the mid-surf zone. Longshore currents transport sediment that has been entrained by wave action, a process referred to as **littoral drift.** Because obliquity of wave incidence tends to retard development of rip currents, longshore currents are generally continuous and have the potential to transport sediment for great distances.

Engineers and other scientists have attempted to construct predictive models to estimate rates of littoral drift. In most cases, their predictions have relied on empirical correlations between the rate of sand movement and some estimate of the wave power expended in the longshore direction (see Komar and Inman 1970; Komar 1976; 1983). In some cases empirical studies show a good correlation between wave variables and transport rates, and parameters used as indicators of the longshore components of wave power or energy are compatible with wave theory (Komar and Inman 1970; Komar 1971a). A complete understanding of the process, however, is complicated by the fact that littoral drift occurs in two different modes of action. When wave steepness is high ($H/L > 0.03$; Bascom 1964) most transport occurs beneath the breaker zone. But when the steepness is low, the drift functions along the beach face in a process known as *swash transport* (beach drift). In this case, waves breaking obliquely to the shoreline push sand grains up the beach face perpendicular to the direction of the wave crests. The backwash, however, affected only by gravity, pulls the water and grains down the beach face in a direction perpendicular to the shoreline. Thus, mobile grains in swash transport move in a spasmodic, zig-zag fashion (fig. 13.16). Besides wave steepness, the angle of wave incidence (α) also is important in the drift rate, maximizing the rate at a 30° angle.

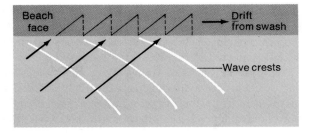

Figure 13.16.
Drift of sediment along a beach face caused by waves striking at an angle to the beach.

The littoral drift process is further complicated because sediment can be moved in suspension or as bedload. Brenninkmeyer (1975) showed that suspended load is important only in a narrow offshore zone. This analysis seems to contrast with earlier studies, which suggested that suspended load was the dominant mode of transport (Fairchild 1973; Thornton 1973). Perhaps some of the difficulty in identifying load types is due to the fact that fine sediment is commonly winnowed from beach debris and carried away from the near-shore environment by rip currents and normal offshore diffusion. The rate and volume of sediment movement is also dependent on the grain size of the transported load (Duane and James 1980).

Considerable doubt remains about the certainty of evaluation of littoral drift rates based on empirical or theoretical analyses. For example, Komar (1983a, p. 17) states, "From the methods described above, the final uncertainty in the evaluation of the net drift could be much greater than the net drift itself, and quite conceivably the direction of the net drift might be incorrectly evaluated."

The best evaluation of littoral drift is based on direct measurements of volumetric loss caused by erosion or beach growth brought on by deposition of drifting sand. These measurements lend themselves to a budget of littoral sediment in which all sediment contributions to a beach (sources) and all losses from a beach (sinks) are taken into account. The main sediment sources are rivers, sea cliff erosion, artificial beach nourishment, and incoming littoral drift. Common sinks are outgoing littoral drift, offshore transport, beach sand mining by humans, and debris moved down submarine canyons. Over a period of years, a particular beach will probably have a positive or negative budget. Obviously, a negative budget indicates more losses than gains, and therefore the beach is eroding. Positive budgets indicate net deposition.

In practice, making reasonable estimates of the components that make up the littoral budget are very difficult. However, the positive or negative character of the budget can be determined by long-term monitoring of erosion and deposition on the beach. This information is very important in predicting the potential impacts that might be caused by humans.

The rate of longshore drifting of sand can be significant when considered on a time scale of human life, and the process is often shown in the accumulation of beach debris against man-made structures such as groins (fig. 13.17). The net littoral drift is the sum total of movements initiated by waves arriving at the shoreline from various directions. Waves may carry debris in one direction for a short period and then, as conditions change, return the material to its original position. Such drift reversals may be random or they may be decidedly seasonal. In California, for example, sand usually drifts southward in the winter and northward in the summer. Net littoral drift can be determined only by long-term budget studies that can detail which direction is the prevailing one. Nonetheless, the process of longshore transport perhaps causes more problems for coastal engineers than any other shoreline phenomenon. It places sand where we do not want it (across harbors and bays) and removes sand from locations where we would like to keep it (resort beaches). Table 13.3 shows net littoral drift along a number of coastlines. The volume of debris involved in the transport demonstrates the tremendous problems engineers face.

Figure 13.17.
Southern coast of Cape Cod, Mass., showing a series of groins constructed to protect the beach from erosion. Groins trap sand that is moving by littoral drift from right to left on photo. Mouth of river is protected by jetties.

Table 13.3 Representative rates of littoral drift along coasts of the United States.

Location	Predominant Direction of Drift	Rate of Drift (cu yd per year)	Method of Measure of Rate of Drift	Years of Record
Atlantic Coast				
Suffolk Co., N.Y.	W	300,000	Accretion	1946–1955
Sandy Hook, N.J.	N	493,000	Accretion	1885–1933
Sandy Hook, N.J.	N	436,000	Accretion	1933–1951
Asbury Park, N.J.	N	200,000	Accretion	1922–1925
Shark River, N.J.	N	300,000	Accretion	1947–1953
Manasquan, N.J.	N	360,000	Accretion	1930–1931
Barneget Inlet, N.J.	S	250,000	Accretion	1939–1941
Absecon Inlet, N.J.	S	400,000	Erosion	1935–1946
Ocean City, N.J.	S	400,000	Erosion	1935–1946
Cold Spring Inlet, N.J.	S	200,000	Accretion	—
Ocean City, Md.	S	150,000	Accretion	1934–1936
Atlantic Beach, N.C.	E	29,500	Accretion	1850–1908
Hillsboro Inlet, Fla.	S	75,000	Accretion	—
Palm Beach, Fla.	S	150,000 to 225,000	Accretion	1925–1939
Gulf of Mexico				
Pinellas Co., Fla.	S	50,000	Accretion	1922–1950
Perdido Pass, Ala.	W	200,000	Accretion	1934–1953
Galveston, Texas	E	437,500	Accretion	1919–1934
Pacific Coast				
Santa Barbara, Calif.	E	280,000	Accretion	1932–1951
Oxnard Plainshore, Calif.	S	1,000,000	Accretion	1938–1948
Port Hueneme, Calif.	S	500,000	Accretion	1938–1948
Santa Monica, Calif.	S	270,000	Accretion	1936–1940
El Segundo, Calif.	S	162,000	Accretion	1936–1940
Redondo Beach, Calif.	S	30,000	Accretion	—
Anaheim Bay, Calif.	E	150,000	Erosion	1937–1948
Camp Pendleton, Calif.	S	100,000	Accretion	1950–1952
Great Lakes				
Milwaukee Co., Wis.	S	8,000	Accretion	1894–1912
Racine Co., Wis.	S	40,000	Accretion	1912–1949
Kenosha, Wis.	S	15,000	Accretion	1872–1909
Ill. State Line to Waukegan	S	90,000	Accretion	—
Waukegan to Evanston, Ill.	S	57,000	Accretion	—
South of Evanston, Ill.	S	40,000	Accretion	—

From J. W. Johnson 1956. Used with permission of the American Association of Petroleum Geologists.

Shoreline Configurations and Landforms

In addition to the modal profile that is oriented perpendicular to the shoreline, many coastal geomorphologists now accept Tanner's (1958) premise that beaches develop a shoreline configuration revealing another type of balance between water energy and sediment supply. This plan-view shape is best established where no long-term unidirectional movement of sediment occurs parallel to the shoreline. Like a graded river, the shoreline configuration is developed "over a period of years" and is adjusted to the prevailing wave characteristics.

It would seem that the most logical environment to preserve a modal configuration would be protected bays where there is no dominant longshore current. In such a locale, waves might produce minor longshore transport of sediment where the wave crests are not completely refracted. The sediment will continue to drift until the shoreline is reoriented parallel to the attacking

Figure 13.18.
Sand barrier off the coast of northern Nantucket Island, Mass., separated from the mainland by a lagoon. Large cuspate features have been formed on the lagoonal side of the barrier.

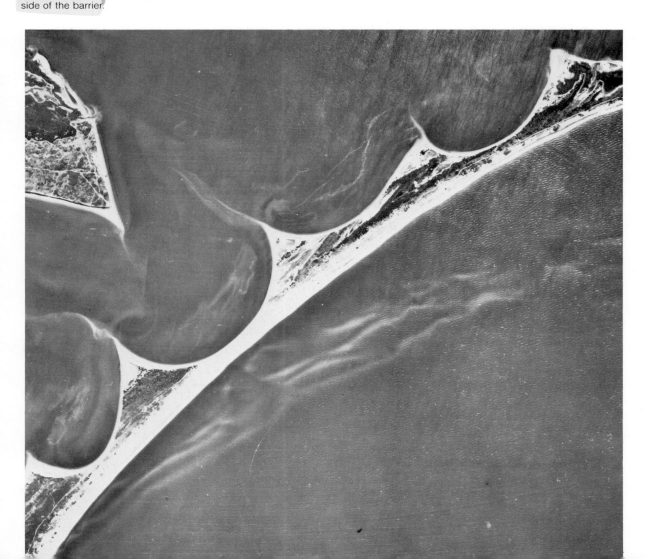

wave crests at every segment. Offshore topography thus determines wave refraction, and the waves in turn establish the shoreline configuration. The complications of even this simple model are staggering. For example, any additional sources of sand to the beach will prevent complete refraction because some longshore transport away from the source will be necessary. In addition, bottom topography is so variable and so susceptible to change that it seems too much to expect that we can precisely describe the form of an equilibrium shoreline.

In light of the above, it is indeed remarkable that many shorelines, open to all types of waves and currents, contain features that are similar in shape and spaced with a regularity that can hardly be attributed to coincidence. Usually crescentic, the features form as periodic seaward projections of the shoreline itself or, on straight shorelines, as narrow pointed accumulations of sediment piled perpendicular to the shore. In either case the seaward projections are separated by a curved embayment (fig. 13.18). A complete hierarchy of these features seems to exist (Dolan and Ferm 1968; Dolan et al. 1974), ranging from minor forms with a wavelength of less than a meter to major cuspate-like indentations of the coastline with spacing measured in hundreds of kilometers. Table 13.4 shows this hierarchy.

Beach Cusps

The most common crescentic forms are **beach cusps.** Cusps develop at the upper part of the beach face and along the outer fringe of the berm. They are usually spaced less than 30 m apart and can form in beach sediment of any size, including boulders and cobbles (Russell and McIntire 1965). Some sorting is produced in the formative mechanics because the cusp projections, or **horns,** are usually more coarse-grained than the intervening embayments.

Table 13.4 Hierarchy of crescentic landforms found on coasts.

Form Characteristic	Cusplet	Cusp	Sand Waves	Secondary Capes	Primary Capes
Spacing	0 to 3 m	3 to 30 m	100 to 3000 m	1 to 100 km	200 km
Material	Fine sand-gravel	Sand-boulders	Sand	Sand	Sand-gravel
Topographic Association	Step	Berm, beach face	Beach berm-offshore bar system	Coastal plains; shores with sufficient sediment	Coastal plain deltas
Rhythmicity	Yes	Yes	Yes	Often	Not always
Motion	Fixed	Normal to beach	Downdrift	Probably downdrift	Slow downdrift
Temporal	Minutes to hours	Hours to days	Weeks to years	Decades	Centuries
Suggested Processes	Swash action on beach face, Groove erosion	Berm deposition and erosion	Wave action, nearshore circulation cells, back eddies of longshore transport currents	Kinematic nature of sediment transport, circulation cells	Wave action, confluence of coastal currents, back-set eddies, and shoals

From Dolan et al. 1974. Used with permission of *Zeitschrift für Geomorphologie*, published by Gebrüder Borntraeger, Stuttgart.

Beach cusps seem to form most readily where wave crests strike parallel to the shoreline. This perhaps explains why the features tend to remain fixed in their positions, although laboratory studies indicate that some longshore migration of the forms is possible if the drift is not excessive (Krumbein 1944). There seems to be some agreement that spacing of cusps is related to the wave height as well as wave direction; the higher the waves, the greater the spacing interval. Even this, however, cannot be pronounced as an inviolate rule; A. T. Williams (1973) found no correlation between wave height and spacing. The spacing of the cusps he studied in a Hong Kong bay were most closely related to the swash distance, i.e., the length between the breaker zone and the highest encroachment of the swash.

Because every study of beach cusps seems to reveal some contradiction of earlier studies, it should come as no surprise that the origin of this feature has been controversial since its earliest description. The simplest explanation of the feature, proposed by D. W. Johnson (1910, 1919) and modified by Kuenen (1948), is based on a process that causes irregular erosion of the beach face. Swash erosion of the beach face initiates cusps, and backwash transports the sediment away. Eroded materials are carried seaward until they deposit as deltaic projections that stand opposite the excavated hollows. Continued swash action progressively transforms the initial hollows into larger embayments until the water crossing the depressions reaches a critical depth. Swash is then refracted in such a way that coarse sediment is deposited on the cusp horns and finer sediment is transported farther offshore. Bays and horns grow until the central area attains a limiting depth, swash action is retarded, and the feature assumes an equilibrium condition.

Actually, the process is more complicated, and it may be that cusps can be generated in several ways. It is now clear, however, that many investigators believe that most cusps are formed in coastal settings that produce a coupling action between incident-surging waves and edge waves (Bowen and Inman 1969, 1971; Bowen 1973; Komar 1973; Guza and Inman 1975; Dolan et al. 1979; Holman 1983). In these settings (primarily reflective state), edge waves augment the breakers systematically in the longshore direction. At positions where the breakers are relatively high, recesses are cut into the beach face or berm by the swash-backwash sequence. The cusp development occurs within minutes or hours after the generating event begins, and following the event the shoreline displays the rhythmic cusp pattern on the beach surface.

It may be possible that cusps can form in other wave climate settings. For example, DuBois (1978, 1981) and Sallenger (1979) suggest that cusps have developed under plunging wave attack. Cusps in these cases form gradually after swash extending over a berm is ponded in low areas on the berm. As the ponded water is allowed to return seaward, it cuts channels through the berm that are molded into cusps by action during tidal variations.

Large-Scale Rhythmic Topography and Capes

Larger features in the hierarchy of crescentic forms have been called **rhythmic topography** by Komar (1976). They are of two main types: (1) crescentic bars, and (2) rhythmic variations that are controlled by rip currents associated with cell circulation. The latter forms have been referred to variously as *sand waves, giant cusps,* or *shoreline rhythms.* The features occur in widely divergent environments including large lakes (Evans 1938; Krumbein and Oshiek 1950), enclosed seas (King and Williams 1949), and open ocean coastlines in many parts of the world.

Rhythmic topography differs from beach cusps in several important ways. (1) Rhythmic topographic features commonly migrate parallel to the shoreline. The rate of the movement varies but can be a kilometer or more a year. (2) Much of rhythmic topography is submerged. The emergent forms are usually observed as giant cusps or sand waves that have considerably greater spacing than beach cusps. The wavelength of these features ranges from 100 to 3000 m, with horns extending oceanward as much as 25 m. The submerged rhythmic topography is usually in the form of crescent-shaped sand bars or longshore bars that are segmented by rip current channels. The crescentic bars are concave toward the shoreline. They commonly stand opposite the horns of large cusps exposed on the beach face itself, but they also exist off straight beaches, especially in protected bays with a small tidal range (Shepard 1952, 1963; King and Williams 1949; Bowen and Inman 1971). (3) Rhythmic topography seems to be more dependent on the bottom configuration in the offshore surf zone than normal beach cusps.

Uncertainty about the origin of rhythmic topography equals the mystery surrounding beach cusps. There is little doubt that rhythmic topography is a function of wave action, cell circulation with associated rip currents, and longshore transport of sediment driven by obliquely striking waves. Precisely how these combine to construct the features and the details of the processes is simply not clear.

In addition to the cuspate features already examined, many coasts of the world display extremely large shoreline crenulations called **capes.** The map in figure 13.19 shows that they are prominently developed on the southeast coast of the United States, where their spacing is roughly one to two orders of magnitude greater than rhythmic topographic features. Capes may be partially relict and therefore not necessarily adjusted to modern conditions. For example, some of the capes shown on the map are fringed by barrier islands that are Holocene in age. Some also coincide with the position of major rivers (W. A. White 1966; Hoyt and Henry 1971), which may indicate that the seaward projections began as ancient deltas.

Many workers feel that these large-scale rhythms are related to rotational cells that are set up as eddy currents along the western edge of the Gulf Stream. Such eddies have not been proven to exist, however, and the direct cause of the capes is unknown.

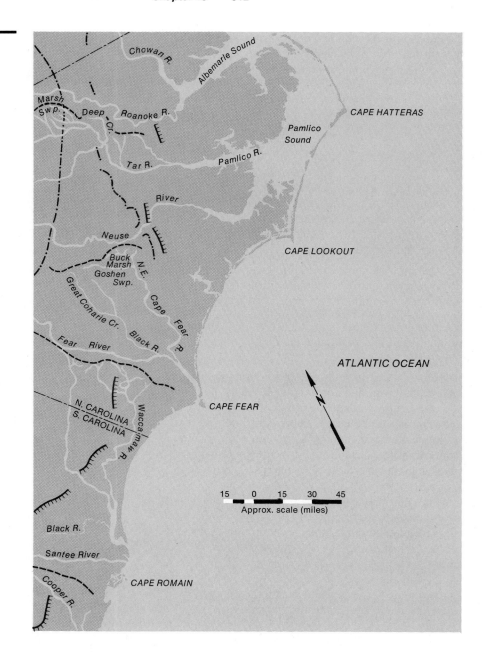

Figure 13.19.
Large capes and crescentic recessions along the south Atlantic coast of the United States.

Erosion of shorelines constitutes one of the major problems facing scientists, engineers, and land managers throughout the world. In the United States alone, approximately 25 percent of our coastlines have been categorized as seriously eroding (U.S. Army Corps of Engineers 1971). The annual cost of prevention techniques is staggering and will probably increase in the future because of the population strain placed on the coastal environment. Shoreline changes along the Great Lakes and along the marine coastlines of the United States have been estimated by May et al. (1983). On a national scale, U.S. shorelines are receding at an average rate of 0.8 m/yr. Rates vary on a regional scale from 0.0 m/yr along the Pacific coast to minus 0.8 m/yr along the Atlantic and minus 1.8 m/yr (negative indicates erosion) along the Gulf coast region (table 13.5). Great Lakes shorelines are retreating at 0.7 m/yr. Significantly, rates vary drastically on a local basis where they depend on geology and wave climate. In fact, net accretion occurs commonly on a local scale over the periods of record (fig. 13.20). Because of this, it is unrealistic to consider average rates as suitable for land management evaluation. Kuhn and Shepard (1983) have shown conclusively that coastal erosion (in this case sea cliff retreat) is episodic, site-specific, strongly related to meteorological conditions, and influenced by man-induced factors. Clearly, then, rates presented here are not meant to be utilized in a predictive manner.

Erosional Landforms and Rates

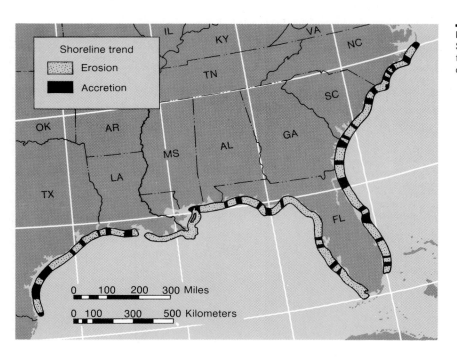

Figure 13.20.
Shoreline erosion and accretion trends along the Atlantic and Gulf coasts of the United States.

Table 13.5   Rate of shoreline change along U.S. marine coasts and bays and lakes. σ indicates within-state standard deviation of rates.

Marine Coasts

Region	$\bar{x}$, m/yr[a]	σ	Total Range[a]		N[b]
Atlantic Coast	−0.8	3.2	25.5	−24.6	510
Maine	−0.4	0.6	1.9	−0.5	16
New Hampshire	−0.5	—	−0.5	−0.5	4
Massachusetts	−0.9	1.9	4.5	−4.5	48
Rhode Island	−0.5	0.1	−0.3	−0.7	17
New York	0.1	3.2	18.8	−2.2	42
New Jersey	−1.0	5.4	25.5	−15.0	39
Delaware	0.1	2.4	5.0	−2.3	7
Maryland	−1.5	3.0	1.3	−8.8	9
Virginia	−4.2	5.5	0.9	−24.6	34
North Carolina	−0.6	2.1	9.4	−6.0	101
South Carolina	−2.0	3.8	5.9	−17.7	57
Georgia	0.7	2.8	5.0	−4.0	31
Florida	−0.1	1.2	5.0	−2.9	105
Gulf of Mexico	−1.8	2.7	8.8	−15.3	358
Florida	−0.4	1.6	8.8	−4.5	118
Alabama	−1.1	0.6	−0.8	−3.1	16
Mississippi	−0.6	2.0	0.6	−6.4	12
Louisiana	−4.2	3.3	3.4	−15.3	106
Texas	−1.2	1.4	0.8	−5.0	106
Pacific Coast	−0.0	1.5	10.0	−5.0	305
California	−0.1	1.3	10.0	−4.2	164
Oregon	−0.1	1.4	5.0	−5.0	86
Washington	0.5	2.2	5.0	−3.9	46
Alaska	−2.4	2.0	2.9	−6.0	69

Bays and Lakes

Region	$\bar{x}$, m/yr[a]	σ	Total Range[a]		N[b]
Delaware Bay					
New Jersey	−1.9	1.3	0.3	−3.0	13
Delaware	−1.3	2.1	5.0	−3.0	12
Chesapeake Bay	−0.7	0.7	1.5	−4.2	136
Western shore	−0.7	0.5	1.5	−1.9	67
Maryland	−0.7	0.3	−0.1	−1.3	35
Virginia	−0.8	0.7	1.5	−1.9	32
Eastern shore	−0.7	0.8	0.1	−4.2	69
Maryland	−0.8	0.9	−0.3	−4.2	47
Virginia	−0.5	0.4	0.1	−1.2	22
Great Lakes	−0.7	0.5	0.6	−2.7	327
Lake Erie	−0.7	0.6	−0.2	−2.4	98
Ohio	−0.6	0.6	−0.2	−2.2	68
Pennsylvania	−0.3	0.1	−0.2	−0.4	14
New York	−1.4	0.6	−0.5	−2.4	20
Lake Ontario	−0.5	0.2	−0.2	−1.2	58
Lake Huron	−0.4	0.3	−0.3	−1.3	28
Lake Michigan	−0.6	0.8	0.6	−9.9	184
Western shore	−0.6	0.4	0.6	−1.5	62
Eastern shore	−0.7	0.9	0.3	−9.9	122
Wisconsin	−0.7	0.3	−0.3	−1.5	46
Illinois	−0.2	0.4	0.6	−0.9	16
Indiana	−0.4	0.5	−0.3	−0.9	12
Michigan	−0.7	0.9	−0.3	−9.9	110
Lake Superior	−1.3	0.7	−0.3	−2.7	35
Minnesota	−0.8	0.4	−0.3	−1.5	16
Wisconsin	−1.8	0.6	−0.9	−2.7	19

[a]Negative values indicate erosion; the positive values indicate accretion.
[b]Total number of 3-min grid cells over which the statistics are calculated.

Table 13.6 Major erosional processes functioning along coasts.

Process	Description
Corrosion	Solution of coastal rocks by chemical action of seawater.
Attrition	Diminution of rock particles as water rolls, bounces, or slides them on a beach or wave-cut platform.
Corrasion	Physical erosion of bedrock caused by the grinding action of rock fragments that are carried in the ocean waves and currents.
Hydraulic Action	Erosion caused by the force of the water itself. Includes wave shock pressure and pneumatic quarrying by air trapped in cracks of the headland rocks.

Much of the erosion included in the rates just provided involves removal of sand from the beach itself. However, because of increased use of the entire coastal zone, engineers and coastal zone managers have placed considerable attention on the processes and rates of sea cliff erosion. When considering sea cliff erosion, several facts become immediately apparent. First, retreat of a sea cliff requires wave erosion at the cliff base. Second, erosion at the base leads to increased mass movement of the sea cliff material because of the resulting increase in slope angle and shear stress. Third, the debris of mass movement collects at the cliff base. No base erosion can occur again until this debris is removed and the toe of the cliff is again exposed to wave attack. The erosion is accomplished by a group of geomorphic processes, listed in table 13.6, that function in a complex of interactions to produce a variety of landforms. The final effect sometimes is controlled by ancillary processes such as burrowing action of marine organisms (see Ahr and Stanton 1973), frost action, and mass movement. As might be expected, the effectiveness of each process varies with the properties of the shore material and with the particular dynamics of the local ocean or lake system. *Corrosion* affects those rocks that are most susceptible to solution. At normal temperatures seawater is saturated with respect to calcium carbonate and does not dissolve limestone and $CaCO_3$ directly. Solution of these materials, however, may be aided by rainwater or by organisms that create local acidic conditions. As chunks of coastal or lakeshore rocks are released, *attrition* decreases the size of the particles and allows subsequent waves to drive the sediment into the cliff face. *Corrasion* then assumes a dominant role in the erosive mechanics. *Hydraulic action* is especially important where the rocks are highly fractured. Not only does the force of the wave exert pressure on the cliff face, but the advancing waves may compress air in the rock cavities, producing a pneumatic effect in the cracks. As the wave recedes, external pressure is instantaneously released, and the compressed air within the rocks exerts an outward stress that may disaggregate the outer zones of the cliff face.

Cliff erosion rates have been documented by a variety of techniques including comparison of sequential ground and aerial photography and maps, exposure of pins inserted into the sea cliff (Hodgkin 1964), instruments designed to measure microerosion (Trudgill 1976; Robinson 1977), and detailed air-photogrammatic maps (Norrman 1980).

Table 13.7 Representative rates of sea cliff retreat.

Location	Material	Rate (m/100 yr)	Reference
New England	Crystalline rock	0	1
U.S.S.R.	Volcanics	0	2
England (Cornish coast)	Crystalline rock	0	3
Northern France	Chalk	25	3
England (Yorkshire)	Sedimentary rocks	9	3
England (Yorkshire)	Glacial drift	28	3
Louisiana (coastal islands)	Sands and clays	800–3800	4
Southern California	Alluvium	30	5
U.S.S.R.	Clay	1200	2
New Jersey	Sand, clay and gravel	180	6
Cape Cod, Mass.	Glacial drift	30	7

1) Johnson 1925
2) in Zenkovich 1967
3) in King 1972
4) Peyronnin 1962
5) Shepard and Grant 1947
6) Rankin 1952
7) Zeigler et al. 1959

Average erosion rates are shown in table 13.7 to demonstrate the variability of sea cliff erosion under different conditions. A much more extensive table of cliff erosion rates is presented in Sunamura (1983). It is clear that rates of sea cliff erosion are extremely variable. This variability derives primarily from differences in geology, wave climate, and time. Lithology and cohesiveness of the coastal material clearly are of paramount importance in determining the rate of cliff retreat (fig. 13.21). Unconsolidated debris is eroded most rapidly and, in some cases, catastrophically. For example, short-term rates resulting from storms can be tens of meters of erosion in one day. Rocks that are nonresistant to wave attack are usually friable sandstones and shales. Rocks that resist sea cliff retreat are massive igneous rocks, high-rank metamorphic rocks, and certain massive carbonates.

In a series of papers, Sunamura (1975, 1976, 1977, 1978, 1982) has examined cliff erosion in theoretical terms of driving and resisting forces. The assailing wave force (f_w) is initially determined by energy derived in deep water, but it is directly influenced by factors such as water level or tide, bottom and beach topography, and beach sediment. These determine the wave type and height, where the waves break, and at what level they strike the sea cliff. Unfortunately, there is no way to directly measure f_w; therefore, some related parameter such as wave height is used as an estimate of the driving force.

Although lithology is the primary determinant of resistance (f_s), it is also influenced by contributing factors such as mechanical strength and geologic structures. Most workers use compressive strength as a surrogate for resistance.

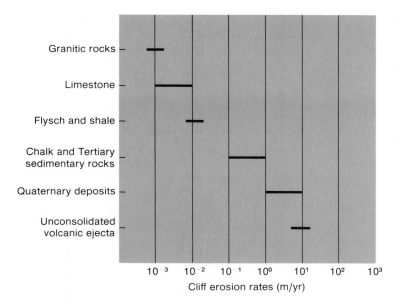

Figure 13.21.
Generalized orders of erosion
rates for sea cliffs composed of
different lithologies and
mechanical strength. (Data from
Sunamura 1983)

Thus, the cliff erosion rate becomes a function of f_w/f_s, which Sunamura (1977, 1983) has expressed as

$$\frac{dx}{dt} = k \left(C + \ln \frac{\rho g H}{S_c} \right),$$

where H is wave height at the cliff base, S_c is compressive strength, k is a dimensionless constant, and C a constant with dimensions of LT^{-1}. This equation suggests that a critical wave height (H_{crit}) exists that is needed to cause sea cliff erosion and can be derived by setting $dX/dt = 0$. This yields $H_{\text{crit}} = \dfrac{S_c}{\rho g} e^{-c}$. In areas of relatively homogeneous parent material and distinct wave climate, this approach is quite significant. For example, at Byobugaura, on the Pacific coast of Japan, a homogeneous Pliocene mudstone forms the lower half of a 10–60 m high cliff. Sunamura (1982) was able to determine k and C and calculate H_{crit} by using data of cliff strength, wave climate, and erosion rates of two time intervals. The plots shown in fig. 13.22 indicate that erosion was caused only under attack of the rarer but larger waves. The smaller, frequent waves produced no sea cliff erosion. In contrast to our earlier remark about average rates, this approach led to the identification of a threshold condition that has direct application in coastal land management. It also illustrates the problem of using average rates because erosion occurs in spasms (fig. 13.23) of extremely rapid erosion rather than continual removal of material at a constant rate from the cliff face. Short-term measurements may be far from reality if measured during unusual times, and long-term analyses

Figure 13.22.
Wave occurrence frequency, wave
duration, erosion rate and
recession distance plotted against
wave height at Byobugaura,
Japan.

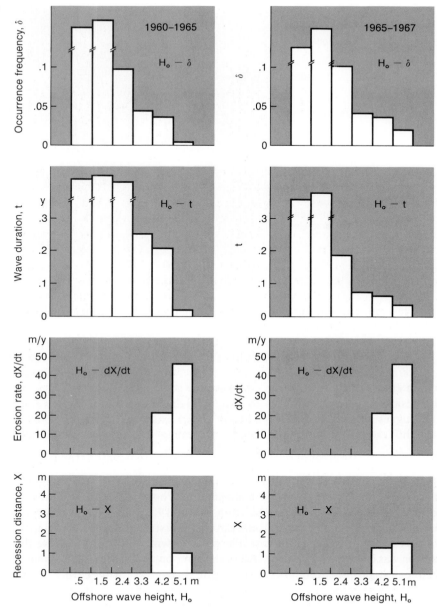

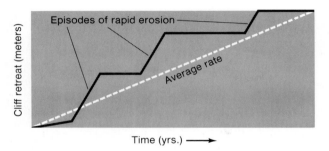

Figure 13.23.
Schematic illustration of cliff
retreat rates using a long-term
average versus the actual
continuum of erosion within the
long-term interval.

may mask the important aspect of prevailing wave climate. Clearly, as Sunamura (1983) suggests, the rate of sea cliff erosion should be determined by
the occurrence frequency of waves exceeding the threshold height value.

The base of any sea cliff is periodically covered by debris. This occurs
during intervals of rapid subaerial erosion when climate change or human activities deliver more debris to the beach surface than can be removed by wave
action (Emery and Kuhn 1982; Kuhn and Shepard 1983). This material prevents further undercutting of the cliff until stronger wave action removes it
and thereby exposes the cliff face to renewed supercritical wave attack. The
implication is that the sea cliff profile will vary with time (Emery and Kuhn
1982; Kuhn and Shepard 1983; Sunamura 1983). These variations tend to
offset one another such that, over a long term, the profile may appear to retreat
in a parallel manner. However, within that long term, the profile will change
according to the amount of debris stored on the beach and/or whether the
beach elevation causes supercritical waves to strike the cliff face at higher or
lower levels.

Emery and Kuhn (1982) have used profiles to classify sea cliffs as *active,
inactive,* or *former.* These apparently indicate whether marine or subaerial
processes are dominant at any given time. A smooth curve at the base (inactive) indicates that subaerial erosion is prevailing. In contrast, a sharp-an-
gled basal contact (active) with the beach suggests that marine erosion is
dominant. A former profile is one that has been removed from the influence
of marine processes. Because inactivity is associated with dry climates, the
incipient climate change in southern California to more humid conditions may
initiate a change to active sea cliff erosion. Areas presently near inactive cliffs
are probably overdeveloped, which will exaccerbate the effects of the oncoming climate shift. Humans, it seems, have primed the pump for serious
environmental problems.

As the cliff retreats, it leaves behind a beveled surface, called the **wave-
cut platform,** that stands slightly below water level at high tide (fig. 13.24).
Although many processes are involved in its creation (see Wentworth 1938),
corrasion at the base of the sea cliff is probably responsible for most of the
planating action. The platform is not flat but slopes gently oceanward with a
declivity that ranges between 0.02 and 0.01. Therefore, under a stationary sea
level the maximum width of the wave-cut platform depends largely on the
depth at which wave abrasion is still a viable process. Bradley (1958) suggests
that wave-cut platforms can be eroded in water no deeper than 10 m; under

Figure 13.24.
Wave-cut platform near MacLeod Harbor, Alaska. Platform has been raised above wave level by recent tectonism.

the common slope range, platforms wider than 500 m probably form only if sea level is continuously rising.

In addition to the effect of water depth, on low-gradient platforms and with constant sea level, the ultimate width is self-controlling because the rate of platform expansion decreases as the sea cliff recedes. Platforms develop rapidly in the early stages of cliff retreat. In time, however, incoming waves lose much of their energy as they interact with the progressively widening platform. Eventually, cliff retreat and platform expansion cease unless deposition on the original platform surface raises the level of incoming waves and changes the frictional component.

As the shoreline retreats and irregularities appear, a group of landforms develop that are characteristic of coastal erosion. **Stacks** are isolated parts of the headland formed when narrow oceanward extensions of the coastal rocks are cut into isolated remnants by wave attack. The process is accentuated when waves are refracted around the headland reach. Sometimes less resistant rock zones are exposed to local corrasion or scouring, and notches or **sea caves** are formed. As sea caves grow, they may extend completely through the headland to produce a feature known as a **sea arch** (fig. 13.25). Any or all of these features indicate a local erosional environment. As the headland rocks undergo attrition, however, more sediment covers the wave-cut platform. Beach profiles develop, and wave energy dissipates farther offshore as waves break over the longshore bars. This sequence may not occur in every situation. For example, in large lakes such as Lake Erie or Lake Michigan, energy loss over longshore bars is of little or no consequence.

Figure 13.25.
Sea arch or window along coast
of Puerto Rico at low tide. Hole is
caused by combination of wave
action and solution of limestone
bedrock.

Depositional Shorelines

In coastal areas where the supply of sand is abundant and local ocean forces are capable of transporting sediment, the shapes of some landforms are determined primarily by depositional events, even though erosional processes are very much involved in the entrainment of the sand before its final deposition. Large depositional features usually occur in the form of **spits, baymouth bars,** or **barrier islands.** The first two are definitely related to longshore transport of sediment (*littoral drift*), with the site of deposition being in tranquil waters of bays, estuaries, or the open ocean. Barrier islands are large elongate features that parallel the shoreline but are not physically connected to it. Barrier islands are very abundant along the Atlantic and Gulf shorelines of the United States and will be treated in a separate section. Other depositional features may include some of the cuspate forms discussed earlier (especially cuspate forelands associated with sand waves). Most of these differ in several ways from the large depositional beaches. First, the cuspate forms show a regularity that is not evident in the large beaches and probably relates to different hydrodynamics. It is known that some crescentic forms (rhythmic topography) will migrate in the direction of the longshore transport. However, very strong longshore drift will at first drastically skew the sand waves and giant cusps and then, with increased transport rates, eliminate the form completely. Second, cuspate features may be superimposed on the larger beach types, indicating that in some cases they are secondary rather than primary features.

Spits and Baymouth Bars

Spits and baymouth bars develop when littoral drift plays a predominant role in the system, provided the drifting sediment enters a zone of slack water where deposition can occur. Possibly the erosional and depositional framework in the littoral system can be considered in terms of a littoral power gradient similar to that proposed by May and Tanner (1973). In their study, a wave power model utilizing E, P_ℓ and dq/dx (q being the quantity of sand transported and x the distance along the beach) as the basic parameters was constructed to predict zones of different littoral transport along a beach. Where dq/dx is greatest, beach erosion is also greatest because a large amount of sediment is being placed in transit. Where dq/dx is negative, deposition is occurring.

Spits and baymouth bars are essentially the same feature. They differ only in that spits extend into the open ocean while baymouth bars cross the gap between two headland reaches. Baymouth bars are also more likely to initiate lagoonal areas shoreward of the bar deposit. These lagoons may gradually change into tidal marshes or swamps as the embayments fill in with fluvially derived sediment.

Although spits and baymouth bars have been attributed to a variety of ocean processes, their origin is clearly and most prominently related to littoral drift. The elongate extensions into open water represent continuations of the beach that rests against the coast. They expand continuously in the direction of littoral drift unless other water motions interfere with the growth process.

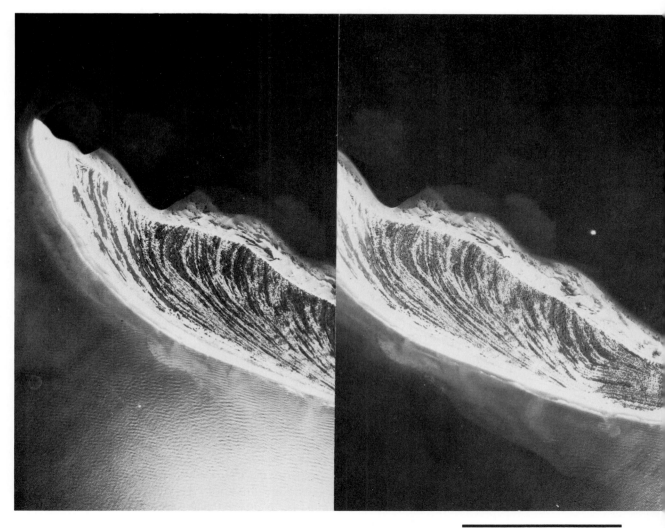

Figure 13.26.
Recurved spit on west coast of
Florida. Series of beach ridges
show accretionary pattern of spit.

For example, it is not uncommon for wave refraction around the free end (Evans 1942) or wave trains approaching from different directions (King and Mc-Cullagh 1971) to reorient the terminal end of the spit. In this case the feature may be called a *recurved spit* or a *hook*. Spits or baymouth bars may also widen by progradation associated with the construction of a series of sandy or pebbly ridges on the oceanward side of the feature (fig. 13.26). These ridges or growth lines, called *beach ridges,* will be discussed in the next section. Finally, spits may link the main coastal region to an offshore island, producing the feature known as a *tombolo*. Tombolos, however, also form on normal coastal beaches when wave refraction around the island shapes the beach into a cuspate foreland that eventually extends to the island itself. This process even functions around breakwaters that have been constructed offshore to prevent beach erosion (Inman and Frautschy 1966).

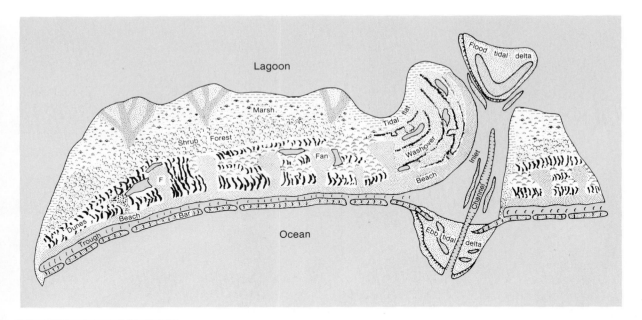

Figure 13.27.
Schematic of idealized barrier
island showing vegetal and
geomorphic facies. (Diagram by
R. Craig Kochel)

Barrier Islands

Distribution and Characteristics   Barrier islands are elongate bodies of sand
that are not attached to the mainland but are separated from it by a **lagoon**
or bay. The islands normally range in width from 2 to 5 km, in length from
10 to 100 km, and are usually less than 6 m in elevation. They are commonly
large enough to support major cities—Atlantic City, Miami Beach, and Galveston are examples. Barrier islands occur on 13 percent of the world's coasts
(King 1972), but they appear to be concentrated where tidal ranges, wave
energies, and offshore gradients are low. Along the boundaries of the United
States, 282 barrier islands exist adjacent to the Gulf coast and the Atlantic
coast between Florida and the middle Atlantic states (Dolan et al. 1980).

The origin of barrier islands has always been steeped in controversy, mainly
because most of the original interpretations were based on morphology alone.
More recent theories rely on subsurface exploration and paleoenvironmental
interpretation of facies both shoreward and seaward of the barrier. Although
differences of opinion exist, most workers seem to believe that barrier islands
have a long growth history associated with postglacial sea level rise. Remember that during the last full glacial episode, sea level was probably 120
m lower than it is today, and the shorelines possibly 60 to 150 km seaward of
their present positions (Curray 1965; Dolan et al. 1980). Beginning as shoreline, depositional nuclei during the low sea level of the late Pleistocene, the
initial ridges migrated landward as sea level rose. Sea level approached its
present level about 4000 to 5000 years ago, at which time our modern barrier
islands developed their distributional and environmental character. The islands are still evolving, and because sea level is still slowly rising, they continue

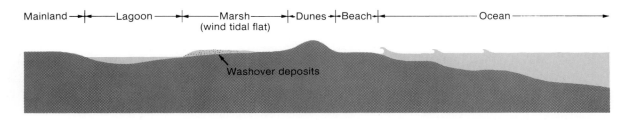

Figure 13.28.
Idealized cross profile of a barrier
island.

to migrate landward. This movement, however, requires the unique combination of environments, sand supply, and sand transport found in the barrier system.

Barrier islands are composed of distinct geomorphic and vegetative zones (figs. 13.27, 13.28). On the ocean side, the islands are characterized by low-gradient beaches that alternately change their configuration during storms and during intervals in which a swell-wave climate prevails. Significantly, the backshore environment of most beaches is characterized by sand dunes rather than a distinct sea cliff. The dune line represents the "backbone" of the island, usually standing up to 6 m in elevation but occasionally reaching much greater heights (fig. 13.29). For example, in a study of southeast African coasts, Orme (1973) reports Holocene dunes climbing to elevations greater than 100 m. Dune elevation and beach properties are closely related to the prevailing wind and wave direction. Where beach orientation is conducive to littoral drift, dunes may be stunted or even absent. These conditions result in low, narrow islands that are extremely vulnerable during storm surges. In contrast, where beaches are perpendicular to the prevailing wind and wave direction, sand is pushed shoreward and eventually is driven by the wind into pronounced dunes.

Behind the dune line is a low, flat zone that is covered with grasses and shrub forests; stands of pine and oak are sometimes found in sheltered areas (fig. 13.27). This zone grades into the salt marshes and tidal flats that abut the lagoon itself. The lagoon, salt marshes, and tidal flats are fed and maintained by tidal inlets that link the open ocean to the sound-side environments.

The Geomorphic Processes   In addition to the wind action associated with dune formation and sand transport, two of the most important processes functioning in barrier island systems are overwash and inlet formation (Dolan et al. 1980). In severe storms, parts of every barrier island are inundated by high water levels and affected by wave action. The process, known as **overwash,** can cover large areas, especially where dune development is minor. Overwash may also occur in narrow washover channels that breach the dune line (fig. 13.27). In such cases, sand is transported from the beach and deposited as distinct geomorphic features known as **washover fans** (Pierce 1970; fig. 13.30). The total transport distance of sand during overwash depends on the tidal range, severity of the storm, island elevation, and storm surge penetration. It is possible, however, for overwash to traverse the entire dune line and place fans on top of older sound-side deposits.

Figure 13.29.
Giant dune at Jockeys Ridge,
N.C.

Figure 13.30.
Outer Banks, N.C. Stabilized dune
breached by washover fan. Old
road surface has been eroded by
ocean and buried by recent
overwash deposits as the barrier
island migrated landward.

Figure 13.31.
Ebb tidal delta and flood tidal delta at Brown Inlet, N.C. Ebb tidal delta marked by offshore shoaling of waves.

Perhaps of equal or greater importance is the fact that major inlets are probably formed during storms. These essentially segment the islands and physically connect the lagoon with the open ocean. The positions of most inlets are ephemeral because they tend to migrate in the direction of littoral drift, extending by spit formation on one side of the inlet and erosion on the other. Some inlets may simply fill in with sediment, thereby destroying the free channel. Nonetheless, when active, inlets serve as avenues for water and sand movement into and out of the lagoonal zone. During high tide, material moves landward through the inlet and deposits a **flood tidal delta** on the inside of the barrier island. During ebb tide, a similar **ebb tidal delta** is formed in the ocean (figs. 13.27, 13.31). During low-tide events, the sound-side shoals are exposed and eventually become the substrate for new salt marshes (Godfrey 1976).

Enormous investments have been made in attempts to fix tidal inlets in a single location. Usually jetties are constructed to prevent filling of the channel and to stabilize its position. In many cases these efforts fail without continuous and costly maintenance. A good example is the Ocean City Inlet, Maryland, which breached a barrier island during a 1933 hurricane. Since construction, the jetty system has interrupted the southward net littoral drift (Dean and Perlin 1977). The effect has been a shoreline advance of about 240 m immediately north of the inlet, where drifting sand has been trapped by the northern jetty. The shoreline immediately south of the inlet has eroded land-ward about 515 m because the beach was starved when its supply of littoral sediment was eliminated (fig. 13.32). The inshore portion of the south jetty is so low and permeable that sand being eroded and transported from the beaches south of the inlet is moving over and through the jetty back into the inlet itself. This is producing a shoal area within the inlet and will require major repair of the south jetty (Dean and Perlin 1977).

There is irrefutable evidence that most mid-Atlantic barrier islands are still migrating landward (Fisher and Simpson 1979). Evidence for this transgression is that shells of lagoonal fauna and salt marsh peats are now found on the beach side of the islands, indicating that the present beach is resting on top of older lagoonal sediment. Clearly, the barrier island system is dynamic. Islands are eroded on the ocean side, and the entrained sediment is transported farther inshore (fig. 13.30). The lagoonal side of the island system grows by deposition of overwash and tidal inlet sediment, and windblown sand. Thus, the landward march of the islands will proceed as long as sea level continues to rise and overwash and inlet processes along with wind action provide sediment to cover the lagoonal facies. This covering simultaneously creates a new substrate for continued growth of the vegetational zones located inshore from the island dunes.

Figure 13.32.
Inlet and jetties at Ocean City, Md., looking north. Barrier island south of inlet (Assateague Island) has migrated landward relative to island north of inlet (Fenwick Island).

Summary

Coastal zones reflect the balance between the driving forces of ocean waters and the resistance offered by the rocks that form the shoreline. Large lakes may experience the same phenomena as ocean coasts. Beaches are the components of coasts that respond most obviously to the dynamics of the system. Energy possessed by waves, currents, and tides is expended on the beach surface. The response to this activity is the creation of a beach profile that is adjusted to the mean values of the wave properties. Waves striking the beach zone obliquely also cause longshore (littoral) drift of sediment.

Coasts and beaches display a wide variety of geomorphic forms that range in size from minor modifications of the beach face to features that encompass kilometers of the shoreline. Commonly the shoreline is indented with a hierarchy of crescentic or rhythmic forms that are related in some way to cellular circulation of the ocean water. Coasts may be typified as erosional or depositional depending on whether the dominant action is causing the shoreline to recede landward or expand oceanward. Each coastal type possesses large-scale features that reflect the prevailing action. In some cases (barrier islands), both erosion and deposition are necessary to maintain the feature.

Suggested Readings

The following references provide greater detail concerning the concepts discussed in this chapter. Each reference cited has an extensive bibliography of topics that may be of particular interest to the reader.

Dolan, R.; Hayden, B.; and Lins, H. 1980. Barrier islands. *Am. Scientist* 68:16–25.

Komar, P. D. 1976. *Beach processes and sedimentation.* Englewood Cliffs, N.J.: Prentice-Hall.

———, ed. 1983. *CRC handbook of coastal processes and erosion.* Boca Raton, Fla.: CRC Press.

May, S.; Dolan, R.; and Hayden, B. 1983. Erosion of U.S. shorelines. *EOS* 64:521–23.

Sunamura, T. 1982. A predictive model for wave-induced erosion, with application to Pacific coasts of Japan. *Jour. Geology* 90:167–78.

Wright, L.; Chappell, J.; Thom, B.; Bradshaw, M.; and Cowell, P. 1979. Morphodynamics of reflective and dissipative beach and inshore systems: Southeastern Australia. *Marine Geol.* 32:105–40.

Bibliography

Abrahams, A. D. 1972. Environmental constraints on the substitution of space for time in the study of natural channel networks. *Geol. Soc. America Bull.* 83:1523–30.

———. 1980. A multivariate analysis of chain lengths in natural channel networks. *Jour. Geology* 88:681–96.

Abrahams, A. D., and Miller, A. J. 1982. The mixed gamma model for channel link lengths. *Water Resour. Res.* 18:1126–36.

Ackers, P., and Charlton, F. G. 1971. The slope and resistance of small meandering channels. *Inst. Civil Engrs. Proc.,* Supp. 15, 1970, paper 73625.

Addington, A. R. 1927. Porter's Cave and recent drainage adjustments in its vicinity. *Indiana Acad. Sci. Proc.* 36:107–16.

Ahlbrandt, T. S., and Fryberger, S. G. 1980. Eolian deposits in the Nebraska Sand Hills. U.S. Geol. Survey Prof. Paper 1120A.

———. 1982. Introduction to eolian deposits. In *Sandstone depositional environments*, edited by P. Scholle and D. Spearing, 11–48. Tulsa, Okla: Am. Assoc. Petroleum Geologists.

Ahlmann, H. W. 1948. *Glaciological research on the North Atlantic.* Royal Geog. Soc. Res. Ser. 1.

Ahnert, F. 1970. Functional relationships between denudation, relief, and uplift in large mid-latitude drainage basins. *Am. Jour. Sci.* 268:243–63.

Ahr, W. M., and Stanton, R. J. 1973. The sedimentologic and paleoecologic significance of Lithotyra, a rock-boring barnacle. *Jour. Sed. Petrology* 43:20–23.

Akagi, Y. 1980. Relations between rock type and the slope form in the Sonora Desert, Arizona. *Zeit. f. Geomorph.* 24:129–140.

Alter, A. J. 1966. Sanitary engineering in Alaska. In *Proc. Permafrost Internat. Conf.* (Lafayette, Ind., 1963). Natl. Acad. Sci. Natl. Research Council Pub. 1287, pp. 407–8.

American Geological Institute. 1972. *Glossary of geological terms.*

Anderson, D. M. 1968. Undercooling, freezing, point depression, and ice nucleation of soil water. *Israel J. Chem.* 6:349–55.

———. 1970. Phase boundary water in frozen soils. U.S. Army Corps Engrs., Cold Regions Res. and Eng. Lab. Research Rept. 274.

Anderson, L. W. 1978. Cirque glacier erosion rates and characteristics of neoglacial tills, Pangnirtung fjord area, Baffin Island, N.W.T., Canada. *Arc. Alp. Res.* 10:749–60.

Anderson, R. S.; Hallet, B.; Walder, J.; and Aubry, B. F. 1982. Observations in a cavity beneath Grinnell Glacier. *Earth Surf. Proc. and Landforms* 7:63–70.

Andrews, D. E. 1980. Glacially thrust bed—An indication of late Wisconsin climate in western New York State. *Geology* 8:97–101.

Andrews, E. D. 1979. Scour and fill in a stream channel, East Fork River, western Wyoming. U.S. Geol. Survey Prof. Paper 1117.

———. 1981. Measurement and computations of bed-material in a shallow sand-bed stream, Muddy Creek, Wyoming. *Water Resour. Res.* 17:131–41.

———. 1983. Entrainment of gravel from naturally sorted riverbed material. *Geol. Soc. America Bull.* 94:1225–31.

Andrews, J. T. 1963. Cross-valley moraines of the Rimrock and Isotoq river valleys, Baffin Island. A descriptive analysis. *Geogr. Bull.* 19:49–77.

———. ed. 1974. *Glacial isostasy.* Stroudsburg, Pa.: Dowden, Hutchinson and Ross.

Andrews, J. T., and Smithson, B. B. 1966. Till fabrics of the cross-valley moraines of north-central Baffin Island. *Geol. Soc. America Bull.* 77:271–90.

Arkley, R. 1963. Calculation of carbonate and water movement in soil from climatic data. *Soil Sci.* 96:239–48.

Arvidson, R. E., and Guinness, E. A. 1982. Clues to tectonic styles in the global topography of Earth, Venus, Mars. *Jour. Geol. Educ.* 30:86–92.

Atkinson, H., and Wright, J. 1957. Chelation and the vertical movement of soil constituents. *Soil Sci.* 84:1–11.

Atkinson, T., and Smith, D. 1976. The erosion of limestones. In *The science of speleology,* edited by T. Ford and C. Cullingford, pp. 151–77. New York: Academic Press.

Atterberg, A. 1911. Die Plastizitat der Tone. *Intern. Mitt. Boden* 1:4–37.

Baas-Becking, L.; Kaplan, I.; and Moore, D. 1960. Limits of the natural environment in terms of pH and oxidation-reduction potentials. *Jour. Geology* 68:243–84.

Bagnold, R. A. 1941. *The physics of blown sand and desert dunes.* London: Methuen and Co.

———. 1960. Some aspects of river meanders. U.S. Geol. Survey Prof. Paper 282-E.

———. 1973. The nature of saltation and of bedload transport in water. *Proc. Royal Soc. London,* ser. A:332:473–504.

———. 1977. Bed-load transport by natural rivers. *Water Resour. Res.* 13:303–12.

Baker, V. R. 1973a. Geomorphology and hydrology of karst drainage basins and cave channel networks in east-central New York. *Water Resour. Res.* 9:695–706.

———. 1973b. *Paleohydrology and sedimentology of Lake Missoula flooding in eastern Washington.* Geol. Soc. America Spec. Paper 144.

———. 1974. Paleohydraulic interpretation of Quaternary alluvium near Golden, Colorado. *Quat. Res.* 4:94–112.

———. 1975. Urban geology of Boulder, Colorado: A progress report. *Environmental Geol.* 1:75–88.

———. 1976a. Hydrogeology of a cavernous limestone terrane and the hydrochemical mechanisms of its formation, Mohawk River basin, New York. *Empire State Geogram* 12, no. 2:2–65.

———. 1976b. Hydrogeomorphic methods for the regional evaluation of flood hazards. *Environ. Geology* 1:261–81.

———. 1977. Stream-channel response to floods, with examples from central Texas. *Geol. Soc. America Bull.* 88:1057–71.

———. 1978. Adjustment of fluvial systems to climate and source terrain in tropical and subtropical environments. In *Fluvial sedimentology,* edited by A. Miall, pp. 211–30. Can. Soc. Petrol. Geol. Mem. 5.

———. 1981. The geomorphology of Mars. *Progress in Phys. Geog.* 5:475–513.

Baker, V. R.; Kochel, R. C.; Patton, P. C.; and Pickup, G. 1983. Palaeohydrologic analysis of Holocene flood slack-water sediments. *Spec. Pubs. Intl. Assoc. Sediment.* 6:229–39.

Baker, V. R.; Pickup, G.; and Polach, H. A. 1983. Desert palaeo floods in central Australia. *Nature* 301:502–4.

Baker, V. R., and Ritter, D. F. 1975. Competence of rivers to transport coarse bedload material. *Geol. Soc. America Bull.* 86:975–78.

Bakker, J. P. 1965. A forgotten factor in the interpretation of glacial stairways. *Zeit. f. Geomorph.* 9:18–34.

Baldwin, M.; Kellogg, C.; and Thorp, J. 1938. Soil classification. In *Soils and Men,* U.S. Dept. Agri. Yearbook, pp. 979–1001.

Ballard, T. M. 1973. Soil physical properties in a sorted stripe field. *Arc. Alp. Res.* 5:127–31.

Bandy, O., and Marincovich, L. 1973. Rates of late Cenozoic uplift, Baldwin Hills, Los Angeles, California. *Science* 181:653–55.

Barnes, H. A. 1968. Roughness characteristics of natural channels. U.S. Geol. Survey Water Supply Paper 1849.

Barnes, P., and Tabor, D. 1966. Plastic flow and pressure melting in the deformation of ice. *Nature* 210:878–82.

Barnett, D. M., and Holdsworth, G. 1974. Origin, morphology, and chronology of sublacustrine moraines, Generator Lake, Baffin Island, Northwest Territories, Canada. *Can. J. Earth Sci.* 11:380–408.

Barsch, D. 1971. Rock glaciers and ice-cored moraines. *Geogr. Annlr.* 53A:203–6.

Barsch, D., and Updike, R. G. 1971. Periglaziale Formung am Kendrick Peak in Nord-Arizona während der letzten kaltzeit. *Geog. Helvetica.* 26:99–114.

Barshad, I. 1964. Chemistry of soil development. In *Chemistry of the soil,* edited by F. Bear, pp. 1–70. New York: Reinhold Book Corp.

Bascom, W. N. 1951. The relationship between sand size and beach face slope. *Am. Geophys. Union Trans.* 32:866–74.

———. 1964. *Waves and beaches.* Garden City, N.Y.: Doubleday.

Bates, C. C. 1953. Rational theory of delta formation. *Am. Assoc. Petroleum Geologists Bull.* 37:2119–62.

Bathurst, J. C.; Thorne, C. R.; and Hey, R. D. 1979. Secondary flow and shear stress at river bends. *Jour. Hydraulics Div., ASCE* 105:1277–95.

Battle, W. R. B. 1960. Temperature observation in bergschrunds and their relationship to frost shattering. In *Norwegian cirque glaciers,* edited by W. V. Lewis, pp. 83–96. Royal Geog. Soc. Res. Ser. 4.

Bauer, B. 1980. Drainage density— An integrative measure of the dynamics and the quality of watersheds. *Zeit. f. Geomorph.* 24:263–72.

Baulig, H. 1957. Peneplains and pediplains. *Geol. Soc. America Bull.* 68:913–30.

Beaty, C. B. 1963. Origin of alluvial fans, White Mountains, California and Nevada. *Ann. Assoc. Am. Geog.* 53:516–35.

———. 1970. Age and estimated rate of accumulation of an alluvial fan, White Mountains, California, U.S.A. *Am. Jour. Sci.* 268:50–77.

———. 1975a. Sublimation or melting observations from the White Mountains, California and Nevada, U.S.A. *Jour. Glaciol.* 14:275–86.

———. 1975b. Coulee alignment and the wind in southern Alberta, Canada. *Geol. Soc. America Bull.* 86:119–28.

Beaumont, P. 1972. Alluvial fans along the foothills of the Elburz Mountain, Iran. *Paleogeogr. Paleoclimatol. Paleoecol.* 12:251–73.

Begin, Z. B. 1981. Stream curvature and bank erosion: A model based on the momentum equation. *Jour. Geology* 89:497–504.

Begin, Z. B.; Meyer, D. F.; and Schumm, S. A. 1980. Knickpoint migration due to base-level lowering. Am. Soc. Civil Engineers, *Jour. Water, Port, Coastal and Ocean Div.* 106:369–87.

———. 1981. Development of longitudinal profiles of alluvial channels in response to base-level lowering. *Earth Surf. Proc. and Landforms* 6:49–68.

Behrendt, J. C. 1965. Densification of snow on the ice sheet of Ellsworth Land and South Antarctic Peninsula. *Jour. Glaciol.* 5:451–60.

Belly, P. V. 1964. Sand movement by wind. U.S. Army Corps Engrs., Coastal Eng. Res. Center Tech. Memo 1.

Bender, M. L.; Fairbanks, R. G.; Taylor, F. W.; Matthews, R. K.; Goddard, J. G.; and Broecker, W. S. 1979. Uranium-series dating of Pleistocene reef tracts of Barbados, West Indies. *Geol. Soc. America Bull.* 90:577–94.

Benedict, J. B. 1970. Downslope soil movement in a Colorado alpine region. Rates, processes, and climatic significance. *Arc. Alp. Res.* 2:165–226.

———. 1976. Frost creep and gelifluction features. A review. *Quat. Res.* 6:55–77.

Benson, M. A. 1950. Use of historical data in flood-frequency analysis. *Am. Geophys. Union Trans.* 31:419–24.

———. 1971. Uniform flood-frequency estimating methods for federal agencies. *Water Resour. Res.* 4:891–908.

Berner, R. A., and Holdren, G. R., Jr. 1977. Mechanism of feldspar weathering: Some observational evidence. *Geology* 5:369–72.

———. 1979. Mechanisms of feldspar weathering—II. Observations of feldspars from soils. *Geochim et Cosmochim Acta* 43:1173–86.

Berner, R. A., and Morse, J. 1974. Dissolution kinetics of calcium carbonate in seawater—IV: Theory of calcite dissolution. *Am. Jour. Sci.* 274:108–34.

Beskow, G. 1947. Soil freezing and frost heaving with specific application to roads and railroads. Evanston, Ill.: Northwestern University Tech. Inst.

Bhowmik, N. G. 1982. Shear stress distribution and secondary currents in straight open channels. In *Gravel-bed rivers,* edited by R. Hey, J. Bathurst, and C. Thorne, pp. 31–62. New York: John Wiley & Sons.

Bhowmik, N. G., and Demissie, M. 1982. Carrying capacity of flood plains. *Am. Soc. Civil Engineers Proc., Jour. Hydraulics Div.* 108, HY3:443–52.

Bindschadler, R. 1983. The importance of pressurized subglacial water in separation and sliding at the glacier bed. *Jour. Glaciol.* 29:3–19.

Bird, E. C. F. 1969. *Coasts.* Cambridge, Mass.: M.I.T. Press.

Birkeland, P. 1974. *Pedology, weathering, and geomorphological research.* London: Oxford Univ. Press.

———. 1984. *Soils and geomorphology.* New York: Oxford Univ. Press.

Black, R. F. 1954. Permafrost, a review. *Geol. Soc. America Bull.* 65:839–55.

———. 1963. Les coins de glace et le gel permanent dans le Nord de l'Alaska. *Annales Géog.* 72:257–71.

———. 1969. Climatically significant fossil periglacial phenomena in north-central United States. *Biuletyn Perygl.* 20:225–38.

———. 1974. Ice-wedge polygons of northern Alaska. In *Glacial geomorphology,* edited by D. R. Coates, pp. 247–75. S.U.N.Y., Binghamton: Pubs. in Geomorphology, 5th Ann. Symposium.

———. 1976. Periglacial features indicative of permafrost: Ice and soil wedges. *Quat. Res.* 6:3–26.

Blackwelder, E. 1927. Fire as an agent in rock weathering. *Jour. Geology* 35:135–40.

———. 1931. The lowering of playas by deflation. *Am. Jour. Sci.* 21:140–44.

———. 1934. Yardangs. *Geol. Soc. America Bull.* 45:159–66.

Blagborough, J. W., and Farkas, S. E. 1968. Rock glaciers in the San Mateo Mountains, south-central New Mexico. *Am. Jour. Sci.* 266:812–23.

Blatt, H., and Jones, R. 1975. Proportions of exposed igneous, metamorphic, and sedimentary rocks. *Geol. Soc. America Bull.* 86:1085–88.

Blissenbach, E. 1954. Geology of alluvial fans in semiarid regions. *Geol. Soc. America Bull.* 65:175–90.

Bloom, A. L. 1965. The explanatory description of coasts. *Zeit. f. Geomorph.* 9(4):422–36.

———. 1967. Pleistocene shorelines: A new test of isostasy. *Geol. Soc. America Bull.* 78:1477–94.

————. 1980. Late Quaternary sea level change on South Pacific coasts: A study in the tectonic diversity. In *Earth rheology, isostasy, and eustasy,* edited by N. A. Morner, pp. 505–16. New York: John Wiley & Sons.

Bloom, A. L.; Broecker, W. S.; Chappell, J. M. A.; Matthews, R. K.; and Mesolella, K. J. 1974. Quaternary sea level fluctuations on a tectonic coast: New ^{230}Th/^{234}U dates from the Huron Peninsula, New Guinea. *Quat. Res.* 4:185–205.

Bluck, B. J. 1964. Sedimentation on an alluvial fan in southern Nevada. *Jour. Sed. Petrology* 34:395–400.

Bögli, A. 1964. Mischungskorrosion—ein Beitrag zur Verkarstungsproblem. *Erdkunde* 18:83–92.

————. 1980. *Karst hydrology and physical speleology.* Berlin: Springer-Verlag.

Boothroyd, J. C., and Ashley, G. M. 1975. Processes, bar morphology, and sedimentary structures on braided outwash fans, northeastern Gulf of Alaska. In *Glaciofluvial and glaciolacustrine sedimentation,* edited by B. McDonald and A. Jopling, pp. 193–222. Tulsa, Okla.: SEPM Spec. Pub. 23.

Born, S. M. 1972. *Late Quaternary history, deltaic sedimentation, and mud-lump formation at Pyramid Lake, Nevada.* Univ. Nevada, Reno, Desert Research Inst.

Born, S. M., and Ritter, D. F. 1970. Modern terrace development near Pyramid Lake, Nevada, and its geologic implications. *Geol. Soc. America Bull.* 81:1233–42.

Bottomley, J. T. 1872. Melting and regelation of ice. *Nature* 5:185.

Boulton, G. S. 1967. The development of a complex supraglacial moraine at the margin of Sorbreen, Ny Friesland, Vestspitsbergen. *Jour. Glaciol.* 6:717–36.

————. 1968. Flow tills and related deposits on some Vestspitsbergen glaciers. *Jour. Glaciol.* 7:391–412.

————. 1970a. On the origin and transport of englacial debris in Svalbard glaciers. *Jour. Glaciol.* 9:213–29.

————. 1970b. On the deposition of subglacial and melt-out tills at the margins of certain Svalbard glaciers. *Jour. Glaciol.* 9:231–46.

————. 1971. Till genesis and fabric in Svalbard, Spitsbergen. In *Till: A symposium,* edited by R. P. Goldthwait, pp. 41–72. Columbus: Ohio State Univ. Press.

————. 1972a. The role of thermal regime in glacial sedimentation—a general theory. In *Polar geomorphology,* edited by R. J. Price and D. E. Sugden, pp. 1–19. Inst. Brit. Geog. Spec. Paper 4.

————. 1972b. Modern Arctic glaciers as depositional models for former ice sheets. *Jour. Geol. Soc. Lond.* 128:361–93.

————. 1974. Processes and patterns of glacial erosion. In *Glacial geomorphology,* edited by D. R. Coates, pp. 41–87. S.U.N.Y., Binghamton: Pubs. in Geomorphology, 5th Ann. Symposium.

————. 1976. The origin of glacially fluted surfaces—observation and theory. *Jour. Glaciol.* 17:287–309.

————. 1978. Boulder shapes and grain-size distribution of debris as indication of transport paths through a glacier and till genesis. *Sedimentology* 25:773–99.

————. 1979. Processes of glacier erosion on different substrata. *Jour. Glaciol.* 23:15–36.

Boulton, G. S., and Jones, A. S. 1979. Stability of temperate ice caps and ice sheets resting on beds of deformable sediment. *Jour. Glaciol.* 24:29–43.

Bowen, A. J. 1969. The generation of longshore currents on a plane beach. *J. Marine Res.* 37:206–15.

————. 1973. Edge waves and the litoral environment. *Proc. 13th Conf. on Coast. Eng.,* pp. 1313–20.

Bowen, A. J., and Inman, D. L. 1969. Rip currents, 2. Laboratory and field observations. *Jour. Geophys. Research* 74:5479–90.

————. 1971. Edge waves and crescentic bars. *Jour. Geophys. Research* 76:8662–71.

Bowen, N. 1928. *The evolution of the igneous rocks.* Princeton, N.J.: Princeton Univ. Press.

Bradley, W. C. 1957. Origin of marine-terrace deposits in the Santa Cruz area, California. *Geol. Soc. America Bull.* 68:421–44.

————. 1958. Submarine abrasion and wave-cut platforms. *Geol. Soc. America Bull.* 69:967–74.

————. 1970. Effect of weathering on abrasion of granitic gravel, Colorado River (Texas). *Geol. Soc. America Bull.* 81:61–80.

Bradley, W. C., and Griggs, G. B. 1976. Form, genesis, and deformation of central California wave-cut platforms. *Geol. Soc. America Bull.* 87:433–49.

Bradley, W. C.; Hutton, J. T.; and Twidale, C. R. 1978. Role of salts in development of granitic tafoni, south Australia. *Jour. Geology* 86:647–54.

Bradley, W. C., and Mears, A. I. 1980. Calculations of floods needed to transport coarse fraction of Boulder Creek alluvium at Boulder, Colorado. *Geol. Soc. America Bull.* 91 (pt. 1):135–38.

Brakenridge, G. R. 1981. Late Quaternary floodplain sedimentation along the Pomme de Terre River, southern Missouri. *Quat. Res.* 15:62–76.

Brenninkmeyer, B. M. 1975. Frequency of sand movement in the surf zone. *Proc. 14th Conf. on Coast. Eng.,* 812–27.

Bretz, J. H. 1942. Vadose and phreatic features of limestone caverns. *Jour. Geology* 50:675–811.

Brice, J. C. 1964. Channel patterns and terraces of the Loup River in Nebraska. U.S. Geol. Survey Prof. Paper 422-D.

————. 1974. Evolution of meander loops. *Geol. Soc. America Bull.* 85:581–86.

Brook, G.; Folkoff, M.; and Box, E. 1983. A world model of soil carbon dioxide. *Earth Surf. Proc. and Landforms* 8:79–88.

Brook, G., and Ford, D. 1982. Hydrologic and geologic control of carbonate water in the subarctic Nahanni karst, Canada. *Earth Surf. Proc. and Landforms* 7:1–16.

Brookfield, M. 1970. Dune trend and wind regime in central Australia. *Zeit. f. Geomorph.,* Suppl. 10:121–58.

Brown, R. J. E. 1967. Permafrost in Canada. Canada Geol. Survey Map 1246A, Natl. Res. Council Publ. NRC 9769.

———. 1969. Permafrost in Canada. Canada Geol. Survey Map 1246A.

———. 1970. *Permafrost in Canada.* Toronto: Univ. Toronto Press.

Browning, J. M. 1973. Catastrophic rock slides, Mount Huascaran, north-central Peru, May 31, 1970. *Am. Assoc. Petroleum Geologists Bull.* 57:1335–41.

Brucker, R. W.; Hess, J. W.; and White, W. B. 1972. Role of vertical shafts in the movement of ground water in carbonate aquifers. *Ground Water* 10(6):5.

Brunsden, D., and Kesel, R. H. 1973. Slope development on a Mississippi River bluff in historic time. *Jour. Geology* 81:576–97.

Brush, L. M. 1961. Drainage basins, channels, and flow characteristics of selected streams in central Pennsylvania. U.S. Geol. Survey Prof. Paper 282-F.

Brush, L. M., and Wolman, M. G. 1960. Knickpoint behavior in noncohesive material: a laboratory study. *Geol. Soc. America Bull.* 71:57–76.

Bruun, P. 1963. Longshore currents and longshore troughs. *Jour. Geophys. Research* 68:1065–78.

Bryan, K. 1922. Erosion and sedimentation in the Papago County, Arizona with a sketch of the geology. *U.S. Geol. Survey Bull.* 730-B:19–90.

Bryan, R. B. 1979. The influence of slope angle on soil entertainment by sheetwash and rainsplash. *Earth Surf. Proc. and Landforms* 4:43–58.

Buckman, H. O., and Brady, N. C. 1960. *The nature and properties of soils.* New York: Macmillan.

Buckman, R. C., and Anderson, R. E. 1979. Estimation of fault-scarp ages from a scarp-height-slope-angle relationship. *Geology* 7:11–14.

Budd, W. F. 1975. A first simple model for periodically self-surging glaciers. *Jour. Glaciol.* 14:3–21.

Büdel, J. 1968. Geomorphology—principles. In *Encyclopedia of geomorphology,* edited by R. W. Fairbridge, pp. 416–22. New York: Reinhold Book Corp.

———. 1982. *Climatic geomorphology.* Princeton, N.J.: Princeton Univ. Press.

Budyko, M. I. 1977. *Climatic changes.* Baltimore: Waverly Press, p. 261.

Bull, W. B. 1963. Alluvial-fan deposits in western Fresno County, California. *Jour. Geology* 71:243–51.

———. 1964a. Alluvial fans and near-surface subsidence in western Fresno County, California. U.S. Geol. Survey Prof. Paper 437-A.

———. 1964b. Geomorphology of segmented alluvial fans in western Fresno County, California. U.S. Geol. Survey Prof. Paper 552-F.

———. 1968. Alluvial fans. *Jour. Geol. Educ.* 16:101–6.

———. 1974. Geomorphic tectonic analysis of the Vidal region. In Woodward-McNeill and Assoc., *Vidal nuclear gen. station, Units 1 and 2,* appendix 2.5B (Geology and seismology). Rosemead, Calif.: So. Calif. Co.

———. 1975a. Allometric change of landforms. *Geol. Soc. America Bull.* 86:1489–98.

———. 1975b. Landforms that do not tend toward a steady state. In *Theories of landform development,* edited by W. N. Melhorn and R. C. Flemal, pp. 111–28. S.U.N.Y., Binghamton: Pubs. in Geomorphology, 6th Ann. Mtg.

———. 1979. Threshold of critical power in streams. *Geol. Soc. America Bull.* 90 (pt. 1):453–64.

———. 1980. Geomorphic thresholds as defined by ratios. In *Thresholds in geomorphology,* edited by D. Coates and J. Vitek, pp. 259–63. London: Allen and Unwin Ltd.

———. 1984. Tectonic geomorphology. *Jour. Geol. Educ.* 32:310–24.

Bull, W. B., and McFadden, L. D. 1977. Tectonic geomorphology north and south of the Garlock Fault, California. In *Geomorphology in arid regions,* edited by D. O. Doehring, pp. 115–38. S.U.N.Y. Binghamton: Proc. 8th Ann. Geomorph. Symposium.

Bullard, F. 1962. *Volcanoes.* Austin: Univ. Texas Press.

Buol, S. W.; Hole, F. D.; and McCracken, R. J. 1973. *Soil genesis and classification.* Ames, Iowa: Iowa State Univ. Press.

Burdon, D. J., and Safadi, C. 1963. Ras-el-ain: The great karst spring of Mesopotamia. *J. Hydrol.* 1:58–95.

Burke, R. M., and Birkeland, P. W. 1979. Reevaluation of multiparameter relative dating techniques and their application to the glacial sequence along the eastern escarpment of the Sierra Nevada, California. *Quat. Res.* 11:21–51.

Burnett, A. W., and Schumm, S. A. 1983. Alluvial river response to neotectonic deformation in Louisiana and Mississippi. *Science* 222:49–50.

Butzer, K. W. 1964. *Environment and archeology.* Chicago: Aldine Publishing.

Cailleaux, A., and Tricart, J. 1956. Le problème de la classification des faits géomorphologiques. *Annales Geog.* 65:162–86.

Caine, N. 1968. *The block fields of northeastern Tasmania.* Australian Natl. Univ., Dept. Geog. Publ. G/6.

————. 1981. A source of bias in rates of surface soil movement as estimated from marked particles. *Earth Surf. Proc. and Landforms* 6:69–75.

————. 1982. Toppling failures from Alpine cliffs on Ben Lomond, Tasmania. *Earth Surf. Proc. and Landforms* 7:133–52.

Calkin, P., and Cailleaux, A. 1962. A quantitative study of cavernous weathering (tafonis) and its application to glacial chronology in Victoria Valley, Antarctica. *Zeit. f. Geomorph.* 6:317–24.

Calkin, P. E., and Rutford, R. H. 1974. The sand dunes of Victoria Valley, Antarctica. *Geogr. Rev.* 64:189–216.

Callander, R. A. 1969. Instability and river channels. *Jour. Fluid Mech.* 36:465–80.

Campbell, W. J., and Rasmussen, L. A. 1969. Three-dimensional surges and recoveries in a numerical glacier model. *Can. J. Earth Sci.* 6:979–86.

Carlston, C. W. 1963. Drainage density and streamflow. U.S. Geol. Survey Prof. Paper 422-C.

————. 1969. Downstream variations in the hydraulic geometry of streams: Special emphasis on mean velocity. *Am. Jour. Sci.* 267:499–509.

Carroll, D. 1958. Role of clay minerals in the transportation of iron. *Geochim. et Cosmochim. Acta* 14:1–27.

————. 1970. *Rock weathering.* New York: Plenum Press.

Carson, M. A. 1969. Models of hillslope development under mass failure. *Geographical Analysis* 1:76–100.

Carson, M. A., and Kirkby, M. 1972. *Hillslope form and process.* London: Cambridge Univ. Press.

Carter, L. D. 1983. Fossil sand wedges on the Alaskan arctic coastal plain and their paleoenvironmental significance. In *Proc. Permafrost 4th Internat. Conf.,* pp. 109–14. Natl. Acad. Sci.

Casagrande, A. 1948. Classification and identification of soils. *Am. Soc. Civil Engineers Trans.* 113:901–91.

CERC (Coastal Engineering Research Center). 1973. *Shore protection manual.* 3 vols. Washington, D.C.: U.S. Army Corps Engrs.

Chadbourne, B. D.; Cole, R. M.; Tootill, S.; and Walford, M. E. R. 1975. The movement of melting ice over rough surfaces. *Jour. Glaciol.* 14:287–92.

Chamberlain, R. T. 1928. Instrumental work on the nature of glacial motion. *Jour. Geology* 36:1–30.

Chambers, M. J. G. 1970. Investigations of patterned ground at Signy Island, South Orkney Islands, IV. Longterm experiments. *British Antarctic Surv. Bull.* 23:93–100.

Chappell, J. M. A. 1974. Geology of coral terraces, Huon Peninsula, New Guinea: A study of Quaternary tectonic movements and sea level changes. *Geol. Soc. America Bull.* 85:553–70.

Cheel, R. 1982. The depositional history of an esker near Ottawa, Canada. *Can. J. Earth Sci.* 19:1417–27.

Chepil, W. S. 1945. Dynamics of wind erosion, II. Initiation of soil movement. *Soil Sci.* 60:397–411.

————. 1959. Equilibrium of soil grains at the threshold of movement by wind. *Soil. Sci. Soc. Am. Proc.* 23:422–28.

Chepil, W. S., and Woodruff, N. P. 1963. The physics of wind erosion and its control. *Advances in Agron.* 15:211–302.

Chorley, R. J. 1957. Illustrating the laws of morphometry. *Geol. Mag.* 94:140–50.

————. 1959. The shape of drumlins. *Jour. Glaciol.* 3:339–44.

————. 1962. Geomorphology and the general systems theory. U.S. Geol. Survey Prof. Paper 500-B.

————. 1969a. *Introduction to fluvial processes.* London: Methuen and Co.

————, ed. 1969b. *Water, earth, and man.* London: Methuen and Co.

————. 1978. The hillslope hydrological cycle. In *Hillslope Hydrology,* edited by M. J. Kirkby, pp. 1–42. New York: John Wiley & Sons.

Chorley, R. J., and Kennedy, B. A. 1971. *Physical geography.* London: Prentice-Hall International.

Chorley, R. J., and Morley, L. S. D. 1959. A simplified approximation for the hypsometric integral. *Jour. Geology* 67:566–71.

Chose, B.; Pandy, S.; and Lal, G. 1967. Quantitative geomorphology of the drainage basins in the central Lumi basin in western Rajasthan. *Zeit. f. Geomorph.* 11:146–60.

Church, M. 1972. Baffin Island sandurs: A study of arctic fluvial processes. *Canada Geol. Survey Bull.* 216.

————. 1978. Paleohydrological reconstructions from a Holocene valley. In *Fluvial sedimentology,* edited by A. Miall, pp. 743–72. Can. Soc. Petrol. Geol. Mem. 5.

Church, M., and Jones, D. 1982. Channel bars in gravel-bed rivers. In *Gravel-bed rivers,* edited by R. Hey, J. Bathurst, and C. Thorne, pp. 291–338. New York: John Wiley & Sons.

Clapperton, C. M. 1975. The debris content of surging glaciers in Svalbard and Iceland. *Jour. Glaciol.* 14:395–406.

Clark, J. A., and Bloom, A. L. 1979. Hydroisostasy and Holocene emergence of South America. In *Proc. Internat. Symposium on Coastal Evolution in the Quaternary,* Sao Paulo, 1978, edited by K. Suguio and others, pp. 41–60.

Clark, S. P., and Jager, E. 1969. Denudation rate in the Alps from geochronological and heat flow data. *Am. Jour. Sci.* 267:1143–60.

Clayton, L. 1967. Stagnant-glacial features of the Missouri Coteau in North Dakota. North Dakota Geol. Survey Misc. Series 30:25–46.

Cleaves, E.; Fisher, D.; and Bricker, O. 1974. Chemical weathering of serpentinite in the eastern piedmont of Maryland. *Geol. Soc. America Bull.* 85:437–44.

Coates, D. R. 1976. *Geomorphology and engineering.* Stroudsburg, Pa.: Dowden, Hutchinson and Ross.

Coates, D. R., and Vitek, J. D., eds. 1980. *Thresholds in geomorphology.* London: Allen and Unwin Ltd.

Colbeck, S. C., ed. 1980. *Dynamics of snow and ice masses.* New York: Academic Press.

Colbeck, S. C., and Evans, R. J. 1973. A flow law for temperate glacier ice. *Jour. Glaciol.* 12:71–86.

Colby, B. 1963. Fluvial sediments: A summary of source, transportation, deposition, and measurement of sediment discharge. *U.S. Geol. Survey Bull.* 1181-A:21.

———. 1964. Scour and fill in sand bed streams. U.S. Geol. Survey Prof. Paper 462-D.

Coleman, J. M. 1968. Deltaic evolution. In *Encyclopedia of geomorphology,* edited by R. W. Fairbridge, pp. 255–60. New York: Reinhold Book Corp.

Colman, S. M. 1982. Clay mineralogy of weathering rinds and possible implications concerning the sources of clay minerals in soils. *Geology* 10:370–75.

———. 1983. Progressive changes in the morphology of fluvial terraces and scarps along the Rappahannock River, Virginia. *Earth Surf. Proc. and Landforms* 8:201–12.

Colman, S. M., and Pierce, K. L. 1981. Weathering rinds on andesitic and basaltic stones as a Quaterny age indicator, western United States. U.S. Geol. Survey Prof. Paper 1210.

Colman, S. M., and Watson, K. 1983. Ages estimated from a diffusion model for scarp degradation. *Science* 221:263–65.

Connell, W., and Patrick, W. 1968. Sulfate reduction in soil. Effects of redox potential and pH. *Science* 159:86–87.

Cook, J. H. 1946. Kame complexes and perforation deposits. *Am. Jour. Sci.* 244:573–83.

Cooke, H. J. 1973. Tropical karst in northeast Tanzania. *Zeit. f. Geomorph.* 17:443–59.

Cooke, R. U. 1970. Morphometric analysis of pediments and associated landforms in the western Mojave Desert, California. *Am. Jour. Sci.* 269:26–38.

Cooke, R. U., and Doornkamp, J. 1974. *Geomorphology in environmental management.* London: Clarendon Press.

Cooke, R. U., and Reeves, R. W. 1972. Relations between debris size and the slope of mountain fronts and pediments in the Mojave Desert, California. *Zeit. f. Geomorph.* 16:76–82.

Cooke, R. U., and Warren, A. 1973. *Geomorphology in deserts.* London: Batsford Ltd.

Cooley, R.; Fiero, G.; Lattman, L.; and Mindling, A. 1973. Influence of surface and near-surface caliche distribution on infiltration characteristics and flooding, Las Vegas area, Nevada. Univ. Nevada, Reno, Desert Research Inst. Proj. Rept. 21.

Cooper, W. S. 1958. Coastal sand dunes of Oregon and Washington. *Geol. Soc. America Mem.* 72.

Corbel, J. 1957. Karsts hauts-Alpins. *Rev. Géogr. Lyon* 32:135–58.

———. 1959. Vitesse de l'erosion. *Zeit. f. Geomorph.* 3:1–28.

———. 1964. L'erosion terrestre, étude quantitative. *Annales Géog.* 73:385–412.

Corte, A. E. 1969. Geocryology and engineering. In *Reviews in engineering geology 2,* edited by D. Varnes and G. Kiersch, pp. 119–85. Boulder, Colo.: Geol. Soc. America.

Costa, J. E. 1974a. Stratigraphic, morphologic, and pedologic evidence of large floods in humid environments. *Geology* 2:301–3.

———. 1974b. Response and recovery of a piedmont watershed from tropical storm Agnes, June 1972. *Water Resour. Res.* 10:106–12.

———. 1978. Holocene stratigraphy in flood frequency analysis. *Water Resour. Res.* 14:626–32.

———. 1983. Paleohydraulic reconstruction of flash-flood peaks from boulder deposits in the Colorado Front Range. *Geol. Soc. America Bull.* 94:986–1004.

Costa, J. E., and Baker, V. R. 1981. *Surficial geology—Building with the Earth.* New York: John Wiley & Sons.

Cox, E. T. 1874. *Fifth annual report of the Geological Survey of Indiana,* pp. 280–305. Indianapolis: Indiana Geol. Survey.

Cramer, H. 1941. Die Systematik der Karstdolinen. *Neues Jb. Miner. Geol. Paläont.* 85:293–382.

Crary, A. P. 1966. Mechanisms for fiord formation indicated by studies of an ice-covered inlet. *Geol. Soc. America Bull.* 77:911–30.

Crittenden, M. 1963. New data on the isostatic deformation of Lake Bonneville. U.S. Geol. Survey Prof. Paper 454-E.

Cronin, J. E. 1983. Design and performance of a liquid natural convection subgrade cooling system for construction on ice-rich permafrost. In *Proc. Permafrost 4th Internat. Conf.,* pp. 198–203. Natl. Acad. Sci.

Cronin, T. M. 1981. Rates and possible causes of neotectonic vertical crustal movements of the emerged southeastern United States Atlantic coastal plain, *Geol. Soc. America Bull.* 92:812–33.

Cronin, T. M.; Szabo, B.; Ager, T. A.; Hazel, J. B.; and Owens, J. P. 1981. Quaternary climates and sea level: US Atlantic coastal plain. *Science* 211:233–40.

Crosby, W. O. 1902. Origin of eskers. *Am. Geologist* 30:1–38.

Crowley, K. D. 1983. Large-scale bed configurations (macroforms), Platte River basin, Colorado and Nebraska: Primary structures and formative processes. *Geol. Soc. America Bull.* 94:117–33.

Crozier, M. J. 1973. Techniques for the morphometric analysis of landslips. *Zeit. f. Geomorph.* 17:78–101.

Cummans, J. 1981. Mudflows resulting from the May 18, 1980, eruption of Mount St. Helens, Washington. U.S. Geol. Survey Circ. 850-B.

Cunningham, F., and Griba, W. 1973. A model of slope development, and its applications to the Grand Canyon, Arizona. *Zeit. f. Geomorph.* 17:43–77.

Curray, J. R. 1960. Sediments and history of Holocene transgression, continental shelf, northwest Gulf of Mexico. In *Recent sediments, northwest gulf of Mexico,* edited by Shepard, F. P., et al. Amer. Assoc. Petroleum Geologists, pp. 221–66.

————. 1961. Late Quaternary sea level; a discussion. *Geol. Soc. America Bull.* 72:1707–12.

————. 1965. Late Quaternary history, continental shelves of the United States. In *The Quaternary of the United States,* edited by H. Wright and D. Frey, pp. 723–35. Princeton, N.J: Princeton Univ. Press.

Cvijíc, J. 1893. Das Karstphanomen. *Geogr. Abh.* 5:217–329.

Czudek, T., and Demek, J. 1970. Thermokarst in Siberia and its influence on the development of lowland relief. *Quat. Res.* 1:103–20.

Dahl, R. 1965. Plastically sculptured detail forms on rock surfaces in northern Nordland. *Geogr. Annlr.* 47:83–140.

Dalrymple, G.; Silver, E.; and Jackson, E. 1973. Origin of the Hawaiian Islands. *Am. Scientist* 61:294–308.

Dalrymple, J. B.; Blong, R. J.; and Conacher, A. J. 1968. A hypothetical nine-unit landsurface model. *Zeit. f. Geomorph.* 12:60–76.

Dalrymple, T. 1960. Flood frequency analysis. *Manual of hydrology,* part 3, Flood flow techniques. U.S. Geol. Survey Water Supply Paper 1543-A.

Dalrymple, T., and Benson, M. A. 1967. Measurement of peak discharge by the slope-area method. *U.S. Geol. Survey Techniques Water Resour. Res. Div.,* Bk. 3, chap. A–2.

Daly, R. 1933. *Igneous rock and the depths of the Earth.* New York: McGraw-Hill.

Daniel, J. R. K. 1981. Drainage density as an index of climatic geomorphology. *J. Hydrol.* 50:147–54.

Davies, J. L. 1973. *Geographical variation in coastal development.* New York: Hafner.

Davies, W. E. 1960. Origin of caves in folded limestone. *Natl. Speleol. Soc. Bull.* 22:5–18.

Davis, D. G. 1980. Cave development in the Guadalupe Mountains, a critical review of recent hypotheses. *Natl. Speleol. Soc. Bull.* 42:42–48.

Davis, J. L. 1964. A morphogenetic approach to world shorelines. *Zeit. f. Geomorph.* 8:127–42.

Davis, R. A., ed. 1978. *Coastal sedimentary environments.* New York: Springer-Verlag.

Davis, W. M. 1902. River terraces in New England. *Mus. Comp. Zoology Bull.* 38:77–111.

————. 1930. Origin of limestone caverns. *Geol. Soc. America Bull.* 41:475–628.

Day, M. 1978. Morphology and distribution of residual limestone hills (mogotes) in the karst of northern Puerto Rico. *Geol. Soc. America Bull.* 89:426–32.

Dean, R. G., and Perlin, M. 1977. Coastal engineering study of Ocean City Inlet, Maryland. In *Coastal sediments 77.* Am. Soc. Civil Engineers, *Symposium Water, Port, Coastal and Ocean Div.* 5:520–42.

Deere, D. U., and Peck, R. B. 1959. Stability of cuts in fine sands and varved clays, Northern Pacific Railroad, Noxon Rapids line change, Montana. *Proc. AREA* 59, pp. 807–15.

Denny, C. S. 1956. Surficial geology and geomorphology of Potter County, Pennsylvania. U.S. Geol. Survey Prof. Paper 288.

————. 1965. Alluvial fans in the Death Valley region, California and Nevada. U.S. Geol. Survey Prof. Paper 466.

————. 1967. Fans and pediments. *Am. Jour. Sci.* 265:81–105.

Derbyshire, E. 1976. *Geomorphology and climate.* New York: John Wiley & Sons.

Dietrich, W. E., and Dunne, T. 1978. Sediment budget for a small catchment in mountainous terrain. *Zeit. f. Geomorph.,* Suppl. Bd. 29:191–206.

Dingman, S. L. 1978. Drainage density and streamflow: A closer look. *Water Resour. Res.* 14:1183–87.

Dionne, J. C. 1974. Polished and striated mud surfaces in the St. Lawrence tidal flats, Quebec. *Can. J. Earth Sci.* 11:860–66.

Dolan, R., and Ferm, J. C. 1968. Crescentic landforms along the mid-Atlantic coast. *Science* 159:627–29.

Dolan, R.; Hayden, B.; and Felder, W. 1979. Shoreline periodicities and edge waves. *Jour. Geology* 87:175–85.

Dolan, R.; Hayden, B.; and Lins, H. 1980. Barrier islands. *Am. Scientist* 68:16–25.

Dolan, R.; Vincent, L.; and Hayden, B. 1974. Crescentic coastal landforms. *Zeit. f. Geomorph.* 18:1–12.

Doornkamp, J. C., and King, C. A. M. 1971. *Numerical analysis in geomorphology: An Introduction*. London: Edward Arnold Ltd.

Douglas, G. R.; McGreevey, J. P.; and Whalley, W. B. 1983. Rock weathering by frost shattering processes. In *Proc. Permafrost 4th Internat. Conf.*, pp. 244–48. Natl. Acad. Sci.

Douglas, I. 1964. Intensity and periodicity in denudation process with special reference to the removal of material in solution by rivers. *Zeit. f. Geomorph.* 8:453–73.

————. 1967. Man, vegetation and sediment yields of rivers. *Nature* 215:925–28.

Drake, J. 1980. The effect of soil activity on the chemistry of carbonate groundwater. *Water Resour. Res.* 16:381–86.

Drake, J. J., and Harmon, R. S. 1973. Hydrochemical environments of carbonate terrains. *Water Resour. Res.* 9:949–57.

Drake, J. J., and Wigley, T. M. L. 1975. The effect of climate on the chemistry of carbonate groundwater. *Water Resour. Res.* 11:958–62.

Drake, L. D. 1968. *Till studies in New Hampshire*. Ph.D. dissertation, Ohio State University.

————. 1974. Till fabric control by clast shape. *Geol. Soc. America Bull.* 85:247–50.

Drake, L. D., and Shreve, R. L. 1973. Pressure melting and regelation of ice by round wires. *Proc. Royal Soc. London,* ser. A:332:51–83.

Dreimanis, A., and Vagners, U. J. 1971. Bimodal distribution of rock and mineral fragments in basal till. In *Till: A symposium*, edited by R. P. Goldthwait. Columbus: Ohio State Univ. Press.

Drever, J. I., and Smith, C. L. 1978. Cyclic wetting and drying of the soil zone as an influence on the chemistry of groundwater in arid terrains. *Am. Jour. Sci.* 278:1448–54.

Duane, D. B., and James, W. R. 1980. Littoral transport in the surf zone elucidated by an Eulerian sediment tracer experiment. *Jour. Sed. Petrology* 50:929–42.

DuBois, R. N. 1972. Inverse relation between foreshore slope and mean grain size as a function of the heavy mineral content. *Geol. Soc. America Bull.* 83:871–76.

————. 1978. Beach topography and beach cusps. *Geol. Soc. America Bull.* 89:1133–39.

————. 1981. Foreshore topography, tides and beach cusps, Delaware. *Geol. Soc. America Bull.* 92 (pt. 1):132–38.

Dunkerley, D. L. 1980. The study of the evolution of slope form over long periods of time: A review of methodologies and some new observational data from Papua, New Guinea. *Zeit. f. Geomorph.* 24:52–67.

Dunne, T. 1978. Field studies of hillslope flow processes. In *Hillslope hydrology,* edited by M. J. Kirby, pp. 227–94. New York: John Wiley & Sons.

————. 1979. Sediment yield and land use in tropical catchments. *J. Hydrol.* 42:281–300.

Dunne, T., and Black, R. D. 1970a. An experimental investigation of runoff production in permeable soils. *Water Resour. Res.* 6:478–90.

————. 1970b. Partial area contribution to storm runoff in a small New England watershed. *Water Resour. Res.* 6:1296–1311.

Dunne, T., and Leopold, L. B. 1978. *Water in environmental planning.* San Francisco: W. H. Freeman.

Dury, G. H. 1964. Principles of underfit streams. U.S. Geol. Survey Prof. Paper 452-A.

————. 1965. Theoretical implications of underfit streams. U.S. Geol. Survey Prof. Paper 452-C.

————. 1966a. Duricrusted residuals on the Barrier and Cobar pediplains of New South Wales. *Jour. Geol. Soc. Australia* 13:299–307.

————. 1966b. Pediment slope and particle size at Middle Pinnacle, near Broken Hill, New South Wales. *Austr. Geog. Studies* 4:1–17.

————. 1969. Relation of morphometry to runoff frequency. In *Introduction to fluvial processes,* edited by R. J. Chorley, pp. 177–88. London: Methuen and Co.

————. 1973. Magnitude-frequency analysis and channel morphology. In *Fluvial geomorphology,* edited by M. Morisawa, pp. 91–121. S.U.N.Y., Binghamton: Pubs. in Geomorphology.

Duval, P. 1981. Creep and fabrics of polycrystalline ice under shear and compression. *Jour. Glaciol.* 27:129–40.

Duval, P., and Hughes, L. G. 1980. Does the permanent creep-rate of polycrystalline ice increase with crystal size? *Jour. Glaciol.* 25:151–57.

Dylik, J. 1964. The essentials of the meaning of the term "Periglacial." *Soc. Sci. et Lettres Łódz Bull.* 15:2:1–19.

Eakin, H. M. 1916. The Yukon-Koyukuk region. Alaska. U.S. Geol. Survey Bull. 631.

Eaton, J., and Murata, K. 1960. How volcanoes grow. *Science* 132:925–38.

Ede, D. P. 1975. Limestone drainage systems. *J. Hydrol.* 27:297–318.

Egemeir, S. J. 1981. Cavern development by thermal waters. *Natl. Speleol. Soc. Bull.* 43:31–52.

Eggler, D. H.; Larson, E. E.; and Bradley, W. C. 1969. Granites, grusses, and the Sherman erosion surface, southern Laramie Range, Colorado-Wyoming. *Am. Jour. Sci.* 267:510–22.

Einstein, H. A. 1950. The bedload function for sediment transportation in open channel flows. U.S. Dept. Agri. Tech. Bull. 1026.

El-Baz, F., and Maxwell, T. A., eds. 1982. *Desert landforms of southwest Egypt: A basis for comparison with Mars.* NASA Sci. and Tech. Information Branch, Washington, D.C., CR–3611.

Elliston, G. R. 1963. Catastrophic glacier advances. *Int. Assoc. Sci. Hydrol. Bull.* 8:65–66.

Elson, J. A. 1968. Washboard moraines and other minor moraine types. In *Encyclopedia of geomorphology,* edited by R. W. Fairbridge, pp. 1213–19. New York: Reinhold Book Corp.

Embleton, C., and King, C. A. M. 1968. *Glacial and periglacial geomorphology.* Edinburgh: Edward Arnold Ltd.

———. 1975a. *Glacial geomorphology.* New York: Halsted Press.

———. 1975b. *Periglacial geomorphology.* New York: Halsted Press.

Emery, K. O., and Kuhn, G. G. 1982. Sea cliffs: Their processes, profiles and classifications. *Geol. Soc. America Bull.* 93:644–54.

Emmett, W. W. 1980. A field calibration of the sediment-trapping characteristics of the Helley-Smith bedload samples. U.S. Geol. Survey Prof. Paper 1139.

Engelhardt, H. F.; Harrison, W. D.; and Kamb, B. 1978. Basal sliding and conditions at the glacier bed as revealed by bore-hole photography. *Jour. Glaciol.* 20:469–508.

Ericksen, G. E.; Pflacker, G.; and Fernandez, J. V. 1970. Preliminary report on the geological events associated with the May 31, 1970, Peru earthquake. U.S. Geol. Survey Circ. 639.

Evans, I. S. 1969. Salt crystallization and rock weathering: A review. *Rev. géomorph. dynamique* 19:157–77.

Evans, O. F. 1938. Classification and origin of beach cusps. *Jour. Geology* 46:615–27.

———. 1940. The low and ball of the east shore of Lake Michigan. *Jour. Geology* 48:476–511.

———. 1942. The origin of spits, bars, and related structures. *Jour. Geology* 50:846–63.

Evans, R. 1976. Observations on a stripe pattern. *Biuletyn Perygl.* 25:9–22.

Everett, D. H. 1961. The thermodynamics of frost damage to porous solids. *Trans. Faraday Soc.* 57:1541–51.

Everitt, B. L. 1968. Use of cottonwood in an investigation of recent history of a flood plain. *Am. Jour. Sci.* 266:417–39.

Eyles, N., and Rogerson, R. J. 1978. Sedimentology of medial moraines in Berendon Glacier, British Columbia, Canada: Implications for debris transport in a glacierized basin. *Geol. Soc. America Bull.* 89:1688–93.

Eynon, G., and Walker, R. G. 1974. Facies relationships in Pleistocene outwash gravels, southern Ontario: A model for bar growth in braided rivers. *Sedimentology* 21:43–70.

Fahey, B. D. 1975. Nonsorted circle development in a Colorado alpine location. *Geogr. Annlr.* 57A:153–64.

———. 1983. Frost action and hydration as rock weathering mechanism on schist: A laboratory study. *Earth Surf. Proc. and Landforms.* 8:535–45.

Fahnestock, R. K. 1961. Competence of a glacial stream. U.S. Geol. Survey Prof. Paper 424-B:211–13.

———. 1963. Morphology and hydrology of a glacial stream— White River, Mount Rainier, Washington. U.S. Geol. Survey Prof. Paper 422-A.

———. 1969. Morphology of the Slims River. In *Icefield Ranges Research Project, Scientific Results #1,* edited by V. C. Bushell and R. H. Ragle, pp. 161–72. Am. Geog. Soc. and Arctic Inst. N. America.

Fairbridge, R. W., and Newman, W. 1968. Postglacial crustal subsidence of the New York area. *Zeit. f. Geomorph.* 12:296–317.

Fairchild, J.C. 1973. Longshore transport of suspended sediment. *Proc. 13th Conf. on Coast. Eng.,* 1069–88.

Ferguson, R. I. 1975. Meander irregularity and wavelength estimation. *J. Hydrol.* 26:315–33.

Ferrians, O. J. 1965. Permafrost map of Alaska. U.S. Geol. Survey Misc. Geol. Inv. Map I–445.

Ferrians, O. J.; Kachadoorian, R.; and Greene, G. W. 1969. Permafrost and related engineering problems in Alaska. U.S. Geol. Survey Prof. Paper 678.

Feth, J.; Robertson, C.; and Polzer, W. 1964. Sources of mineral constituents in water from granitic rocks, Sierra Nevada, California and Nevada. U.S. Geol. Survey Water Supply Paper 1535-I.

Finkel, H. J. 1959. The barchans of southern Peru. *Jour. Geology* 67:614–47.

Finlayson, B. 1981. Field measurements of soil creep. *Earth Surf. Proc. and Landforms* 6:35–48.

Fisher, J. E. 1963. Two tunnels in cold ice at 4000 m on the Breithorn. *Jour. Glaciol.* 4:513–20.

Fisher, J. J., and Simpson, E. J. 1979. Washover and tidal sedimentation rates as environmental factors in development of a transgressive barrier shoreline. In *Barrier islands,* edited by S. Leatherman. New York: Academic Press.

Fisk, H. N. 1944. *Geological investigation of the alluvial valley of the lower Mississippi River.* Vicksburg, Miss.: Mississippi River Comm.

————. 1951. Loess and Quaternary geology of the lower Mississippi Valley. *Jour. Geology* 59:333–56.

Fiske, R. S.; Hopson, C. A.; and Waters, A. C. 1963. Geology of Mount Ranier National Park. U.S. Geol. Survey Prof. Paper 444.

Fitze, P. 1971. Messungen von Bodenbewegungen auf West-Spitzbergen. *Geog. Helvetica* 26:148–52.

Flemal, R. C. 1976. Pingos and pingo scars: Their characteristics, distribution, and utility in reconstructing former permafrost environments. *Quat. Res.* 6:37–53.

Flemal, R. C.; Hinkley, K. C.; and Hesler, J. L. 1976. DeKalb mounds: A possible Pleistocene (Woodfordian) pingo field in north-central Illinois. *Geol. Soc. America Mem.* 136:229–50.

Fleming, R. W., and Johnson, A. M. 1975. Rates of seasonal creep of silty clay soil. *Quart. Jour. Engr. Geol.* 8:1–29.

Flint, R. F. 1963. Altitude, lithology, and the Fall Zone in Connecticut. *Jour. Geology* 71:683–97.

————. 1971. *Glacial and Quaternary geology.* New York: John Wiley & Sons.

Flint, R. F., and Bond, G. 1968. Pleistocene sand ridges and pans in western Rhodesia. *Geol. Soc. America Bull.* 79:299–314.

Foley, M. G. 1978. Scour and fill in steep, sand-bed ephemeral streams. *Geol. Soc. America Bull.* 89:559–70.

————. 1980a. Bed-rock incision by streams. *Geol. Soc. America Bull.* 91 (pt. 2):2189–213.

————. 1980b. Quaternary diversion and incision, Dearborn River, Montana. *Geol. Soc. America Bull.* 91 (pt. 2):2152–88.

Folk, R. L. 1971a. Longitudinal dunes of the northwestern edge of the Simpson Desert, Northern Territory, Australia, 1. Geomorphology and grain size relationships. *Sedimentology* 16:5–54.

————. 1971b. Genesis of longitudinal and oghurd dunes elucidated by rolling upon grease. *Geol. Soc. America Bull.* 82:3461–68.

————. 1975. Glacial deposits identified by chattermark trails in detrital garnets. *Geology* 3:473–75.

Folk, R. L., and Patton, E. B. 1982. Buttressed expansion of granite and development of grus in central Texas. *Zeit. f. geomorph.* 26:17–32.

Foose, R. M. 1967. Sinkhole formation by groundwater withdrawal: Far West Rand, South Africa. *Science* 157:3792:1045–48.

Forbes, J. D. 1843. *Travels through the Alps of Savoy.* Edinburgh: Oliver and Boyd.

Ford, D., and Ewers, R. 1978. The development of limestone cave systems in length and depth. *Can. J. Earth Sci.* 15:1783–98.

Fournier, M. F. 1960. *Climat et érosion.* Paris: Presses Univ. France.

Francis, J. R. D. 1973. Experiments on the motion of solitary grains along the bed of a water stream. *Proc. Royal Soc. London,* ser. A:332:443–71.

Frazee, C. J.; Fehrenbacher, J. B.; and Krumbein, W. C. 1970. Loess distribution from a source. *Soil Sci. Soc. Am. Proc.* 34:296–301.

Free, G. R. 1960. Erosion characteristics of rainfall. *Agri. Engineering* 41:447–49, 455.

Freeze, R. A. 1980. A stochastic-conceptual analysis of rainfall-runoff processes on a hillslope. *Water Resour. Res.* 16:391–408.

French, H. M. 1974. Mass-wasting at Sachs Harbour, Barks Island, N.W.T., Canada. *Arc. Alp. Res.* 6:77–78.

Friedkin, J. F. 1945. A laboratory study of the meandering of alluvial rivers. U.S. Army Corps Engrs., U.S. Waterways Eng. Exp. Sta.

Friese-Greene, T. W., and Pert, G. J. 1965. Velocity fluctuations of Berksackerbrae, east Greenland. *Jour. Glaciol.* 5:739–47.

Fryberger, S. G. 1979. Dune forms and wind regime. In *A study of global sand seas,* edited by E. McKee. U.S. Geol. Survey Prof. Paper 1052:137–70.

Fryberger, S. G., and Ahlbrandt, T. S. 1979. Mechanisms for the formation of eolian sand seas. *Zeit. f. Geomorph.* 23:440–60.

Frye, J. C., and Leonard, A. R. 1954. Some problems of alluvial terrace mapping. *Am. Jour. Sci.* 252:242–51.

Fuller, M. L. 1914. The geology of Long Island, New York. U.S. Geol. Survey Prof. Paper 82.

Galvin, C. J. 1967. Longshore current velocity: A review of theory and data. *Revs. in Geophys.* 5:3:287–304.

————. 1968. Breaker type classification on three laboratory beaches. *Jour. Geophys. Research* 73:3651–59.

Galvin, C. J., and Eagleson, P. S. 1965. Experimental study of longshore currents on a plane beach. U.S. Army Corps Engrs., Coastal Eng. Res. Center Tech. Memo 10.

Gamper, M. W. 1983. Control and rates of movement of solifluction lobes in the eastern Swiss Alps. In *Proc. Permafrost 4th Internat. Conf.,* pp. 328–33. Natl. Acad. Sci.

Gams, I. 1969. Some morphological characteristics of the Dinaric karst. *Geogr. Jour.* 135:563–72.

Gardner, J. H. 1935. Origin and development of limestone caverns. *Geol. Soc. America Bull.* 46:1255–74.

Gardner, J. S. 1979. The movement of material on debris slopes in the Canadian Rocky Mountains. *Zeit. f. Geomorph.* 23:45–57.

Gardner, T. W. 1973. A model study of river meander incision. M.S. thesis, Colorado State Univ., Fort Collins.

———. 1983. Experimental study of knickpoint and longitudinal profile evolution in cohesive, homogeneous material. *Geol. Soc. America Bull.* 94:664–72.

Garrels, R. M., and Mackenzie, F. T. 1971. *Evolution of sedimentary rocks.* New York: Norton and Co.

Garwood, N. C.; Janos, D. P.; and Brokaw, N. 1979. Earthquake-caused landslide: A major disturbance to tropical forests. *Science* 205:997–99.

Gentilli, J. 1968. Exfoliation. In *Encyclopedia of geomorphology,* edited by R. W. Fairbridge, pp. 336–39. New York: Reinhold Book Corp.

Gerrard, A. J. 1981. *Soils and landforms.* London: Allen and Unwin Ltd.

Gibbs, R. J. 1967. The geochemistry of the Amazon River system, part I. *Geol. Soc. America Bull.* 78:1203–32.

Gilbert, G. K. 1877. *Geology of the Henry Mountains (Utah).* U.S. Geog. and Geol. Survey of the Rocky Mtn. Region. Washington, D.C.: U.S. Govt. Printing Office.

———. 1917. Hydraulic-mining debris in the Sierra Nevada. U.S. Geol. Survey Prof. Paper 105.

Gile, L. H. 1966. Cambic and certain non-cambic horizons in desert soils of southern New Mexico. *Soil Sci. Soc. Am. Proc.* 30:773–81.

———. 1975. Holocene soils and soil-geomorphic relations in an arid region of southern New Mexico. *Quat. Res.* 5:321–60.

Gile, L., and Grossman, R. 1979. *The Desert Project soil monograph.* U.S. Dept. Agri., Soil Conserv. Serv.

Gile, L.; Hawley, J.; and Grossman, R. 1981. Soils and geomorphology in the Basin and Range area of Southern New Mexico—Guidebook to the Desert Project. New Mexico Bur. Mines and Min. Res. Memo 39.

Gile, L. H.; Peterson, F.; and Grossman, R. 1965. The K horizon. A master soil horizon of carbonate accumulation. *Soil Sci.* 99:74–82.

———. 1966. Morphological and genetic sequences of carbonate accumulation in desert soils. *Soil Sci.* 101:347–60.

Gilluly, J. 1937. Physiography of the Ajo region, Arizona. *Geol. Soc. America Bull.* 43:323–48.

———. 1949. The distribution of mountain-building in geologic time. *Geol. Soc. America Bull.* 60:561–90.

———. 1955. *Geologic contrasts between continents and ocean basins.* Geol. Soc. America Spec. Paper 62:7–18.

———. 1964. Atlantic sediments, erosion rates, and the evolution of the Continental Shelf—Some speculations. *Geol. Soc. America Bull.* 75:483–92.

———. 1969. Geological perspectives and the completeness of the geologic record. *Geol. Soc. America Bull.* 80:2303–12.

Gilpin, R. R. 1980. A model for the prediction of ice lensing and frost heave in soils. *Water Resour. Res.* 16:918–30.

Gjessing, J. 1967. On plastic scouring and subglacial erosion. *Norsk. Geogr. Tidsskr.* 20:1–37.

Glen, J. W. 1952. Experiments on the deformation of ice. *Jour. Glaciol.* 2:111–14.

———. 1955. The creep of polycrystalline ice. *Proc. Royal Soc. London,* ser. A: 228:519–38.

———. 1958. Mechanical properties of ice. I. The plastic properties of ice. *Philos. Mag.,* Suppl. 7:254–65.

Glen, J. W., and Lewis, W. V. 1961. Measurements of side-slip at Austerdalsbreen, 1959. *Jour. Glaciol.* 3:1121.

Glennie, K. W. 1970. *Desert sedimentary environments.* Amsterdam: Elsevier.

Godfrey, P. J. 1976. Barrier beaches of the East Coast. *Oceanus* 19:27–40.

Gold, L. W., and Lachenbruch, A. H. 1973. Thermal conditions in permafrost—A review of North American literature. In *Proc. Permafrost 2nd Internat. Conf.* pp. 3–23. Natl. Acad. Sci.-Natl. Res. Council, Yakutsk, U.S.S.R., 1973.

Goldich, S. 1938. A study of rock weathering. *Jour. Geology* 46:17–58.

Goldthwait, R. P. 1951. Development of end moraines in east central Baffin Island. *Jour. Geology* 59:567–77.

———. 1969. Patterned soils and permafrost on the Presidential Range (abs). Paris. 8th INQUA Cong. *Résumés des Communications.* 150.

———, ed. 1971. *Till: A symposium.* Columbus: Ohio State Univ. Press.

———. 1973. Jerky glacier motion and meltwater. *Int. Assoc. Sci. Hydrol. Bull.* 95:183–88.

———. 1976. Frost sorted patterned ground: A review. *Quat. Res.* 6:27–35.

———. 1979. Giant grooves made by concentrated basal ice streams. *Jour. Glaciol.* 23:297–307.

Goodman, D. J.; King, G. C. P.; Millar, D. H. M.; and Robin, G. DeQ. 1979. Pressure-melting effects in basal ice of temperate glaciers: Laboratory studies and field observations under Glaciere D'Argentiere. *Jour. Glaciol.* 23:259–70.

Gordon, J. E. 1977. Morphometry of cirques in the Kintail Affric-Cannich area of northwest Scotland. *Geogr. Annlr.* 59A:177–94.

———. 1981. Ice-scoured topography and its relationship to bedrock structure and ice movements in parts of northern Scotland and West Greenland. *Geogr. Annlr.* 63A:55–65.

Gordon, M.; Tracey, J.; and Ellis, M. 1958. Geology of the Arkansas bauxite region. U.S. Geol. Survey Prof. Paper 299.

Gorsline, D. S. 1966. Dynamic characteristics of west Florida Gulf Coast beaches. *Marine Geol.* 4:187–206.

Gow, A. J., and Williamson, T. 1976. Rheological implications of the internal structure and crystal fabrics of the West Antarctic ice sheet as revealed by deep core drilling at Byrd Station. *Geol. Soc. America Bull.* 87:1665–77.

Graf, W. H. 1971. *Hydraulics of sediment transport.* New York: McGraw-Hill.

Graf, W. L. 1970. The geomorphology of the glacial valley cross section. *Arc. Alp. Res.* 2:303–12.

———. 1976. Cirques as glacier locations. *Arc. Alp. Res.* 8:79–90.

Gravenor, C. P. 1953. The origin of drumlins. *Am. Jour. Sci.* 251:674–81.

———. 1982. Chattermarked garnets in Pleistocene glacial sediments. *Geol. Soc. America Bull.* 93:751–58.

Gravenor, C. P., and Kupsch, W. O. 1959. Ice disintegration features in western Canada. *Jour. Geology* 67:48–64.

Gravenor, C. P., and Meneley, W. A. 1958. Glacial flutings in central and northern Alberta. *Am. Jour. Sci.* 256:715–28.

Gray, J. M. 1982. Unweathered, glaciated bedrock on an exposed lake bed in Wales. *Jour. Glaciol.* 28:483–97.

Gray, W. M. 1965. Surface spalling by thermal stresses in rocks. In *Rock mechanics symposium.* Toronto: Proc. Ottawa, Can. Dept. Mines and Tech. Surveys.

Green, J., and Short, N. 1971. *Volcanic landforms and surface features.* New York: Springer-Verlag.

Gregory, K. J., and Walling, D. E. 1973. *Drainage basin form and process.* New York: Halsted Press.

Griggs, D. 1936a. The factor of fatigue in rock exfoliation. *Jour. Geology* 44:783–96.

———. 1936b. Deformation of rocks under high confining pressures. *Jour. Geology* 44:541–77.

Grim, R. 1962. *Applied clay mineralogy.* New York: McGraw-Hill.

Grove, J. M. 1960. The bands and layers of Vesl-Skautbreen. In *Norwegian cirque glaciers,* edited by W. V. Lewis, pp. 11–23. Royal Geog. Soc. Res. Ser. 4.

Guilcher, A. 1958. *Coastal and submarine morphology.* London: Methuen and Co.

Gunn, J. 1981. Hydrological processes in karst depressions. *Zeit. f. Geomorph.* 25:313–31.

———. 1983. Point recharge of limestone aquifers—A model from New Zealand karst. *J. Hydrol.* 61:19–29.

Gupta, A., and Fox, H. 1974. Effects of high-magnitude floods on channel form: A case study in Maryland Piedmont. *Water Resour. Res.* 10:499–509.

Gutenberg, B. 1941. Changes in sea level, postglacial uplift, and mobility of the earth's interior. *Geol. Soc. America Bull.* 52:721–72.

Guza, R. T., and Inman, D. L. 1975. Edge waves and beach cusps. *Jour. Geophys. Research* 80:21:2997–3012.

Haan, C. T., and Johnson, H. P. 1966. Rapid determination of hypsometric curves. *Geol. Soc. America Bull.* 77:123–25.

Hack, J. T. 1941. Dunes of the western Navajo country. *Geogr. Rev.* 31:240–63.

———. 1957. Studies of longitudinal stream profiles in Virginia and Maryland. U.S. Geol. Survey Prof. Paper 294-B:45–97.

———. 1960a. Relation of solution features to chemical character of water in the Shenandoah Valley, Virginia. U.S. Geol. Survey Prof. Paper 400-B:387–90.

———. 1960b. Interpretation of erosional topography in humid temperate regions. *Am. Jour. Sci.* (Bradley Vol.) 258-A:80–97.

———. 1965. Postglacial drainage evolution in the Ontonagan area, Michigan. U.S. Geol. Survey Prof. Paper 504-B:1–40.

———. 1966. Circular patterns and exfoliation in crystalline terrane, Grandfather Mountain area, North Carolina. *Geol. Soc. America Bull.* 77:975–86.

———. 1973. Stream-profile analysis and stream-gradient index. U.S. Geol. Survey Jour. Research 1:421–29.

Hack, J. T., and Goodlett, J. C. 1960. Geomorphology and forest ecology of a mountain region in the central Appalachians. U.S. Geol. Survey Prof. Paper 347.

Hadley, R. F. 1961. Influence of riparian vegetation on channel shape, northeastern Arizona. U.S. Geol. Survey Prof. Paper 424-C:30–31.

———. 1967. Pediments and pediment-forming processes. *Jour. Geol. Educ.* 15:83–89.

Hadley, R. F., and Schumm, S. A. 1961. Sediment sources and drainage basin characteristics in upper Cheyenne River basin. U.S. Geol. Survey Water Supply Paper 1531-B:137–96.

Haefeli, R., and Brentani, F. 1955. Observations in a cold ice cap. *Jour. Glaciol.* 2:571–80.

Hagerty, D. J. 1980. Multifactor analysis of bank caving along a navigable stream. In *Natl. Waterways Roundtable Proc.* U.S. Army Engr. Water Res. Support Ctr., Inst. for Water Resources, IWR–80–1:463–92.

Haig, M. 1979. Ground retreat and slope evolution on regraded surface-mine dumps, Waunafon, Gwent. *Earth Surf. Proc. and Landforms* 4:183–89.

Haigh, M. J., and Wallace, W. L. 1982. Erosion of strip-mine dumps in LaSalle County, Illinois: Preliminary results. *Earth Surf. Proc. and Landforms* 7:79–84.

Hallberg, G. R. 1979. Wind-aligned drainage in loess in Iowa. *Iowa Acad. Sci. Proc.* 86:4–9.

Hallet, B. 1976a. Deposits formed by subglacial precipitation of $CaCO_3$. *Geol. Soc. America Bull.* 87:1003–15.

———. 1976b. The effect of subglacial chemical processes on glacier sliding. *Jour. Glaciol.* 17:209–21.

———. 1979. A theoretical model of glacial abrasion. *Jour. Glaciol.* 23:39–50.

Hammad, H. Y. 1972. River-bed degradation after closure of dams. Am. Soc. Civil Engineers, *Jour. Hydraulics Div.* 98:591–607.

Handy, R. L. 1976. Loess distribution by variable winds. *Geol. Soc. America Bull.* 87:915–27.

Hanshaw, B., and Back, W. 1980. Chemical mass-wasting of the northern Yucatan Peninsula by groundwater dissolution. *Geology* 8:222–24.

Happ, S. C.; Rittenhouse, G.; and Dobson, G. C. 1940. Some principles of accelerated stream and valley sedimentation. U.S. Dept. Agri. Tech. Bull. 695.

Harpstead, M., and Hole, R. 1980. *Soil science simplified.* Ames: Iowa State Univ. Press.

Harris, C. 1973. Some factors affecting the rates and processes of periglacial mass movement. *Geogr. Annlr.* 55A:24–58.

Harris, S. A. 1983. Comparison of the climatic and geomorphic methods of predicting permafrost distribution in western Yukon Territory. In *Proc. Permafrost 4th Internat. Conf.,* pp. 450–55. Natl. Acad. Sci.

Harris, S. E. 1943. Friction cracks and the direction of glacial movement. *Jour. Geology* 51:244–58.

Harrison, A. E. 1964. Ice surges on the Muldrow Glacier, Alaska. *Jour. Glaciol.* 5:365–68.

Harrison, W. D. 1975. Temperature measurements in a temperate glacier. *Jour. Glaciol.* 14:23–30.

Hartshorn, J. H. 1958. Flowtill in southeastern Massachusetts. *Geol. Soc. America Bull.* 69:477–82.

Harvey, A. M.; Hitchcock, D. H.; and Hughes, D. J. 1979. Event frequency and morphological adjustment of fluvial systems. In *Adjustments of the fluvial system,* edited by D. D. Rhodes and G. P. Williams, pp. 139–67. Dubuque, Iowa: Kendall Hunt.

Harwood, T. A. 1969. Some possible problems with pipelines in permafrost regions. *Proc. 3rd Canadian Conf. on Permafrost,* pp. 79–84. Natl. Res. Council of Canada, Tech. Memo 96.

Hastenrath, S. L. 1967. The barchans of the Arequipa region, southern Peru. *Zeit. f. Geomorph.* 11:300–331.

Hastenrath, S., and Kruss, P. 1982. On the secular variation of ice flow velocity at Lewis Glacier, Mount Kenya, Kenya. *Jour. Glaciol* 28:333–39.

Hausenbuiller, R. 1972. *Soil science: Principles and practices.* Dubuque, Iowa: Wm. C. Brown Company Publishers.

Haynes, V. C. 1982. The Darb El-Arba'in Desert: A product of Quaternary climatic change. In *Desert landforms of southwest Egypt: A basis for comparison with Mars,* edited by F. El-Baz and T. Maxwell, pp. 91–118. NASA, CR–3611.

Heim, A. 1932. *Bergsturz und Menschenleben.* Zurich: Fretz and Wasmuth Verlag.

Heller, P. L. 1981. Small landslide types and controls in glacial deposits: Lower Skagit River drainage, northern Cascade Range, Washington. *Environ. Geol.* 3:221–28.

Helley, E. J., and Smith, W. 1971. Development and calibration of a pressure difference bedload sampler. U.S. Geol. Survey, Water Resources Div., Open-File Rpt.

Hembree, C., and Rainwater, F. 1961. Chemical degradation on opposite flanks of the Wind River Range, Wyoming. U.S. Geol. Survey Water Supply Paper 1535-E.

Hennion, F. B., and Lobacz, E. F. 1973. Corps of engineers technology related to design of pavements in areas of permafrost. In *Proc. Permafrost 2nd Internat. Conf.,* pp. 426–29. Natl. Acad. Sci.-Natl. Res. Council, Yakutsk, U.S.S.R., 1973.

Herak, M., and Stringfield, V. T. 1972. *Karst. Important karst regions of the northern hemisphere.* Amsterdam: Elsevier.

Hey, R. D.; Bathurst, J. C.; and Thorne, C. R. 1982. *Gravel-bed rivers.* New York: John Wiley & Sons.

Hey, R. D., and Thorne, C. R. 1975. Secondary flows in river channels. *Area* 7:191–95.

Hickin, E. J. 1974. The development of meanders in natural river channels. *Am. Jour. Sci.* 274:414–42.

Hickin, E. J., and Nanson, G. C. 1975. The character of channel migration on the Beatton River, northeast British Columbia, Canada. *Geol. Soc. America Bull.* 86:487–94.

Hill, A. R. 1971. The internal composition and structure of drumlins in north Down and south Antrim, northern Ireland. *Geogr. Annlr.* 53:14–31.

———. 1973. Erosion of river banks composed of glacial till near Belfast, Northern Ireland. *Zeit. f. Geomorph.* 17:428–42.

Hill, C. A. 1981. Speleogenesis of Carlsbad Caverns and other caves of the Guadalupe Mountains. *Proc. 8th Intl. Cong. Speleol.* Bowling Green, Ky., pp. 143–44.

Hillaire-Marcel, C., and Fairbridge, R. W. F. 1978. Isostasy and eustasy of Hudson Bay. *Geology* 6:117–22.

Hino, M. 1975. Theory on formation of rip current and cuspidal coast. *Proc. 14th Conf. on Coast. Eng.,* pp. 901–19.

Hjulström, F. 1939. Transportation of detritus by moving water. In *Recent marine sediments: A symposium,* edited by P. Trask. Tulsa, Okla.: Am. Assoc. Petroleum Geologists.

Hodge, S. M. 1974. Variations in the sliding of a temperate glacier. *Jour. Glaciol.* 13:349–69.

———. 1976. Direct measurement of basal water pressures: A pilot study. *Jour. Glaciol.* 16:205–17.

Hodgkin, E. P. 1964. Rate of erosion of intertidal limestone. *Zeit. f. Geomorph.* 8:385–92.

Holdsworth, G. 1973. Ice calving into the proglacial Generator Lake Baffin Island, N.W.T., Canada. *Jour. Glaciol.* 12:235–50.

Holland, H. D.; Kirsipu, T. V.; Huebner, J. S.; and Oxburgh, V. M. 1964. On some aspects of the chemical evolution of cave waters. *Jour. Geology* 72:36–67.

Holman, R. A. 1983. Edge waves and the configuration of the shoreline. In *CRC handbook of coastal processes and erosion,* edited by P. Komar, pp. 21–33. Boca Raton, Fla.: CRC Press.

Holmes, C. D. 1947. Kames. *Am. Jour. Sci.* 245:240–49.

———. 1960. Evolution of till-stone shapes, central New York. *Geol. Soc. America Bull.* 71:1645–60.

Holmes, G. W.; Hopkins, D. M.; and Foster, H. L. 1968. Pingos in central Alaska. U.S. Geol. Survey Bull. 1241-H.

Holtedahl, H. 1967. Notes on the formation of fjords and fjord valleys. *Geogr. Annlr.* 49:188–203.

Hooke, R. LeB. 1967. Processes on arid-region alluvial fans. *Jour. Geology* 75:438–60.

———. 1968. Steady-state relationships on arid-region alluvial fans in closed basins. *Am. Jour. Sci.* 266:609–29.

———. 1972. Geomorphic evidence for Late Wisconsin and Holocene tectonic deformation, Death Valley, California. *Geol. Soc. America Bull.* 83:2073–97.

———. 1977. Basal temperatures in polar ice sheets: A qualitative review. *Quat. Res.* 7:1–13.

Hooke, R. LeB.; Alexander, E. C., Jr.; and Gustafson, R. J. 1980. Temperature profiles in the Barnes Ice Cap, Baffin Island, Canada, and heat flux from the subglacial terrane. *Can. J. Earth Sci.* 17:1174–88.

Hooke, R. LeB., and Hudleston, P. J. 1980. Ice fabrics in a vertical flow plane, Barnes Ice Cap, Canada. *Jour. Glaciol.* 25:195–214.

———. 1981. Ice fabrics from a borehole at the top of the south dome, Barnes Ice Cap, Baffin Island. *Geol. Soc. America Bull.* 92 pt. 1:274–81.

Hooke, R. LeB., and Rohrer, W. L. 1977. Relative erodibility of source-area rock types, as determined from second-order variations in alluvial-fan size. *Geol. Soc. America Bull.* 88:1177–82.

———. 1979. Geometry of alluvial fans: Effect of discharge and sediment size. *Earth Surf. Proc. and Landforms* 4:147–66.

Hopkins, D. M., and Sigafoos, R. S. 1951. Frost action and vegetation patterns on Seward Peninsula, Alaska. U.S. Geol. Survey Bull. 974-C:51–100.

Hoppe, G., and Schytt, V. 1953. Some observations on fluted moraine surfaces. *Geogr. Annlr.* 35:105–15.

Horton, R. E. 1933. The role of infiltration in the hydrological cycle. *Am. Geophys. Union Trans.* 14:446–60.

———. 1945. Erosional development of streams and their drainage basins: Hydrophysical approach to quantitative morphology. *Geol. Soc. America Bull.* 56:275–370.

Howard, Alan D. 1971. Simulation model of stream capture. *Geol. Soc. America Bull.* 82:1355–76.

———. 1977. Effect of slope on the threshold of motion and its application to orientation of wind ripples. *Geol. Soc. America Bull.* 88:853–56.

Howard, Alan D.; Morton, J. B.; Gal-el-Hak, M.; and Pierce, D. 1977. Simulation model of erosion and deposition on a barchan dune. NASA, CR–2838.

Howard, Arthur D. 1959. Numerical systems of terrace nomenclature: A critique. *Jour. Geology* 67:239–43.

———. 1967. Drainage analysis in geologic interpretation: A summation. *Am. Assoc. Petroleum Geologists Bull.* 51:2246–59.

Howard, Arthur D.; Fairbridge, R. W.; and Quinn, J. H. 1968. Terraces, fluvial—Introduction. In *Encyclopedia of geomorphology,* edited by R. W. Fairbridge, pp. 1117–23. New York: Reinhold Book Corp.

Hoyt, J. H. 1966. Air and sand movements in the lee of dunes. *Sedimentology* 7:137–44.

———. 1967. Barrier island formation. *Geol. Soc. America Bull.* 78:1125–36.

Hoyt, J. H., and Henry, V. J. 1971. Origin of capes and shoals along the southeastern coast of the United States. *Geol. Soc. America Bull.* 82:59–66.

Hsu, K. J. 1965. Isostasy, crustal thinning, mantle changes and the disappearance of ancient land masses. *Am. Jour. Sci.* 263:97–109.

———. 1975. Catastrophic debris streams (sturzstroms) generated by rockfalls. *Geol. Soc. America Bull.* 86:129–40.

Hubbert, M. K. 1940. The theory of groundwater motion. *Jour. Geology* 48:785–944.

Huddart, D., and Lister, H. 1981. The origin of ice marginal terraces and contact ridges of East Kangerdluarssuk Glacier, SW Greenland. *Geogr. Annlr.* 63A:31–39.

Hunt, C. B.; Averitt, P.; and Miller, R. L. 1953. Geology and geography of the Henry Mountains region, Utah. U.S. Geol. Survey Prof. Paper 228.

Hunt, C. B., and Mabey, D. R. 1966. Stratigraphy and structure. Death Valley, California. U.S. Geol. Survey Prof. Paper 494-A.

Huntley, D. A., and Bowen, A. J. 1973. Field observations of edge waves. *Nature* 243:160–61.

Huntoon, P. W. 1974. The karstic groundwater basins of the Kaibab Plateau, Arizona. *Water Resour. Res.* 10:579–90.

Hursh, C. R. 1936. Storm-water and absorption. *Am. Geophys. Union Trans.* 17:301–2.

Hursh, C. R., and Brater, E. F. 1941. Separating storm hydrographs from small drainage areas into surface and subsurface flow. *Am. Geophys. Union Trans.* 22:863–70.

Hutchinson, J. N. 1968. Mass movement. In *Encyclopedia of geomorphology,* edited by R. W. Fairbridge, pp. 688–96. New York: Reinhold Book Corp.

Hutter, K. 1982. Glacier flow. *Am. Scientist* 70:26–34.

Hutton, C. E. 1947. Studies of loess-derived soils in southwestern Iowa. *Soil Sci. Soc. Am. Proc.* 12:424–31.

Iken, A.; Rothlisberger, H.; Flotron, A.; and Haeberli, W. 1983. The uplift of Unteraargletscher at the beginning of the melt season—A consequence of water storage at the bed? *Jour. Glaciol.* 19:28–47.

Inglis, C. C. 1949. The behavior and control of rivers and canals. *Research Pub.* Poona, India, no. 13, 2 vols.

Inman, D. L., and Bagnold, R. A. 1963. Littoral processes. In *The sea,* edited by M. N. Hill, 3:529–53. New York: Interscience.

Inman, D. L., and Brush, B. M. 1973. The coastal challenge. *Science* 181:20–32.

Inman, D. L.; Ewing, G. C.; and Corliss, J. B. 1966. Coastal sand dunes of Guerrero Negro, Baja, California, Mexico. *Geol. Soc. America Bull.* 77:787–802.

Inman, D. L., and Frautschy, J. D. 1966. Littoral processes and the development of shoreline. *Proc. Coast. Eng. Speciality Conf.,* Am. Soc. Civil Engineers (Santa Barbara, Calif.), pp. 511–36.

Inman, D. L., and Nordstrom, C. E. 1971. On the tectonic and morphologic classification of coasts. *Jour. Geology* 79:1–21.

Ippen, A. T., ed. 1966. *Estuary and coastline hydrodynamics.* New York: McGraw-Hill.

Isherwood, D., and Street, A. 1976. Biotite-induced grussification of the Boulder Creek Granodiorite, Boulder County, Colorado. *Geol. Soc. America Bull.* 87:366–70.

Ives, J. D., and Fahey, B. D. 1971. Permafrost occurrence in the Front Range, Colorado Rocky Mountains, U.S.A. *Jour. Glaciol.* 10:105–11.

Jackson, J. A.; Gagnepain, J.; Houseman, G.; King, G. C. P.; Papadimitriou, P.; Soufleris, C.; and Virieux, J. 1982. Seismicity, normal faulting, and the geomorphological development of the Gulf of Corinth (Greece): The Corinth earthquakes of February and March 1981. *Earth and Planat. Sci. Letters* 57:377–97.

Jackson, M.; Hseung, Y.; Corey, R.; Evans, E.; and Heuval, R. 1952. Weathering sequence of clay size minerals in soils and sediments. *Soil Sci. Soc. Am. Proc.* 16:3–6.

Jackson, T., and Keller, W. 1970. A comparative study of the role of lichens and "inorganic" processes in the chemical weathering of recent Hawaiian lava flows. *Am. Jour. Sci.* 269:446–66.

Jacobel, R. W. 1982. Short-term variations in velocity of South Cascade Glacier, Washington, U.S.A. *Jour. Glaciol* 28:325–32.

Jahn, A. 1960. Some remarks on evolution of slopes on Spitsbergen. *Zeit. f. Geomorph.,* Suppl. 1:49–58.

Jansen, J. M. L., and Painter, R. B. 1974. Predicting sediment yield from climate and topography. *J. Hydrol.* 21:371–80.

Jarvis, G. T., and Clarke, G. K. C. 1975. The thermal regime of Trapridge Glacier and its relevance to glacier surging. *Jour. Glaciol.* 14:235–49.

Jarvis, R. S., and Sham, C. H. 1981. Drainage network structure and the diameter-magnitude relation. *Water Resour. Res.* 17:1019–27.

Jennings, J. N. 1967. Some karst areas of Australia. In *Landform studies from Australia and New Guinea,* edited by J. N. Jennings and J. A. Mabbutt, pp. 256–92. Canberra.

———. 1971. *Karst.* Cambridge, Mass.: M.I.T. Press.

———. 1983. Karst landforms. *Am. Scientist* 71:578–86.

Jennings, J. N., and Bik, M. J. 1962. Karst morphology in Australian New Guinea. *Nature* 194:1036–38.

Jenny, H. 1941. *Factors of soil formation.* New York: McGraw-Hill.

———. 1950. Origin of soils. In *Applied sedimentation,* edited by P. Trask, pp. 41–61. New York: John Wiley & Sons.

Jenny, H., and Leonard, C. 1939. Functional relationships between soil properties and rainfall. *Soil Sci.* 38:363–81.

Johnson, A. 1970. *Physical process in geology.* San Francisco: Freeman, Cooper and Co.

Johnson, A. M., and Rahn, P. H. 1970. Mobilization of debris flows. *Zeit. f. Geomorph.,* Suppl. 9:168–86.

Johnson, D. W. 1910. Beach cusps. *Geol. Soc. America Bull.* 21:604–21.

———. 1919. *Shore processes and shoreline development.* New York: John Wiley & Sons. Facsimile edition: Hafner, New York, 1965.

———. 1925. *New England-Acadian shoreline.* New York: John Wiley & Sons.

———. 1932. Rock fans of arid regions. *Am. Jour. Sci.* 23 (5th ser.):389–416.

———. 1944. Problems of terrace correlation. *Geol. Soc. America Bull.* 55:793–818.

Johnson, J. W. 1956. Dynamics of nearshore sediment movement. *Am. Assoc. Petroleum Geologists Bull.* 40:2211–32.

Johnson, W. D. 1904. The profile of maturity in alpine glacial erosion. *Jour. Geology* 12:7:569–78.

Johnston, G. H. 1963. Pile construction in permafrost. *Proc. Permafrost Internat. Conf.,* (Lafayette, Ind., 1963). Natl. Acad. Sci.-Natl. Res. Council Pub. 1287, pp. 477–81.

———. 1983. Performance of an insulated roadway on permafrost, Inuvik, N.W.T. In *Proc. Permafrost 4th Internat. Conf.,* pp. 548–51. Natl. Acad. Sci.

Jopling, A. V. 1966. Some application of theory and experiment to the study of bedding genesis. *Sedimentology* 7:71–102.

Judson, S. 1968a. Erosion rates near Rome, Italy. *Science* 160:1444–46.

———. 1968b. Erosion of the land. *Am. Scientist* 56:356–74.

Judson, S., and Ritter, D. F. 1964. Rates of regional denudation in the United States. *Jour. Geophys. Research* 69:3395–401.

Kachadoorian, R., and Ferrians, O. J., Jr. 1973. Permafrost-related engineering problems posed by the Trans-Alaskan Pipeline. *Permafrost 2nd Internat. Conf.,* pp. 684–87. Natl. Acad. Sci.-Natl. Res. Council, Yakutsk, U.S.S.R., 1973.

Kamb, B. 1964. Glacier mechanics. *Science* 146:353–65.

———. 1970. Sliding motion of glaciers: Theory and observation. *Rev. Geophys. and Space Phys.* 8:673–728.

Kamb, B., and LaChapelle, E. 1964. Direct observation of the mechanism of glacier sliding over bedrock. *Jour. Glaciol.* 5:159–72.

Kaye, C. A. 1957. The effect of solvent motion on limestone solutions. *Jour. Geology* 65:34–47.

———. 1964a. Outline of Pleistocene geology of Martha's Vineyard, Massachusetts. U.S. Geol. Survey Prof. Paper 501-C:134–39.

———. 1964b. Illinoian and early Wisconsin moraines of Martha's Vineyard, Massachusetts. U.S. Geol. Survey Prof. Paper 501-C:140–43.

Keller, E. A. 1971. Areal sorting of bedload material. *Geol. Soc. America Bull.* 82:753–56.

———. 1972. Development of alluvial stream channels. A five-stage model. *Geol. Soc. America Bull.* 83:1531–36.

Keller, E. A.; Bonkowski, M. S.; Korsch, R. J.; and Shlemon, R. J. 1982. Tectonic geomorphology of the San Andreas fault zone in the southern Indio Hills, Coachella Valley, California. *Geol. Soc. America Bull.* 93:46–56.

Keller, E. A., and Melhorn, W. 1973. Bedforms and fluvial processes on alluvial stream channels: selected observations. In *Fluvial geomorphology,* edited by M. Morisawa, pp. 253–83. S.U.N.Y., Binghamton: Pubs. in Geomorphology.

———. 1978. Rhythmic spacing and origin of pools and riffles. *Geol. Soc. America Bull.* 89:723–30.

Keller, E. A., and Swanson, F. J. 1979. Effects of large organic material on channel form and fluvial processes. *Earth Surf. Proc. and Landforms* 4:361–80.

Keller, E. A., and Tally, T. 1979. Effects of large organic debris on channel form and fluvial processes in the coastal redwood environment. In *Adjustments of the fluvial system,* edited by D. Rhodes and G. Williams, pp. 169–97. Dubuque, Iowa: Kendall/Hunt Publishing Company.

Keller, W. 1954. Bonding energies of some silicate minerals. *Am. Mineralogist* 39:783–93.

———. 1976. Scan electron micrographs of kaolins collected from diverse environments of origin. Pt. 1. *Clays and Clay Minerals* 24:107–13.

———. 1978. Kaolinization of feldspar as displayed in scanning electron micrographs. *Geology* 6:184–88.

———. 1982. Kaolin—A most diverse rock in genesis, texture, physical properties and uses. *Geol. Soc. America Bull.* 93:27–36.

Kellerhals, R. 1967. Stable channels with gravel-paved beds. Am Soc. Civil Engineers Proc., *Jour. Waterways and Harbors* 93:63–84.

Kelsey, H. M. 1980. A sediment budget and analysis of geomorphic process in the Van Duzen River basin, north coastal California, 1941–1975. *Geol. Soc. America Bull.* 91:190–95.

Kemmerly, P. R. 1976. Definitive doline characteristics in the Clarksville quadrangle, Tennessee. *Geol. Soc. America Bull.* 87:42–46.

———. 1980a. Sinkhole collapse in Montgomery County, Tennessee. Tenn. Div. Geol., Environ. Geol. Ser. no. 6.

———. 1980b. A time-distribution study of doline collapse: Framework for prediction. *Environ. Geol.* 3:123–30.

———. 1982. Spatial analysis of a karst depression population: Clues to genesis. *Geol. Soc. America Bull.* 93:1078–86.

Kemmerly, P., and Towe, S. 1978. Karst depressions in a time context. *Earth Surf. Proc. and Landforms* 3:355–61.

Kesel, R. H. 1973. Inselberg landform elements: Definition and synthesis. *Rev. géomorph. dynamique* 22:97–108.

———. 1977. Some aspects of the geomorphology of inselbergs in central Arizona, U.S.A. *Zeit. f. Geomorph.* 21:119–46.

Kesel, R. H.; Dunne, K.: McDonald, R.; Allison, K.; and Spicer, B. 1974. Lateral erosion and overbank deposition on the Mississippi River in Louisiana caused by 1973 flooding. *Geology* 2:461–64.

Kesseli, J. E. 1941. Rock streams in the Sierra Nevada, California. *Geogr. Rev.* 31:203–27.

Keyes, C. R. 1912. Deflative scheme of the geographic cycle in an arid climate. *Geol. Soc. America Bull.* 23:537–62.

Kiersch, G. A. 1964. Vaiont reservoir disaster. *Civil Engineering* 34:32–39.

Kilpatrick, F. A., and Barnes, H. H. 1964. Channel geometry of piedmont streams as related to frequency of floods. U.S. Geol. Survey Prof. Paper 422E:1–10.

King, C. A. M. 1972. *Beaches and coasts.* New York: St. Martin's Press.

King, C. A. M., and Buckley, J. T. 1968. The analysis of stone size and shape in Arctic environments. *Jour. Sed. Petrology* 38:200–214.

King, C. A. M., and Lewis, W. V. 1961. A tentative theory of ogive formation. *Jour. Glaciol.* 3:913–39.

King, C. A. M., and McCullagh, M. J. 1971. A simulation model of a complex recurved spit. *Jour. Geology* 79:22–37.

King, C. A. M., and Williams, W. W. 1949. The formation and movement of sandbars by wave action. *Geogr. Jour.* 107:70–84.

King, D. 1956. The Quaternary stratigraphic record at Lake Eyre North and the evolution of existing topographic forms. *Trans. Royal Soc. Australia* 79:93–103.

King, L. C. 1953. Canons of landscape evolution. *Geol. Soc. America Bull.* 64:751–52.

Kirkby, M. J. 1967. Measurement and theory of soil creep. *Jour. Geology* 75:359–78.

———. 1969. Infiltration, throughflow, and overland flow; and erosion by water on hillslopes. In *Water, earth, and man,* edited by R. J. Chorley, pp. 215–38. London: Methuen and Co.

Kirkby, M. J., and Chorley, R. J. 1967. Throughflow, overland flow and erosion. *Int. Assoc. Sci. Hydrol. Bull.* 12:5–21.

Kirkby, M. J., and Kirkby, A. V. 1969. Erosion and deposition on a beach raised by the 1964 earthquake, Montague Island, Alaska. U.S. Geol. Survey Prof. Paper 543-H:1–41.

Kirkby, R. P. 1969. Variation in glacial deposition in a subglacial environment: An example from Midlothian. *Scott. J. Geol.* 5:49–53.

Kneale, W. R. 1982. Field measurements of rainfall drop-size distribution, and the relationship between rainfall parameters and soil movement by rainsplash. *Earth Surf. Proc. and Landforms* 7:499–502.

Knighton, A. D. 1974. Variation in width-discharge relation and some implications for hydraulic geometry. *Geol. Soc. America Bull.* 85:1069–76.

———. 1977. Alternative derivation of the minimum variance hypothesis. *Geol. Soc. America Bull.* 88:364–66.

Knox, J. C. 1972. Valley alluviation in southwestern Wisconsin. *Ann. Assoc. Am. Geog.* 62:401–10.

———. 1976. Concept of the graded stream. In *Theories of landform development,* edited by W. Melhorn and R. Flemal, pp. 168–98. S.U.N.Y., Binghamton: Pubs. in Geomorphology.

Kochel, R. C.; Baker, V. R.; and Patton, P. C. 1982. Paleohydrology of Southwestern Texas. *Water Resour. Res.* 18:1165–83.

Kochel, R. C., and Johnson, R. A. 1984. Geomorphology and sedimentology of humid-temperate alluvial fans, central Virginia. In *Gravels and conglomerates,* edited by E. Koster and R. Steel, pp. 109–22. Can. Soc. Petrol. Geol. Mem. 10.

Kolb, C. R., and Van Lopik, J. R. 1958. Geology of the Mississippi River deltaic plain, southeastern Louisiana. U.S. Army Corps Engrs., Waterway Exp. Sta. Rept. 3–483, Vicksburg.

Komar, P. D. 1971a. The mechanics of sand transport on beaches. *Jour. Geophys. Research* 76:3:713–21.

———. 1971b. Nearshore cell circulation and the formation of giant cusps. *Geol. Soc. America Bull.* 82:2643–50.

———. 1973. Observations of beach cusps at Mono Lake, California. *Geol. Soc. America Bull.* 84:3593–3600.

———. 1975. Nearshore currents: Generation by obliquely incident waves and longshore variations in breaker height. In *Proc. symposium on nearshore sediment dynamics,* edited by J. R. Hails and A. Carr. pp. 17–45. London: John Wiley & Sons.

———. 1976. *Beach processes and sedimentation.* Englewood Cliffs, N.J.: Prentice-Hall.

———. 1979. Comparisons of the hydraulics of water flows in Martian outflow channels with flows of similar scale on Earth. *Icarus* 37:156–81.

———. 1983a. Beach processes and erosion—an introduction. In *CRC handbook of coastal processes and erosion,* edited by P. Komar, pp. 1–20. Boca Raton, Fla.: CRC Press.

———, ed. 1983b. *CRC handbook of coastal processes and erosion.* Boca Raton, Fla.: CRC Press.

Komar, P. D., and Inman, D. L. 1970. Longshore sand transport on beaches. *Jour. Geophys. Research* 75:30:5914–27.

Komar, P. D., and Reimers, C. E. 1978. Grain shape effects on settling rates. *Jour. Geology* 86:193–209.

Konrad, J. M., and Morgenstern, N. R. 1983. Frost susceptibility of soils in terms of their segregation potential. In *Proc. Permafrost 4th Internat. Conf.,* pp. 660–65. Natl. Acad. Sci.

Kottlowski, F.; Cooley, M.; and Ruhe, R. 1965. Quaternary geology of the southwest. In *The Quaternary of the United States,* edited by H. Wright and D. Frey. Princeton, N.J.: Princeton Univ. Press.

Krigstrom, A. 1962. Geomorphological studies of sandar plains and their braided rivers in Iceland. *Geogr. Annlr.* 44:328–46.

Krinsley, D. H., and Donahue, J. 1968. Environmental interpretation of sand grain surface textures of electron microscopy. *Geol. Soc. America Bull.* 79:743–48.

Kroethe, N., and Libra, R. 1983. Sulfur isotopes and hydrochemical variations in spring waters of southern Indiana, U.S.A. *J. Hydrol.* 61:267–83.

Krumbein, W. C. 1944. Shore currents and sand movement on a model beach. U.S. Army Corps Engrs., Beach Erosion Board Tech. Memo 7.

Krumbein, W. C., and Oshiek, L. E. 1950. Pulsation transport of sand by shore agents. *Am. Geophys. Union Trans.* 31:216–20.

Kuenen, P. H. 1948. The formation of beach cusps. *Jour. Geology* 56:34–40.

———. 1960. Experiment abrasion, 4. Eolian action. *Jour. Geology* 68:427–49.

Kuhn, G. G., and Shepard, F. P. 1983. Beach processes and sea cliff erosion in San Diego County, California. In *CRC handbook of coastal processes and erosion,* edited by P. Komar, pp. 267–84. Boca Raton, Fla.: CRC Press.

Kuno, H. 1969. Plateau basalts. In *The Earth's crust and upper mantle,* edited by P. Hart, pp. 495–500. Am. Geophys. Union, Geophys. Monograph 13.

Kupsch, W. O. 1955. Drumlins with jointed boulders near Dollard, Saskatchewan. *Geol. Soc. America Bull.* 66:327–38.

Lachenbruch, A. H. 1966. Contraction theory of ice wedge polygons: A qualitative discussion. *Proc. Permafrost Internat. Conf.* (Lafayette, Ind. 1963). Natl. Acad. Sci.-Natl. Res. Council Pub. 1287, pp. 63–71.

———. 1970. Some estimates of the thermal effects of a heated pipeline in permafrost. U.S. Geol. Survey Circ. 632.

Lachenbruch, A. H., and Marshall, B. V. 1969. Heat flow in the Arctic. *Arctic* 22:300–311.

Lagally, M. 1934. *Mechanik und Thermodynamik des stationären Gletschers.* Leipzig.

Lambe, T. 1953. The structure of inorganic soils. *Am. Soc. Civil Engineers Proc.* 79: Separate 315.

Lancaster, N. 1980. The formation of seif dunes from barchans—Supporting evidence for Bagnold's model from the Namib Desert. *Zeit. f. Geomorph.* 24:160–67.

———. 1982. Dunes on the Skeleton Coast, Namibia (South West Africa): Geomorphology and grain size relationships. *Earth Surf. Proc. and Landforms* 7:575–87.

Lane, E. W. 1937. Stable channels in erodible materials. *Am. Soc. Civil Engineers Trans.* 102:123–94.

———. 1955. Design of stable channels. *Am. Soc. Civil Engineers Trans.* 120:1234–79.

———. 1957. A study of the shape of channels formed by natural streams in erodible material. *M.R.D. Sediments Series no. 9,* U.S. Army Corps Engrs., Eng. Div., Missouri River, Omaha, Neb.

Langbein, W. B. 1964. Geometry of river channels. Am. Soc. Civil Engineers, *Jour. Hydraulics Div.* 90:301–13.

Langbein, W. B., and Leopold, L. B. 1964. Quasi-equilibrium states in channel morphology. *Am. Jour. Sci.* 262:782–94.

———. 1966. River meanders: Theory of minimum variance. U.S. Geol. Survey Prof. Paper 422-H.

Langbein, W. B., and Others. 1949. Annual runoff in the United States. U.S. Geol. Survey Circular 52.

Langbein, W. B., and Schumm, S. A. 1958. Yield of sediment in relation to mean annual precipitation. *Am. Geophys. Union Trans.* 39:1076–84.

Langford-Smith, T., and Dury, G. H. 1964. A pediment at Middle Pinnacle, near Broken Hill, New South Wales. *Jour. Geol. Soc. Australia* 11:79–88.

Lattman, L. H. 1960. Cross section of a flood plain in a moist region of moderate relief. *Jour. Sed. Petrology* 30:275–82.

———. 1968. Structural control in geomorphology. In *Encylopedia of geomorphology,* edited by R. W. Fairbridge, pp. 1074–79. New York: Reinhold Book Corp.

———. 1973. Calcium carbonate cementation of alluvial fans in Southern Nevada. *Geol. Soc. America Bull.* 84:3013–28.

Laury, R. L. 1971. Stream bank failure and rotational slumping. Preservation and significance in the geologic record. *Geol. Soc. America Bull.* 82:1251–66.

LaValle, P. 1967. Some aspects of linear karst depression development in south central Kentucky. *Ann. Assoc. Am. Geog.* 57:49–71.

———. 1968. Karst depression morphology in south-central Kentucky. *Geogr. Annlr.* 50A:94–108.

Lawson, A. C. 1915. The epigene profiles of the desert. *Univ. of Calif. Dept. Geol. Bull.* 9:23–48.

Lee, W., and Uyeda, S. 1965. Review of heat flow data. In *Terrestrial heat flow,* edited by W. Lee. Am. Geophys. Union, Geophys. Monograph 8.

Legget, R. 1967. Soil: Its geology and use. *Geol. Soc. America Bull.* 78:1433–60.

LeGrand, H. 1983. Perspective on karst hydrology. *J. Hydrol.* 61:343–55.

Lehman, D. 1963. Some principles of chelation chemistry. *Soil Sci. Soc. Am. Proc.* 27:167–70.

Lehmann, H. 1936. Morphologische Studien auf Java. Stuttgart: *Geogr. Abh.* 3:9.

Lehre, A. K. 1982. Sediment budget of a small coast range drainage basin in north-central California. In *Sediment budgets and routing in forested drainage basins,* edited by F. J. Swanson et al., pp. 67–77. U.S.D.A., Forest Serv. Genl. Tech. Rpt. PNW–141.

Leigh, C. 1982. Sediment transport by surface wash and throughflow at the Pasoh Forest Reserve, Negri Sembilan, Peninsular Malaysia. *Geogr. Annlr.* 64A:171–80.

Leighton, M. W., and Pendexter, C. 1962. Carbonate rock types. In *Classification of carbonate rocks,* edited by W. E. Ham, pp. 33–61. Am. Soc. Petroleum Geologists Mem. 1.

Leliavsky, S. L. 1966. *An introduction to fluvial hydraulics.* New York: Dover Publications.

Lemke, R. W. 1958. Narrow linear drumlins near Velva, North Dakota. *Am. Jour. Sci.* 256:270–83.

Leopold, L. B. 1953. Downstream change of velocity in rivers. *Am. Jour. Sci.* 251:606–24.

———. 1982. Water surface topography in river channels and implications for meander development. In *Gravel-bed rivers,* edited by R. Hey, J. Bathurst, and C. Thorne, pp. 359–88. New York: John Wiley & Sons.

Leopold, L. B., and Bull, W. B. 1979. Base level aggradation and grade. *Proc. Am. Phil. Soc.* 123:168–202.

Leopold, L. B., and Emmett, W. W. 1976. Bedload measurements, East Fork River, Wyoming. *Natl. Acad. Sci. Proc.* 73:1000–1004.

———. 1977. 1976 bedload measurements, East Fork River, Wyoming. *Natl. Acad. Sci. Proc.* 74:2644–48.

Leopold, L. B., and Langbein, W. B. 1962. The concept of entropy in landscape evolution. U.S. Geol. Survey Prof. Paper 500-A:20.

———. 1963. Association and indeterminancy in geomorphology. In *The fabric of geology,* edited by C. C. Albritton, pp. 184–92. Reading, Mass.: Addison-Wesley.

Leopold, L. B., and Maddock, T., Jr. 1953. The hydraulic geometry of stream channels and some physiographic implications. U.S. Geol. Survey Prof. Paper 252.

Leopold, L. B., and Miller, J. P. 1956. Ephemeral streams— hydraulic factors and their relation to the drainage net. U.S. Geol. Survey Prof. Paper 282-A.

Leopold, L. B., and Wolman, M. G. 1957. River channel patterns; braided, meandering and straight. U.S. Geol. Survey Prof. Paper 282-B.

———. 1960. River meanders. *Geol. Soc. America Bull.* 71:769–794.

Leopold, L. B.; Wolman, M. G.; and Miller, J. P. 1964. *Fluvial processes in geomorphology.* San Francisco: W. H. Freeman.

Lewin, J. 1976. Initiation of bed forms and meanders in coarse-grained sediment. *Geol. Soc. America Bull.* 87:281–85.

Lewis, W. V. 1947. Valley steps and glacial valley erosion. *Inst. Brit. Geog. Trans.* 14:19–44.

———. 1954. Pressure release and glacial erosion. *Jour. Glaciol.* 2:417–22.

———, ed. 1960. *Norwegian cirque glaciers.* Royal Geog. Soc. Res. Ser. 4.

Li, Y. H. 1976. Denudation of Taiwan Island since the Pliocene Epoch. *Geology* 4:105–7.

Likens, G. E.; Bormann, F. H.; Pierce, R. S.; Eaton, J. S.; and Johnson, N. M. 1977. *Biogeochemistry of a forested ecosystem.* New York: Springer-Verlag.

Linell, K. A., and Johnston, G. H. 1973. Engineering design and construction in permafrost regions. *Proc. Permafrost 2nd Internat. Conf.,* pp. 553–75. Natl. Acad. Sci.-Natl. Res. Council, Yakutsk, U.S.S.R., 1973.

Livesey, R. H. 1965. Channel armoring below Fort Randall dam. *Proc. Fed. Inter-Agency Sedimentation Conf.* U.S. Dept. of Agri. Pub. 970:461–70.

Livingstone, D. A. 1963. Chemical composition of rivers and lakes. U.S. Geol. Survey Prof. Paper 440-G.

Lliboutry, L. 1964. Subglacial "supercavitation" as a cause of the rapid advances of glaciers. *Nature* 202:77.

———. 1968. General theory of subglacial cavitation and sliding of temperate glaciers. *Jour. Glaciol.* 7:21–58.

Lliboutry, L., and Reynaud, L. 1981. "Global dynamics" of a temperature valley glacier. Mer de Glace and past velocities deduced from Forbes bands. *Jour. Glaciol.* 27:207–26.

Lobacz, E. F., and Quinn, W. F. 1963. Thermal regime beneath buildings constructed on permafrost. *Proc. Permafrost Internat. Conf.,* (Lafayette, Ind., 1963). Natl. Acad. Sci.-Natl. Res. Council Pub. 1287, pp. 159–64.

Lockwood, J. G. 1979. *Causes of climate.* London: Edward Arnold Ltd.

Lohnes, R. A., and Handy, R. L. 1968. Slope angles in friable loess. *Jour. Geology* 76:247–58.

Lombard, R. E.; Miles, M. B.; Nelson, L. M.; Kresch, D. L.; and Carpenter, P. J. 1981. Channel conditions in the lower Toutle and Cowlitz Rivers resulting from the mudflows of May 18, 1980. U.S. Geol. Survey Circ. 850-C.

Long, J. T., and Sharp, R. P. 1964. Barchan-dune movement in the Imperial Valley, California. *Geol. Soc. America Bull.* 75:149–56.

Longuet-Higgins, M. S. 1970. Longshore currents generated by obliquely incident sea waves. *Jour. Geophys. Research* 75:6778–6801.

Longwell, C. R. 1960. Interpretation of the leveling data. U.S. Geol. Survey Prof. Paper 295:33–38.

Loughnan, F. 1969. *Chemical weathering of the silicate minerals.* New York: American Elsevier.

Loughnan, F., and Bayliss, P. 1961. The mineralogy of the bauxite deposits near Weipa, Queensland. *Am. Mineralogist* 46:209–17.

Loziński, W. 1912. Die periglaziale Fazies der mechanischen Verwitterung. Internat. Geol. Cong., 11th, Stockholm, 1910, *Compte rendu,* pp. 1039–53.

Lugn, A. L. 1962. *The origin and sources of loess.* Lincoln: Univ. Nebraska Studies, new series, no. 26.

————. 1968. The origin of loesses and their relation to the Great Plains in North America. In *Loess and related eolian deposits of the world,* edited by C. B. Schultz and J. C. Frye, p. 139. Lincoln: Univ. Nebraska Press.

Luk, S. H. 1979. Effect of soil properties on erosion by wash and splash. *Earth Surf. Proc. and Landforms* 4:241–55.

Lumley, J. L., and Panofski, H. A. 1964. *The structure of atmospheric turbulence.* New York: John Wiley & Sons.

Lustig, L. K. 1965. Clastic sedimentation in Deep Springs Valley, California. U.S. Geol. Survey Prof. Paper 352-F.

————. 1969. Trend surface analysis of the Basin and Range province and some geomorphic implications. U.S. Geol. Survey Prof. Paper 500-D.

Lynch, D. K. 1982. Tidal bores. *Sci. Amer.* 247:146–56.

Mabbutt, J. A. 1966. Mantle-controlled planation of pediments. *Am. Jour. Sci.* 264:78–91.

————. 1971. The Australian arid zone as a prehistoric environment. In *Aboriginal man and environment in Australia,* edited by D. Mulvaney and J. Golson, pp. 66–79. Canberra: Australian Natl. Univ. Press.

Macdonald, G. 1972. *Volcanoes.* Englewood Cliffs, N.J.: Prentice-Hall.

Mackay, J. R. 1970. Disturbances to the tundra and forest tundra environment of the western Arctic. *Can. Geotechnical Jour.* 7:420–32.

————. 1973. The growth of pingos, western Arctic coast, Canada. *Can. J. Earth Sci.* 10:979–1004.

————. 1974. Ice-wedge cracks, Garry Island, Northwest Territories. *Can. J. Earth Sci.* 11:1366–83.

————. 1975a. The closing of ice-wedge cracks in permafrost, Garry Island, Northwest Territories. *Can. J. Earth Sci.* 12:1668–74.

————. 1975b. The stability of permafrost and recent climatic change in the Mackenzie Valley, NWT. Canada Geol. Survey Paper 75-1B:173–76.

Mackay, J. R., and Matthews, W. H. 1974. Movement of sorted stripes, the Cinder Cone, Garibaldi Park, B.C., Canada. *Arc. Alp. Res.* 6:347–59.

Mackin, J. H. 1936. The capture of the Greybull River. *Am. Jour. Sci.* 31:373–85.

————. 1937. Erosional history of the Big Horn Basin, Wyoming. *Geol. Soc. America Bull.* 48:813–93.

————. 1948. Concept of the graded river. *Geol. Soc. America Bull.* 59:463–512.

————. 1963. Rational and empirical methods of investigation in geology. In *The fabric of geology,* edited by C. Albritton. Reading, Mass.: Addison-Wesley.

Maddock, T., Jr. 1969. The behavior of straight open channels with movable beds. U.S. Geol. Survey Prof. Paper 622-A:70.

Malott, C. A. 1921. A subterranean cut-off and other subterranean phenomena along Indian Creek, Laurence Co., Indiana. *Indiana Acad. Sci. Proc.* 31:203–10.

————. 1938. Invasion theory of cavern development (abs.). *Geol. Soc. America Proc. 1937,* p. 323.

————. 1939. Karst valleys. *Geol. Soc. America Bull.* 50:1984.

Mammerickx, J. 1964. Quantitative observations on pediments in the Mojave and Sonoran deserts (southwestern United States). *Am. Jour. Sci.* 262:417–35.

Manley, S., and Manley, R. 1968. *Beaches; their lives, legends and lore.* Philadelphia: Chilton.

Mark, D. M. 1974. Line intersection method for estimating drainage density. *Geology* 2:235–36.

Marrs, R. W., and Kolm, K. E., eds. 1982. *Interpretation of windflow characteristics from eolian landforms.* Geol. Soc. America Spec. Paper 192.

Martel, E. A. 1921. *Nouveau traité des eaux souterraines.* Paris: Delagrave.

Martini, I. P. 1978. Tafoni weathering, with examples from Tuscany, Italy. *Zeit. f. Geomorph.* 22:44–67.

Matschinski, M. 1968. Alignment of dolines northwest of Lake Constance, Germany. *Geol. Mag.* 105(1):56–61.

Matsumota, T. 1967. Fundamental problems in the circum-Pacific orogenesis. *Tectonophysics* 4:595–613.

Matthes, F. E. 1900. Glacial sculpture of the Bighorn Mountains, Wyoming. *U.S. Geol. Survey 21st Ann. Rept., 1899–1900,* pt. 2, pp. 167–90.

————. 1930. Geologic history of the Yosemite Valley. U.S. Geol. Survey Prof. Paper 160.

Matthews, R. K. 1973. Relative elevation of late Pleistocene high sea level stands; Barbados uplift rates and their implications. *Quat. Res.* 3:147–53.

May, J. P., and Tanner, W. F. 1973. The littoral power gradient and shoreline changes. In *Coastal geomorphology,* edited by D. R. Coates, pp. 43–60. S.U.N.Y., Binghamton: 3rd Ann. Geomorph. Symposium.

May, S. K.; Dolan, R.; and Hayden, B. P. 1983. Erosion of U.S. shorelines. *EOS* 64:521–23.

McCall, J. G. 1952. The internal structure of a cirque glacier: Report on studies of the englacial movements and temperatures. *Jour. Glaciol.* 2:122–31.

————. 1960. The flow characteristics of a cirque glacier and their effect on glacial structure and cirque formation. In *Norwegian cirque glaciers,* edited by W. V. Lewis, pp. 39–62. Royal Geog. Soc. Res. Ser. 4.

McComas, M.; Hinkley, K.; and Kempton, J. 1969. Coordinated mapping of geology and soils for land-use planning. Illinois Geol. Survey Environ. Geol. Note 29.

McCoy, R. M. 1971. Rapid measurement of drainage density. *Geol. Soc. America Bull.* 82:757–62.

McDonald, R. C. 1975. Observations on hillslope erosion in tower karst topography of Belize. *Geol. Soc. America Bull.* 86:255–56.

———. 1979. Tower karst geomorphology in Belize. *Zeit f. Geomorph.*, Suppl. 32:35–45.

McDowall, I. C. 1960. Particle size reduction of clay minerals by freezing and thawing. *New Zealand J. Geol. and Geophys.* 3:337–43.

McGee, W. J. 1897. Sheetflood erosion. *Geol. Soc. America Bull.* 8:87–112.

McGinnis, L. D. 1966. Crustal tectonics and Precambrian basement in northeastern Illinois. Illinois Geol. Survey Rept. of Inv. 219.

McGowen, J. H., and Garner, J. H. 1970. Physiographic features and stratification types of coarse-grained point bars. Modern and ancient examples. *Sedimentology* 14:77–111.

McGreevey, J. P. 1981. Some perspectives on frost shattering. *Proc. in Phys. Geog.* 5:56–75.

McHattie, R. L., and Esch, D. C. 1983. Benefits of a peat underlay used in road construction on permafrost. In *Proc. Permafrost 4th Internat. Conf.*, pp. 826–31. Natl. Acad. Sci.

McKee, E. D. 1966. Structures of dunes at White Sands National Monument, New Mexico, and a comparison with structures of dunes from other selected areas. *Sedimentology* 7:1–69.

———. 1979. Introduction to a study of global sand seas. In *A study of global sand seas,* edited by E. McKee. U.S. Geol. Survey Prof. Paper 1052:1–20.

McKee, E. D., and Tibbitts, G. C., Jr. 1964. Primary structures of a seif dune and associated deposits in Libya. *Jour. Sed. Petrology* 34:5–17.

McKenzie, G. D. 1969. Observations on a collapsing kame terrace in Glacier Bay National Monument, S.E. Alaska. *Jour. Glaciol.* 8:413–25.

McLean, R. F., and Kirk, R. M. 1969. Relationship between grain size, size-sorting and foreshore slope on mixed sand-shingle beaches. *New Zealand J. Geol. and Geophys.* 12:138–55.

McPherson, H. J., and Rannie, W. F. 1967. Geomorphic effects of the May, 1967, flood in Graburn watershed, Cypress Hills, Alberta, Canada. *J. Hydrol.* 9:307–21.

McPherson, M. B. 1974. *Hydrological effects of urbanization.* Paris: UNESCO Press.

Meade, R. H. 1969. Errors in using modern stream-load data to estimate natural rates of denudation. *Geol. Soc. America Bull.* 80:1265–74.

———. 1982. Sources, sinks and storage of river sediment in the Atlantic drainage of the United States. *Jour. Geology* 90:235–52.

Meier, M. F. 1960. Mode of flow of Saskatchewan glacier, Alberta, Canada. U.S. Geol. Survey Prof. Paper 351.

Meier, M. F., and Johnson, A. 1962. The kinematic wave on Nisqually Glacier, Washington. *Jour. Geophys. Research* 67:886.

Meier, M. F.; Kamb, W. B.; Allen, C. R.; and Sharp, R. P. 1974. Flow of Blue Glacier, Olympic Mountains, Washington, U.S.A. *Jour. Glaciol.* 13:187–212.

Meier, M. F., and Post, A. S. 1969. What are glacier surges? *Can. J. Earth Sci.* 6:807–17.

Meier, M. F., and Tangborn, W. V. 1965. Net budget and flow of South Cascade Glacier, Washington. *Jour. Glaciol.* 5:547–66.

Mellor, M. 1970. Phase composition of pore water in cold rocks. U.S. Army Corps Engrs., Cold Regions Res. and Eng. Lab. Research Rept. 292.

Melton, F. A. 1940. A tentative classification of sand dunes. *Jour. Geology* 48:113–73.

Melton, M. A. 1958. Correlation structure of morphometric properties of drainage systems and their controlling agents. *Jour. Geology* 66:442–60.

———. 1965a. The geomorphic and paleoclimatic significance of alluvial deposits in southern Arizona. *Jour. Geology* 73:1–38.

———. 1965b. Debris-covered hillslopes of the southern Arizona desert—consideration of their stability and sediment contribution. *Jour. Geology* 73:715–29.

Menard, H. W. 1961. Some rates of regional erosion. *Jour. Geology* 69:155–61.

Menzies, J. 1979. The mechanics of drumlin formation with particular reference to the change in pore-water content of the till. *Jour. Glaciol.* 22:373–84.

———. 1981. Temperatures within subglacial debris—A gap in our knowledge. *Geology* 9:271–73.

Meyer-Peter, E., and Muller, R. 1948. Formulas for bed-load transport. Intnal. Assoc. for Hydr. Structures Res. Proc., 2nd Meeting, Stockholm, pp. 39–65.

Mickelson, D. M. 1973. Nature and rate of basal till deposition in a stagnating ice mass, Burroughs Glacier, Alaska. *Arc. Alp. Res.* 5:17–27.

Mickelson, D. M., and Berkson, J. M. 1974. Till ridges presently forming above and below sea level in Wachusett Inlet, Glacier Bay, Alaska. *Geogr. Annlr.* 56A;111–19.

Mielenz, R., and King, M. 1955. Physical-chemical properties and engineering performance of clays. *California Div. of Mines Bull.* 169:196–254.

Milanović, P. 1981. *Karst hydrology.* Littleton, Colo.: Water Resources Publications.

Miller, J. P. 1958. High mountain streams; effects of geology on channel characteristics and bed material. New Mexico State Bur. Mines and Min. Res. Memo 4.

Miller, R. D. 1966. Phase equilibria and soil freezing. *Proc. Permafrost Internat. Conf.* (Lafayette, Ind., 1963). Natl. Acad. Sci.-Natl. Res. Council Pub. 1287, pp. 193–97.

Millette, J. F. G., and Higbee, H. W. 1958. Periglacial loess, I. Morphological properties. *Am. Jour. Sci.* 256:284–93.

Milliman, J. D., and Meade, R. H. 1983. World-wide delivery of river sediment to the oceans. *Jour. Geology* 91:1–22.

Mills, H. C., and Wells, P. D. 1974. Ice-shove deformation and glacial stratigraphy of Port Washington, Long Island, New York. *Geol. Soc. America Bull.* 85:357–64.

Mills, H. H. 1977. Textural characteristics of drift from some representative Cordilleran glaciers. *Geol. Soc. America Bull.* 88:1135–48.

———. 1980. An analysis of drumlin forms in the northeastern and north-central United States. *Geol. Soc. America Bull.* 91:2214–89.

———. 1981. Boulder deposits and the retreat of mountain slopes or "Gully Gravure" revisited. *Jour. of Geology* 89:649–60.

Mills, H., and Starnes, D. 1983. Sinkhole morphometry in a fluviokarst region: Eastern Highland Rim, Tennessee, U.S.A. *Zeit f. Geomorph.* 27:39–54.

Miotke, F. D. 1973. The subsidence of the surfaces between mogotes in Puerto Rico east of Arecibo. *Caves and Karst* 15:1–12.

Monroe, W. H. 1969. Evidence of subterranean sheet solution under weathered detrital cover in Puerto Rico. In *Problems of karst denudation.* Internat. Speleol. Cong., 5th, Stuttgart.

———. 1970. A glossary of karst terminology. U.S. Geol. Survey Water Supply Paper 1899 K.

———. 1976. The karst landforms of Puerto Rico. U.S. Geol. Survey Prof. Paper 899.

Moon, J. W. 1980. On the expected diameter of random channel networks. *Water Resour. Res.* 16:1119–20.

Moore, G., ed. 1960. Origin of limestone caves: A symposium with discussion. *Natl. Speleol. Soc. Bull.* 22.

Moore, J. 1970. Relationship between subsidence and volcanic load, Hawaii. *Bull. Volcanol.* 34:562–76.

Moore, T. R. 1979. Land use and erosion in the Machakos Hills. *Ann. Assoc. Am. Geographers,* 69:419–31.

Moran, S.; Clayton, L.; Hooke, R. LeB.; Fenton, M.; and Andriashek, L. 1980. Glacier-bed landforms of the prairie region of North America. *Jour. Glaciol.* 25:457–76.

Morehouse, D. F. 1968. Cave development via the sulfuric acid reactions. *Natl. Speleol. Soc. Bull.* 30:1–10.

Morey, G.; Fournier, R.; and Rowe, J. 1962. The solubility of quartz in water in the temperature interval from 25°C to 300°C. *Geochim. et Cosmochim. Acta* 26:1029–43.

———. 1964. The solubility of amorphous silica at 25°C. *Jour. Geophys. Research* 69:1995–2002.

Morgan, J. P. 1970. Deltas—a résumé. *Jour. Geol. Educ.* 18:107–17.

Morisawa, M. E. 1962. Quantitative geomorphology of some watersheds in the Appalachian Plateau. *Geol. Soc. America Bull.* 73:1025–46.

———. 1964. Development of drainage systems on an upraised lake floor. *Am. Jour. Sci.* 262:340–54.

———. 1968. *Streams, their dynamics and morphology.* New York: McGraw-Hill.

Morner, N. A. 1980. *Earth rheology, isostasy, and eustasy.* New York: John Wiley & Sons.

Morris, E. M. 1976. An experimental study of the motion of ice past obstacles by the process of regelation. *Jour. Glaciol.* 17:79–98.

Morton, R. A., and Donaldson, A. C. 1978. Hydrology, morphology, and sedimentology of the Guadalupe fluvial-deltaic system. *Geol. Soc. America Bull.* 89:1030–36.

Mosley, M. P. 1979. Streamflow generation in a forested watershed, New Zealand. *Water Resour. Res.* 15:795.

———. 1982. The effect of a New Zealand beech forest canopy on the kinetic energy of water drops and on surface erosion. *Earth Surf. Proc. and Landforms* 7:103–7.

Moss, J. H., and Bonini, W. 1961. Seismic evidence supporting a new interpretation of the Cody terrace near Cody, Wyo. *Geol. Soc. America Bull.* 72:547–56.

Moss, J. H., and Kochel, R. C. 1978. Unexpected geomorphic effects of the hurricane Agnes storm and flood, Conestoga drainage basin, southeastern Pennsylvania. *Jour. Geology* 86:1–11.

Mueller, J. E. 1972. Re-evaluation of the relationship of master streams and drainage basins. *Geol. Soc. America Bull.* 83:3471–74.

Mugridge, S. J., and Young, H. R. 1983. Disintegration of shale by cyclic wetting and drying and frost action. *Can. J. Earth Sci.* 20:568–76.

Muir Wood, A. M. 1969. *Coastal hydraulics.* London: Macmillan.

Mukerji, A. B. 1976. Terminal fans of inland streams in Sutlej-Yamuna plain, India. *Zeit. f. Geomorph.* 20:190–204.

Mullenders, W., and Gullentops, F. 1969. The age of the pingos of Belgium. In *The periglacial environment,* edited by T. Péwé. Montreal: McGill-Queens Univ. Press.

Muller, E. H. 1974. Origin of drumlins. In *Glacial geomorphology,* edited by D. R. Coates, pp. 187–204. S.U.N.Y., Binghamton: Pubs. in Geomorphology, 5th Ann. Symposium.

Müller, F. 1962. Zonation in the accumulation areas of the glaciers of Axel Heiberg Island, N.W.T., Canada. *Jour. Glaciol.* 4:302–11.

———. 1963. *Observations on pingos (Beobachtungen über Pingos).* Can. Natl. Res. Council Tech. Translation 1073.

Muller, S. W. 1947. *Permafrost or permanently frozen ground and related engineering problems.* Ann Arbor, Mich.: J. W. Edwards.

Munk, W. H. 1949. Surf beats. *Am. Geophys. Union Trans.* 30:849–54.

Murphey, J. B.; Wallace, D. E.; and Lane, L. J. 1977. Geomorphic parameters predict hydrograph characteristics in the southwest. *Water Resour. Res. Bull.* 13:25–38.

Myrick, R. M., and Leopold, L. B. 1963. Hydraulic geometry of a small tidal estuary. U.S. Geol. Survey Prof. Paper 422-B.

Nanson, G. C., and Young, R. W., 1981. Overbank deposition and floodplain formation on small coastal streams of New South Wales. *Zeit. f. Geomorph.* 25:332–45.

Nash, D. 1980a. Forms of bluffs degraded for different lengths of time in Emmet County, Michigan, U.S.A. *Earth Surf. Proc. and Landforms* 5:331–45.

———. 1980b. Morphologic dating of degraded normal fault scarps. *Jour. Geology* 88:353–60.

National Academy of Sciences. 1983. *Proc. Permafrost 4th Internat. Conf.* Washington, D.C.: National Academy Press.

National Research Council of Canada. 1978. *Proc. Permafrost 3rd Internat. Conf.* Ottawa, Ont.: National Research Council of Canada.

Norris, R. M. 1966. Barchan dunes of Imperial Valley, California. *Jour. Geology* 74:292–306.

Norrman, J. O. 1980. Coastal erosion and slope development in Surtsey Island. *Zeit. f. Geomorph.,* Suppl. 34:20–38.

Nunn, K. R., and Rowell, D. M. 1967. Regelation experiments with wires. *Philos. Mag.* 16:1281–83.

Nye, J. F. 1952a. The mechanics of glacier flow. *Jour. Glaciol.* 2:82–93.

———. 1952b. A comparison between the theoretical and the measured long profiles of the Unteraar glacier. *Jour. Glaciol.* 2:103–7.

———. 1957. The distribution of stress and velocity in glaciers and ice-sheets. *Proc. Royal Soc. London,* ser. A:239:113–33.

———. 1960. The response of glaciers and ice-sheets to seasonal and climatic changes. *Proc. Royal Soc. London,* ser. A:256:559–84.

———. 1965. The flow of a glacier in a channel of rectangular, elliptic, or parabolic cross-section. *Jour. Glaciol.* 5:661–90.

Nye, J. F., and Martin, P. C. S. 1967. *Glacial erosion,* pp. 78–83. Int. Assoc. Sci. Hydrol., Comm. Snow and Ice, Bern.

Oberlander, T. M. 1972. Morphogenesis of granitic boulder slopes in the Mojave Desert, California. *Jour. Geology* 80:1–20.

———. 1974. Landscape inheritance and the pediment problem in the Mojave Desert of southern California. *Am. Jour. Sci.* 274:849–75.

Officer, C. B., and Drake, C. L. 1982. Epeirogenic plate movements. *Jour. Geology* 90:139–54.

Ollier, C. D. 1963. Insolation weathering: Examples from central Australia. *Am. Jour. Sci.* 261:376–81.

———. 1969. *Weathering.* Edinburgh: Oliver and Boyd.

———. 1976. Catenas in different climates. In *Geomorphology and climate,* edited by E. Derbyshire, London: John Wiley & Sons.

Olyphant, G. 1981a. Allometry and cirque evolution. *Geol. Soc. America Bull.* 92:697–85.

———. 1981b. Interaction among controls of cirque development: Sangre Cristo Mountains, Colorado, U.S.A. *Jour. Glaciol.* 27:449–58.

Orme, A. R. 1973. Barrier and lagoon systems along the Zululand coast, South Africa. In *Coastal geomorphology,* edited by D. R. Coates, pp. 181–217. S.U.N.Y., Binghamton: 3rd Ann. Geomorph. Symposium.

———. 1974. Quaternary deformation of marine terraces between Ensenada and El Rosario, Baja California, Mexico. In *Geology of peninsular California, Pacific sections,* pp. 67–79. AAPG, SEPM, and SEG.

Osborn, G. D. 1975. Advancing rock glaciers in the Lake Louise area, Banff National Park, Alberta. *Can. J. Earth Sci.* 12:1060–62.

Osterkamp, W. R. 1978. Gradient, discharge and particle-size relations of alluvial channels in Kansas, with observations on braiding. *Am. Jour. Sci.* 278:1253–68.

Østrem, G. 1964. Ice-cored moraines in Scandinavia. *Geogr. Annlr.* 46:282–337.

Outcalt, S. I., and Benedict, J. B. 1965. Photo-interpretation of two types of rock glaciers in the Colorado Front Range, U.S.A. *Jour. Glaciol.* 5:849–56.

Owens, E. H., and Harper, J. R. 1977. Frost-table and thaw depths in the littoral zone near Pearl Bay, Alaska. *Arctic* 30:155–68.

Owens, L. B., and Watson, J. P. 1979. Rates of weathering and soil formation on granite in Rhodesia. *Soil Sci. Soc. Am. Proc.* 43:160–66.

Paige, S. 1912. Rock-cut surfaces in the desert ranges. *Jour. Geology* 20:442–50.

Pakiser, L. C., and Robinson, R. 1966. Composition of the continental crust as estimated from seismic observations. In *The Earth beneath the continents,* edited by J. Steinhart and T. Smith, pp. 620–26. Am. Geophys. Union, Geophys. Monograph 10.

Palmer, A. C. 1972. A kinematic wave model of glacier surges. *Jour. Glaciol.* 11:65–72.

Palmer, A. N. 1975. Origin of maze caves. *Natl. Speleol. Soc. Bull.* 37:57–76.

———. 1981a. Hydrochemical factors in the origin of limestone caves. *Proc. 8th Intl. Cong. Speleol.,* Bowling Green, Ky., pp. 120–22.

———. 1981b. *A geological guide to Mammoth Cave National Park.* Teaneck, N.J.: Zephyrus Press.

———. 1984. Recent trends in karst geomorphology. *Jour. Geol. Educ.* 32:247–53.

Palmer, V. E., and Palmer, A. N. 1975. Landform development of the Mitchell Plain of southern Indiana: Origin of a partially karstic plain. *Zeit. f. Geomorph.* 19:1–39.

Palmquist, R. C. 1975. Preferred position model and subsurface symmetry of valleys. *Geol. Soc. America Bull.* 86:1391–98.

———. 1979. Geological controls on doline characteristics in mantled karst. *Zeit. f. Geomorph.,* Suppl. 32:90–106.

Panoś, V., and Stelcl, O. 1968. Physiographic and geologic control in development of Cuban mogotes. *Zeit. f. Geomorph.* 12:117–65.

Paredes, J. R. and Buol, S. W. 1981. Soils in an aridic, ustic, udic, climosequence in the Maracaibo Lake Basin, Venezuela. *Soil Sci. Soc. Am. Proc.* 45:385–91.

Parham, W. E., 1969. Formation of halloysite from feldspar: Low temperature, artificial weathering versus natural weathering. *Clays and Clay Minerals* 17:13–22.

Parizek, R. 1969. *Glacial ice-contact rings and ridges.* Geol. Soc. America Spec. Paper 123:49–102.

Park, C. 1977. World-wide variations in hydraulic geometry exponents of streams channels: An analysis and some observations. *J. Hydrol.* 33:133–46.

Parker, G. 1976. On the cause and characteristic scale of meandering and braiding in rivers. *Jour. Fluid Mech.* 76:459–80.

Paterson, W. S. B. 1964. Variations in velocity of Athabasca Glacier with time. *Jour. Glaciol.* 5:277–85.

———. 1969. *The physics of glaciers.* Oxford: Pergamon Press.

———. 1981. *The physics of glaciers,* 2d ed. Oxford: Pergamon Press.

Patton, H. B. 1910. Rockstreams of Veta Park, Colorado. *Geol. Soc. America Bull.* 22:663–76.

Patton, P. C., and Baker, V. R. 1976. Morphometry and floods in small drainage basins subject to diverse hydrogeomorphic controls. *Water Resour. Res.* 12:941–52.

———. 1977. Geomorphic response of central Texas stream channels to catastrophic rainfall and runoff. In *Geomorphology of arid and semiarid regions,* edited by D. O. Doehring, pp. 189–217. S.U.N.Y., Binghamton: Pubs. in Geomorphology.

Patton, P. C.; Baker, V. R.; and Kochel, R. C. 1979. Slack-water deposits: A geomorphic technique for the interpretation of fluvial paleohydrology. In *Adjustments of the fluvial system,* edited by D. D. Rhodes and G. P. Williams, pp. 225–53: Dubuque, Iowa: Kendall Hunt.

Patton, P. C., and Dibble, D. S. 1982. Archeological and geomorphic evidence for the paleohydrologic record of the Pecos River in west Texas. *Am. Jour. Sci.* 282:97–121.

Patton, P. C., and Schumm, S. A. 1975. Gulley erosion, northwestern Colorado: A threshold phenomenon. *Geology* 3:88–90.

Peltier, L. 1950. The geographical cycle in periglacial regions as it is related to climatic geomorphology. *Ann. Assoc. Am. Geog.* 40:214–36.

Perutz, M. F. 1940. Mechanism of glacier flow. *Proc. Royal Soc. London,* ser. A:52:132–35.

Pesci, M. 1968. Loess. In *Encyclopedia of geomorphology,* edited by R. W. Fairbridge. New York: Reinhold Book Corp.

Petrie, G., and Price, R. J. 1966. Photogrammetric measurements of the ice wastage and morphological changes near the Casement Glacier, Alaska. *Can. J. Earth Sci.* 3:827–40.

Petterssen, S. 1964. Meteorology. In *Handbook of applied hydrology,* edited by V. T. Chow, pp. 3–39. New York: McGraw-Hill.

Pettijohn, F. 1941. Persistence of heavy minerals and geologic age. *Jour. Geology* 49:610–25.

Péwé, T. L. 1955. Origin of the upland silt near Fairbanks, Alaska. *Geol. Soc. America Bull.* 66:699–724.

———. 1959. Sand-wedge polygons (Tesselations) in the McMurdo Sound region, Antarctica—A progress report. *Am. Jour. Sci.* 257:545–52.

———. 1966. Ice-wedges in Alaska—Classification, distribution and climatic significance. In *Proc. Permafrost Internat. Conf.* (Lafayette, Ind., 1963). Natl. Acad. Sci.-Natl. Res. Council Pub. 1287, pp. 76–81.

———. 1969. *The periglacial environment.* Montreal: McGill-Queens Univ. Press.

———. 1973. Ice wedge casts and past permafrost distribution in North America. *Geoform* 15:15–26.

———. 1981. Desert dust: Origin, characteristics, and effect on man. Geol. Soc. America Spec. Paper 186.

Péwé, T. L., and Journaux, A. 1983. Origin and character of loesslike silt in unglaciated south-central Yakutia, Siberia, U.S.S.R. U.S. Geol. Surv. Prof. Paper 1262.

Peyronnin, C. A., Jr. 1962. Erosion of Isles Dernieres and Timbalier Islands. Am. Soc. Civil Engineers, *Jour. Waterways and Harbors,* 1:57–69.

Phillip, H. 1920. Geologische Untersuchungen über den Mechanismus der Gletscher Bewegung und die Entstehung der gletschertextur. *Neuer Jb. Miner. Geol. Paläont.* 43:439–556.

Pickup, G. 1977. Simulation modelling of river channel erosion. In K. J. Gregory, *River channel changes,* pp. 47–60. London: John Wiley & Sons.

Pierce, J. W. 1970. Tidal inlets and washover fans. *Jour. Geology* 78:230–34.

Pike, R. J., and Wilson, S. E. 1971. Elevation-relief ratio, hypsometric integral, and geomorphic area-altitude analysis. *Geol. Soc. America Bull.* 82:1079–84.

Pillans, B. 1983. Upper Quaternary marine terrace chronology and deformation, South Taranaki, New Zealand. *Geology* 11:292–97.

Piper, A. M. 1932. Ground water in north-central Tennessee. U.S. Geol. Survey Water Supply Paper 640:69–89.

Pissart, A. 1970. *The pingos of Prince Patrick Island (76°N–120°W).* Natl. Res. Council of Canada, Tech. Trans. 1401.

Plummer, L. 1975. Mixing of seawater with calcium carbonate groundwater. In *Quantitative studies in the geological sciences,* edited by E. Whitten, pp. 219–36. Geol. Soc. America Mem. 142.

Plummer, L.; Vacher, H.; Mackenzie, F.; Bricker, O.; and Land, L. 1976. Hydrochemistry of Bermuda: A case history of groundwater diagenesis of biocalcarenites. *Geol. Soc. America Bull.* 87:301–16.

Plummer, N.; Wigley, T.; and Parkhurst, D. 1978. The kinetics of calcite dissolution in CO_2 water systems at 5° to 60°C and 0.0 to 1 atm. CO_2. *Am. Jour. Sci.* 278:179–216.

Pohl, E. R. 1955. *Vertical shafts in limestone caves.* Natl. Speleol. Soc. Occasional Paper 2.

Pohl, E. R., and White, W. B. 1965. Sulfate minerals: Their origin in the central Kentucky karst. *Am. Mineralogist* 50:1461–65.

Poldervaart, A., ed. 1955. Chemistry of the Earth's crust. In *Crust of Earth,* Geol. Soc. America Spec. Paper 62:119–44.

Pomeroy, J. S. 1980. Storm-induced debris avalanching and related phenomena in the Johnstown area, Pennsylvania, with references to other studies in Appalachians. U.S. Geol. Survey Prof. Paper 1191.

———. 1982. Landslides in the greater Pittsburgh region, Pennsylvania. U.S. Geol. Survey Prof. Paper 1229.

Porslid, A. E. 1938. Earth mounds in unglaciated Arctic northwestern America. *Geogr. Rev.* 28:46–58.

Porter, S. C. 1977. Present and past glaciation threshold in the Cascade Range, Washington, U.S.A. Topographic and climatic controls, and paleoclimatic implications. *Jour. Glaciol.* 18:101–16.

Porter, S. C., and Orombelli, G. 1980. Catastrophic rockfall of September 12, 1717, on the Italian flank of the Mont Blanc Massif. *Zeit f. Geomorph.* 24:200–18.

Post, A. S. 1960. The exceptional advances of the Muldrow, Black Rapids and Sustina glaciers. *Jour. Geophys. Research* 65:3703–12.

———. 1966. The recent surge of Walsh glacier, Yukon and Alaska. *Jour. Glaciol.* 6:375–81.

———. 1967. Effects of the March 1964 Alaskan earthquake on glaciers. U.S. Geol. Survey Prof. Paper 544-D.

———. 1969. Distribution of surging glaciers in western North America. *Jour. Glaciol.* 8:229–40.

Potter, N., Jr. 1972. Ice-cored rock glacier, Galena Creek, northern Absaroka Mountains, Wyoming. *Geol. Soc. America Bull.* 83:3025–57.

Potter, N., Jr., and Moss, J. H. 1968. Origin of the Blue Rocks block field deposits, Berks County, Pennsylvania. *Geol. Soc. America Bull.* 79:255–62.

Potts, A. S. 1970. Frost action in rocks: Some experimental data. *Inst. Brit. Geog. Trans.* 49:109–24.

Powers, R. W. 1962. Arabian Upper Jurassic carbonate reservoir rocks. In *Classification of carbonate rocks,* edited by W. E. Ham. Am. Assoc. Petroleum Geologists Mem 1:122–92.

Prestegaard, K. L. 1983a. Bar resistance in gravel bed streams at bankfull stage. *Water Resour. Res.* 19:472–76.

———. 1983b. Variables influencing water-surface slopes in gravel-bed streams at bankfull stage. *Geol. Soc. America Bull.* 94:673–78.

Price, L. W. 1972. *The periglacial environment, permafrost, and man.* Assoc. Am. Geog., Comm. on College Geog. Resource Paper 14.

Price, R. J. 1966. Eskers near the Casement glacier, Alaska. *Geogr. Annlr.* 48:111–25.

———. 1969. Moraines, sandar, kames, and eskers near Breidamerkurjökull, Iceland. *Inst. Brit. Geog. Trans.* 46:17–43.

———. 1970. Moraines at Fjallsjökull, Iceland. *Arc. Alp. Res.* 2:27–42.

———. 1973. *Glacial and fluvioglacial landforms.* New York: Hafner.

Pritchard, G. B. 1962. Inuvik, Canada's new Arctic town. *Polar Record* 11:71:145–54.

Ragan, R. M. 1968. An experimental investigation of partial area contributions. *Intl. Assoc. Sci. Hydrol.* Pub. 76:241–51.

Rahn, P. H. 1966. Inselbergs and nickpoints in southwestern Arizona. *Zeit. f. Geomorph.* 10:217–25.

———. 1967. Sheetfloods, streamfloods, and the formation of pediments. *Ann. Assoc. Am. Geog.* 57:593–604.

————. 1976. Coulee alignment and the wind in southern Alberta, Canada: Discussion. *Geol. Soc. America Bull.* 87:157.

Rains, R., and Shaw, J. 1981. Some mechanisms of controlled moraine development, Antarctica. *Jour. Glaciol* 27:113–28.

Rankin, J. K. 1952. Development of the New Jersey shore. In *3rd Coastal Engr. Conf. Proc.,* edited by J. W. Johnson, pp. 306–17. Cambridge, Mass.: Council of Wave Research.

Rapp, A. 1960. Recent development of mountain slopes in Karkevagge and surroundings, northern Scandinavia. *Geogr. Annlr.* 42:65–206.

Raudkivi, A. J. 1967. *Loose boundary hydraulics.* Oxford: Pergamon Press.

Rawitz, E.; Engman, E. T.; and Cline, G. D. 1970. Use of the mass balance method for examining the role of soils in controlling watershed performance. *Water Resour. Res.* 6:115–23.

Ray, L. L. 1951. Permafrost. *Arctic* 4:196–203.

Ray, R. J.; Krantz, W. B.; Caine, T. N.; and Gunn, R. D. 1983a. A mathematical model for patterned ground: Sorted polygons and stripes, and underwater polygons. In *Proc. Permafrost 4th Internat. Conf.,* pp. 1036–41. Natl. Acad. Sci.

————. 1983b. A model for sorted patterned ground regularity. *Jour. Glaciol.* 29:317–37.

Raymond, C. F. 1971. Flow in a transverse section of Athabasca Glacier, Alberta, Canada. *Jour. Glaciol.* 10:55–84.

Reed, B.; Galvin, C. J.; and Miller, J. P. 1962. Some aspects of drumlin geometry. *Am. Jour. Sci.* 260:200–210.

Reeve, I. J. 1982. A splash transport model and its application to geomorphic measurement. *Zeit. f. Geomorph.* 26:55–71.

Reheis, M. J. 1975. Source transportation and deposition of debris on Arapaho Glacier, Front Range, Colorado, U.S.A., *Jour. Glaciol.* 14:407–20.

Reiche, P. 1943. Graphic representation of chemical weathering. *Jour. Sed. Petrology* 13:58–68.

Reid, I.; Layman, J.; and Frostick, L. 1980. The continuous measurement of bedload discharge. *Jour. Hydraul. Res.* 18:243–49.

Reid, J. M.; MacLeod, D. A.; and Cresser, M. 1981. The assessment of chemical weathering rates within an upland catchment in North-East Scotland. *Earth Surf. Proc. and Landforms* 6:447–57.

Rendell, H. 1982. Clay hillslope erosion rates in Basento Valley, S. Italy, *Geogr. Annlr.* 64A:141–47.

Renwick, W.; Brumbaugh, R.; and Loeher, L. 1982. Landslide morphology and processes on Santa Cruz Island, California: *Geogr. Annlr.* 64A:149–59.

Revue de géomorphologie dynamique. 1967. Field methods for the study of slope and fluvial processes. *Rev géomorph. dynamique* 17:145–88.

Rhodes, D. D. 1977. The b-f-m diagram: Graphical representation and interpretation of at-a-station hydraulic geometry. *Am. Jour. Sci.* 277:73–96.

Rice, A. 1976, Insolation warmed over. *Geology,* pp. 61–62.

Rich, J. L. 1935. Origin and evolution of rock fans and pediments. *Geol. Soc. America Bull.* 46:999–1024.

Richards, K. S. 1976a. Channel width and the riffle-pool sequence. *Geol. Soc. America Bull.* 87:883–90.

————. 1976b. The morphology of riffle-pool sequences. *Earth Surf. Proc. and Landforms* 1:71–88.

————. 1979. Prediction of drainage density from surrogate measures. *Water Resour. Res.* 15:435–42.

————. 1982. *Rivers.* London: Methuen and Co.

Richardson, H. W. 1942. Alcan-America's glory road, parts I and II. *Engr. News Record* 129:25:81–96 and 27:35–42.

————. 1943. Alcan-America's glory road, part III. *Engr. News Record* 130:1:131–38.

————. 1944. Controversial Canol. *Engr. News Record* 132:2:78–84.

Rieke, R. D.; Vinson, T. S.; and Mageau, D. W. 1983. The role of specific surface area and related index properties in the frost heave susceptibility of soils. In *Proc. Permafrost 4th Internat. Conf.,* pp. 1066–71. Natl. Acad. Sci.

Rigsby, G. P. 1960. Crystal orientation in glacier and in experimentally deformed ice. *Jour. Glaciol.* 3:589–606.

Ritter, D. F. 1967a. Terrace development along the front of the Beartooth Mountains, southern Montana. *Geol. Soc. America Bull.* 78:467–84.

————. 1967b. Rates of denudation. *Jour. Geol. Educ.* 15, C.E.G.S. short rev. 6:154–59.

————. 1972. The significance of stream capture in the evolution of a piedmont region, southern Montana. *Zeit. f. Geomorph.* 16:83–92.

————. 1975. Stratigraphic implications of coarse-grained gravel deposited as overbank sediment, southern Illinois. *Jour. Geology* 83:645–50.

————. 1982. Complex river terrace development in the Nenana Valley near Healy, Alaska. *Geol. Soc. America Bull.* 93:346–56.

Ritter, D. F.; Kinsey, W. F.; and Kauffman, M. E. 1973. Overbank sedimentation in the Delaware River valley during the last 6,000 years. *Science* 179:374–75.

Rittman, A. 1962. *Volcanoes and their activity.* New York: Wiley-Interscience.

Robin, G. deQ. 1976. Is the basal ice of a temperate glacier at the pressure-melting point? *Jour. Glaciol.* 16:183–95.

Robin, G. deQ., and Weertman, J. 1973. Cyclic surging of glaciers. *Jour. Glaciol.* 12:3–18.

Robinson, G. 1966. Some residual hillslopes in the Great Fish River Basin, South Africa. *Geogr. Jour.* 132:386–90.

Robinson, L. A. 1977. Marine erosive processes at the cliff foot. *Marine Geol.* 23:257–71.

Ronov, A., and Yaroshevsky, A. 1969. Chemical composition of the Earth's crust. In *The Earth's crust and upper mantle,* edited by P. Hart, pp. 37–57. Am. Geophys. Union, Geophys. Monograph 13.

Rossby, C. G. 1941. The scientific basis of modern meteorology. In *Climate and man,* U.S. Dept. Agri. Yearbook, pp. 599–655.

Röthlisberger, H. 1972. Water pressure in intra- and sub-glacial channels. *Jour. Glaciol.* 11:177–203.

Rouse, L. J., Jr.; Roberts, H. H.; and Cunningham, R. 1978. Satellite observation of subaerial growth of the Atchafalaya Delta, Louisiana. *Geology* 6:405–8.

Rubey, W. W. 1938. The force required to move particles on a stream bed. U.S. Geol. Survey Prof. Paper 189-E.

———. 1952. Geology and mineral resources of the Hardin and Brussels quadrangles (in Illinois). U.S. Geol. Survey Prof. Paper 218.

Ruhe, R. V. 1952. Topographic discontinuities of the Des Moines lobe. *Am. Jour. Sci.* 250:46–56.

———. 1964. Landscape morphology and alluvial deposits in southern New Mexico. *Ann. Assoc. Am. Geog.* 54:147–59.

———. 1965. Quaternary paleopedology. In *The Quaternary of the United States,* edited by H. Wright and D. Frey, pp. 735–64. Princeton, N.J.: Princeton Univ. Press.

———. 1967. Geomorphic surfaces and surficial deposits in southern New Mexico. New Mexico State Bur. Mines and Min. Res. Memo 18.

———. 1969. *Quaternary landscapes in Iowa.* Ames: Iowa State Univ. Press.

———. 1975. *Geomorphology.* Boston: Houghton Mifflin Co.

Runnells, D. D. 1969. Diagenesis, chemical sediments, and the mixing of natural waters. *Jour. Sed. Petrology* 39:1188–1201.

Russell, R. 1943. Freeze-thaw frequencies in the United States. *Am. Geophys. Union Trans.* 24:125–33.

Russell, R. J. 1967a. Aspects of coastal morphology. *Geogr. Annlr.* 49A:299–309.

———. 1967b. *River and delta morphology.* Louisiana State Univ., Coastal Studies Inst. Tech. Rept. 52.

Russell, R. J., and McIntire, W. G. 1965. Beach cusps. *Geol. Soc. America Bull.* 76:307–20.

Russell-Head, D. S., and Budd, W. F. 1979. Ice-sheet flow properties derived from bore-hole shear measurements combined with ice-core studies. *Jour. Glaciol.* 24:117–30.

Rust, B. R. 1972. Structure and process in a braided river. *Sedimentology* 18:221–45.

———. 1977. Mass flow deposits in a Quaternary succession near Ottawa, Canada: Diagnostic criteria for subaqueous outwash. *Can. J. Earth Sci.* 14:175–84.

Ruxton, B. P., and Berry, L. 1961. Weathering profiles and geomorphic position on granite in two tropical regions. *Rev. géomorph. dynamique* 12:16–31.

Ruxton, B. P., and McDougall, I. 1967. Denudation rates in northeast Papua from potassium-argon dating of lavas. *Am. Jour. Sci.* 265:545–61.

Ryckborst, H. 1975. On the origin of pingos. *J. Hydrol.* 26:303–14.

Ryder, J. M. 1971a. The stratigraphy and morphology of paraglacial alluvial fans in south-central British Columbia. *Can. J. Earth Sci.* 8:279–98.

———. 1971b. Some aspects of the morphometry of paraglacial alluvial fans in south-central British Columbia. *Can. J. Earth Sci.* 8:1252–64.

St. Amand, P. 1957. Geological and geophysical synthesis of the tectonics of portions of British Columbia, the Yukon Territory, and Alaska. *Geol. Soc. America Bull.* 68:1343–70.

Sakamoto-Arnold, C. M. 1981. Eolian features produced by the December 1977 windstorm in southern San Joaquin Valley, California. *Jour. Geology* 89:129–37.

Sallenger, A. D. 1979. Beach-cusp formation. *Marine Geol.* 29:23–37.

Salter, P., and Williams, J. 1965. The influence of texture on the moisture characteristics of soils, I. A critical comparison of techniques for determining the available-water capacity and moisture characteristic curve of a soil. *Jour. Soil Sci.* 16:1–5.

Saucier, R. T., and Fleetwood, A. R. 1970. Origin and chronologic significance of Late Quaternary terraces, Ouachita River, Arkansas and Louisiana. *Geol. Soc. America Bull.* 81:869–90.

Savage, J. C., and Paterson, W. S. B. 1963. Borehole measurements in the Athabasca Glacier. *Jour. Geophys. Research* 68:4521–36.

Savat, J. 1981. Work done by splash: Laboratory experiments. *Earth Surf. Proc. and Landforms* 6:275–83.

Savigear, R. 1952. Some observations on slope development in South Wales. *Inst. Brit. Geog. Trans.* 18:31–51.

Sawkins, J. 1869. Report on the geology of America. *Mem. Geol. Survey.*

Schalscha, E.; Appelt, H.; and Schatz, A. 1967. Chelation as a weathering mechanism, I. Effect of complexing agents on the solubilization of iron from minerals and granodiorite. *Geochim. et Cosmochim. Acta* 31:587–96.

Schatz, A. 1963. Chelation in nutrition, soil microorganisms and soil chelation. The pedogenic action of lichens and lichen acids. *Jour. Agri. and Food Chem.* 11:112–18.

Schatz, A.; Cheronis, N.; Schatz, V.; and Trelawney, G. 1954. Chelation (sequestration) as a biological weathering factor in pedogenesis. *Pennsylvania Acad. Sci. Proc.* 28:44–57.

Schowengerdt, R. A., and Glass, C. E. 1983. Digitally processed topographic data for regional tectonic evaluation. *Geol. Soc. America Bull.* 94:549–56.

Schumm, S. A. 1956. Evolution of drainage systems and slopes in badlands at Perth Amboy, New Jersey. *Geol. Soc. America Bull.* 67:597–646.

———. 1960. The shape of alluvial channels in relation to sediment type. U.S. Geol. Survey Prof. Paper 352-B.

———. 1962. Erosion of miniature pediments in Badlands National Monument, South Dakota. *Geol. Soc. America Bull.* 73:719–24.

———. 1963a. A tentative classification of alluvial river channels. U.S. Geol. Survey Circ. 477.

———. 1963b. Sinuosity of alluvial channels on the Great Plains. *Geol. Soc. America Bull.* 74:1089–1100.

———. 1963c. Disparity between present rates of denudation and orogeny. U.S. Geol. Survey Prof. Paper 454-H.

———. 1965. Quaternary paleohydrology. In *The Quaternary of the United States,* edited by H. E. Wright and D. G. Frey, pp. 783–94. Princeton, N.J.: Princeton Univ. Press.

———. 1967a. Rates of surficial rock creep on hillslopes in western Colorado. *Science* 155:560–61.

———. 1967b. Meander wavelength of alluvial rivers. *Science* 157:1549–50.

———. 1968. River adjustment to altered hydrologic regimen, Murrumbidgee River and paleochannels, Australia. U.S. Geol. Survey Prof. Paper 598.

———. 1969. River metamorphosis. Am. Soc. Civil Engineers, *Jour. Hydraulics Div.,* HY1:255–73.

———. 1971. Fluvial geomorphology. In *River mechanics,* edited by H. W. Shen. chs. 4 and 5. Ft. Collins, Colo.: Colorado State Univ.

———, ed. 1972. *River morphology.* Stroudsburg, Pa.: Dowden, Hutchinson, and Ross.

———. 1973. Geomorphic thresholds and complex response of drainage systems. In *Fluvial geomorphology,* edited by M. Morisawa, pp. 299–310. S.U.N.Y., Binghamton: Pubs. in Geomorphology, 4th Ann. Mtg.

———. 1977. *The fluvial system.* New York: John Wiley & Sons.

———. 1980. Some applications of the concept of geomorphic thresholds. In *Thresholds in geomorphology,* edited by D. Coates and J. Vitek, pp. 473–86. London: Allen and Unwin Ltd.

Schumm, S. A.; Bean, D. W.; and Harvey, M. D. 1982. Bed-form-dependent pulsating flow in Medano Creek, Southern Colorado. *Earth Surf. Proc. and Landforms* 7:17–28.

Schumm, S. A., and Chorley, R. J. 1964. The fall of Threatening Rock. *Am. Jour. Sci.* 262:1041–54.

———. 1966. Talus weathering and scarp recession in the Colorado Plateaus. *Zeit. f. Geomorph.* 10:11–36.

Schumm, S. A., and Hadley, R. F. 1957. Arroyos and the semiarid cycle of erosion. *Am. Jour. Sci.* 255:161–74.

Schumm, S. A., and Khan, H. R. 1972. Experimental study of channel patterns. *Geol. Soc. America Bull.* 83:1755–70.

Schumm, S. A., and Lichty, R. W. 1963. Channel widening and floodplain construction along Cimarron River in southwestern Kansas. U.S. Geol. Survey Prof. Paper 352-D.

———. 1965. Time, space and causality in geomorphology. *Am. Jour. Sci.* 263:110–19.

Schumm, S. A., and Parker, R. S. 1973. Implications of complex response of drainage systems for Quaternary alluvial stratigraphy. *Nat. Phys. Sci.* 243:99–100.

Scott, A. J., and Fisher, W. L. 1969. Delta systems and deltaic deposition. In *Delta systems in the exploration for oil and gas,* edited by W. Fisher, L. Brown, A. Scott, and J. McGowen. Univ. Texas, Austin, Bureau Econ. Geology.

Scott, K. M., and Gravlee, G. C., Jr., 1968. Flood surge on the Rubicon River, California—Hydrology, hydraulics, and boulder transport. U.S. Geol. Survey Prof. Paper 422-M.

Seed, H.; Woodward, R.; and Lundgren, R. 1964. Clay mineralogical aspects of the Atterberg limits. *Am. Soc. Civil Engineers Proc.* 90:SM4:107–31.

Selby, M. J. 1966. Methods of measuring soil creep. *J. Hydrol.* 5:54–63.

———. 1967. Aspects of the geomorphology of the Greywacke ranges bordering the lower and middle Waikato basins. *Earth Sci. Jour.* 1:1–22.

———. 1980. A rock mass strength classification for geomorphic purposes: With tests from Antarctica and New Zealand. *Zeit. f. Geomorph.* 24:31–51.

———. 1982. *Hillslope materials and processes.* New York: Oxford Univ. Press.

Senstius, M. 1958. Climax forms of chemical rock-weathering. *Am. Scientist* 46:355–67.

Sevon, W. D. 1969. Sedimentology of some Mississippian and Pleistocene deposits of northeastern Pennsylvania. In *Geology of selected areas in New Jersey and eastern Pennsylvania.* New Brunswick, N.J.: Rutgers Univ. Press.

Sharp, R. P. 1940. Geomorphology of the Ruby-East Humboldt Range, Nevada. *Geol. Soc. America Bull.* 51:337–72.

————. 1951. Features of the firn on upper Seward Glacier, St. Elias Mountains, Canada. *Jour. Geology* 59:599–621.

————. 1953. Deformation of a vertical bore hole in a piedmont glacier. *Jour. Glaciol.* 2:182–84.

————. 1960. *Glaciers.* Eugene, Ore.: Univ. Oregon Press.

————. 1963. Wind ripples. *Jour. Geology* 71:617–36.

————. 1964. Wind-driven sand in Coachella Valley, California. *Geol. Soc. America Bull.* 75:785–804.

————. 1966. Kelso Dunes, Mojave Desert, California. *Geol. Soc. America Bull.* 77:1045–74.

————. 1979. Intradune flats of the Algodones chain, Imperial Valley, California. *Geol. Soc. America Bull.* 90:908–16.

————. 1980. Wind-driven sand in Coachella Valley, California: Further data. *Geol. Soc. America Bull.* 91:724–30.

Sharp, R. P., and Noble, L. H. 1953. Mudflow of 1941 at Wrightwood, southern California. *Geol. Soc. America Bull.* 64:547–60.

Sharpe, C. F. S. 1938. *Landslides and related phenomena.* New York: Columbia Univ. Press.

Shaw, J. 1972. Sedimentation in the ice-contact environment, with examples from Shropshire (England). *Sedimentology* 18:23–62.

————. 1980. Drumlins and large-scale flutings related to glacier folds. *Arc. Alp. Res.* 12:287–98.

Shawe, D. R. 1963. Possible wind-erosion origin of linear scarps on the Saga Plain, southwestern Colorado. U.S. Geol. Survey Prof. Paper 475-C:138–42.

Shepard, F. P. 1950. Beach cycles in southern California. U.S. Army Corps Engrs., Beach Erosion Board Tech. Memo 20.

————. 1952. Revised nomenclature for depositional coastal features. *Am. Assoc. Petroleum Geologists Bull.* 36:1902–12.

————. 1963. *Submarine geology.* New York: Harper & Row.

Shepard, F. P.; Emery, K. O.; and LaFond, E. C. 1941. Rip currents: A process of geological importance. *Jour. Geology* 49:337–69.

Shepard, F. P., and Grant, U. S., IV. 1947. Wave erosion along the southern California coast. *Geol. Soc. America Bull.* 58:919–26.

Shepard, F. P., and Inman, D. L. 1950. Nearshore circulation related to bottom topography and wave refraction. *Am. Geophys. Union Trans.* 31:2:196–212.

Shepard, F. P., and Wanless, H. R. 1971. *Our changing coastlines.* New York: McGraw-Hill.

Shepherd, R. G., and Schumm, S. A. 1974. Experimental study of river incision. *Geol. Soc. America Bull.* 85:257–68.

Shlemon, R. J. 1975. Subaqueous delta formation—Atchafalaya Bay, Louisiana. In *Deltas,* edited by M. Brousard, pp. 209–21. Houston: Houston Geologic Society.

Shoji, S.; Yamada, I.; and Kurashima, K. 1981. Mobilities and related factors of chemical elements in the topsoils of andosols in Tohuku, Japan: 2. Chemical and mineralogical compositions of size fractions and factors influencing the mobilities of major chemical elements. *Soil Sci.* 132:330–46.

Short, A. D. 1979. Three-dimensional beach stage model. *Jour. Geology* 87:553–71.

Short, A. D., and Wright, L. D. 1981. Beach systems of the Sydney Region. *Aust. Geog.* 15:8–16.

Shreve, R. L. 1966a. Statistical law of stream numbers. *Jour. Geology* 74:17–37.

————. 1966b. Sherman landslide, Alaska. *Science* 154:1639–43.

————. 1967. Infinite topologically random channel networks. *Jour. Geology* 75:178–86.

————. 1968. *The Blackhawk landslide.* Geol. Soc. America Spec. Paper 108.

————. 1974. Variation of mainstream length with basin area in river networks. *Water Resour. Res.* 10:1167–77.

Shumskii, P. A. 1964. *Principles of structural glaciology.* New York: Dover Publications.

Shuster, E. T., and White, W. B. 1971. Seasonal fluctuations in the chemistry of limestone springs: A possible means for characterizing carbonate aquifers. *J. Hydrol.* 14:93–128.

Sidle, R., and Swanston, D. 1982. Analysis of a small debris slide in coastal Alaska. *Can. Geotechnical Jour.* 19:167–74.

Silar, J. 1965. Development of tower karst of China and North Vietnam. *Natl. Speleol. Soc. Bull.* 27(2):35–46.

Silvester, R. 1966. Wave refraction. In *Encyclopedia of oceanography,* edited by R. W. Fairbridge, pp. 975–76. New York: Reinhold Book Corp.

Simmons, G., and Richter, D. 1976. Microcracks in rocks. In *The Physics and chemistry of minerals and rocks,* edited by R. G. J. Strens, pp. 105–37. London: John Wiley & Sons.

Simons, D. B., and Richardson, E. V. 1962. Resistance to flow in alluvial channels. *Am. Soc. Civil Engineers Trans.* 127:927–52.

————. 1963. Forms of bed roughness in alluvial channels. *Am. Soc. Civil Engineers Trans.* 128:284–302.

————. 1966. Resistance to flow in alluvial channels. U.S. Geol. Survey Prof. Paper 422-J.

Simpson, D. 1964. Exfoliation in the upper Pocahontas Sandstone, Mercer County, West Virginia. *Am. Jour. Sci.* 262:545–51.

Skempton, A. W. 1953. Soils mechanics in relation to geology. *Yorkshire Geol. Soc. Proc.* 29:33–62.

————. 1964. The long-term stability of clay slopes. *Geotechnique* 2:75–102.

Small, R.; Clark, M.; and Cawse, T. J. 1979. The formation of medial moraines on Alpine glaciers. *Jour. Glaciol.* 22:43–52.

Smalley, I. J. 1966. Drumlin formation. A rheological model. *Science* 151:1379.

———. 1970. Cohesion of soil particles and the intrinsic resistance of simple soil systems to wind erosion. *Jour. Soil Sci.* 21:154–61.

———. 1981. Conjectures, hypotheses, and theories of drumlin formation. *Jour. Glaciol.* 27:503–5.

Smalley, I. J., and Unwin, D. J. 1968. The formation and shape of drumlins and their distribution and orientation in drumlin fields. *Jour. Glaciol.* 7:377–90.

Smart, J. S., and Wallis, J. R. 1971. Cis and trans links in natural channel networks. *Water Resour. Res.* 7:1346–48.

Smith, D. D., and Wischmeier, W. H. 1962. Rainfall erosion. *Advances in Agron.* 14:109–48.

Smith, D. G. 1976. Effect of vegetation on lateral migration of anastomosed channels of a glacier meltwater river. *Geol. Soc. America Bull.* 87:857–60.

Smith, D. G., and Smith, N. D. 1976. Sedimentation in anastamosed river systems: Examples from alluvial valleys near Banff, Alberta. *Jour. Sed. Petrology* 50:157–64.

Smith, D. I. 1969. The solution erosion of limestone in an arctic morphogenetic region. In *Problems of the karst denudation,* edited by O. Stelcl, pp. 99–110. Brno.

Smith, D. I., and Atkinson, T. 1976. Process landforms and climate in limestone regions. In *Geomorphology and climate,* edited by E. Derbyshire. London: John Wiley & Sons.

Smith, G. D. 1942. Illinois loess-variations in its properties and distribution; a pedologic interpretation. *Univ. Illinois Agr. Exp. Sta. Bull.* 490:137–84.

Smith, H. T. U. 1948. Giant glacial grooves in northwest Canada. *Am. Jour. Sci.* 246:503–14.

———. 1953. The Hickory Run boulder field, Carbon County, Pennsylvania. *Am. Jour. Sci.* 251:625–42.

Smith, N. D. 1970. The braided stream depositional environment: Comparison of the Platte River with some Silurian clastic rocks, north-central Appalachians. *Geol. Soc. America Bull.* 81:2993–3014.

———. 1971. Transverse bars and braiding in the Lower Platte River, Nebraska. *Geol. Soc. America Bull.* 82:3407–20.

———. 1974. Sedimentology and bar formation in the upper Kicking Horse River, a braided outwash stream. *Jour. Geology* 82:205–23.

Snodgrass, D.; Groves, G.; Hasselmann, K.; Miller, G.; Munk, W.; and Powers, W. 1966. Propagation of ocean swell across the Pacific. *Phil. Trans. Royal Soc. London,* ser. A:259:431–97.

Soil Survey Staff. 1951. *Soil survey manual.* U.S. Dept. Agri. Handbook 18, Soil Conserv. Serv.

———. 1960. *Soil classification, a comprehensive system—7th approximation.* U.S. Dept. Agri., Soil Conserv. Serv.

———. 1975. *Soil taxonomy.* U.S. Dept. Agri. Handbook 436, Soil Conserv. Serv.

———. 1981. Replacement chapter to Handbook 18, *Soil Survey Manual,* released May 1981. U.S. Dept. Agri., Soil Conserv. Serv.

Sonu, C. J. 1972. Field observation of nearshore circulation and meandering currents. *Jour. Geophys. Research* 77:18:3232–47.

Soucie, G. 1973. Where beaches have been going: Into the ocean. *Smithsonian* 4:3:55–61.

Springer, M. E. 1958. Desert pavement and vesicular layer of some desert soils in the desert of the Lahontan Basin, Nevada. *Soil Sci. Am. Proc.* 22:63–66.

Stalker, A. MacS. 1960. Ice-pressed drift forms and associated deposits in Alberta. *Canada Geol. Survey Bull.* 57.

Stanley, D. J.; Krinitzsky, E. L.; and Compton, J. R. 1966. Mississippi River bank failure, Fort Jackson, Louisiana. *Geol. Soc. America Bull.* 77:850–66.

Stanley, J. M., and Cronin, J. E. 1983. Investigations and implications of subsurface conditions beneath the Trans Alaska Pipeline in Atigun Pass. In *Proc. Permafrost 4th Internat. Conf.,* pp. 1188–93. Natl. Acad. Sci.

Stanley, S. R., and Ciolkosz, E. J. 1981. Classification and genesis of spodosols in the central Appalachians. *Soil Sci. Soc. Am. Proc.* 45:912–17.

Stearns, S. R. 1966. Permafrost (perenially frozen ground). U.S. Army Corps Engrs., Cold Regions Res. and Eng. Lab, Cold Regions Sci. and Eng. 1 (A2).

Steers, J. A. 1962. *The sea coast.* London: Collins.

Steinemann, S. 1954. Results of preliminary experiments on the plasticity of ice crystals. *Jour. Glaciol.* 2:404–12.

———. 1958. Flow and recrystallization of ice. *Intl. Assoc. Sci. Hydrol. Pub.* 39:449–62.

Steinen, R. P.; Harrison, R. S.; and Matthews, R. K. 1973. Eustatic low stand of sea level between 125,000 and 105,000 B.P.: Evidence from the subsurface of Barbados, West Indies. *Geol. Soc. America Bull.* 84:63–70.

Stevens, R., and Carron, M. 1948. Simple field test for distinguishing minerals by abrasion pH. *Am. Mineralogist* 33:31–49.

Stewart, J. H., and LaMarche, V. C., Jr. 1967. Erosion and deposition produced by the flood of Dec. 1964 on Coffee Creek, Trinity County, California. U.S. Geol. Survey Prof. Paper 422-K:1–22.

Stoddart, D. 1969. Climatic geomorphology. In *Introduction to fluvial processes,* edited by R. Chorley, pp. 189–201. London: Methuen and Co.

Stokes, W. L. 1964. Incised, wind-aligned stream patterns of the Colorado Plateau. *Am. Jour. Sci.* 262:808–16.

Stone, R. 1968. Deserts and desert landforms. In *Encyclopedia of geomorphology,* edited by R. W. Fairbridge, pp. 271–79. New York: Reinhold Book Corp.

Strahler, A. N. 1950. Equilibrium theory of slopes approached by frequency distribution analysis. *Am. Jour. Sci.* 248:800–814.

———. 1952a. Dynamic basis of geomorphology. *Geol. Soc. America Bull.* 63:923–38.

———. 1952b. Hypsometric (area-altitude) analysis of erosional topography. *Geol. Soc. America Bull.* 63:1117–42.

———. 1957. Quantitative analysis of watershed geomorphology. *Am. Geophys. Union Trans.* 38:913–20.

———. 1958. Dimensional analysis applied to fluvially eroded landforms. *Geol. Soc. America Bull.* 69:279–99.

———. 1964. Quantitative geomorphology of drainage basins and channel networks. In *Handbook of applied hydrology,* edited by V. T. Chow, pp. 4-39–4-76. New York: McGraw-Hill.

———. 1965. *Introduction to physical geography.* New York: John Wiley & Sons.

———. 1966. Tidal cycle of changes on an equilibrium beach. *Jour. Geology* 74:247–68.

———. 1968. Quantitative geomorphology. In *Encyclopedia of geomorphology,* edited by R. W. Fairbridge, pp. 898–912. New York: Reinhold Book Corp.

Strakhov, N. 1967. *Principles of lithogenesis.* London: Oliver and Boyd.

Stringfield, V. T., and LeGrand, H. E. 1969. Hydrology of carbonate rock terranes—A review. *J. Hydrol.* 8:349–413.

Sugden, D. E., and John, B. S. 1976. *Glaciers and landscape.* London: Edward Arnold Ltd.

Sunamura, T. 1975. A laboratory study of wave-cut platform formation. *Jour. Geology* 83:389–97.

———. 1976. Feedback relationship in wave erosion of laboratory rocky coast. *Jour. Geology* 84:427–37.

———. 1977. A relationship between wave-induced cliff erosion and erosive force of waves. *Jour. Geology* 85:613–18.

———. 1978. Mechanisms of shore platform formation on the southeast coast of the Izu peninsula, Japan. *Jour. Geology* 86:211–22.

———. 1982. A predictive model for wave-induced erosion, with application to Pacific coasts of Japan. *Jour. Geology* 90:167–78.

———. 1983. Processes of sea cliff and platform erosion. In *CRC handbook of coastal processes and erosion,* edited by P. Komar, pp. 233–65. Boca Raton, Fla.: CRC Press.

Sundborg, A. 1956. The river Klarälven, a study of fluvial processes. *Geogr. Annlr.* 38:280–91.

Suzuki, T., and Takahashi, K. 1981. An experimental study of wind abrasion. *Jour. Geology* 89:509–22.

Svasek, J. N., and Terwindt, J. H. J. 1974. Measurements of sand transport by wind on a natural beach. *Sedimentology* 21:311–22.

Swanson, D. 1972. Magma supply rate at Kilauea Volcano, 1925–1971. *Science* 175:169–70.

Sweeting, M. M. 1950. Erosion cycles and limestone caverns in the Ingleborough District of Yorkshire. *Geogr. Jour.* 115:63–78.

———. 1958. The karstlands of Jamaica. *Geogr. Jour.* 124:184–99.

———. 1973. *Karst landforms.* New York: Columbia Univ. Press.

———, ed. 1981. *Karst geomorphology.* New York: Academic Press.

Swineford, A., and Frye, J. C. 1951. Petrography of the Peorian loess in Kansas. *Jour. Geology* 59:306–22.

Swinnerton, A. C. 1932. Origin of limestone caverns. *Geol. Soc. America Bull.* 43:662–93.

Swinzow, G. K. 1969. Certain aspects of engineering geology in permafrost. *Eng. Geol.* 3:177–215.

Taber, S. 1929. Frost heaving. *Jour. Geology* 37:428–61.

———. 1930. The mechanics of frost heaving. *Jour. Geology* 38:303–17.

———. 1943. Perennially frozen ground in Alaska: Its origin and history. *Geol. Soc. America Bull.* 54:1433–1548.

———. 1953. Origin of Alaska silts. *Am. Jour. Sci.* 251:321–36.

Tan, K. H. 1980. The release of silicon, aluminum, and potassium during decomposition of soil minerals by humic acid. *Soil Sci.* 129:5–11.

Tanner, W. F. 1958. The equilibrium beach. *Am. Geophys. Union Trans.* 39:889–91.

———. 1974. The incomplete flood plain. *Geology* 2:105–6.

Tator, B. A. 1952. Pediment characteristics and terminology. *Ann. Assoc. Am. Geog.* 42:295–317.

———. 1953. Pediment characteristics and terminology. *Ann. Assoc. Am. Geog.* 43:47–53.

Ten Brink, N. W. 1974. Glacio-isostasy: New data from West Greenland and geophysical implications. *Geol. Soc. America Bull.* 85:219–28.

Terzaghi, K. 1936. The shearing resistance of saturated soils. *Proc. 1st Internat. Conf. on Soils Mech. and Foundation Eng.* 1:54–66.

————. 1943. *Theoretical soils mechanics.* New York: John Wiley & Sons.

————. 1950. Mechanism of landslides. In *Application of geology to engineering practice,* edited by S. Paige. Geol. Soc. America Berkey Vol., pp. 83–123.

————. 1962. Stability of steep slopes on hard unweathered rock. *Geotechnique* 12:251–70.

Thie, J. 1974. Distribution and thawing of permafrost in the southern part of the discontinuous permafrost zone in Manitoba. *Arctic* 27:189–200.

Thomas, H. P., and Ferrell, J. E. 1983. Thermokarst features associated with buried sections of the Trans Alaska pipeline. In *Proc. Permafrost 4th Internat. Conf.,* pp. 1245–50. Natl. Acad. Sci.

Thomas, R. H. 1979. West Antarctic Ice Sheet: Present-day thinning and Holocene retreat of margins. *Science* 205:1257–58.

Thomas, R. H.; MacAyeal, D. R.; Bentley, C. R.; and Clapp, J. L. 1980. The creep of ice, geothermal heat flow, and Roosevelt Island, Antarctica. *Jour. Glaciol.* 25:47–60.

Thorn, C. E. 1976. Quantitative evaluation of nivation in the Colorado Front Range. *Geol. Soc. America Bull.* 87:1169–78.

————. 1979. Bedrock freeze-thaw weathering regime in an alpine environment, Colorado Front Range. *Earth Surf. Proc. and Landforms* 4:211–28.

Thorn, C. E., and Hall, K. 1980. Nivation: An arctic-alpine comparison and reappraisal. *Jour. Glaciol.* 25:109–24.

Thorne, C. R. 1982. Processes and mechanisms of river bank erosion. In *Gravel-bed rivers,* edited by R. Hey, J. Bathurst, and C. Thorne, pp. 227–72. New York: John Wiley & Sons.

Thorne, C. R., and Lewin, J. 1979. Bank processes, bed material movement, and platform development in a meandering river. In *Adjustments of the fluvial system,* edited by D. D. Rhodes and G. P. Williams, pp. 117–37. Dubuque, Iowa: Kendall-Hunt.

Thorne, C. R., and Tovey, N. K. 1981. Stability of composite river banks. *Earth Surf. Proc. and Landforms* 6:469–84.

Thornes, J. B. 1970. Hydraulic geometry of stream channels in the Xingu-Araguaia headwaters. *Geogr. Jour.* 136:376–82.

Thornton, E. B. 1973. Distribution of sediment transport across the surf zone. *Proc. 13th Conf. on Coast. Eng.,* pp. 1049–68.

Thorp, J., and Smith, G. 1949. Higher categories of soil classifications. Order, Suborder, and Great Soil Groups. *Soil Sci.* 67:117–26.

Thrailkill, J. 1968. Chemical and hydrologic factors in the excavation of limestone caves. *Geol. Soc. America Bull.* 79:19–45.

————. 1972. Carbonate chemistry of aquifer and stream water in Kentucky. *J. Hydrol.* 16:93–104.

Thrailkill, J., and Robl, T. 1981. Carbonate geochemistry of vadose water recharging limestone aquifers. *J. Hydrol.* 54:195–208.

Tinkler, K. J. 1982. Avoiding error when using the Manning equation. *Jour. Geology* 90:326–28.

Todd, D. K. 1959. *Groundwater hydrology.* New York: John Wiley & Sons.

————, ed. 1970. *The water encyclopedia.* Port Washington, N.Y.: Water Information Center.

Townsend, D. W., and Vickery, R. P. 1967. An experiment in regelation. *Philos. Mag.* 16:1275–80.

Toy, T. J. 1977. Hillslope form and climate. *Geol. Soc. America Bull.* 88:16–22.

————. 1982. Accelerated erosion: Process, problems, and prognosis. *Geology* 10:524–29.

Trainer, F., and Heath, R. 1976. Bicarbonate content of groundwater in carbonate rock in eastern North America. *J. Hydrol.* 31:37–55.

Trenhaile, A. S. 1971. Drumlins: Their distribution, orientation and morphology. *Can. Geog.* 15:113–26.

————. 1977. Cirque elevation and Pleistocene snowlines. *Zeit. f. Geomorph.* 21:445–59.

————. 1979. The morphology of valley steps in the Canadian Cordillera. *Zeit. f. Geomorph.* 23:27–44.

Tricart, J. 1967. Le modelé des régions périglaciaires. In *Traité de géomorphologie 2,* edited by J. Tricart and A. Cailleux. Paris: SEDES.

————. 1968. Notes géomorphologiques sur la karstification en Barbade (Antilles): *Mém. docums. cent. docum. cartogr. géogr.* 4:329–34.

————. 1969. *Geomorphology of cold environments,* trans. by Edward Watson. New York: St. Martin's Press.

Tricart, J., and Cailleux, A. 1972. *Introduction to climatic geomorphology.* London: Longman.

Trimble, S. W. 1977. The fallacy of stream equilibrium in contemporary denudation studies. *Am. Jour. Sci.* 277:876–87.

Trimble, S. W., and Lund, S. W. 1982. Soil conservation and the reduction of erosion and sedimentation in the Coon Creek Basin, Wisconsin. U.S. Geol. Survey Prof. Paper 1234.

Troll, C. 1958. Structure soils, solifluction, and frost climates of the Earth. U.S. Army Corps Engrs., Snow, Ice, Permafrost Res. Est. Trans. 43.

Trudgill, S. 1976. The marine erosion of limestones on Aldabra Atoll, Indian Ocean. *Zeit. f. Geomorph.,* Suppl. 26:164–200.

Tuan, Ti-Fu. 1959. *Pediments in southeastern Arizona.* Berkeley, Calif.: Univ. Calif. Pubs. in Geog. 13.

————. 1962. Structure, climate and basin land forms in Arizona and New Mexico. *Ann. Assoc. Am. Geog.* 52:51–68.

Tucker, M. J. 1950. Surf beats: Sea waves of 1 to 5 minute period. *Proc. Royal Soc. London,* ser. A, 202:565–73.

Twidale, C. R. 1962. Steepened margins of inselbergs from northwestern Eyre Peninsula, South Australia. *Zeit. f. Geomorph.* 6:51–69.

————. 1964. Erosion of an alluvial bank at Birdwood, South Australia. *Zeit. f. Geomorph.* 8:189–211.

————. 1967. Origin of the piedmont angle as evidenced in South Australia. *Jour. Geology* 75:393–411.

————. 1968. Weathering. In *Encyclopedia of geomorphology,* edited by R. W. Fairbridge, pp. 1228–32. New York: Reinhold Book Corp.

————. 1972. Evolution of sand dunes in the Simpson Desert, central Australia. *Inst. Brit. Geog. Trans.* 56:77–110.

————. 1978. On the origin of pediments in different structural settings. *Am. Jour. Sci.* 278:1138–76.

Twidale, C. R., and Bourne, J. A. 1975. Episodic exposure of inselbergs. *Geol. Soc. America Bull.* 86:1473–81.

Ueta, H. T., and Garfield, D. E. 1968. Deep core drilling program at Byrd Station 1967–68. *U.S. Arctic Journal* 3:111–12.

U.S. Army Corps of Engineers. 1971. National shoreline study. Washington, D.C.: U.S. Army Corps Engr.

Ursic, S. J., and Dendy, F. E. 1965. Sediment yields from small watersheds under various land uses and forest covers. *Proc. Fed. Inter-Agency Sedimentation Conf.* (1963). U.S. Dept. Agri. Misc. Publ. 970:47–52.

Vagners, V. J. 1966. Lithologic relationship of till to carbonate bedrock in southern Ontario. M.S. thesis, Geology Dept., Univ. Western Ontario.

Van Arsdale, R. 1982. Influence of calcrete on the geometry of arroyos near Buckeye, Arizona. *Geol. Soc. America Bull.* 93:20–26.

Van Dorn, W. G. 1965. Tsunamis. In *Hydroscience advances 2.* New York: Academic Press.

————. 1966. Tsunamis. In *Encyclopedia of oceanography,* edited by R. W. Fairbridge, pp. 941–43. New York: Reinhold Book Corp.

Van Heukon, T. K. 1977. Distant source of 1976 dustfall in Illinois and Pleistocene weather models. *Geology* 5:693–95.

Vanoni, V. A. 1941. Some experiments on the transportation of suspended load. *Am. Geophys. Union Trans.,* 22nd Ann. Mtg., pt. 3:608–20.

————. 1946. Transportation of suspended sediment by water. *Am. Soc. Civil Engineers Trans.* 3:67–133.

Vanoni, V. A., and others. 1966. Sediment transportation mechanics. Initiation of motion. Am. Soc. Civil Engineers, *Jour. Hydraulics Div.* 92:HY2:291–313.

Varnes, D. J. 1958. Landslide types and processes. In *Landslides and engineering practice,* edited by E. Eckel, pp. 20–47. Washington, D.C.: Highway Research Board Spec. Rept. 29.

————. 1978. Slope movement types and processes. In *Landslides,* edited by R. Schuster and R. Krizak, pp. 11–33. Washington, D.C.: Trans. Res. Board, Natl. Acad. Sci.

Vernon, P. 1966. Drumlins and Pleistocene ice flow over the Ards Peninsula. *Jour. Glaciol.* 6:401–9.

Verstappen, H. Th. 1964. Karst morphology of the Star Mountains (central New Guinea) and its relation to lithology and climate. *Zeit. f. Geomorph.* 8:40–49.

Vitek, J. D. 1983. Stone polygons: Observations of surficial activity. In *Proc. Permafrost 4th Internat. Conf.,* pp. 1326–31. Natl. Acad. Sci.

Vivian, R. 1970. Hydrologie et erosion sous-glaciaires. *Rev. géog. alp.* 58:241–64.

————. 1980. The nature of the ice-rock interface: The results of investigation on 20,000 m² of the rock bed of temperate glaciers. *Jour. Glaciol.* 25:267–77.

Vivian, R., and Bouquet, G. 1973. Subglacial cavitation phenomena under Glacier d' Argentiere, Mont Blanc, France. *Jour. Glaciol.* 12:439–52.

Wahrhaftig, C. 1965. Stepped topography of the southern Sierra Nevada. *Geol. Soc. America Bull.* 76:1165–90.

Wahrhaftig, C., and Cox, A. 1959. Rock glaciers in the Alaska Range. *Geol. Soc. America Bull.* 70:383–436.

Waitt, R. B., Jr. 1980. About forty last-glacial Lake Missoula jökulhlaups through southern Washington. *Jour. Geology* 88:653–79.

Walcott, R. I. 1972. Late Quaternary vertical movements in eastern North America. Quantitative evidence of glacio-isostatic rebound. *Rev. Geophys. and Space Physics* 10:849–84.

Walker, A. S. 1982. Deserts of China. *Am. Scientist* 70:366–76.

Wallace, R. E. 1977. Profiles and ages of young fault scarps, north-central Nevada. *Geol. Soc. America Bull.* 88:1267–81.

————. 1978. Geometry and rates of change of fault-generated range fronts, north-central Nevada. *U.S. Geol. Survey Jour. of Res.* 6:637–49.

Walters, J. C. 1978. Polygonal patterned ground in central New Jersey. *Quat. Res.* 10:42–54.

Ward, W. H. 1945. The stability of natural slopes. *Geogr. Jour.* 105:170–97.

Warnke, D. A. 1969. Pediment evolution in the Halloran Hills, central Mojave Desert, California. *Zeit. f. Geomorph.* 13:357–89.

Warren, A. 1970. Dune trends and their implications in the central Sudan. *Zeit. f. Geomorph.*, Suppl. 10:154–80.

———. 1971. Dunes in the Ténéré Desert. *Geogr. Jour.* 137:458–61.

Washburn, A. L. 1956. Classification of patterned ground and review of suggested origins. *Geol. Soc. America Bull.* 67:823–66.

———. 1967. Instrumental observations of mass-wasting in the Mesters Vig district, Northeast Greenland. *Medd. om Grønland* 166:4.

———. 1970. An approach to a genetic classification of patterned ground. *Acta Geogr. Lødz.* 24:437–46.

———. 1973. *Periglacial processes and environments.* London: Edward Arnold Ltd.

———. 1980. *Geocryology.* New York: John Wiley & Sons.

Wasson, R. J. 1977. Catchment processes and the evolution of alluvial fans in the lower Derwent Valley, Tasmania. *Zeit. f. Geomorph.*, Suppl. 21:147–68.

Wasson, R. J., and Hall, G. 1982. A long record of mudslide movement at Waerenga-O-Kuri, New Zealand. *Zeit f. Geomorph.* 26:73–85.

Watts, S. H. 1979. Some observations on rock weathering, Cumberland Peninsula, Baffin Island. *Can. J. Earth Sci.* 16:977–83.

Wayne, W. J. 1967. Periglacial features and climatic gradient in Illinois, Indiana and western Ohio, east-central United States. In *Quaternary paleoecology,* edited by E. Cushing and H. Wright, pp. 393–414. New Haven, Conn.: Yale Univ. Press.

———. 1981. Ice segregation as an origin for lenses of nonglacial ice in "ice-cemented" rock glaciers. *Jour. Glaciol.* 27:506–10.

Wear, J., and White, J. 1951. Potassium fixation in clay mineral studies as related to crystal structure. *Soil Sci.* 71:1–14.

Weertman, J. 1957. On the sliding of glaciers. *Jour. Glaciol.* 3:33–38.

———. 1964. The theory of glacier sliding. *Jour. Glaciol.* 5:287–303.

———. 1967. An examination of the Lliboutry theory of glacier sliding. *Jour. Glaciol.* 6:489–94.

———. 1979. The unsolved general glacier sliding problem. *Jour. Glaciol.* 23:97–111.

Weertman, J., and Weertman, J. B. 1964. *Elementary dislocation theory.* New York: Macmillan.

Weinert, H. 1961. Climate and weathered Karroo dolerites. *Nature* 191:325–29.

———. 1965. Climatic factors affecting the weathering of igneous rocks. *Agri. Meterol.* 2:27–42.

Welder, F. A. 1959. *Processes of deltaic sedimentation in the lower Mississippi River.* Louisiana State Univ., Coastal Studies Inst. Tech. Rept. 12.

Wellman, P. 1982. Surging of Fisher Glacier, eastern Antarctica: Evidence from geomorphology. *Jour. Glaciol.* 28:23–28.

Wells, J. T.; Prior, D. B.; and Coleman, J. M. 1980. Flowslides in muds on extremely low angle tidal flats, northeastern South America. *Geology* 8:272–75.

Wells, S. 1976. Sinkhole plain evolution in the central Kentucky karst. *Natl. Speleol. Soc. Bull.* 38:103–6.

———. 1977. Geomorphic controls of alluvial fan deposition in the Sonoran Desert, southwestern Arizona. In *Geomorphology in arid regions,* edited by D. Doehring, pp. 27–50. S.U.N.Y. Binghamton: 8th Geomorph. Symposium.

Wentworth, C. K. 1938. Marine beach formation: Water level weathering. *Jour. Geomorphology* 1:6–32.

Wescott, W. A., and Ethridge, F. G. 1980. Fan-delta sedimentology and tectonic-Hallahs Fan delta, southeast Jamaica. *Am. Assoc. Petroleum Geologists Bull.* 64:374–99.

Westgate, J. A. 1968. Linear sole markings in Pleistocene till. *Geol. Mag.* 105:501–5.

Weyl, P. K. 1958. The solution kinetics of calcite. *Jour. Geology* 66:163–76.

Whalley, W. B. 1983. Rock glaciers—Permafrost features or glacial relics. In *Proc. Permafrost 4th Internat. Conf.,* pp. 1396–1401. Natl. Acad. Sci.

Whalley, W. B.; Douglas, G. R.; and McGreevey, J. P. 1982. Crack propogation and associated weathering in igneous rocks. *Zeit. f. Geomorph.* 26:33–54.

Whillans, I. M. 1978. Inland ice sheet thinning due to Holocene warmth. *Science* 201:1014–16.

Whitaker, R. H.; Buol, S. W.; Niering, W. A.; and Havens, Y. H. 1968. A soil and vegetation pattern in the Santa Catalina Mountains, Arizona. *Soil Sci.* 105:440–51.

White, E. L., and Reich, B. M. 1970. Behaviour of annual floods in limestone basins in Pennsylvania. *J. Hydrol.* 10:193–98.

White, E., and White, W. 1979. Quantitative morphology of landforms in carbonate rock basins in the Appalachian Highlands. *Geol. Soc. America Bull.* 90:385–96.

———. 1983. Karst landforms and drainage basin evolution in the Obey River basin, north-central Tennessee, U.S.A. *J. Hydrol.* 61:69–82.

White, S. E. 1976a. Is frost action really only hydration shattering? A review. *Arc. Alp. Res.* 8:1–6.

———. 1976b. Rock glaciers and block fields, review and new data. *Quat. Res.* 6:77–98.

White, W. A. 1966. Drainage asymmetry and the Carolina capes. *Geol. Soc. America Bull.* 77:223–40.

White, W. B. 1960. Termination of passages in Appalachian caves as evidence for a shallow phreatic origin. *Natl. Speleol. Soc. Bull.* 22:43–53.

———. 1969. Conceptual models for carbonate aquifers. *Ground Water* 7:15–21.

————. 1976. Geology and biology of Pennsylvania caves. Pennsylvania Geol. Survey Gen. Rept. 66:1–71.

————. 1977. Conceptual models for carbonate aquifers: revisited. In *Hydrologic problems in karst terrains,* edited by R. Dilamarter and S. Csallany, pp. 176–87. Bowling Green: Western Kentucky Univ. Press.

White, W. B., and Schmidt, V. A. 1966. Hydrology of a karst area in east central West Virginia. *Water Resour. Res.* 2:549–60.

Whitney, M. I. 1978. The role of vorticity in developing lineation by wind erosion. *Geol. Soc. America Bull.* 89:1–18.

Whitney, M. I., and Dietrich, R. V. 1973. Ventifact sculpture by windblown dust. *Geol. Soc. America Bull.* 84:2561–82.

Whittecar, G. R., and Mickelson, D. 1979. Composition, internal structures, and a hypothesis of formation for drumlins, Waukesha County, Wisconsin, U.S.A. *Jour. Glaciol.* 22:357–71.

Wiegel, R. L. 1964. *Oceanographical engineering.* Englewood Cliffs, N.J.: Prentice-Hall.

Wilcock, D. N. 1971. Investigation into the relations between bedload transport and channel shape. *Geol. Soc. America Bull.* 82:2159–76.

————. 1975. Relations between planimetric and hypsometric variables in third- and fourth-order drainage basins. *Geol. Soc. America Bull.* 86:47–50.

Wilford, G. E., and Wall, J. R. D. 1965. Karst topography in Sarawak. *J. Trop. Geogr.* 21:44–70.

Williams, A. T. 1973. The problem of beach cusp development. *Jour. Sed. Petrology* 43:857–66.

Williams, G. 1964. Some aspects of the eolian saltation load. *Sedimentology* 3:257–87.

Williams, G. P. 1978. Hydraulic geometry of river cross sections— Theory of minimum variance. U.S. Geol. Survey Prof. Paper 1029.

Williams, G. P., and Guy, H. P. 1973. Erosional and depositional aspects of Hurricane Camille in Virginia, 1969. U.S. Geol. Survey Prof. Paper 804.

Williams, G. P., and Wolman, M. G. 1984. Downstream effects of dams on alluvial rivers. U.S. Geol. Survey Prof. Paper 1286.

Williams, J. R. 1970. Groundwater in the permafrost region of Alaska. U.S. Geol. Survey Prof. Paper 696.

Williams, L. 1964. Regionalization of freeze-thaw activity. *Ann. Assoc. Am. Geog.* 54:597–611.

Williams, P. J. 1966. Downslope soil movement at a sub-Arctic location with regard to variations with depth. *Canadian Geotech. Jour.* 3:191–203.

Williams, P. W. 1963. An initial estimate of the speed of limestone solution in County Clare. *Ir. Geogr.* 4:432–41.

————. 1966. Morphometric analysis of temperate karst landforms. *Ir. Speleol.* 1:23–31.

————. 1971. Illustrating morphometric analyses of karst with examples from New Guinea. *Zeit. f. Geomorph.* 15(1):40–61.

————. 1972a. The analysis of spatial characteristics of karst terrains. In *Spatial analysis in geomorphology,* edited by R. J. Chorley. London: Methuen and Co.

————. 1972b. Morphometric analysis of polygonal karst in New Guinea. *Geol. Soc. America Bull.* 83:761–96.

————. 1983. The role of the subcutaneous zone in karst hydrology. *J. Hydrol.* 61:45–67.

Willman, H. B., and Frye, J. C. 1970. Pleistocene stratigraphy of Illinois. *Illinois Geol. Survey Bull.* 94.

Wilson, B. W. 1966. Seiche. In *Encyclopedia of oceanography,* edited by R. W. Fairbridge, pp. 804–11. New York: Reinhold Book Corp.

Wilson, I. G. 1972. Aeolian bedforms—their development and origins. *Sedimentology* 19:173–210.

————. 1973. Ergs. *Sed. Geol.* 10:77–106.

Wilson, L. 1968. Morphogenetic classification. In *Encyclopedia of geomorphology,* edited by R. W. Fairbridge, pp. 717–28. New York: Reinhold Book Corp.

————. 1969. Les relations entre les processus géomorphologiques et le climat moderne comme méthode de paleoclimatologie. *Rev. géog. phys. et de géologie dyn.* 11:303–14.

————. 1972. Seasonal sediment yield patterns of United States rivers. *Water Resour. Res.* 8:1470–79.

————. 1973. Variations in mean annual sediment yield as a function of mean annual precipitation. *Am. Jour. Sci.* 273:335–49.

Winkler, E. M. 1965. Weathering rates as exemplified by Cleopatra's Needle in New York City. *Jour. Geol. Educ.* 13:50–52.

————. 1975. *Stone: Properties, durability in man's environment,* 2d ed. New York: Springer-Verlag.

Winkler, E. M., and Wilhelm, E. J. 1970. Salt bursts by hydration pressures in architectural stone in urban atmosphere. *Geol. Soc. America Bull.* 81:567–72.

Woldenberg, M. J. 1969. Spatial order in fluvial systems. Horton's laws derived from mixed hexagonal hierarchies of drainage basin areas. *Geol. Soc. America Bull.* 80:97–112.

Wolfe, T. E. 1964. Cavern development in the Greenbrier Series, West Virginia. *Natl. Speleol. Soc. Bull.* 26:37–60.

Wollast, R. 1967. Kinetics of the alteration of K-feldspar in buffered solutions at low temperature. *Geochim. et Cosmochim. Acta* 31:635–48.

Wolman, M. G. 1955. The natural channel of Brandywine Creek, Pennsylvania. U.S. Geol. Survey Prof. Paper 271.

————. 1959. Factors influencing erosion of a cohesive river bank. *Am. Jour. Sci.* 257:204–16.

————. 1967. A cycle of sedimentation and erosion in urban river channels. *Geogr. Annlr.* 49-A:385–95.

Wolman, M. G., and Brush, L. M., Jr. 1961. Factors controlling the size and shape of stream channels in coarse noncohesive sands. U.S. Geol. Survey Prof. Paper 282-G.

Wolman, M. G., and Gerson, R. 1978. Relative scales of time and effectiveness of climate in watershed geomorphology. *Earth Surf. Proc. and Landforms* 3:189–208.

Wolman, M. G., and Leopold, L. B. 1957. River flood plains; some observations on their formation. U.S. Geol. Survey Prof. Paper 282-C.

Wolman, M. G., and Miller, J. P. 1960. Magnitude and frequency of forces in geomorphic processes. *Jour. Geology* 68:54–74.

Womack, W. R., and Schumm, S. A. 1977. Terraces of Douglas Creek, northwestern Colorado: An example of episodic erosion. *Geology* 5:72–76.

Woo, M., and Heron, R. 1981. Occurrence of ice layers at the base of High Arctic snowpacks. *Arc. Alp. Res.* 13:225–30.

Woo, M., and Marsh, P. 1977. Effect of vegetation as limestone solution in a small High Arctic basin. *Can. J. Earth Sci.* 14:571–81.

Wood, A. 1942. The development of hillside slopes. *Proc. Geol. Assoc.* 53:128–40.

Wood, F. J. 1978. The strategic role of perigean spring tides. Washington, D.C.: U.S. Dept. of Commerce.

Woodcock, A. H. 1974. Permafrost and climatology of a Hawaii volcano crater. *Arc. Alp. Res.* 6:49–62.

Woodcock, A. H.; Furumoto, A. S.; and Woollard, G. P. 1970. Fossil ice in Hawaii. *Nature* 226:873.

Woodruff, N. P., and Siddoway, F. H. 1965. A wind erosion equation. *Soil Sci. Soc. Am. Proc.* 29:602–8.

Wright, H. E., and Frey, D. G., eds. 1965. *The Quaternary of the United States.* Princeton, N.J.: Princeton Univ. Press.

Wright, J., and Schnitzer, M. 1963. Metallo-organic interactions associated with podsolisation. *Soil Sci. Soc. Am. Proc.* 27:171–76.

Wright, L. D. 1977. Sediment transport and deposition at river mouths: A synthesis. *Geol. Soc. America Bull.* 88:857–68.

Wright, L. D.; Chappell, J.; Thom, B. G.; Bradshaw, M. P.; and Cowell, P. 1979. Morphodynamics of reflective and dissipative beach and inshore systems: Southeastern Australia. *Marine Geol.* 32:105–40.

Wright, L. D.; Guza, R. T.; and Short, A. D. 1982a. Dynamics of a high-energy dissipative surf zone. *Marine Geol.* 45:41–62.

Wright, L. D.; Nielson, P. N.; Short, A. D.; and Green, M. O. 1982b. Morphodynamics of a macrotidal beach. *Marine Geol.* 50:97–128.

Wright, L. D., and Short, A. D. 1983. Morphodynamics of beaches and surf zones in Australia. In *CRC handbook of coastal processes and erosion,* edited by P. Komar, pp. 35–64. Boca Raton, Fla.: CRC Press.

Wyllie, P. 1971. *The dynamic Earth.* New York: John Wiley & Sons.

Yaalon, D. H. 1975. Conceptual models in pedogenesis. Can soil-forming functions be solved? *Geoderma* 14:189–205.

Yair, A.; Lavee, H.; Bryan, R. B.; and Adar, E. 1980. Runoff and erosion processes and rates in the Zin Valley badlands, northern Negev, Israel. *Earth Surf. Proc. and Landforms* 5:205–25.

Yang, C. T. 1973. Incipient motion and sediment transport. Am. Soc. Civil Engineers, *Jour. Hydraulics Div.* 99:HY10:1679–1704.

Yeats, R. S. 1978. Neogene acceleration of subsidence rates in southern California. *Geology* 6:456–60.

Yefimov, A. I., and Dukhin, I. E. 1968. Some permafrost thicknesses in the Arctic. *Polar Record* 14:68.

Young, A. 1960. Soil movement by denudational processes on slopes. *Nature* 188:120–22.

————. 1961. Characteristic and limiting slope angles. *Zeit. f. Geomorph.* 5:126–31.

————. 1972a. *Slopes.* Edinburgh: Oliver and Boyd.

————. 1972b. The soil catena: A systematic approach. In *International geography 1972,* Congr. Int. Geogr. Commun., n. 22, v. 1:287–89.

Young, R. W. 1983. The tempo of geomorphological change: Evidence from southeastern Australia. *Jour. Geology* 91:221–30.

Zeigler, J. M.; Hayes, C. R.; and Tuttle, S. D. 1959. Beach changes during storms on outer Cape Cod, Massachusetts. *Jour. Geology* 67:318–36.

Zenkovich, V. P. 1967. *Processes of coastal development,* translated by D. G. Fry, edited by J. A. Steers. Edinburgh: Oliver and Boyd.

Zirjacks, W. L., and Hwang, C. T. 1983. Underground utilidors at Barrow, Alaska: A two-year history. In *Proc. Permafrost 4th Internat. Conf.,* pp. 1513–17. Natl. Acad. Sci.

Credits

Illustrations

Chapter 1

Figure **1.6**: "From *Handbook of Applied Hydrology* by Ven T. Chow. Copyright © 1964 by McGraw-Hill, Inc. Used with permission of McGraw-Hill Book Company." Figure **1.11**: From Patton and Schumm 1975, with permission of The Geological Society of America. Reprinted by permission.

Chapter 2

Figure **2.1**: From Wyllie, P., (ed.), *The Dynamic Earth*. © 1971 John Wiley & Sons, Inc., New York. Reprinted by permission. Figure **2.4**: Adapted from Ten Brick in *Geological Society of America Bulletin*, vol. 85, page 224. © 1974 The Geological Society of America. Reprinted by permission. Figure **2.7**: From Burnett, A. W. and S. A. Schumm, "Alluvial-River Response to Neotectonis Deformation in Louisiana and Mississippi," in *Science,* Vol. 222, pp. 49–50, figure, October 7, 1983. Copyright 1983 by the American Association for the Advancement of Science. Reprinted by permission. Figure **2.10**: Copyright © 1968 by Dowden, Hutchinson & Ross, Inc. Figure **2.11**: H. E. Wright and D. G. Frey, eds., *The Quaternary of the United States.* Copyright © 1965 by Princeton University Press. Reprinted by permission of Princeton University Press. Figure **2.12**: From Bloom et al., 1974. Used with permission of *Quaternary Research*. Figure **2.13**: From Saucier and Fleetwood in *Geological Society of America Bulletin*, Volume 81, figure 7, page 879.

© 1970 The Geological Society of America. Reprinted by permission. Figure **2.14**: From Brakenridge, G. R., 1981. Used with permission of *Quaternary Research*. Figures **2.15** and **2.16**: H. E. Wright and D. G. Frey, eds., *The Quaternary of the United States.* Copyright © 1965 by Princeton University Press. Reprinted with permission of Princeton University Press. Figure **2.17**: "From Langbein and Schumm, *Transactions of the American Geophysical Union,* Vol. 39, page 1077, 1958, copyrighted by American Geophysical Union."

Chapter 3

Figure **3.3**: From Bass-Becking et al., 1960. © by University of Chicago Press. Reprinted by permission. Figures **3.4** and **3.7**: From Loughnan, *Chemical Weathering of the Silicate Minerals.* © 1969 American Elsevier Publishing Co. Reprinted by permission. Figure **3.9**: Reprinted with permission of Macmillan Publishing Company from *The Nature and Properties of Soils,* 6th ed., by Harry O. Buckman and Nyle C. Brady. Copyright © 1960 by Macmillan Publishing Company. Figure **3.12**: From Senstius, M. W., "Climax Forms of *Chemical* Rock-Weathering," in *American Scientist,* Vol. 46. © 1958 *American Scientist.* Reprinted by permission. Figure **3.16**: From Hausenbuiller, R. L., *Soil Science: Principles & Practices.* © 1972 Wm. C. Brown Publishers, Dubuque, Iowa. All Rights Reserved. Reprinted by permission. Figure **3.17**: From Jenny and Leonard in *Soil Science,* Vol. 36:367. © 1939 The Williams & Wilkins Co.

Reprinted by permission. Figure **3.19**: From *Soils and Geomorphology* by Peter W. Birkeland. Copyright © 1984 by Oxford University Press, Inc. Reprinted by permission. Figure **3.20**: From Gile, 1975. Used with the permission of *Quaternary Research*. Figures **3.21** and **3.22**: From Gile et al., 1981. Used with the permission of New Mexico Bureau of Mines and Mineral Resources. Figure **3.23**: From Gile, 1975. Used with the permission of *Quaternary Research*.

Chapter 4

Figure **4.7**: From Peltier, L., *Annals of the Association of American Geographers,* vol. 40. © 1950 Association of American Geographers. Reprinted by permission. Figures **4.11** and **4.15**: From Carson, M. A., and M. J. Kirkby, *Hillslope Form and Process.* © 1972 Cambridge University Press. Reprinted by permission. Figure **4.12**: From Legget in *Geological Society of America Bulletin,* vol. 78, page 1444. © 1967 The Geological Society of America. Reprinted by permission. Figure **4.18**: From Kirkby, M. J., "Measurement and Theory of Soil Creep," in *Journal of Geology,* vol. 75. © 1967 The University of Chicago Press. Reprinted by permission. Figures **4.20** and **4.21a**: After Varnes 1958 in *Landslides and Engineering Practice, Special Report 29.* Used with permission of the Transportation Research Board. Figure **4.24**: From Ritter, Dale F., *Zeitschrift Für Geomorphologie,* Vol. 16, page 91, figure 4. © 1972 Gebrüder Borntraeger. Reprinted by permission. Figure **4.26**: From Dalrymple, et al., *Zeitschrift Für Geomorphologie,* Vol. 12, figure 1. Used with permission of *Zeitschrift Für*

Geomorphologie published by Gebrüder Borntraeger Verlagsbuchhandlung Berlin. Stuttgart, Johannesstraβe 3A, 7000 Stuttgart 1. Figure **4.27:** From Selby, M. J., *Zeitschrift Für Geomorphologie,* Vol. 24, page 48, figure 4. Used with permission of *Zeitschrift Für Geomorphologie* published by Gebrüder Borntraeger Verlagsbuchhandlung Berlin. Stuttgart, Johannesstraβe 3A, 7000 Stuttgart 1. Figure **4.31:** From "Forms of bluffs degraded for different lengths of time in Emmet County, Michigan, U.S.A." in *Earth Surface Processes and Landforms* by D. Nash. Copyright © 1980 John Wiley & Sons, Ltd. Reprinted by permission of John Wiley & Sons, Ltd.

Chapter 5
Figure **5.1:** From *Bulletin AAPG: 1967.* Used with permission of the American Association of Petroleum Geologists. Figure **5.28:** "From Kochel, et al., *Water Resources Research,* vol. 18, page 1168, 1982, copyright by the American Geophysical Union." Figure **5.32:** From Milliman and Meade 1983, *Journal of Geology,* © The University of Chicago Press. Used with permission of the University of Chicago Press. Figure **5.36:** After Wolman in *Geografiska Annaler,* vol. 49-A, figure 1, page 386. © 1967 Almqvist & Wiksell International. Reprinted by permission.

Chapter 6
Figure **6.2:** From Simons and Richardson, *Transactions of the American Society of Civil Engineers,* vol. 128, figure 2, page 289, 1963. Used with permission by the American Society of Civil Engineers. Figure **6.3:** From Vanoni, *Transactions of the American Society of Civil Engineers,* vol. 3, figure 24, page 127, 1946. Used with permission by the American Society of Civil Engineers. Figure **6.7:** From Hjulstrom 1939. Used with permission of American Association of Petroleum Geologists. Figure **6.19:** From Hey & Thorne in *Area,* Vol. 7, figure 4, page 193. © 1975 The Institute of British Geographers. Reprinted by permission. Figure **6.25:** From Schumm and Khan in *Geological Society of America Bulletin,* vol. 83, page 1768. © 1972 The Geological Society of America. Reprinted by permission. Figure **6.27:** From Gardner, Thomas in *Geological Society of America Bulletin,* vol. 94, page 671. © 1982 The Geological Society of America. Reprinted by permission.

Chapter 7
Figure **7.3:** From Hickin, in *American Journal of Science,* Vol. 274, figure 2a, page 416. © 1974 Kline Geology Laboratory-Yale University. Reprinted by permission. Figure **7.4:** From Hickin, in *GSA Bulletin,* vol. 86, figure 6, page 491. © 1975 The Geological Society of America. Reprinted by permission. Figure **7.8:** From Everitt, in *American Journal of Science,* vol. 266, figure 5, page 429. © 1968 Kline Geological Laboratory-Yale University. Reprinted by permission. Figure **7.20:** From Bull, in *Journal of Geological Education,* vol. 16, figure 2, page 105. © 1968 National Association of Geology Teachers. Reprinted by permission. Figure **7.23:** After Cooke, in *American Journal of Science,* vol. 269, figure 1, page 27. © 1970 Kline Geology Laboratory-Yale University. Reprinted by permission. Figure **7.25:** After Ruxton and Berry, in *Revue de Geomorphologie Dynamique,* vol. 12, 1961. Reprinted by permission. Figure **7.27:** From Scott and Fisher, in *Delta Systems in the Exploration of Oil and Gas,* figure 37, 1969. Reprinted by permission of the Texas Bureau of Economic Geology. Figure **7.28:** From Russell, in *River and Delta Morphology,* Coastal Studies Series #20, figure 8, page 44. © 1967 Louisiana State University Press. Reprinted by permission. Figure **7.29:** Reprinted with permission, from *Encyclopedia of Geomorphology,* Rhodes W. Fairbridge, ed., Dowden, Hutchinson & Ross, Stroudsburg, Pa. Figure **7.30:** From Rouse et al., in *Geology,* vol. 6, page 407. © 1978 The Geological Society of America. Reprinted by permission. Figure **7.31:** From Morgan, in *Journal of Geological Education,* vol. 18, page 113. © 1970 National Association of Geology Teachers. Reprinted by permission.

Chapter 8
Figure **8.1:** From Mabbutt, in *Aboriginal Man and Environment in Australia,* D. J. Mulvaney and J. Golson, editors. © 1971 Australian National University Press. Reprinted by permission. Figures **8.4, 8.5,** and **8.8:** From R. A. Bagnold, 1941, *"The Physics of Blown Sand and Desert Dunes."* Used with permission of Methuen and Co., Ltd. Figure **8.9:** From Williams, G., *Sedimentology,* vol. 3, page 282, figure 19. © 1964 Blackwell Scientific Publications. Reprinted by permission. Figure **8.10:** From Wilson, I. G., *Sedimentology,* vol. 19, page 193,

figure 2. © 1972 Blackwell Scientific Publications. Reprinted by permission. Figures **8.11** and **8.15:** From R. A. Bagnold, 1941, *"The Physics of Blown Sand and Desert Dunes."* Used with permission of Methuen and Co., Ltd. Figure **8.18:** *The Quaternary of Illinois:* Univ. Ill. Coll. Agr. Spec. Pub. 14.

Chapter 9
Figures **9.4, 9.12, 9.15, 9.16,** and **9.17:** "Reproduced from the *Journal of Glaciology* by permission of the International Glaciological Society." Figure **9.20:** From Sharp, R. P., *Glaciers.* © 1960 University of Oregon Press. Reprinted by permission.

Chapter 10
Figures **10.1** and **10.11:** After Boulton 1974. Used with permission of Donald R. Coates, *Glacier Geomorphology.* Figure **10.4:** From Lewis, W. V., *RGS Research Series: 4.* © 1960 Royal Geographical Society. Reprinted by permission. Figure **10.8:** After McCall, *RGS Research Series: 4.* © 1960 Royal Geographical Society. Reprinted by permission. Figure **10.16:** After Goldthwait 1951. © The University of Chicago. Used with permission. Figure **10.22:** "Reproduced from the *Journal of Glaciology* by permission of the International Glaciological Society."

Chapter 11
Figure **11.5:** From Gold and Lachenbruch in "Permafrost: 2d International Conference Proceedings," 1973, with permission of the National Academy of Sciences, Washington, D.C.

Chapter 12
Figures **12.1** and **12.2:** From A. N. Palmer, 1984, *Journal of Geological Education,* v. 32, p. 248. Figure **12.8:** After Williams 1972b in *Geological Society of America Bulletin.* © The Geological Society of America. Reprinted by permission. Figure **12.9:** "From Baker, *American Geophysical Union Water Resources Research,* 1973, copyrighted by American Geophysical Union." Reprinted by permission. Figure **12.22:** From the *Geological Society of America Bulletin,* vol. 79, figure 14, page 39. © 1968 The Geological Society of America. Reprinted by permission.

Chapter 13

Figure **13.7:** "Adapted from Gavin, *Journal of Geophysical Research,* 1968, copyrighted by American Geophysical Union." Reprinted by permission. Figure **13.11:** "Adapted from Shepard and Inman, *AGU Transactions,* 1950, copyrighted by American Geophysical Union." Reprinted by permission. Figure **13.19:** From W. A. White, in *Geological Society of America Bulletin,* vol. 77. © 1966 The Geological Society of America. Reprinted by permission. Figure **13.20:** From May et al., issue 35, 1983, page 522. Copyright by the American Geophysical Union. Reprinted by permission. Figure **13.22:** From Sunamura 1982. © by The University of Chicago Press. Used by permission.

Photographs

Frontispiece

G. K. Gilbert, USGS

Chapter 1

Figure **1.1:** © John Shelton. Figure **1.2:** NASA. Figure **1.7:** T. S. Lovering, USGS. Figure **1.8:** G. W. Stose, USGS. Figure **1.9:** W. T. Lee, USGS. Figure **1.10:** R. R. Dutcher.

Chapter 2

Figure **2.5:** NASA. Figure **2.6:** R. Guillemette. Figure **2.8:** USGS. Figure **2.9:** H. Miller Cowling.

Chapter 3

Figure **3.2:** D. F. Ritter.

Chapter 4

Figure **4.1:** R. R. Dutcher. Figure **4.3:** R. R. Dutcher. Figure **4.4:** F. C. Calkins. Figure **4.5:** W. T. Schaller, USGS. Figure **4.6:** D. F. Ritter. Figure **4.16:** G. W. Stose, USGS. Figure **4.19 a, b:** USGS. Figure **4.21 b:** USGS. Figure **4.23 a:** USGS. Figure **4.23 b:** J. B. Hadley, USGS. Figure **4.23 c:** C. F. Erskine, USGS.

Chapter 5

Figure **5.13:** USGS. Figure **5.24:** D. F. Ritter.

Chapter 6

Figure **6.4 a, b:** W. W. Emmett, USGS. Figure **6.9:** D. F. Ritter. Figure **6.22:** D. F. Ritter. Figure **6.24:** USGS.

Chapter 7

Figure **7.1:** D. F. Ritter. Figure **7.4 b:** USDA. Figure **7.9 b:** R. Guillemette. Figure **7.17:** USGS. Figure **7.18 a:** USGS. Figure **7.22:** R. R. Dutcher. Figure **7.24:** H. E. Malde, USGS.

Chapter 8

Figure **8.3:** D. F. Ritter. Figure **8.7:** M. R. Campbell, USGS. Figure **8.14 a:** G. K. Gilbert, USGS. Figure **8.14 b:** C. E. Erdmann, USGS. Figure **8.14 c:** USDA. Figure **8.17:** D. F. Ritter.

Chapter 9

Figure **9.3:** US Coast Guard and Geodetic Survey. Figure **9.18:** M. F. Meier, USGS. Figure **9.19:** M. F. Meier, USGS. Figure **9.21:** D. F. Ritter.

Chapter 10

Figure **10.2:** D. F. Ritter. Figure **10.5:** A. Werner. Figure **10.6:** Austin Post, USGS. Figure **10.9 a:** F. E. Mathes, USGS. Figure **10.12:** R. R. Dutcher. Figure **10.13 a:** J. H. Moss, USGS. Figure **10.15 a:** R. R. Dutcher. Figure **10.15 b:** J. H. Moss, USGS. Figure **10.15 c:** J. Travis. Figure **10.17:** D. F. Ritter. Figure **10.18:** R. R. Dutcher. Figure **10.19:** J. Travis. Figure **10.20:** C. William Horrell. Figure **10.21 b:** W. C. Alden, USGS. Figure **10.23:** D. F. Ritter.

Chapter 11

Figure **11.1:** O. J. Ferrians, USGS. Figure **11.11:** O. J. Ferrians, USGS. Figure **11.12:** G. W. Holmes, USGS. Figure **11.13:** O. J. Ferrians, USGS. Figure **11.14:** R. Craig Kochel. Figure **11.15:** Noel Potter, Jr. Figure **11.16:** O. J. Ferrians, USGS. Figure **11.18:** D. F. Ritter.

Chapter 12

Figure **12.4:** D. F. Ritter. Figure **12.10:** C. William Horrell. Figure **12.12:** D. F. Ritter. Figure **12.16:** USGS. Figure **12.17:** W. H. Monroe, USGS. Figure **12.18:** W. H. Monroe, USGS.

Chapter 13

Figures **13.1** and **13.2 a, b:** R. Craig Kochel; Figure **13.10:** UPI; Figure **13.12:** P. Michaels; Figure **13.17:** US Coast Guard and Geodetic Survey; Figure **13.18:** USGS. Figure **13.24:** USGS. Figure **13.25:** W. H. Monroe, USGS. Figure **13.26** (both): USGS. Figures **13.29–13.32:** R. Craig Kochel.

Index